Nanocantilever Beams

Nanocantilever Beams

Modeling, Fabrication, and Applications

edited by

Ioana Voiculescu

Mona Zaghloul

PAN STANFORD PUBLISHING

Published by

Pan Stanford Publishing Pte. Ltd.
Penthouse Level, Suntec Tower 3
8 Temasek Boulevard
Singapore 038988

Email: editorial@panstanford.com
Web: www.panstanford.com

British Library Cataloguing-in-Publication Data
A catalogue record for this book is available from the British Library.

Nanocantilever Beams: Modeling, Fabrication, and Applications

Cover image: Courtesy of Profs. Takahito Ono and Masaya Toda, Department of Mechanical Systems and Design, Tohoku University, Sendai, Japan. The image shows a single nanocantilever beam with a magnetic microsphere and an array of bimaterial nanocantilever beams.

ISBN 978-981-4613-23-1 (Hardcover)
ISBN 978-981-4613-24-8 (eBook)

Printed in the USA

Contents

Part 3: Applications of Nanocantilever Beams

Preface

The cantilever beam is an important structure of microelectromechanical system (MEMS) devices. This simple structure, fixed at one side and not connected at the opposite side, was inspiring Kurt Peterson, in the classic paper "Silicon as Mechanical Material." He was the first scientist to think of the fabrication of this structure in silicon. Since this pioneering study to fabricated cantilever beam in silicon was performed, a large number or research groups have joined this field and greatly expanding the scope of microcantilever beam structure. At this moment, the cantilever beam is an active research topic and there are many interesting applications based on this concept especially in the areas of chemical and biological sensors. The special advantages of these types of sensors rely on their versatile qualities such as microscale dimensions of the sensor area, label-free detection, high sensitivity, simple integration with electric circuits, and the potential for simultaneous detection of tens or even hundreds of targets by using arrays of cantilever beams. The cantilever beam arrays could be mass fabricated using the semiconductor and MEMS technologies. The performance of these devices has been dramatically improved through the development of new materials and micro and nanotechnologies.

In the past decade, microtechnology has explored the submicron regime, and nanomechnical systems (NEMS) have started to attract interest for a wide range of applications due to their unique properties. Micro- and nanosized cantilever beams are robust devices whose high sensitivity and selectivity allow them to detect physical, chemical, and biological components by measuring changes in cantilever bending or in resonant frequency. Several types of optical and electrical signal transductions have been reported for diverse application in vacuum, air, or liquid. Nanocantilever beams have an intrinsically high mass sensitivity and thus evolved into a powerful tool for absolute mass detection of molecules.

This book focuses on the nanocantilever beam. Nanometer-size mechanical structures show exceptional properties generated by their reduced dimensions. These properties enable new sensing concepts and transduction mechanisms that will allow the enhancement of the performance of the actual devices to their fundamental limits. The book is divided into three parts and covers important research that was performed in this area in the past decade. The first part is about nanocantilever beam fabrication using silicon or polymer as substrate materials. The second part studies the nonlinearity of nanocantilever beam resonators. At small vibrational amplitudes, nanocantilever beams behave as mechanical devices. The nonlinear effects are manifested when the vibration amplitude increases. Euler–Bernoulli beam theory is widely used to successfully predict the linear dynamics of micro- and nanocantilever beams. However, its capacity to characterize the nonlinear dynamics of these devices has not yet been rigorously assessed, despite its use in nanoelectromechanical systems (NEMS) development. The study of nonlinear dynamics of nanocantilever beam provides an analytical tool in order to optimize the resonant nanosensor design and enhance its performance for precision measurements applications. The third part focuses on important applications of diverse nanostructures such as carbon nanotubes, nanomembrane structures, lead zirconate titanate (PZT) nanofibers, and nanomechanical beams. Nanocantilever beams have important applications as optical transducers and actuators. Fully integrated cavity optomechanical transducers for mechanical position and motion sensing with high precision, high bandwidth, and a small footprint are described in the book. Nanophotonics has emerged as a viable option for addressing the transduction issues of nanomechnical beams decreasing sizes. Integrated nano-optomechanical systems have demonstrated exceptional displacement sensitivity. Nanobridge structures fabricated from zinc oxide (ZnO) nanowires and used as gas sensors are also presented in the book. The fabrication techniques for the ZnO nanowire chemical sensor and the experimental testing of the sensor are discussed. An important application of nanocantilever beams is in the area of biotechnology. When the nanocantilever beam is fabricated from materials with different thermal expansion coefficients, it could act as a calorimeter used to detect the temperature of mammalian cells. The nanocantilever beam functionalized with an antibody was demonstrated as a highly sensitive sensor for the detection of diverse biomolecules.

This book offers very diverse information about the nanocantilever beam, nanobridge, and nanomembrane structures whose high sensitivity allow them to detect physical, chemical, and biological components. A variety of materials could be used for the fabrication of nanocantilever beams. Several types of optical and electrical signal transductions methods have been reported for diverse applications in the air, liquid, and vacuum. This book presents the state of the art of all presented subjects, and the editors hope that this book will be very useful for researchers in this field. We would like to thank the authors for their efforts to contribute to this collection of research topics focused on nanocantilever beams. We were fortunate to meet each author and discussed each chapter at the NMC 2013 Workshop on Nanomechanical Sensing, at Stanford University. We hope that this collected work will provide excellent scientific reference for an audience with a diversity of backgrounds and interests, including students, academic researchers, industry specialists, policy-makers, and enthusiasts.

Ioana Voiculescu
The City College of New York, New York, NY, USA
Mona Zaghloul
The George Washington University, Washington, DC, USA

Part 1

Techniques for the Fabrication of Nanocantilever Beams

Chapter 1

Nanocantilever Beam Fabrication for CMOS Technology Integration

Gemma Rius[a] **and Francesc Perez-Murano**[b]

[a]*Centre for Fostering Young and Innovative Researchers, Nagoya Institute of Technology, Gokiso cho, Showa ku, 466-8555 Nagoya, Japan*
[b]*Institut de Microelectronica de Barcelona, Centro Nacional de Microelectronica, CNM-CSIC, Campus UAB, 08193 Bellaterra, Spain*

1.1 From Micromechanics to Nanomechanics

In the past years, advances in nanofabrication methods have led to a size reduction of microelectromechanical systems (MEMS) originating the nanoelectromechanical systems (NEMS). NEMS may provide significant improvements in terms of integration, power consumption, and production costs.

Nanometer-size mechanical structures show exceptional properties rooted precisely on their reduced dimensions. These properties can result in novel sensing and transduction mechanisms that will allow to push the performance of existing devices to their fundamental limits or, more importantly, to enable completely new sensing concepts with an unprecedented impact on applications.[1,2] Probably, the most representative example of the

Nanocantilever Beams: Modeling, Fabrication, and Applications
Edited by Ioana Voiculescu and Mona Zaghloul

ISBN 978-981-4613-23-1 (Hardcover), 978-981-4613-24-8 (eBook)
www.panstanford.com

benefits of downscaling effects in nanomechanical properties is the mechanical resonance. Nanomechanical resonators are freestanding nanostructures with well-defined mechanical resonant modes. They are used to construct high-resolution mass sensors[3–7] and very high frequency oscillators for telecommunications.[8,9]

Nanomechanical sensors are based on the principle that external effects, ranging from simple mass loading to complex molecular recognition events on functionalized surfaces, result in measurable bending (static mode) or changes in the resonant frequency (dynamic mode) of nanomechanical structures. Remarkably, the sensitivity and detection limits of nanomechanical sensors increase dramatically for decreasing dimensions. For instance, miniaturization has pushed the detection limits of nanomechanical mass detection from the picogram to the zeptogram range in less than a decade.

Application of NEMS for developing high-efficiency telecommunication systems is being addressed by several groups.[5,10–22] Nanomechanical resonators are called to replace quartz crystals in the field of the RF communications due to their capability to be fabricated with standard integrated circuit (IC) process. The reason derives from the high(er) frequencies that they can achieve and the small area that they would require. The primary building blocks for any telecommunication system are oscillators. Self-sustaining oscillator with feedback can be implemented by employing a NEMS resonator as the frequency determining-element for the feedback oscillator. Some other relevant examples of the use of NEMS for telecommunications are the mechanical implementation of filters[23] and frequency convertors.[15]

As it applies for testing or establishing of device production changes, when developing novel nanofabrication methods, the first phase is to realize discrete structures or devices. Once it has been demonstrated that one single nanostructure or nanodevice is useful by itself, the next stage is to optimize the fabrication method in order to improve its robustness and reliability, as well as the throughput, understood as the number of structures per unit time to be produced. However, most of the time, nanostructure and nanodevices are not useful by themselves, alone. To get a functional system, it is needed to have a set of devices interconnected or working simultaneously, or to combine them with other structures and devices, eventually of different critical size. In the latter case, it is requisite to combine and adapt different fabrication methods.

The highest level of system complexity, and in turn of functionality, can be achieved by the combination of nanodevices with microelectronic circuits. Its most straightforward example is to use a nanodevice for the sensing purposes and interface it to a microelectronic circuit for the signal conditioning and amplification. This approach has been followed during many years in microsystem technology, and it is now being expanded to incorporate novel nanotechnology concepts.

While microelectronic fabrication (and specifically, CMOS technology) is an extremely reliable and productive technology, novel nanofabrication methods are under development and, for most of the cases, far to reach the maturity to provide commercially viable applications. Yet, the field of nanofabrication is viral and progressing rapidly, and some recent developments unveil the high potential for achievement of breakthrough concepts. As a paradigmatic example, directed self-assembly of block co-polymers is seriously considered by the integrated circuit industries as a complementary lithography method for fulfilling the future needs of CMOS (further) miniaturization.[24]

The so-called top-down fabrication may involve techniques such as UV optical lithography (OL), electron beam lithography (EBL), focused ion beam (FIB) lithography, or nanoimprint lithography (NIL). Nanopatterning based on all these techniques has made impressive progress in terms of both minimum achievable lateral dimensions and large-scale reproducibility and process repeatability. However, NEMS fabrication challenges go far beyond nanolithography patterning, as they include many other aspects of processing technology such as material deposition and etching, release of the free standing structures, etc. In fact, while top-down fabrication of discrete nanomechanical devices had been demonstrated long time ago,[25] only very recently the concept of large-scale integration (LSI) applied to NEMS has been raised up.[26] LSI of NEMS is indeed an important but highly non-trivial issue for the practical application of devices. A basic trend is to push top-down fabrication limits to meet the standards of LSI.

In this chapter, we will review recent advances in the development of fabrication methods for nanomechanical structures and to integrate nanomechanical devices into CMOS circuits.

Nanomechanical devices present a lot of interest for developing multisensing platforms and for building up telecommunication systems. Therefore, it constitutes an ideal example to study the combination of novel and emerging nanofabrication methods with well-known and established CMOS technology.

1.2 Lithography-Based Fabrication of Nanomechanical Structures

The manufacture of the mechanical devices in MEMS and NEMS is done by a modified version of the semiconductor device fabrication technologies. The characteristic feature is precisely the materialization of the 3D movable element, in contrast to the planarity pursued in microelectronics.[1]

MEMS were originally produced by the so-called bulk micromachining, where the whole thickness of the silicon wafer is used to constitute the micromechanical structure. For NEMS, the processing sequence generally relies on the patterning of the movable structure on a thin structural layer and its release by sacrificial etching of an underlying layer. This technology is commonly referred as surface micromachining and is also applied to MEMS fabrication (Fig. 1.1).[27]

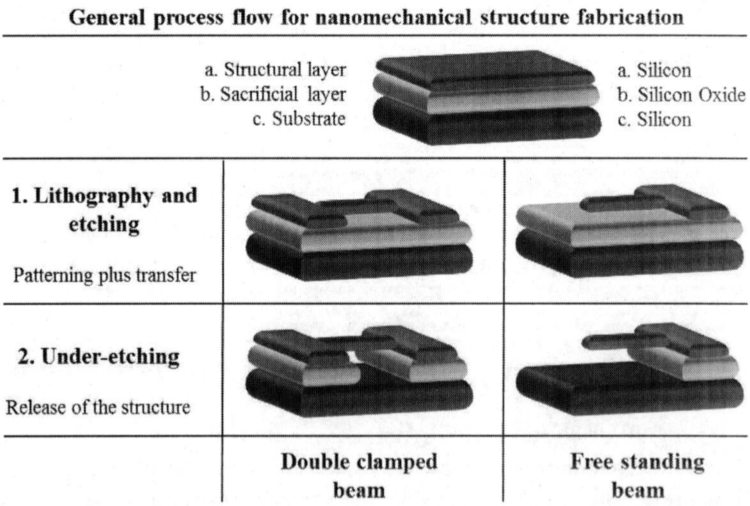

Figure 1.1 Fabrication sequence for single and double clamped beams.

1.2.1 Process Flow for the Fabrication of (Nano)Mechanical Devices

A relevant aspect of the design of mechanical structures is the choice of the structural material. This aspect has also important implications in terms of fabrication, the processing conditions, the capability to effectively realize, especially with tight control at the nanoscale, the mechanical devices, and obviously for the applications. Polymers can be produced in large amounts and with a great variety of interesting material characteristics. Metals, such as gold, nickel, aluminum, etc. may provide high reliability under their mechanically limited operational range. Ceramics can have added value properties, such as piezoelectricity, which widens the structure sensitivity to both normal and shear forces, thus useful for sensing.[28]

In this chapter, the focus on fabrication of nanomechanical devices is based exclusively on single-crystal and polycrystalline silicon. Complementarily, we briefly refer to relevant alternate examples in Section 1.4. Silicon material of high quality is readily available at very competitive costs. On top of it, single-crystal Si has exceptional elastic and mechanical properties; no flexural hysteresis means minimal energy dissipation and very little fatigue implies that repeatable motion is very reliable, hence prolonging device operation lifetime. The possibility to incorporate electronic functionality by doping makes for Si a combination of material properties deal for MEMS and NEMS.[29]

In terms of Si mechanical device fabrication, the manufacturing strongly benefits from the well-established tools and processes developed for the set of silicon-based materials used in microelectronics industry. The possibility of low-defect single-crystal growth and sophisticate wafer technology allows the processing cost optimization and makes possible precise-production complex architectures, such as the CMOS integration that will be considered here. These are main merits for the success of silicon technologies.

The basic fabrication sequence of Si mechanical devices is depicted in Fig. 1.1. The substrate, silicon, is covered by two layers, the sacrificial layer (silicon dioxide) and the structural layer (silicon). Patterning techniques are applied to the structural layer to define the mechanical element (Step 1 in Fig. 1.1), and then etching techniques are used to remove the sacrificial layer (Step 2 in

Fig. 1.1). As a result, the mechanical device is shaped in the desired 3D morphology and the mechanical structure is free to move in determined directions and vibration modes.[30]

Very commonly, the patterning involves two methodologies: lithography and pattern transfer. Lithography is used to define the surface (~2D) projection of the mechanical element and selective material etching is applied for the pattern transfer to the structural layer. In both cases and depending on the fabrication strategy, it can involve a number of intermediate processing steps. On the other hand, sometimes lithography and mask manufacture constitute a single processing step (direct patterning). We will particularly discuss (nano)lithography in detail in Section 1.2.2 and variations of the (nano)patterning basic approach, in Section 1.3, applied to nanobeams.

Worth to mention and for completeness, we summarize some of the features related to etching in Table 1.1, as we are not able to allocate space to discuss this important step of the fabrication processing sequence and the integration of the nanomechanical devices into CMOS circuitry. Etching has a major role in the definition of nanoscaled 3D movable element as it is critical in terms of both the transfer to the structural layer (guaranteeing patterning fidelity) and the release of the structure (inducing minimal stress and preventing collapse). More details on the latter issue are discussed elsewhere.[31]

Table 1.1 Basics of etching treatments applicable to silicon (nano) mechanical structures fabrication

Purpose	Method	Feature	Shaping	Material/ etchant	Principle	Limitations
Pattern transfer	Dry etching	Anisotropy		Si/ion etching (RIE)	Efficiency	Trenches, profile
Structure release	Wet etching	Isotropy		SiO_2/ buffered HF	Selectivity	Structure collapse

1.2.2 Nanolithographies for the Patterning of Nanomechanical Devices

As a universal definition, lithography is the ability to transfer a pattern from one media to another. In semiconductor electronics fabrication, OL is the sovereign method for batch patterning and microfabrication.[32] It is based on selective light irradiation of a photosensitive material (resist) according to shape projection by means of a shadowing mask. The exposure of the resist media to the photons causes its chemical changes, so that selective etching of either the irradiated or non-irradiated areas can be applied (the so-called resist development). Although the pattern dimensions (and shape) are primarily determined by the mask (dimensions) and exposure (contact/projection approach, focus...) as well as the resist (thickness, molecular chains length, etc.), shrinking the features beyond the micron scale it is mainly limited by diffraction phenomenon, i.e., as a result of the light wavelength. Pushing the diffraction limits of OL faces the challenge of increasing tool expenses (Table 1.2). While some examples of fabrication of nanometer-scale beams by a combination of UV lithography and oxidation have been presented,[33] the use of state of the art DUV lithography seems to be mandatory to achieve good dimensional control. In consequence, due to the extremely high cost of DUV lithography tools, the definition of nanomechanical structures by optical lithography have been realized by using standard CMOS processes in dedicated foundries. The activity in this regard will be presented in Section 1.3.3.4.

Table 2.2 Main nanolithographies applicable to MEMS/NEMS fabrication and characteristic features

Technique	Principle	Resolution	Merits	Limitations
Extreme optical lithography	Light-photons Resist media	<100 nm	Throughput–parallel Compatibility	Resolution Mask, ~costs
Electron beam lithography	Focused energetic electron probe Resist media	<5 nm	Pattern flexibility Resolution	Throughput–serial Charging, circuit damage
Ion beam lithography	Focused energetic ion probe Direct patterning	>10 nm Depending on approach	Multiple processing Resolution	Throughput–serial Complexity
Nanostencil lithography	Masked material deposition Direct patterning	>10 nm Edge roughness	Throughput–parallel Compatibility	Mask, ~cost Complexity

1.2.2.1 Electron beam lithography

Speaking of downscaling the dimensions of the lithography features, the use of a fine focused (reasonably) energetic electron beam overcomes the wavelength limiting factor encountered for light. The technique is called EBL[34] and is the preferred method for nanomechanical device patterning. The feature patterning transfer is similar to OL in the sense that both are rooted in the use of a resist media. Pattern flexibility is one of the merits of the technique, as the charged particles can not only be focused to extremely narrow probes, but also be driven electromagnetically with high accuracy across big areas (direct writing), being very convenient for prototyping without the cost and constraint of (expensive) masks. However, the scan capability and local delivery of the electron beam is at the same time a serious limitation in terms of throughput and eventual electronic damage of the substrate materials.

In both cases, OL and EBL, the patterned resist is used as a template for subsequent material deposition and etching. A very common post-lithography processing sequence for mechanical structures fabrication is the deposition of a thin metal layer plus resist lift-off. The patterned thin metal layer will act as a mask for subsequent etching. Pattern transfer to the structural layer is preferably done by any of the anisotropic etching techniques and conditions (for example, reactive ion etching, RIE), which in principle allows preserving the feature fidelity.

Obviously, resist-based fabrication, although widely consolidated, involves a significant number of processing steps. Nonetheless, and increasingly acute for fabrication at the nanoscale, each process step is susceptible of compromising feature dimensions and risking the success rate. At this point, simplified nanopatterning methods are proposed. For direct nanopatterning of nanomechanical structures, we will introduce two techniques, processing based on (focused) ion beam lithography (IBL)[35] and nanostencil lithography (NSL).[36]

1.2.2.2 Focused ion beam patterning

The invention of the liquid metal ion sources and the capability of focusing their charged particle beams, similarly to the electron beams, below submicron dimensions enabled multiple processing steps in a single tool. About the middle 1980s, the FIB apparatuses became an essential instrument for inspection, metrology, and failure analysis, in particular applied to materials science and microsystems. However, an important application of FIB relies on its capability to perform both subtractive and additive treatments.[37]

In other words, by using ion beams etching and deposition techniques, two of the three routines required for nanomechanical structure fabrication, as depicted in Fig. 1.1, are made possible. This is a result of a number of ion interactions with matter (collisions, chemical reactions, etc.) through a range of (rather complex) phenomena. As the FIB can be driven electromagnetically similar to EBL, FIB-based processing acquires complete mask-less lithography and patterning capabilities interestingly at the nanoscale.

Energetic FIB can be used for the local removal of material, as in conventional ion beam milling. Instead, under proper irradiation conditions the FIB can be used for substrate doping by the ion beam species. This alternate process resembles implantation technique as the implanted region is penetration depth dependent upon the beam energy (therefore, thickness-limited process) and often introduces etching selectivity upon the original substrate material and the irradiated areas. We will particularly illustrate how to exploit this approach in Section 1.3.3.2. Additional competences for etching are enabled by the introduction of selected gas-species during the FIB exposure, so-called FIB induced etching, which usually provides, for example, higher etching rate and selectivity as well as anisotropy. Gas-assisted FIB techniques are also applied to deposition of materials, known as FIB-induced deposition, which allows surface pattern definition in a single-process step that can be used as an etching mask.[38]

In summary, commercially available FIB tools provide direct writing, direct patterning based 3D nanofabrication. The set of FIB capabilities conform FIB-IBL as a versatile route for prototyping nanomechanical structures with high processing flexibility. Compared to EBL, proximity effect and electron damage are avoided while significantly reducing the amount of steps in the fabrication sequence (processing efficiency), yet it is limited in terms of standardization and throughput (serial patterning) among other disadvantages.

Following the trend of addressing processing reduction, throughput, and consolidation of nanolithography techniques for nanomechanical device technologies, NSL is worth mentioning.

1.2.2.3 Nanostencil lithography

NSL relies on the use of a shadow mask (stencil) consisting of nanoscaled features for the direct deposition of patterned thin layers of selected materials. The principle is simple and follows the lithography concept of OL in the sense that it is based on a single re-usable mask and parallel processing. However, the nanofeature deposition is performed directly, without the use of a resist media (resist-less), thus, the nanopatterning is accomplished in a single processing step. Not involving thermal and chemical treatments, NSL avoids resist-related contamination, overcomes processing limitations such as resist lift-off and prevents eventual damage of sensitive substrate materials. Performed under a good vacuum, high purity materials can be deposited.[39]

Similar to FIB-IBL techniques, ideally NSL can be implemented not only as deposition but also for implantation and etching routine; however, the integrity of the (nano)stencil, the mask, may be compromised. Actually, the main limitations of NSL as a nanopatterning technique derive from the uncertainty and actual problems encountered experimentally, such as pattern distortion, spatial resolution accuracy (blurring) and alignment (static and dynamic modes can be applied), clogging, mask break, etc.

The most common application of NSL for nanomechanical structure patterning is its application for the deposition of thin metal patterned features, as it will be described in Section 1.3.3.3.[40] Unless the metal conforms a structural part of the nanomechanical device, the metal layer (nano)pattern acts as a mask for the pattern transfer step to the structural material by etching before nanomechanical structure release. Therefore, at this point the approach becomes similar to the route described above for EBL. Depending on the device design and within the variations of thefabrication strategies, the metal mask is finally removed prior todevice operation.

For compactness and consistence with the contents of next section, we omit a discussion on scanning probe lithography (SPL) and NIL applied to the nanopatterning and fabrication of nanomechanical structures. However, as the patterning capabilities at the nanoscaled of both approaches are remarkable we provide a list of related works in the literature. Examples of AFM-based and NIL-based nanomechanical devices fabrication can be found elsewhere.[41–43]

To conclude this section and considering again the design of the nanoscaled structure and device, and related to its characteristic nanoscaled dimensionality, which makes it interesting in terms of (electro)mechanics, we should consider the two main morphological aspects in terms of the fabrication of the nanomechanical element: (i) the top-view shape and dimensions, and (ii) the thickness. Whether one or the other determines the (nano)mechanical response will constraint not only the resolution tolerance of the fabrication but also the most suitable processing sequence for the device. As seen in the introduction, the mechanical performance relevant to be considered in the nanodomain has the most great contribution resulting from these morphological attributes, which are designed as a function of the operational mode (static or dynamic) and desired functionality (bending or vibrational modes).

Concerning to the top-view nanomechanical structure and dimensions, pattern shape complexity and required nanoscaled precision is characteristic of some NEMS, as it will be shown, for example, in Section 1.3.3.1. A smart way to improve the performance of NEMS from their design is the use of coupled nanomechanical structures. The overall response of the system depends not only on the individual response of each structure, but also on the interaction between them. An illustrative example of this is the use of coupled nanocantilevers to increase the mass sensitivity.[44]

In other cases, the structural layer thickness (a few hundred nanometers) or the combination of thin layers of various materials usually represent a challenge during the etching steps. It requires precision and optimal conditions in terms of both pattern transfer and structure release to avoid lithography pattern distortion and stress or collapse of the nanomechanical element.

A paradigmatic example of this demanding fabrication accuracy are the arrays of cantilevers, which are found convenient when working in parallel for sensing applications (Fig. 1.2).[46,47] As mentioned, not only permanent bending at the end of the fabrication sequence should be avoided, but exactitude in the cantilevers dimensions is the only guarantee for reliable operation and measurements. If using the EBL plus metal deposition and anisotropic etching route, final cantilever dimensions are a result of each of the fabrication stages. Inaccuracy in the lithography (initial) step will propagate in the subsequent processing steps. In this approach nanofabrication process compatibility with microfabricated part often only implies alignment capabilities of the nanopatterning technique.

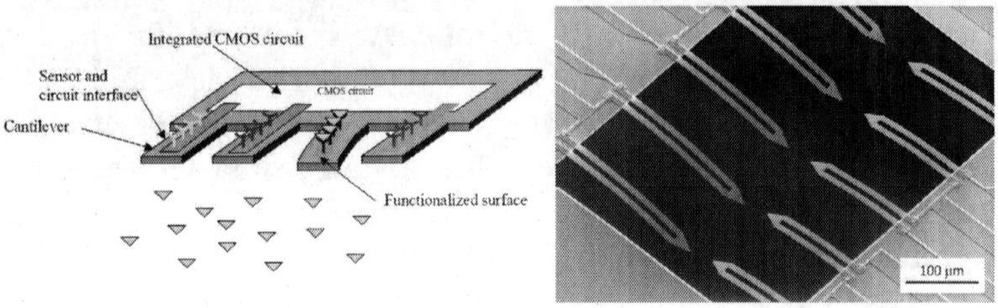

Figure 1.2 (Left) Concept of selective sensing based on specific functionalization of cantilever nanobeams. (Right) SEM images of an array cantilever nanobeams fabricated for enabling selective sensing on integrated circuit architecture.[45]

As an additional remark, arrays of cantilevers are used for multiplexed sensing. Simultaneous, selective sensing of different species is possible by the functionalization of each cantilever surface with an appropriate marker. Important operational and performance aspects are the actuation and readout of the nanomechanical element. Integration of the nanomechanical device into CMOS circuits is the smart solution that has been proposed, as presented next.

1.3 Integration of NEMS into CMOS Technology

1.3.1 Interest of NEMS/CMOS Integration

In Section 1.2, we described the basic process flow for nanomechanical structure fabrication and restricted the manufacture analysis on the basis of a few nanopatterning strategies. Specifically, we have mentioned in the last part of Section 1.2.2 some fabrication issues for arrays of single-clamped cantilever nanobeams, which were linked to their real operation such as stress and morphological accuracy.

Continuing with the cantilever array instance, in terms of fabrication these arrays are embedded into the entire fabrication process of the microstructure that typically interfaces the nanosensor for its macroscopic access. The patterning of the nanomechanical structure is just intercalated into the wafer whole fabrication sequence as an additional lithography level (multilevel lithography). As the lithography technique for the nanostructures patterning often differs from the OL, the use of combined methods is commonly known as mix and match.

Many efforts are directed toward the combination of NEMS with diverse devices and structures: integration of field-effect transistor for electrical readout,[48,49] integration of a Schottky diode for optical detection,[50] integration of waveguides[51,52] and integration of nanofluidics channels.[53–55] In this section, we focus on an even more advanced aspect of making compatible standard microfabrication/patterning techniques with nanopatterning methods. The complexity of the approach consists of the following: how to introduce the fabrication of the 3D movable nanoscaled element (nanocantilever beam) into the yet intrinsically complex-design, precision-demanding and multiple-processing of CMOS technology. Particularly, we will present relevant examples of NEMS integration into pre-fabricated CMOS circuitry based on the previously discussed nanopatterning techniques. Some main features of each fabrication strategy applied to cantilever beams will be highlighted.

Combination of nanomechanical devices into CMOS circuits provides a powerful route toward system integration, as it incorporates on-chip the functionalities of signal readout (mainly using capacitive methods), multiplexing, and amplification. Dedicated discussion of the relevance of this aspect would deserve a separate, profuse analysis beyond the scope of this chapter.

1.3.2 CMOS Integration: Hybrid versus Monolithic Approach

Over the past decades, progress in silicon planar technologies has allowed miniaturized systems (microsystems, MEMS) to be performed by exploiting the sensing properties of IC materials (silicon, polysilicon, aluminum, silicon oxide, and silicon nitride) or additional deposited materials (such as piezoelectric zinc oxide, applicable polymers, and supplementary metallization layers). When MEMS are fabricated using IC technologies and materials, it is possible to integrate the interface circuit and several functional non-CMOS structures or devices on the same chip. Alternatively, they can be embedded in the same package, leading to the concept of systems-on-chip (SoC).[56] The potential advantages of this approach are numerous: the cost is reduced due to batch fabrication of both the structures and the interface circuits; its size and interconnections are minimized; and its reliability is improved. Different approaches have been developed over the years for the systems integration. Basically, they can be classified into two groups: monolithic integration and hybrid integration.

In the monolithic integration, the MEMS and the IC are integrated on the same chip. The whole system is implemented using a fabrication process optimized for integrated circuits, including a few compatible post-processing steps when necessary. The MEMS structure and fabrication must be designed by taking into account the material characteristics and design rules given by the standard IC process used. Following are some advantages of this approach: (i) The parasitic signals due to the interconnections between the sensors and the interface circuitry are minimized and therefore become well-defined and reproducible; (ii) the system assembly is simple, inexpensive, and independent of the number of connections needed; and finally (iii) the use of the same technology allows to achieve a good matching between the sensor and the interface circuitry allowing accurate compensation of many parasitic effects. R. T. Howe and R. S. Muller were the first to publish on the fabrication of polysilicon microstructures integrated monolithically with NMOS circuitry.[57,58] Another pioneering work has been performed by H. Baltes and co-workers, who, in 1989, demonstrated a new approach based on sensor SoCs, consisting in integrating MEMS with CMOS by releasing the microstructures from the front of the wafer.[59]

Technology for CMOS/MEMS monolithical integration can be divided in three types according to the stage in which the MEMS structure is defined: before (pre-CMOS), during (intermediate-CMOS) or after the fabrication of the CMOS circuit (post-CMOS).

In the pre-CMOS approach, the MEMS structures or part of them are formed before the CMOS fabrication process sequence. In this case, the CMOS IC is not altered by the micromachining process, but the MEMS devices have to be protected during the whole CMOS process. An example of this approach is the fabrication of polysilicon structures using the iMEMS technology of Sandia National Laboratories[60] or recently the ModMEMS technology.[61] In the intermediate-CMOS approach, standard CMOS processing is used to define the MEMS structure or alternatively the CMOS process sequence is interrupted for additional material deposition or including new process steps. Intermediate-CMOS is the one used in commercial existing products.[62]

In the post-CMOS approach, two general fabrication strategies can be applied. As a first strategy, the MEMS structures are entirely built on top of a finished CMOS substrate, leaving the CMOS layers intact.[63,64] As a second strategy, the MEMS are obtained by machining CMOS layers after the completion of the regular CMOS process sequence.[65]

Alternatively, in the hybrid integration the MEMS and the IC are integrated on different chips. They are included in the same package or mounted on the same substrate. The interconnections between the sensor chip and the interface circuit chip can be performed with bonding wires or other techniques, such as flip-chip or wafer-bonding methods. With this approach, the two chips can be fabricated using two different technologies, which are optimized for the sensors and the circuitry, independently.

However, this approach has some serious limitations. The assembling can be quite expensive, limited and a source of possible failures. The parasitic effects due to the interconnections are some orders of magnitude larger, more unpredictable, and less repeatable than with the monolithic approach. As a result, in many cases, these effects may hinder any improvements obtained in sensor performance by technology optimization.

Finally, matching between MEMS and IC cannot always be guaranteed. A relevant example of this approach, closer to the area of NEMS, is the Millipede project developed for many years by IBM.[66] The Millipede concept stands for an array of hundreds of atomic force microscopy cantilevers with integrated heaters that is fabricated using advanced MEMS technologies. By means of a dedicated CMOS circuit, each cantilever in the array can be individually addressed (actuation) and controlled (readout, feedback loop). Millipede was originally proposed for developing a massive data storage system. Nowadays, it is still considered a useful concept for massive fabrication at the nanoscale, then relevant for industry. In any case, it remains one of the most advanced developments and a major achievement to date in terms of combination of nanotechnology and microelectronics.

1.3.3 NEMS–CMOS Fabrication Based on Nanolithography Methods

We base this section mainly on some of our works on NEMS–CMOS integrative approaches which have been developed totally or partially at the IMB-CNM-CSIC,[67] but we complement with and refer to works found in the literature where needed. Most of the results shown here are part or developments based on several thesis works[31,45,68–70] where more detailed gathered information can be found.

We have seen in Section 1.3.1 the basic process flow for discrete fabrication of (nano)mechanical devices and some general aspects related to the processing steps. In the case of the nanomechanical device integration on CMOS using the monolithic approach that we show here, the nanomechanical structure is defined by using a surface micromachining strategy. Particularly a dedicated area (termed as nanoarea) has been included in the chip layout to be used for the fabrication of the nanomechanical structure after the CMOS circuitry fabrication is complete, i.e., as a post-CMOS processing module (Fig. 1.3).

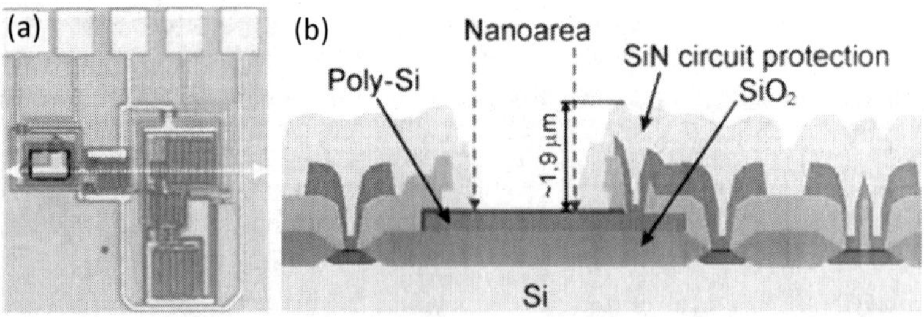

Figure 1.3 Schematic drawing depicts the cross section of the nanoarea within the CMOS (right) and its location respect to the circuitry in top view (left). Adapted from Ref. 71.

The same Si-based layers that form the CMOS chip are used as the structural and sacrificial layer of the nanomechanical device. Particularly, 0.6 μm-thick polysilicon layer and 1 μm are used, respectively, as seen in the cross section of the nanoarea in Fig. 1.3.[72] This structure will be common for the three approaches shown next, which vary in terms of the applied nanopatterning strategy (and consequently involve certain variations of the rest of the fabrication sequence).

1.3.3.1 Electron beam lithography

Beyond the OL-based fabrication method for the NEMS-on-CMOS integration (Section 1.3.2), EBL is the most straightforward lithography technique for the nanomechanical device manufacture.

The most common EBL-based approach for nanomechanical structure fabrication follows the standard process sequence that is applied when a resist-media is used, as in OL. The main fabrication steps are given in Ref. 45. The entirety can be listed as follows (material or treatment purpose, in brackets): resist deposition, EBL (feature nanopatterning), resist development, thin metal layer deposition (etching mask), resist lift-off, Si etching (nanoelement structuring), SiO_2 etching (nanoelement release). Sometimes an additional OL level is used, before the underetching, to preserve the integrity of the CMOS circuitry during the etching for the nanomechanical structure release. Certain wet etching to remove the metal mask is often used as well to get the devices ready for operation.

In addition to superior potential for spatial resolution lithography, the main difference that the use of EBL introduces to the NEMS-on-CMOS respect to OL is the pattern flexibility. EBL is a direct writing technique featuring high scan precision and ultrafast beam blanking capability, automated routines available over large wafer ranges, etc. Resulting from the absence of mask, the possibility of defining any pattern feature but, especially, of being able to modify wisely when repeating the whole fabrication sequence is of great advantage. The benefit of EBL is particularly important when used for prototyping and also during testing and optimization of the process conditions, such as critical anisotropic etching.

Patterns of high complexity can be exposed. For NEMS, this is especially useful, as done in reference,[73] where the structure patterned in the nanoarea include a set of two perpendicular beams with a common anchor, whose vibrating movement is simultaneously driven and monitored by the nanometer-order close electrode. The actuator/readout electrode is placed in the nanoarea by the very nanofabrication (Fig. 1.4). The structure allows differential analysis of the mass sensing in one of the cantilever beams (the second acts as reference), therefore, ideally providing higher sensitivity. The direct connection to the CMOS allows optimal signal treatment.

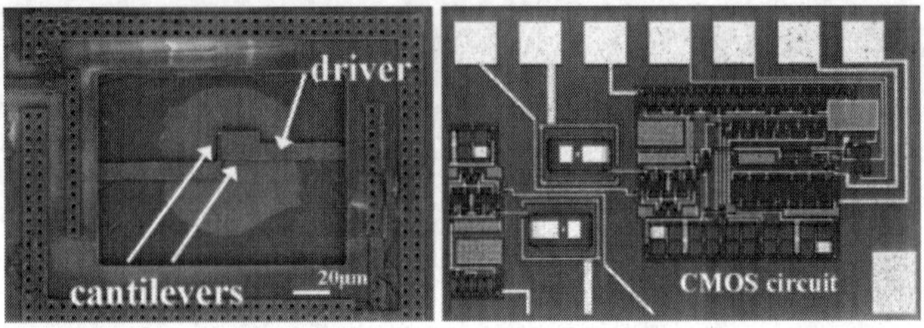

Figure 1.4 Orthogonal cantilever beams for differential measurements integrated into CMOS circuitry by EBL method. Reprinted from Ref. 76.

To enable compatibility with the pre-fabricated CMOS morphology, that is to be able to perform the (blind) pattern irradiations in the selected locations, the nanoareas, a key aspect is the alignment of the EBL respect to the pre-structured topology. This can simply be addressed including proper alignment marks as routinely included in any multilevel lithography fabrication sequence. As mentioned, automated mark recognition can be applied. However, the time allocated for alignment does not overcome but accumulates to the scanning time, which is intrinsically long due to being executed in a serial fashion. Throughput would be a serious drawback of the EBL approach in batch fabrication.

Although NEMS-on-CMOS based on EBL has repeatedly demonstrated to be successfully accomplished, nanomechanical structure manufacture using the post-CMOS fabrication approach is not a routine. Summarizing the multiple steps processing, the nanofabrication is long and complex and requires high precision. Moreover, many potentially aggressive treatments are used, especially for the CMOS circuit and in spite of the final passivation layer that is typically used, such as the dry and wet etchings and resist heat and chemical treatments. As mentioned earlier, additional protection during etching is an option.

More specifically on the nanolithography, the notorious topography after the whole CMOS makes EBL difficult in a basic point. EBL is typically performed upon thin resists layers (commonly PMMA), which are deposited on the substrate by spin coating. As a result resist of variable thickness in certain points are found, as for example in the edges of the nanoarea where the anchoring part and electrical contact of the nanomechanical structure are defined. The inhomogeneity of the resist layer thickness is of great importance not only in terms of EBL resolution and pattern definition, but can really compromise the metal deposition and resist lift-off success rate, and therefore the final NEMS.[45]

Last but not least, and also in terms of CMOS circuitry damage, the EBL irradiation can degrade the electronic performance of the electronic circuits. The higher the beam energy, the higher the spatial resolution, i.e., smaller features can be in principle achieved, which is a result of scattering processes. Charged beam-solid interactions cause effective beam broadening, which also manifests as proximity effect in dense patterns. However, at the same time, increasing electron beam energy also increases electron penetration depth and lateral scattering (and charge-induced cascade), which may lead to the formation of trapped charges in the various interfaces of the CMOS layers. Eventually, this can lead to permanent CMOS circuit damage.[74,75]

Based on this reason and the other difficulties presented above, this scenario promotes the pursuit of an alternative: direct patterning techniques that reduce the number of steps and overcome the vulnerabilities of EBL-based fabrication process, while ensuring the capability of nanoscaled feature definition and providing CMOS compatibility.

1.3.3.2 Ion beam patterning

Most original methods can be addressed by ion beam-based nanopatterning strategies. The all-in-one capabilities for micro-nanomachining of the FIB tools allow multiple approaches for the nanopatterning of NEMS embedded into CMOS circuits. A few examples of FIB-based possibilities for NEMS on CMOS technology are presented here; more details can be found in Refs. 76 and 77.

The diversity of FIB nanofabrication sequences mostly derive from the possibility of applying direct nanopatterning, but multiply on the basis of being available in etching, deposition and implantation modes. The strategies can be classified as (i) strictly based on FIB, (ii) incorporating FIB-based processing to the conventional nanomechanical device fabrication, and (iii) using FIB strategically as local action.

Simplest strictly FIB-based approach would involve direct structuring of the Si layer according to the desired device geometry by FIB milling. Ideally, even release (underetching) could be addressed on the basis of FIB capabilities. However, following this approach would not only be very time consuming but also involves risks from FIB-intrinsic re-deposition phenomena, the accuracy needed to adjust the conditions to the layer thickness, pattern dimensions, etc.

Among the fabrication routes by combination of FIB-based processing with the conventional flow of treatments, patterned deposition of a thin material is one of the options. Any materials can be patterned by FIBID, essentially, depending on the volatile precursor used, but most common metals include platinum, carbon or tungsten. We tested TEOS as mask for RIE from certain pentasiloxane precursor molecules, which confirmed its suitability for sustaining the pattern transfer treatment (Fig. 1.5).

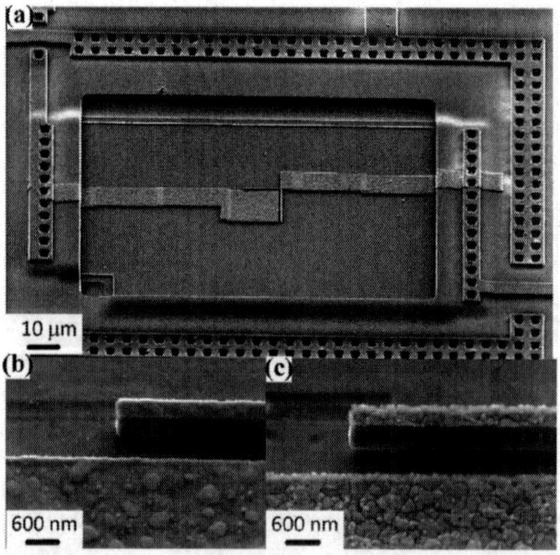

Figure 1.5 (a) SEM images of a polysilicon nanomechanical device integrated into a CMOS circuit defined by IBID of TEOS and RIE in the nanoarea. (b) and (c) are tilted SEM images of the beam tip before and after the wet etching of the underlying oxide to release the cantilever from the substrate. Reprinted with permission from Ref. 76.

Alternative single-step nanopatterning for the fabrication of a RIE shadow mask can be done by the local amorphization of the Si surface in the nanoarea. The exposure of a (Si) crystal to the energetic (Ga) ion beam under specific conditions (doses, beam energy, etc.) causes the amorphization of the crystalline Si according to the feature design. Pattern distortion is not significant due to the limited travel length of Ga ions inside Si. As the resulting ultrathin amorphous Si layer has demonstrated high resistance to etching, it can be used for masking purposes (Fig. 1.6).[78]

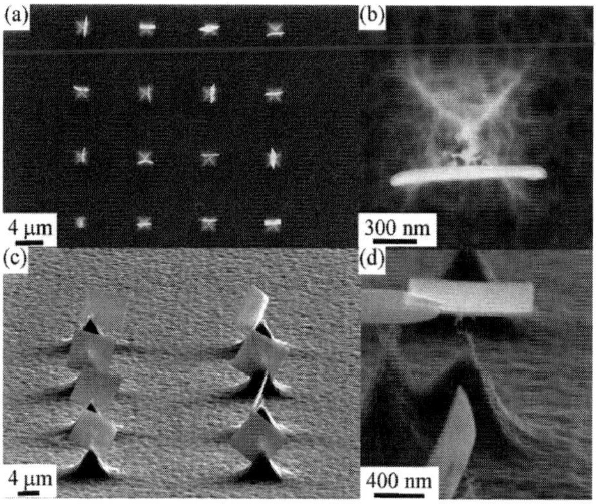

Figure 1.6 SEM images in (a) and (c) top-view and (b) and (d) tilted view of patterns made by local FIB exposure of squared areas on Si, plus RIE in isotropic etching conditions. Reprinted with permission from Ref. 76.

Remarkably, the penetration depth of the ions in the crystal is determined by the incident ion beam energy, being ~25 nm on crystalline Si for 30 kV Ga^+. Based on this, we are testing the application of this phenomenon to form a thickness-limited material to be used as the structural layer of ultrathin nanomechanical devices and NEMS.

Finally and just briefly mentioned, a third application of FIB-based processing is the use of the capabilities of the fine FIB probe to pattern strategic point of the NEMS. For example, formation of an extremely narrow gap between the cantilever beam and the electrode can be in principle realized by (i) milling of a prepatterned thin metal layer plus RIE[79] or (ii) trench milling of the Si structural layer once pre-patterned by RIE.[45] In both cases, precision and side effects, such as derived from the amorphization, make this route not very reliable. At the same time, the advantage of a reduced processing flow is lost.

A summary of the main features for the different FIB strategies discussed above is compiled in Table 1.3.

Table 1.3 Summary of FIB-based processes tested for NEMS-on-CMOS integration

Process	Purpose	Merits	Limitations
Direct etching of Si	Single step patterning and transfer	Simplicity	Time consuming
FIBID of TEOS	Patterning of etching shadow mask	Etching selectivity	Spatial resolution
Ion induced Si amorphization	Structural layer (or etching mask)	Simplicity Ultrathin layer	Operational properties?
Strategic FIB patterning	Critical device-shape patterning (gap, etc.)	Resolution	Side effects Processing complexity

Not only purity and contamination are an issue when ion beam is used, but the use of charged particles beam may have additional effects, as explained for EBL. We have evaluated the effect of FIB-based processing on CMOS and concluded that, as compared to EBL, the use of ion beams may have a lower impact.[80]

FIB tools are not yet made as compatible as EBL for wafer-scale lithography and automation of routines yet, where gas-based processes may be increasingly complex to be implemented. Undeniably, the route following direct patterning strategies is very promising, as we present next for NSL. FIB-based processing appears as an excellent option for prototyping and very flexible tool for preliminary testing of alternate nanofabrication sequences.

1.3.3.3 Nanostencil lithography

To conclude the description of main nanolithography strategies we have tested for the integration of NEMS into CMOS circuits, we briefly present the development based on NSL. As mentioned in Section 1.2.2, the concept of NSL conveniently follows an approach based on direct patterning, therefore advantageous in two aspects: resist-less lithography and reduction of steps in the fabrication sequence. Noteworthy, and in this sense different from FIB-based routines, it relies on the use of a mask, which means less patterning flexibility but represents an important improvement in terms of fabrication throughput.

Figure 1.7 shows the results of wafer-scale patterning of the pre-fabricated CMOS nanoareas by using NSL. Each of the dedicated areas has been patterned with a specific nanoelectromechanical device design according to the mask. Particularly, the bottom left optical image in Fig. 1.7, shows a single clamped beam structure in capacitive scheme for signal monitoring, as described for example in Ref. 50.

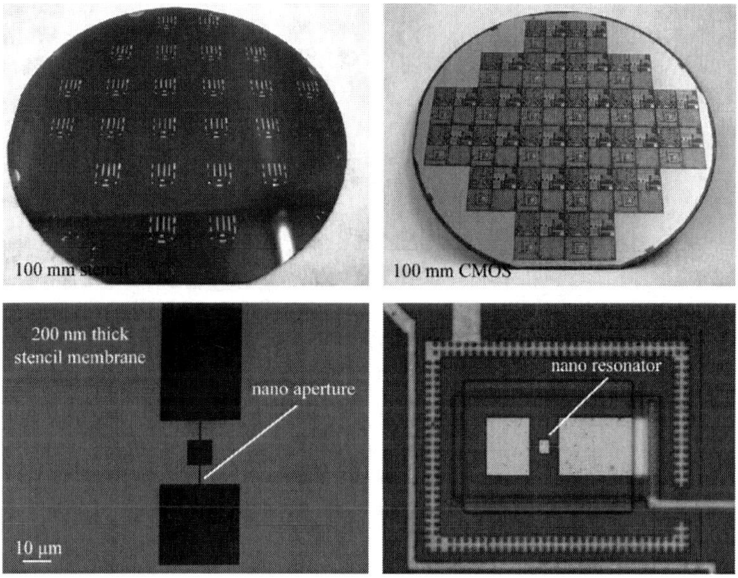

Figure 1.7 Example of monolithical integration of NEMS by using NSL.[81]

In the case of consolidation of NEMS-on-CMOS batch fabrication, including the selection of optimal NEMS design, and aim for its commercial exploitation, NSL would probably be the preferred nanopatterning technique for massive production at competitive cost following NEMS monolithic integration on pre-fabricated CMOS circuits. Another relevant aspect is that the use of masked lithography provides maximum match with the OL used for the

CMOS circuitry fabrication, as it is equivalent method. Special mark requirements and mask alignment are not an issue, so NSL tools adapted to wafer manipulation would do. From the viewpoint of each of the processing conditions and involved treatments it is probably the most compatible of the three alternatives evaluated here.

1.3.3.4 Deep UV optical lithography

For completeness, we describe recent efforts on defining sub-micromechanical devices into CMOS circuits using OL. As the evolution of optical lithography has brought sub-100 nm resolution by deep UV and extreme UV technologies, they should be effective enough for CMOS/NEMS integration. However, the complexity and cost of these techniques makes that they can only be performed in specialized foundries, and, in consequence, the NEMS definition has to be realized as part of the CMOS process in a compatible way as well.

Monolithic CMOS/NEMS integration using standard CMOS technologies has been developed by the Barcelona group during the last years.[5,11,82–84] Nanomechanical resonators are fully defined during the CMOS process, and the structural material is one of the existing layers of the CMOS circuit. A major advantage of this approach is the use of an established technology, based on conventional OL-based patterning. The second and remarkable benefit is the capability to fully exploit the added functionalities by the CMOS circuit (multiplexing, amplification, signal processing, etc). The main disadvantage is the lack of flexibility in terms of material composition and material layer thicknesses, which affects mechanical performance of the structures. A way to overcome this limitation is to provide additional layers to the CMOS in a post-processing approach.

1.4 Concluding Remarks

We have tried to show how and why most of the fabrication methods for NEMS and CMOS/NEMS integration rely on silicon-based technology. The main reason is the combination of good mechanical properties and well-established processing methods. The added-value of NEMS/CMOS integration is its superior performance as a practical system.

However, while most of the works that have been reviewed in this chapter are made of silicon, silicon nitride,[85] silicon carbide,[86–90] or related materials such SiCN[91] are also of high interest. In parallel, an increasing attention is being played to piezoelectric materials[92–95] due to the possibility of an easy implementation of self-transduction (sensing and actuation).

On the other hand, carbon-based nanomechanical devices are gaining more and more interest, due to the possibilities of achieving higher resonance frequencies and more sensitive devices. Fabrication processes for carbon nanotube-based[6,96–99] and graphene-based[100,101] nanomechanical resonators are now well demonstrated for prototyping and proof-of-concept purposes.

EBL is still the most used method to define nanomechanical devices from an experimental point of view, as it is a simple, proven, and reliable method. Nonetheless, EBL is not yet made convenient for massive fabrication and future industrial application. New prototyping approaches have been recently reported, including the use of FIB-based processing or emerging nanopatterning methods such as NSL. Anyway, little progress has been made in finding methodologies that would allow scale-up for the fabrication of NEMS. Some activities include the technology being developed by the so-called Nanosystems alliance between Leti and Caltech.[102] They are working together on OL-based technologies for fabricating single silicon nanowire devices[103] and the extension of CMOS technology to integrated nanomechanical resonators in microelectronic circuits, taking advantage of the high resolution provided by deep UV optical lithography.

To conclude, there has been a sustained research and development activity during the past two decades focused toward the development of specific nanofabricaction methods for nano-mechanical structures, devices, and systems. While the progress is evident and has provided some tools to investigate the performance limits of nanomechanical devices, there are still issues to be clarified before a real NEMS technology would be available for industrial applications. In the following years, there is the need to state if NEMS would play a relevant role in the areas of sensing, telecommunications or even nanoelectronics. For this to be the case, incorporation of NEMS into integrated systems is a requisite.

References

1. H. G. Craighead. Nanoelectromechanical systems (2000). *Science*, **290**, 1532.
2. K. L. Ekinci, and M. Roukes (2005). Nanoelectromechanical systems. *Rev. Sci. Instrum.*, **76**, 061101.
3. K. L. Ekinci, Y. T. Yang, and M. L. (2004).Roukes. Ultrasensitive nano-electromechanical mass detection. *J. Appl. Phys.*, **95**, 2682.
4. J. Verd, A. Uranga, G. Abadal, J. Teva, F. Torres, F. Pérez-Murano, and J. Fraxedas (2007). Monolithic mass sensor fabricated using a conventional technology with attogram resolution in air conditions. *Appl. Phys. Lett.*, **91**, 013501.
5. J. Arcamone, M. Sansa, J. Verd, A. Uranga, G. Abadal, N. Barniol, M. van den Boogaart, J. Brugger, and F. Perez-Murano (2009). Nanomechanical mass sensor for spatially resolved ultrasensitive monitoring of deposition rates in stencil lithography. *Small*, **5**, 176–180.
6. B. Lassagne, D. Garcia-Sanchez, A. Aguasca, and A. Bachtold (2008). Ultrasensitive mass sensing with a nanotube electromechanical resonator. *Nano Lett.*, **8**(11), 3735–3738.
7. K. Jensen, K. Kim, and A. Zettl (2008). An atomic-resolution nanomechanical mass sensor. *Nat. Nanotechnol.*, **3**, 5333.
8. X. Huang, C. Zorman, M. Mehregany, and M. Roukes (2003). Nanoelectromechanical systems: Nanodevice motion at microwave frequencies. *Nature*, **496**, 421.
9. J. Verd, A. Uranga, J. Teva, J. L. Lopez, F. Torres, J. Esteve, G. Abadal, F. Perez-Murano, and N. Barniol (2006). Integrated CMOS-MEMS with on chip read-out electronics for high frequency applications. *IEEE Electron Device Letters*, **27**, 495.
10. I. Martin-Fernandez, X. Borrisé, E. Lora-Tamayo, P. Godignon, and F. Perez-Murano (2010). Batch wafer scale fabrication of passivated carbon nanotube transistors for electrochemical sensing applications. *J. Vac. Sci. Technol. B*, **28**, C6P1
11. J. L. Lopez, J. Verd, J. Teva, G. Murillo, J. Giner, F. Torres, A. Uranga, G. Abadal, and N. Barniol (2009). Integration of RF-MEMS resonators on submicrometric commercial CMOS technologies. *J. Micromech. Microeng.*, **19**, 015002.
12. J. L. Lopez, J. Verd, A. Uranga, J. Giner, G. Murillo, F. Torres, G. Abadal, and N. Barniol (2009). A CMOS–MEMS RF-tunable bandpass filter based on two high-Q 22-MHz polysilicon clamped-clamped beam resonators. *IEEE Electron Device Lett.*, **30**, 718.

13. J. Verd, M. Sansa, A. Uranga, F. Perez-Murano, J. Segura, and N. Barniol (2011). Metal microelectromechanical oscillator exhibiting ultra-high water vapour resolution. *Lab Chip*, **11**, 2670.

14. D. V. Scheible and R. H. Blick (2010). A mode-locked nanomechanical electron shuttle for phase-coherent frequency conversion. *New J. Phys.*, **12**, 023019.

15. X. L. Feng, M. H. Matheny, C. A. Zorman, M. Mehregany, and M. L. Roukes (2010). Low voltage nanoelectromechanical switches based on silicon carbide nanowires. *Nano Lett.*, **10**, 2891.

16. A. Ayari, P. Vincent, S. Perisanu, M. Choueib, M. Gouttenoire, V. Bechelany, D. Cornu, and S. Purcell (2007). Self-oscillations in field emission nanowire mechanical resonators: a nanometric DC-AC conversion. *Nano Lett.*, **7**, 2252.

17. S. Perisanu, T. Barois, P. Poncharal, T. Gaillard, A. Ayari, S. T. Purcell, and P. Vincent (2011). The mechanical resonances of electrostatically coupled nanocantilevers. *Appl. Phys. Lett.*, **98**, 063110.

18. E. Colinet, L. Duraffourg, S. Labarthe, P. Andreucci, S. Hentz, and P. Robert (2009). Self-oscillation conditions of a resonant nanoelectromechanical mass sensor. *J. Appl. Phys.*, **105**, 124908.

19. Y.-C. Liu, M.-H. Tsai, W.-C. Chen, S.-S. Li, and W. Fang (2011). High-*Q*, large stop-band rejection integrated CMOS-MEMS oxide resonators with embedded metal electrodes. *Transducers*, 2011, 934–937.

20. W.-L. Huang, Z. Ren, Y.-W. Lin, H.-Y. Chen, J. Lahann, and C. T.-C. Nguyen (2008). Fully monolithic CMOS nickel micromechanical resonator oscillator. *Tech. Digest, 21st IEEE International Conference on Micro Electro Mechanical Systems* (MEMS'08), Tucson, Arizona, 10–13.

21. J. Verd, A. Uranga, G. Abadal, J. L. Teva, F. Torres, J. Lopez, F. Perez-Murano, J. Esteve, and N. Barniol (2008). Monolithic CMOS MEMS oscillator circuit for sensing in the attogram range. *IEEE Electron Device Lett.*, **29**, 146–148.

22. M. K. Zalalutdinov, J. D. Cross, J. W. Baldwin, B. R. Ilic, W. Zhou, B. H. Houston, and J. M. Parpia (2010). CMOS-integrated RF MEMS resonators. *J. Microelectromech. Syst.*, **19**, 807–815.

23. T.-J. King Liu, J. Jeon, R. Nathanael, H. Kam, V. Pott, and E. Alon (2010). Prospects for MEM logic switch technology. *IEDM10*, 424.

24. X. Chevalier, C. Nicolet, R. Tiron, A. Gharbi, M. Argoud, J. Pradelles, and C. Navarro (2013). Scaling-down lithographic dimensions with block-copolymer materials: 10 nm sized features with PS-*b*-PMMA. *Proc. SPIE*, 8680, 868006.

25. D. W. Carr and H. G. Craighead (1997). Fabrication of nanoelectromechanical systems in single crystal silicon using SOI substrates and electron beam lithography. *J. Vac. Sci. Technol. B*, **15**, 2760.
26. Ollier, E., et al. NEMS based on top-down technologies: from stand-alone NEMS to VLSI NEMS & NEMS-CMOS integration. *Electron Devices and Solid-State Circuits, 2008. EDSSC 2008. IEEE International Conference on*. IEEE, 2008.
27. M. Madou. *Fundamentals of Microfabrication: The Science of Miniaturization* (3rd ed.). Taylor & Francis Limited (2011) ISBN0849331803, 9780849331800.
28. B. Bhushan (2007). Nanotribology and nanomechanics of MEMS/NEMS and BioMEMS/BioNEMS materials and devices Materials for MEMS/NEMS, *Microelectronic Engineering*, **84**(3), 387–412.
29. S. E. Lyshevski. MEMS and NEMS: Systems, Devices, and Structures Nano- and Microscience, Engineering, Technology and Medicine. CRC Press, 2002 ISBN 0849312620, 9780849312625.
30. J. A. Pelesko, and D. H. Bernstein. Modeling MEMS and NEMS, CRC Press, 2002, ISBN 1420035290, 9781420035292.
31. J. Arcamone (2007). Integration of nanomechanical sensors on CMOS by nanopatterning methods, PhD thesis, Universitat Autonoma de Barcelona.
32. B. J. Lin, *Optical Lithography*, SPIE Press, Bellingham, WA, 2009, p. 136.
33. A. Koumela, S. Hentz, D. Mercier, C. Dupre, E. Ollier, P. X.–L. Feng, S. T. Purcell, and L. Duraffourg (2013). High frequency top-down junction-less silicon nanowire resonators. *Nanotechnology*, **24**, 435203.
34. A. Constancias, S. Landis, S. Manakli, L. Martin, L. Pain, and D. Rio (2013). Electron beam lithography, in *Lithography* (ed. S. Landis), John Wiley & Sons, Inc., Hoboken, NJ USA.
35. S. Reyntjens and R. Puers (2001). A review of focused ion beam applications in microsystem technology. *J. Micromech. Microeng.*, **11**, 287–300.
36. G. M. Kim, M. A. F. van den Boogaart, and J. Brugger (2003). Fabrication and application of a full wafer size micro/nanostencil for multiple length-scale surface patterning. *Microelectron. Eng.*, **67–68**, 609–614.
37. J. Melngailis (1987). Focused ion beam technology and applications. *J. Vac. Sci. Technol. B*, **5**, 469.
38. S. J. Randolph, J. D. Fowlkes, and P. D. Rack (2006). Focused, nanoscale electron-beam-induced deposition and etching. *Crit. Rev. Solid State Mater. Sci.*, **31**, 3.

39. P. Zahl, M. Bammerlin, G. Meyer, and R. R. Schlittler (2005). All-in-one static and dynamic nanostencil atomic force microscopy/scanning tunneling microscopy system, *Rev. Sci. Instrum.*, **76**, 023707.

40. J. Brugger, J. W. Berenschot, S. Kuiper, W. Nijdam, B. Otter, M. Elwenspoek (2000). Resistless patterning of sub-micron structures by evaporation through nanostencils. *Microelectron. Eng.*, **53**(1–4), 403–405.

41. G. Abadal, Z. J. Davis, A. Boisen, F. Pérez-Murano, N. Barniol, and X. Borrisé (2001). Combining laser and jumping mode AFM lithography on aluminum for the fabrication of nanoelectromechanical devices, *Microscopy*, **2**, 121–128.

42. M. Villarroya, F. Pérez-Murano, C. Martín, Z. Davis, A. Boisen, J. Esteve, E. Figueras, J. Montserrat, and N. Barniol. AFM lithography for the definition of nanometre scale gaps: application to the fabrication of a cantilever-based sensor with electrochemical current detection. *Nanotechnology*, **15**, 771–776 (2004).

43. G. Luo. Nanoimprint lithography based nanoelectromechanical device fabrication. PhD Thesis. Lund University (2009).

44. E. Gil-Santos, D. Ramos, A. Jana, M. Calleja, A. Raman, and J. Tamayo. Mass sensing based on deterministic and stochastic responses of elastically coupled nanocantilevers. *Nano Lett.*, **9**, 4122–4127 (2009).

45. G. Rius. Electron beam lithography for nanofabrication. PhD Thesis, Universitat Autonoma de Barcelona (2008).

46. G. Villanueva, J. Montserrat, F. Pérez-Murano, G. Rius, J. Bausells. Submicron piezoresistive cantilevers in a CMOS-compatible technology for intermolecular force detection. *Microelectron. Eng.*, **73–74**, 480–486 (2004).

47. J. Arcamone, G. Rius, G. Abadal, J. Teva, N. Barniol, and F. Pérez-Murano. Micro/nanomechanical resonators for distributed mass sensing with capacitive detection. *Microelectron. Eng.*, **83**, 4–9, 1216–1220 (2006).

48. G. Tosolini, G. Villanueva, F. Perez-Murano, and J. Bausells. Silicon microcantilevers with MOSFET detection. *Microelectron. Eng.*, **87**, 1245 (2010).

49. J.-S. Wenzler, T. Dunn, S. Erramilli, and P. Mohanty. Nanoelectromechanical system-integrated detector with silicon nanomechanical resonator and silicon nanochannel field effect transistor. *J. Appl. Phys.*, **105**, 094308 (2009).

50. Q. P. Unterreithmeier, T. Faust, S. Manus, and J. P. Kotthaus. On-chip interferometric detection of nanomechanical motion. *Nano Lett.*, **10**, 887 (2010).

51. K. Y. Fong, W. H. P. Pernice, Mo Li, and H. X. Tang (2010). High-*Q* optomechanical resonators in silicon nitride nanophotonic circuits. *Appl. Phys. Lett.*, **97**, 073112.
52. M. Li, W. H. P. Pernice, and H. X. Tang (2010). Ultrahigh-frequency nano-optomechanical resonators in slot waveguide ring cavities. *Appl. Phys. Lett.*, **97**, 183110.
53. R. A. Barton, B. Ilic, S. S. Verbridge, B. R. Cipriany, J. M. Parpia, and H. G. Craighead (2010). Fabrication of a nanomechanical mass sensor containing a nanofluidic channel. *Nano Lett.*, **10**(6), 2058–2063.
54. J. Lee, W. Shen, K. Payer, T. P. Burg, and S. R. Manalis (2010). Toward attogram mass measurements in solution with suspended nano-channel resonators. *Nano Lett.*, **10**(7), 2537–2542.
55. V. Agache, G. Blanco-Gomez, M. Cochet, and P. Caillat (2011). Suspended nanochannel in MEMS plate resonator for mass sensing in liquid. *MEMS*, 2011, **157–160.**
56. A. Brand (2006). Microsensor integration into systems-on-chip. *Proc. IEEE*, **94**(6), 1160–1176.
57. R. T. Howe and R. S. Muller (1983). Polycrystalline silicon micro-mechanical beams. *J. Electrochem. Soc.*, **130**, 1420–1423.
58. R. T. Howe and R. S. Muller (1986). Resonant microbridge vapor sensor. *IEEE Trans. Electron Devices*, ED-**33**, 499–507.
59. M. Parameswaran, H. P. Baltes, L. Ristic, A. C. Dhaded, and A. M. Robinson (1989). A new approach for the fabrication of microelectro-mechanical structures. *Sens. Actuators*, **19**(3), 289–307.
60. J. Smith, S. Montague, J. Sniegowski, J. Murray, and P. McWhorter (1995). Embedded micromechanical devices for the monolithic integration of MEMS with CMOS. *Proc. IEEE IEDM*, **95**, 609–612.
61. J. Yasaitis, M. Judy, T. Brosnihan, P. Garone, N. Pokrovskiy, D. Sniderman, S. Limb, R. Howe, B. Boser, M. Palaniapan, X. Jiang, and S. Bhave (2003). A modular process for integrating thick polysilicon MEMS devices with sub-micron CMOS. *Proc. SPIE*, **4979**, 145–154.
62. C. Hierold. Intelligent CMOS sensors (2000). *Proc. IEEE MEMS*, 2000, **1–6**.
63. P. F. van Kessel, J. J. Hornbeck, R. E. Meier, and M. R. Douglas (1998). A MEMS-based projection display. *Proc. IEEE*, **86**(8), 1687–1704.
64. G. K. Fedder, S. Santhanam, M. L. Reed, S. C. Eagle, D. F. Guillou, M. S.-C. Lu, and L. R. Carley (1996). Laminated high-aspect-ratio micro-structures in a conventional CMOS process. *Sens. Actuators A*, **57**, 103–110.

65. T. Ishihara, K. Suzuki, S. Suwazono, M. Hirata, and H. Tanigawa (1987). CMOS integrated silicon pressure sensor. *IEEE J. Solid State Circ.*, **22**(2), 151–155.

66. P. Vettiger, G. Cross, M. Despont, U. Drechsler, U. Durig, B. Gotsmann, W. Haberle, M. A. Lantz, H. E. Rothuizen, R. Stutz, and G. K. Binnig (2002). The "millipede"—nanotechnology entering data storage. *IEEE Trans. Nanotechnol.*, **1**, 39–55.

67. Institut de Microelectronica de Barcelona, IMB-CNM-CSIC, http://www.imb-cnm.csic.es/.

68. J. Verd (2008). Monolithic CMOS-MEMS resonant beams for ultra-sensitive mass detection. PhD Thesis, Universitat Autonoma de Barcelona.

69. M. Villarroya (2006). Diseño y fabricación de sistemas micro/nano electromecánicos integrados monolíticamente para aplicaciones de sensores de masa y sensores biológicos con palancas como elementos transductores. PhD Thesis, Universitat Autonoma de Barcelona.

70. M. Sansa (2008). Caracterització d'oscil•ladors basats en CMOS-M/NEMS i la seva aplicació com a sensors de massa. PFC Universitat Autonoma de Barcelona.

71. E. Forsen, P. Carlberg, S. G. Nilsson, L. Montelius, F. Pérez-Murano, J. Esteve, J. Montserrat, E. Figueras, F. Campabadal, J. Verd, G. Abadal, N. Barniol, and A. Boisen (2004). Fabrication of cantilever based mass sensors; integrated with CMOS using laser lithography on resist. *Nanotechnology*, **15**, 628–633.

72. J. Verd, G. Abadal, M. Villarroya, J. Teva, A. Uranga, F. Campabadal, J. Esteve, E. Figueras, F. Pérez-Murano, Z. Davis, E. Forsen, A. Boisen, and N. Barniol (2005). Design, fabrication, and characterization of a submicroelectromechanical resonator with monolithically integrated CMOS readout circuit. *J. Microelectromechan. Syst.*, **14**, 508–519.

73. J. Arcamone, G. Rius, J. Llobet, X. Borrisé, and F. Pérez-Murano (2008). Mass measurements based on nanomechanical devices: differential measurements. *J. Phys. Conf. Ser.*, **100**, 052031.

74. F. Campabadal, S. Ghatnekar-Nilsson, G. Rius, C. Fleta, J. M. Rafi, E. Figueras, and J. Esteve (2005). CMOS degradation effects due to electron beam lithography in smart NEMS fabrication. *Smart Sens. Actuators, MEMS II*, 5836, **667–674**.

75. F. Campabadal, S. Ghatnekar-Nilsson, G. Rius, E. Figueras, and J. Esteve (2005). Nanocantilevers with integrated CMOS: Effects of electron

beam lithography on NMOS transistors. 2005 Spanish conference on electron devices proceedings, 213–216.

76. G. Rius, J. Llobet, X. Borrisé, N. Mestres, A. Retolaza, S. Merino, and F. Perez-Murano (2009). Fabrication of complementary metal-oxide-semiconductor integrated nanomechanical devices by ion beam patterning. *J. Vac. Sci. Technol. B*, **27**(6), 2691.

77. G. Rius, J. Llobet, X. Borrise, and F. Perez-Murano (2011). Fabrication of nanomechanical devices integrated in CMOS circuits by ion beam exposure of silicon, *AIP Conference Proceedings-American Institute of Physics*, **1336**, 239–242.

78. N. Chekurov, K. Grigoras, A. Peltonen, A. Franssila, and I. Tittonen (2009). The fabrication of silicon nanostructures by local gallium implantation and cryogenic deep reactive ion etching. *Nanotechnology*, **20**, 065307.

79. M. Villarroya, N. Barniol, C. Martin, F. Pérez-Murano, J. Esteve, L. Bruchhaus, R. Jede, E. Bourhis, and J. Gierak (2007). Fabrication of nanogaps for MEMS prototyping using focused ion beam as a lithographic tool and reactive ion etching pattern transfer. *Microelectron. Eng.*, **84**, 1215–1218.

80. G. Rius, J. Llobet, J. Arcamone, X. Borrise, and F. Perez-Murano (2009). Electron and ion beam lithography for the fabrication of nanomechanical devices integrated on CMOS circuits. *Microelectron. Eng.*, **86**(4–6) 1046–1049.

81. J. Arcamone, A. Sánchez-Amores, J. Montserrat, M. A. F. van den Boogaart, J. Brugger, and F. Pérez-Murano (2007). Dry etching for the correction of gap-induced blurring and improved pattern resolution in nanostencil lithography. *J. Micro/Nanolith. MEMS MOEMS*, **6**, 013005.

82. J. L. Lopez, J. Verd, J. Teva, G. Murillo, J. Giner, F. Torres, A. Uranga, G. Abadal, and N. Barniol (2009). Integration of RF-MEMS resonators on submicrometric commercial CMOS technologies. *J. Micromech. Microeng.*,**19**, 015002.

83. J. L. Lopez, J. Verd, A. Uranga, J. Giner, G. Murillo, F. Torres, G. Abadal, and N. Barniol (2009). A CMOS–MEMS RF-tunable bandpass filter based on two high-*Q* 22-MHz polysilicon clamped-clamped beam resonators. *IEEE Electron Device Lett.*, **30**, 718.

84. W.-C. Chen, W. Fang, and S.-S. Li (2011). A generalized CMOS-MEMS platform for micromechanical resonators monolithically integrated with circuits. *J. Micromech. Microeng.*, **21**, 065012.

85. Q. P. Unterreithmeier, E. M. Weig, and J. P. Kotthaus (2009). Universal transduction scheme for nanomechanical systems based on dielectric forces. *Nature*, **458**, 1001–1004.

86. R. B. Karabalin, X. L. Feng, and M. L. Roukes (2009). Parametric nanomechanical amplification at very high frequency. *Nano Lett.*, **9**(9), 3116–3123.

87. K. Naik, M. S. Hanay, W. K. Hiebert, X. L. Feng, and M. L. Roukes (2009). Towards single-molecule nanomechanical mass spectrometry. *Nat. Nanotechnol.*, **4**, 445–450.

88. T.-H. Lee, S. Bhunia, and M. Mehregany (2010). Electromechanical computing at 500°C with silicon carbide. *Science*, **329**(5997), 1316–1318.

89. M. Li, E. B. Myers, H. X. Tang, S. J. Aldridge, H. C. McCaig, J. J. Whiting, R. J. Simonson, N. S. Lewis, and M. L. (2010). Roukes nanoelectromechanical resonator arrays for ultrafast, gas-phase chromatographic chemical analysis. *Nano Lett.*, **10**(10), 3899–390.

90. Y. T. Yang, C. Callegari, X. L. Feng, and M. L. Roukes (2011). Surface adsorbate fluctuations and noise in nanoelectromechanical systems. *Nano Lett.*, **11**(4), 1753–1759.

91. C. Guthy, R. M. Das, B. Drobot, and S. Evoy (2010). Resonant characteristics of ultranarrow SiCN nanomechanical resonators. *J. Appl. Phys.*, **108**, 014306.

92. R. B. Karabalin, M. H. Matheny, X. L. Feng, E. Defaÿ, G. Le Rhun, C. Marcoux, S. Hentz, P. Andreucci, and M. L. Roukes (2009). Piezoelectric nanoelectromechanical resonators based on aluminum nitride thin films. *Appl. Phys. Lett.*, **95**, 103111.

93. M. Faucher, B. Grimbert, Y. Cordier, N. Baron, A. Wilk, H. Lahreche, P. Bove, M. François, P. Tilmant, T. Gehin, C. Legrand, M. Werquin, L. Buchaillot, C. Gaquière, and D. Théron (2009). Amplified piezoelectric transduction of nanoscale motion in gallium nitride electromechanical resonators. *Appl. Phys. Lett.*, **94**, 233506.

94. Ü. Sökmen, A. Stranz, A. Waag, A. Ababneh, H. Seidel, U. Schmid, and E. Peiner (2010). Evaluation of resonating Si cantilevers sputter-deposited with AlN piezoelectric thin films for mass sensing applications. *J. Micromech. Microeng.*, **20**, 064007.

95. P. Ivaldi et al. (2011). 50 nm thick AlN film-based piezoelectric cantilevers for gravimetric detection. *J. Micromech. Microeng.*, **21**, 085023.

96. G. A. Steele, A. K. Hüttel, B. Witkamp, M. Poot, H. B. Meerwaldt, L. P. Kouwenhoven, and H. S. J. van der Zant. Strong coupling between

single-electron tunneling and nanomechanical motion. *Science*, **28**, 1103–1107 (2009).

97. Eichler, J. Chaste, J. Moser, and A. Bachtold. Parametric amplification and self-oscillation in a nanotube mechanical resonator. *Nano Lett.*, **11**(7), 2699–2703 (2011).
98. H. Cho, M.-F. Yu, and A. F. Vakakis, L. A. Bergman, and D. M. McFarland. Tunable, broadband nonlinear nanomechanical resonator. *Nano Lett.*, **10**(5), 1793–1798 (2010).
99. A. Eichler, J. Moser, J. Chaste, M. Zdrojek, I. Wilson-Rae, and A. Bachtold. Nonlinear damping in mechanical resonators made from carbon nanotubes and graphene. *Nat. Nanotechnol.*, **6**, 339–342 (2011).
100. A. Chen, S. Rosenblatt, K. I. Bolotin, W. Kalb, P. Kim, I. Kymissis, H. L. Stormer, T. F. Heinz, and J. Hone. Performance of monolayer graphene nanomechanical resonators with electrical readout. *Nat. Nanotechnol.*, **4**, 861–867 (2009).
101. S. Shivaraman, R. A. Barton, X. Yu, J. Alden, L. Herman, M. V. S. Chandrashekhar, J. Park, P. L. McEuen, J. M. Parpia, H. G. Craighead, and M. G. Spencer. Free-standing epitaxial graphene. *Nano Lett.*, **9**, 3100–3105 (2009).
102. wwww.nanovlsi.com.
103. A. Koumela, et al. (2013). High frequency top-down junction-less silicon nanowire resonators. *Nanotechnology*, **24**(43), 435203.

Chapter 2

Polymer Cantilevers and Novel Transduction Techniques for Nanoelectromechanical Sensing

Prasenjit Ray,[a] V. Seena,[b] and V. Ramgopal Rao[a]

[a]*Department of Electrical Engineering, IIT Bombay, Mumbai, India*
[b]*Department of Avionics, IIST, Thiruvananthapuram, India*

rrao@ee.iitb.ac.in

Polymer-based micro/nanoelectromechanical systems open a new area of research interest for their possible use in a variety of domains such as physical sensing, chemical sensing, bio-sensing, and inertial sensing applications. Polymer micro-electro-mechanical system (MEMS) devices have potential to find their place in mainstream sensing area because of their low processing cost, versatility to adapt to different sensing applications such as from thin films to high-aspect-ratio structures, low stiffness, and biocompatibility. In a broader view, micro-electromechanical systems also encompass micro-actuators.

2.1 Different Transduction Mechanisms

In this section, a few fundamental concepts and methods about transduction are reviewed. Most widely used transduction

Nanocantilever Beams: Modeling, Fabrication, and Applications
Edited by Ioana Voiculescu and Mona Zaghloul

ISBN 978-981-4613-23-1 (Hardcover), 978-981-4613-24-8 (eBook)
www.panstanford.com

mechanisms are piezoresistive, piezoelectric, capacitive, thermal, and tunneling resonant frequency–based techniques.

2.1.1 Piezoresistive Transduction

Piezoresistive effect is a widely used transduction mechanism for sensors. This is the effect of change in resistance in an electrical resistor upon application of an external strain. Such an effect is seen in various materials from metals to semiconductors and composites. The change in resistance can happen mainly in two ways: (1) geometrical change due to strain, which is particularly dominant in the case of metal strain gauges and (2) change in the resistivity of the material due to internal atomic position and their motion because of strain.[1,2] These days, piezoresistive effect is used in the MEMS field for a wide variety of sensing applications, including accelerometers, pressure sensors, cantilever-based sensors, tactile sensors, gyros, flow sensors, and many other chemical and biological sensors.

2.1.1.1 Metal strain gauge

The most well-known piezoresistor for sensing purposes is a metal strain gauge; metal strain gauges were used even before semiconducting materials such as silicon came into picture. The origin of piezoresistivity in metals lies in the geometry changes and in a change of the internal atomic positions upon application of stress. Due to stress, the energy band of metals undergoes a slight distortion resulting in a change in conduction, leading to the piezoresistive effect. In MEMS fabrication processes, metal thin-film resistors are generally deposited using sputter or evaporation. Elemental metal thin films used as strain gauges in MEMS usually have their gauge factors ranging from 0.8 to 3. Strain gauges made of thin-film metals do not compare favorably with semiconductor strain gauges in terms of their gauge factor values. However, the main reason for use of metal strain gauges instead of semiconductor materials is that this would eliminate the need for doping. Metals can also generally sustain a much greater elongation before fracture. As such metal resistors can be integrated with polymer materials for polymer MEMS device applications.[3,4] To meet the requirements of aeronautics and aerospace applications, where the stress gradients encountered are high, thin-film strain gauges (TFSGs) using palladium-13% rhodium as piezoresistive materials have been reported.[5] Gold piezoresistors have also been used for fabrication of an SU-8 three axis sensor for measuring tactile sensitivity and force exerted during locomotion of small biological organisms.[6] Doped silicon piezoresistors are preferred over metal strain gauges for sensor applications because of the low gauge factor for the metals. However, metal thin-film piezoresistors are desirable in devices fabricated using polymer MEMS technology, since low processing temperatures are involved.

2.1.1.2 Doped silicon piezoresistor

The piezoresistivity in silicon arises in the same way as in metal strain gauges, though the dominant mechanism is often different. When a stress is applied to a semiconductor, the band energies change by a small amount and unlike in metals, this small shift in energy band causes a significant change in conductivity in semiconductors. This gives a large piezoresistive gauge factor. Relative orientation of crystallographic direction of silicon affects the piezoresistive coefficients. Considering a rectangular coordinate system having arbitrary orientation with respect to the crystallographic axes of a homogeneous semiconductor, the electric field components E_i and current density components i_i are related by a symmetric resistivity matrix Eq. 2.1.

$$\begin{pmatrix} E_x \\ E_y \\ E_z \end{pmatrix} = \begin{bmatrix} \rho_1 & \rho_6 & \rho_5 \\ \rho_6 & \rho_2 & \rho_4 \\ \rho_5 & \rho_4 & \rho_3 \end{bmatrix} \begin{pmatrix} i_x \\ i_y \\ i_z \end{pmatrix} \tag{2.1}$$

There are six independent stress components in 3D-space, three normal stresses (σ_{xx}, σ_{yy}, σ_{zz}) and three shear stresses (τ_{xy}, τ_{yz}, τ_{zx}). These notations are simplified and unified accordingly as $\sigma_{xx} \rightarrow T_1, \sigma_{yy} \rightarrow T_2, \sigma_{zz} \rightarrow T_3, \tau_{yz} \rightarrow T_4, \sigma_{zx} \rightarrow T_5$ and $\tau_{xy} \rightarrow T_6$. Six independent components of resistivity matrix, ρ_1 to ρ_6 are related to six stress components by Eq. 2.2. If *x*, *y*, and *z* axes are aligned to <100> crystal axis in silicon, the relation between resistivity and stress expressed in matrix equation format (see Eq. 2.2).

$$\begin{pmatrix} \Delta\rho_1/\rho_0 \\ \Delta\rho_2/\rho_0 \\ \Delta\rho_3/\rho_0 \\ \Delta\rho_4/\rho_0 \\ \Delta\rho_5/\rho_0 \\ \Delta\rho_6/\rho_0 \end{pmatrix} = [\pi][T] = \begin{bmatrix} \pi_{11} & \pi_{12} & \pi_{12} & 0 & 0 & 0 \\ \pi_{12} & \pi_{11} & \pi_{12} & 0 & 0 & 0 \\ \pi_{12} & \pi_{12} & \pi_{11} & 0 & 0 & 0 \\ 0 & 0 & 0 & \pi_{44} & 0 & 0 \\ 0 & 0 & 0 & 0 & \pi_{44} & 0 \\ 0 & 0 & 0 & 0 & 0 & \pi_{44} \end{bmatrix} \begin{pmatrix} T_1 \\ T_2 \\ T_3 \\ T_4 \\ T_5 \\ T_6 \end{pmatrix}, \quad (2.2)$$

where ρ_0 is the isotropic resistivity of the unstressed *d* crystal and the terms Π_{ij} the component of piezoresistor tensor. There are three independent piezoresistor co-efficient matrices: Π_{11}, Π_{12}, and Π_{44}. Doping and temperature of substrate determine the piezoresistive coefficients of single-crystal silicon. For both p- and n-type silicon, the value of the piezoresistive coefficient decreases with increasing temperature and doping concentrations. The Tables 2.1 and 2.2 show the piezoresistivity components for single-crystal silicon for different doping values and transverse and longitudinal gauge factors, respectively.

Table 2.1 Piezoresistivity components for single-crystal silicon for different doping values

Piezoresistance co-efficient (10^{-11}Pa^{-1})	n-type (resistivity = 11.7 Ω-cm)	p-type (resistivity = 7.8 Ω-cm)
Π_{11}	–102.2	6.6
Π_{12}	53.4	–1.1
Π_{44}	–13.6	138.1

Table 2.2 Transverse and longitudinal gauge factors for various commonly encountered resistor configurations

Direction of strain	Direction of current	Configuration	Piezoresistive co-efficient
<100>	<100>	Longitudinal	Π_{11}
<100>	<010>	Transverse	Π_{12}
<110>	<110>	Longitudinal	$(\Pi_{11} + \Pi_{12} + \Pi_{44})/2$
<110>	<110>	Transverse	$(\Pi_{11} + \Pi_{12} - \Pi_{44})/2$
<111>	<111>	Longitudinal	$(\Pi_{11} + 2\Pi_{12} + 2\Pi_{44})/2$

2.1.1.3 Polymer composite piezoresistors

Another new and important piezoresistor element is based on conducting polymers. Plastic materials are usually considered as insulating. With the increasing demand for flexible electronics, polymer conducting materials find wide spread use in sensor and MEMS areas. When conductive fillers such as metal nanoparticles, carbon nanoparticles, and graphite fibers are dispersed into a polymer matrix above a certain threshold concentration value, a percolation path will be formed initiating conduction through the nanoparticles. Depending on the concentration of nanoparticles inside the polymer matrix, resistivity of these polymer composites changes. Carbon black is also often used as a filler material. Incomplete decomposition of aromatic hydrocarbons at high temperatures through rupturing of C-H bonds leads to the formation of carbon black. Then, the carbon atoms and the aromatic radicals react to form layered structures with hexagonal carbon rings. They stack in layers to form crystallographic structures and hence impart conductivity to the material. Conducting polymers/conducting polymer composites have several applications in the field of sensors such as composite thin film of polymethylmethacrylate (PMMA) with multiwall carbon nanotube for gas sensing application,[8] MWNT-polydimethylsiloxane (PDMS) composites used to fabricate heaters integrated into microfluidic channels, and even as capacitive pressures sensors.[9] Carbon black–filled polymers often exhibit a large piezoresistive behavior. The mechanism for piezoresistance in this composite is not the same as with metal thin films or doped silicon. Piezoresistance in composite films comes from the particle separation in polymer matrix. The transition from insulator to conductor for a polymer takes place abruptly at a threshold concentration of the filler particles known as the percolation threshold. Any process in this composite, which is able to change the distance between particles and contact area between nanoparticles, can also change the resistivity of the material. The main reason for this piezoresistivity is thought to be because of the difference between the filler and matrix compressibility. A variety of sensors have been fabricated using the piezoresistive property of conductive filler polymer composites. An all polymer two-axis artificial hair cell flow sensor using carbon black-polyurethane composite with 30% loading has been demonstrated.[10] A gauge factor as high as 10 has been reported for 10% CB in PDMS.[11] A much higher gauge factor has been shown in 10% CB in SU-8 polymer.[12–14]

2.1.2 Piezoelectricity

Because of application of a mechanical stress, certain class of crystals show potential generation. This crystal also shows a mechanical deformation when an external electric field is applied to them. The first effect is called the *direct effect* of piezoelectricity and second one is referred to as the *inverse effect*. These properties of the material enable such materials to be used as both sensors and actuators. Using molecular models, it is possible to explain the generation of electric charge due to the application of force. In the case of piezoelectric material, the centers of gravity of positive and negative charges of each molecule coincide and thereby cancelling the external effects of these charges. While exerting pressure on such a material, its internal structure get deformed causing the separation of positive and negative charge centres and formation of dipoles. Thus the material gets polarized and with this the mechanical energy can be transformed in to electrical energy.

2.1.3 Capacitive Sensing Methods for MEMS

Popular MEMS sensors that are in existence today in mainstream commercial products use capacitive sensing. The capacitance based sensing methods are relatively simpler compared to the other techniques. Usually the structure consists of a capacitor with one or more fixed set of electrodes and one or more movable set of electrodes. The capacitance for simple parallel plate configuration can be represented by the Eq. 2.3.

$$C = \frac{\varepsilon_0 \varepsilon_r A}{d}, \tag{2.3}$$

where ε_0 and ε_r are permittivity of free space and relative permittivity of material between plates, respectively. A is the area overlap between two plates and d is distance between the plates. Due to mechanical motion, either the overlap area or distance between the two plates changes, which in turn changes the capacitance. One of the most widely used products based on capacitive technique is an accelerometer. In such a device, the a proof mass connected between fixed springs moves when the structure undergoes acceleration. This leads to change in gap between two different electrodes thereby creating a differential capacitance. This is one of the most popular scheme that are widely used in all commercial airbag sensors because of the ruggedness and high sensitivity of such accelerometers. In another approach, the capacitive based accelerometer can be implemented where proof mass moves in Z direction and changes the gap between two conducting plates. Temperature changes do not affect the performance of these devices. However, the capacitance techniques are often difficult to implement for sensing in liquids.

2.1.4 Tunneling MEMS

Several groups used the high displacement sensitivity of tunneling processes for high sensitivity transduction applications.[18–21] Operation of this sensor depends on the exponential dependence of tunneling current I_t, with electrode gap separation h given by Eq. 2.4.

$$I_t \propto V_B \exp(-\alpha(\Phi)^{1/2} h), \tag{2.4}$$

where V_B is the tunneling bias across electrodes, α is constant (α = 1.025 eV$^{-1/2}$ Angstrom^{-1}) and Φ is the tunneling barrier height. Tunneling sensor has the advantage over the more common capacitive, piezoresistive, and piezoelectric transducers, in that it has smaller critical sensing area and the tunneling current is largely temperature independent. Without sacrificing on the signal to noise ratio of the sensor, it can be scaled down to extremely small areas. The tunneling sensor is illustrated in Fig. 2.1.

Figure 2.1 Schematic of tunneling-based sensor.

2.1.5 Thermal MEMS Sensor

This is a less complex MEMS sensor based on temperature difference. The main part of this device is a micro heater.[11,12] In this system due to the temperature difference between a heated element and the surrounding gas inside the sensor, a convective flow of surrounding gas is generated. When any acceleration or force is applied, the change in convective flow of gas causes the temperature difference between the sides of the heated element, which is proportional to applied acceleration. Temperature sensors placed on either side of the heat source measure a differential output corresponding to the applied acceleration. Temperature sensing can be based on thermocouple or thermistor effect. The device is packaged in a sealed chamber to prevent external air flow from disturbing the device operation.

2.1.6 MOSFET-Embedded MEMS Sensor

Another new promising mechanical sensor based on MEMS technology is a MOSFET and a thin-film transistor embedded micro-cantilever.[22,23] In this approach, one metal-oxide-semiconductor field effect transistor was embedded in the most high stress region. When any bio-molecule (target molecule) binds with the receptor molecule on the cantilever surface, a surface stress is generated, which changes the channel mobility underneath the gate region because of the resultant strain. Any change in mobility of the carriers inside channel will change the drain current. With optimized design and process parameters, the MOSFET-embedded microcantilever has shown a high sensitivity. Apart from the high sensitivity, another major advantage of this MOSFET embedded micro-cantilever is its possibility for integration with on-chip radio-frequency element and its low power dissipation.

2.2 Polymer Microcantilever Sensors with Novel Transduction Mechanisms

In the past one decade, polymer MEMS technologies gained lot of attention from research community worldwide. Most of the earlier MEMS devices were made out of silicon, silicon nitride, or silicon dioxide. These materials have a very high Young's modulus and are brittle in nature. Due to the lower Young's modulus value of polymers, they would have advantages over silicon-based devices in terms of their surface stress sensitivity. The standard process flow followed for fabricating silicon based MEMS devices can be used for the fabrication of polymer MEMS devices as well. However, these existing processes are supplemented by a few other polymer specific processes such as spin coating, spray coating, molding, low-temperature vapor deposition, stereo lithography, screen printing, etc. Lower process cost is another advantage with polymer-based MEMS processes. In addition to these obvious advantages, it is easy to integrate novel sensing mechanisms into polymer devices by embedding different kinds of nanostructures inside polymer matrix thereby bringing lot of functionality to the polymer structural material.

2.2.1 SU-8/MWNT Nanocomposite Film–Based Polymeric Piezoresistive Microcantilever

SU8-based polymer cantilevers with gold (Au) and polysilicon as piezoresistive layers were reported earlier.[24,25,43] However, due to the lower gauge factor of gold, of the order of 2, there is a limit to the achievable sensitivity with such devices for applications in sensors. Carbon black/SU-8 composite is reported recently as another option for piezoresistive polymer cantilevers with gauge factor close to 90 with 8–9 vol% CB in SU-8.[26] These cantilevers have been successfully used for bio applications and explosive application.[14,26] Instead of carbon black, the use of multiwall carbon nanotubes (MWNT) is expected to show improvement in dispersion after chemical modification as has been shown in the next section resulting in a higher gauge factor for MWNT/SU-8 composite compared to CB/SU-8 composite. SU-8/MWNT composite would be another option for polymeric piezoresistive sensor.[28]

2.2.1.1 SU-8/MWNT composite film preparation

SU-8/MWNT films are ideally suited for piezoresistive sensing applications. To get a proper piezoresistive behavior, CNTs need to be uniformly distributed all over the polymer matrix and the composite film should achieve some conductivity.

First, this NT (NC-3100, purified multiwall carbon nanotubes, L/D 100–1000, purity > 95%), which was procured from Nanocyl CA, Belgium, was modified using octadecyl triphenyl phosphonium bromide (OTPB) as reported earlier.[27] The purpose of modification is to reduce MWNT aggregate size in SU-8 matrix and assist in the de-bundling of MWNT. The de-bundling leads to an improved dispersion of MWNT in SU-8. Subsequently, this modified MWNT powder was kept in a baking oven for a dehydration bake at 120°C to remove the moisture.

$$\text{CNT\%} = \frac{M_{\text{MWNT}}}{M_{\text{MWNT}} + V_{\text{SU-8}} \times \rho_{\text{SU-8}}} \times 100$$

where M_{MWNT} is the mass of MWNT powder and $V_{\text{SU-8}}$ and $\rho_{\text{SU-8}}$ are the volume and density of SU-8, respectively.

Several mixing methods have been tried to mix the modified MWNT with SU-8. Probe sonication and bath sonication techniques were tried out. However, the probe sonication method gave better dispersion compared to the other method. The probe sonication energy was fixed at 500 J to avoid the breakage in the MWNT structure. To avoid solvent evaporation due to the heat generated during the sonication process, an ice bath was used during the sonication. In addition, nanothinner (cyclopentanone) from Microchem was added in 1:1 ratio with SU-8. MWNT was initially sonicated in nanothinner followed by sonication after adding SU-8 to the MWNT/nanothinner. This resulted in a uniformly dispersed spin coatable SU-8 polymer composite with MWNT concentration between 0.05 and 0.2 wt%.

The spin coating parameters were optimized for this SU-8/MWNT composite. The photopatternable composite film was exposed to UV rays in a standard mask aligner system to cross link the SU-8. UV dose required was found to be slightly larger than that required for pure SU-8 2002. The optimized UV-dose for the film thickness of 1 μm is around 60 mJ/cm^2. Rest of the processes are same as that of SU-8 lithography. After development the residual developer should be removed using Isopropyl alcohol (IPA). Bath-sonication was performed to remove MWNT residues from the unexposed area. The spin coated and UV-exposed SU-8/MWNT composite film is shown in Fig. 2.2.

Figure 2.2 SEM micrograph of (a) patterned SU-8/MWNT composite. SU-8/MWNT composite film at (b) 1 um scale and (c) 100 nm scale.

2.2.1.2 Electrical characterization

The conduction inside polymer composite happens through percolation path formation. There are mainly three possibilities for the distribution of nanoparticles inside the polymer matrix: (i) The nanoparticles in complete isolation inside composite, (ii) Nanoparticles in close proximity to each other, and (iii) the nanoparticle in contact with each other. In the first case when the nanoparticles are in isolation there will be no conduction through the composite matrix with only a change in the mechanical or dielectric properties. For the second case, when the nanoparticles are in close proximity to each other, conduction can happen through mechanisms such as tunneling or hopping. In tunneling, the electrons can cross the energy barrier without activation energy. In hopping, a higher activation energy is required to raise the energy of electron for it to cross the barrier. The traditional hopping probability can be represented by Eq. 2.5.[29]

$$P = a\tau e^{-\Delta E/kT} + e^{-\Delta E_{\mathrm{p}}/kT}, \tag{2.5}$$

where a is constant, τ is the tunneling factor, ΔE is the activation energy and ΔE_p is the polarization energy and T is temperature. In the third case, when nanoparticles are in contact with each other, the conduction mechanism depends on the inherent conduction properties of nanoparticles.

The conduction through polymer matrix depends on the following factors: (i) nanoparticle concentration (ii) nanoparticle conductivity, and (iii) aspect ratio of the nanoparticles.

Percolation threshold value for conductivity of polymer composite is really sensitive to the aspect ratio of the nanoparticles. Aspect ratio is the ratio of the long dimension to the short dimension of the nanoparticle. With increasing aspect ratio, the percolation threshold shifts towards lower value, i.e., the film starts conducting at a lower concentration of nanoparticles. In this work, multiwall carbon nanotube (MWNT) has been used as filler nanoparticle material with a very high aspect ratio, so it helps achieve a higher conduction at lower concentration values. Also, the conductivity of MWNT is high compared to other nanoparticles such as carbon black.

The piezoresistors with different OTPB modified MWNT (1:1) concentrations of 0.05 wt% in SU-8, 0.075 wt% in SU-8, 0.1 wt% in SU-8, and 0.2 wt% in SU-8 were tested without any bending. A programmable voltage source is used to vary the voltage applied to the piezoresistor and current is measured using a Keithley 4200 parametric analyzer system. The results are shown in Fig. 2.3, and it is observed that reasonable conductivity is obtained only beyond 0.1 wt% of MWNT in MWNT/SU-8 piezoresistors.

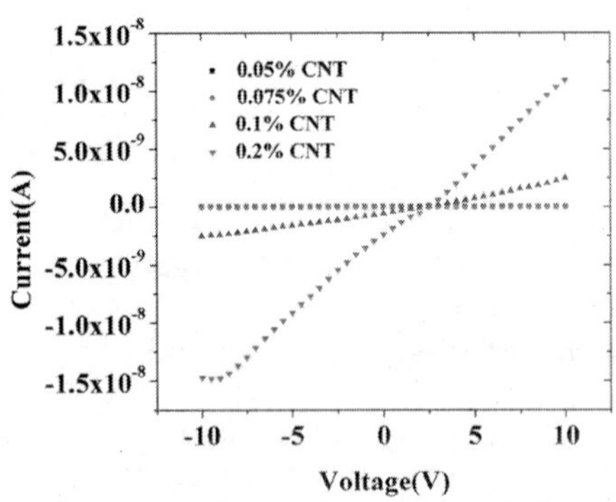

Figure 2.3 Current versus voltage plot for four different concentrations of MWNT in SU-8.[28]

2.2.1.3 Polymer micro-cantilever fabrication with SU-8/MWNT composite as piezoresistor

Flip-chip technique was followed to fabricate piezoresistive cantilevers with SU-8/MWNT polymer composite as the piezoresistive layer.[26,27,44,47] The process starts with an oxidized silicon wafer with thickness of 1000 nm. This oxide acts as a sacrificial layer for the final release of the device from the silicon wafer.

To protect the piezoresistive composite layer, a polymer layer needs to be patterned on top of the cantilever beam. This layer will also form the first level of SU-8 die with contact vias. The thickness of this layer should be thin enough so that the piezoresistive layer will not be closer to the neutral axis of the cantilever beam. The minimum thickness we can get from the SU-8 2002 is 900 nm by spinning at 6000 rpm and from SU-8 2000. 5 it is 500 nm at a spin speed of 3000 rpm. However, the film quality of the first layer is better compared to the second one in terms of porosity. To avoid this porosity problem, SU-8 2002 was chosen and was spin coated at 6000 rpm and prebaked at 70 and 90°C both for 3 min each to get around 0.9 μm thickness of SU-8. Using required dose of UV light, this SU-8 layer was patterned. This patterned SU8-2002 will act as an encapsulation layer for SU-8/MWNT composite layer. A post exposure bake is given with the same baking parameters such as used for the prebaking. The SU-8 was developed using SU-8 developer PGMEA from Microchem. A hard bake is given at a temperature of 120°C to cross link the structure completely and to release any residual stress in the SU-8 film. In the next step, contact via and electrode need to be fabricated. For that purpose a Cr/Au layer of 15 nm/100 nm was deposited using RF sputtering method. In this process wet etching technique has been followed instead of lift off, because the adhesion of Cr layer with SU-8 is very poor and it may be possible that the electrode layer will come off during liftoff. For this reason, after deposition the contact via and electrode layer were patterned using etching of Cr/Au layer after standard photolithography process with S1813 resist. At this stage, the sample is ready to form the piezoresistive layer of SU-8/MWNT. This layer was fabricated as per the process described in previous section using piezoresistor mask. Now the bottom encapsulation layer needs to be fabricated to protect the piezoresistor layer. For this SU8-2005 was spin coated and patterned to cover the

MWNT/SU-8 composite piezoresistor. Here the achievable thickness using SU-8 2005 was 5 μm. An anchor layer of thickness of 150 μm using SU8-2100 is coated and patterned using standard photolithography. This much thickness is required for easy handling of the SU-8 die using tweezer after release from the silicon wafer. A hard bake at 110°C is given to release the residual stress. As a final stage of release, the sacrificial layer of oxide is etched by buffered hydrofluoric acid (5:1). Average etching time depends on the oxide thickness and quality, which can last for a few minutes to hours. After release, the cantilever die can be found to be floating in the etchant. The die is then taken from the etchant and was rinsed in deionized water and isopropyl alcohol. Then it was dip in Cr etchant (22% Ceric ammonium nitrate, 8% acetic acid, and 70% DI water) to etch the Cr layer on top of the contact vias. The entire process sequence is shown in Fig. 2.4. SEM microphotographs of fabricated devices are shown in Fig. 2.5.

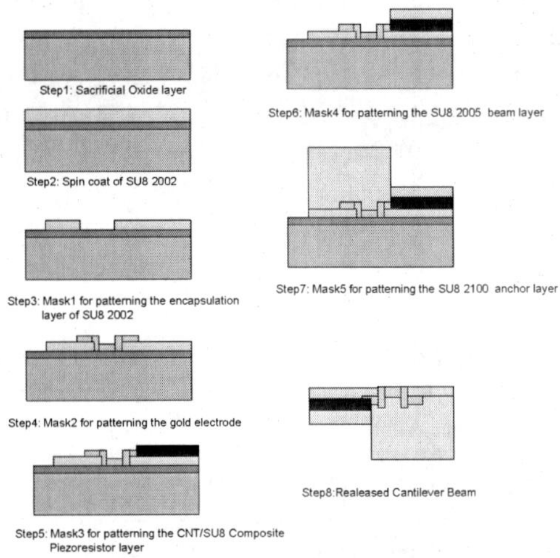

Figure 2.4 Process flow for Polymer/CNT composite cantilever.[28]

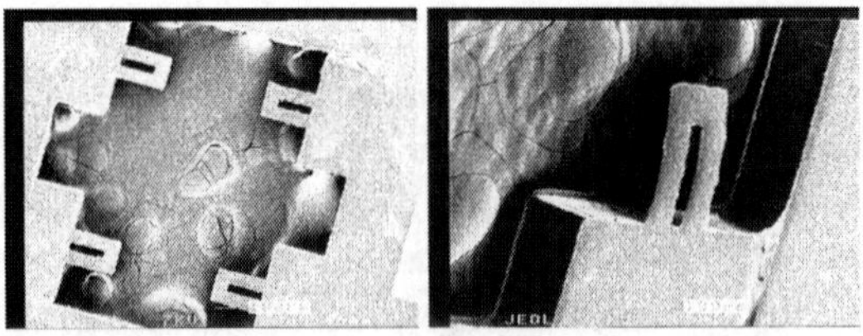

Figure 2.5 SEM images for fabricated polymer micro-cantilever.[28]

2.2.1.4 Electromechanical characterization of the fabricated micro-cantilever

The fabricated Polymer Microcantilever with SU-8/MWNT composite was characterized electromechanically to explore the piezoresistive behavior for sensing purposes. This electromechanical characterization was performed by bending the micro-cantilever by calibrated micromanipulator needle using a Suss Microtech probe station. The minimum possible deflection was 10 μm in vertical direction. Simultaneously the current was measured using Keithley 4200 SMU by varying the voltage through the resistor. The Polymer micro-cantilever with composite was fabricated for 0.1% and 0.2% weight percentage of MWNT in SU-8. First, the reading of current-versus-voltage has been taken for resistor without bending it. The cantilever beam is then deflected from 10 to 50 μm with 10 μm steps. The *I*–*V* results for 0.1 and 0.2 wt% MWNT are shown in Fig. 2.6. The current is an order of magnitude higher for 0.2 wt% MWNT than 0.1 wt% sample.

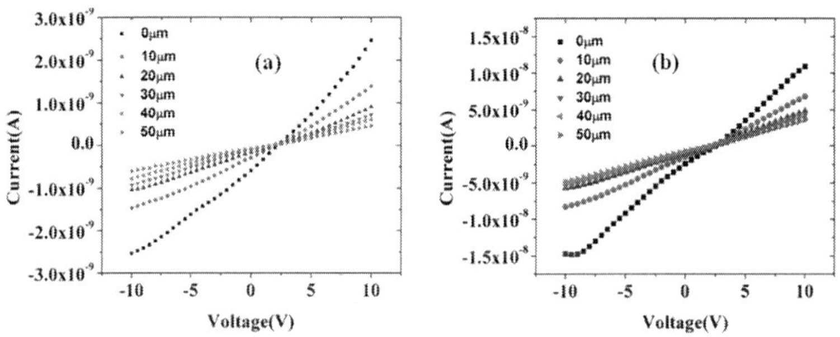

Figure 2.6 Current versus voltage for different value of deflection of cantilever beam. (a) With 0.1 wt% MWNT in SU-8 and (b) 0.2 wt% CNT in SU-8.[28]

The relative change of resistance ($\Delta R/R$) with deflection of cantilever beam is shown in Fig. 2.7. In the SU-8/MWNT based composite, the conduction is due to the tunneling between the carbon nanotubes. So when any stress is generated due to the bending of cantilever beam the separation between the gap changes. This distance between the carbon nanotubes is related exponentially with the tunneling factor (τ). So with change of gap between MWNT the tunneling factor changes exponentially, which in turn helps to change the current. This sensor is extremely attractive for ultrasensitive sensing applications. Linearity, however, needs to be improved by further experiments by varying the MWNT concentration.

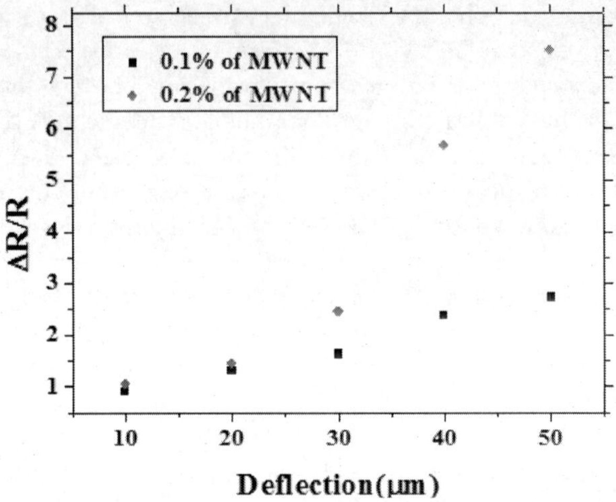

Figure 2.7 Relative change of resistance as a change in deflection for two different cantilevers with 0.1 and 0.2 wt% MWNT in SU-8.[28]

2.2.2 Organic CantiFET

Polymer nanomechanical sensor platforms with integrated strain-sensitive materials having lower Young's modulus values similar to the structural polymer layer would be ideal for developing ultra-sensitive sensors. Pentacene, which is a well-understood organic material, is commonly used as a channel material in organic thin-film transistors (OTFT). Pentacene is reported to exhibit good strain sensitivity and also has a Young's modulus nearly matching that of SU-8.[30,31] The most popular deposition process for Pentacene is vacuum sublimation. This low-temperature vacuum sublimation process of deposition of pentacene ensures uniform deposition of pentacene layers suitable for polymer substrates.[32] OTFTs are used in realizing large-area, flexible and low-cost devices and circuits such as paper-like displays, radio frequency identification tags, etc. However, the effect of mechanical strain on the electrical behavior of pentacene-based OTFTs has always been a concern while discussing the reliability of OFETs for flexible electronics applications.[33–45] There are a few reports available in literature that investigated the effect of bending induced strain on change in current of pentacene OTFTs.[32–36,37] This negative aspect of dependence of electrical behavior on bending induced strain in pentacene OFETs for flexible electronics could be utilized to realize strain sensor applications using pentacene.

The new device concept is to embed a pentacene OTFT inside an SU-8 microcantilever as shown in Fig. 2.8. This novel device, named "organic CantiFET," can be considered a low-cost polymer counterpart of a similar device in silicon, a MOSFET-embedded microcantilever reported earlier.[22,38] When SU-8 microcantilevers with embedded OTFT undergo nanomechanical motion during sensing events, the organic semiconductor pentacene responds to this by changing its mobility and hence the drain current of the FET. The change in drain current can be measured and recorded using appropriate signal conditioning circuitry in order to perform the sensing operation. A simple differential amplifier circuit with sensing and reference transistors could be used in a complete sensor configuration.

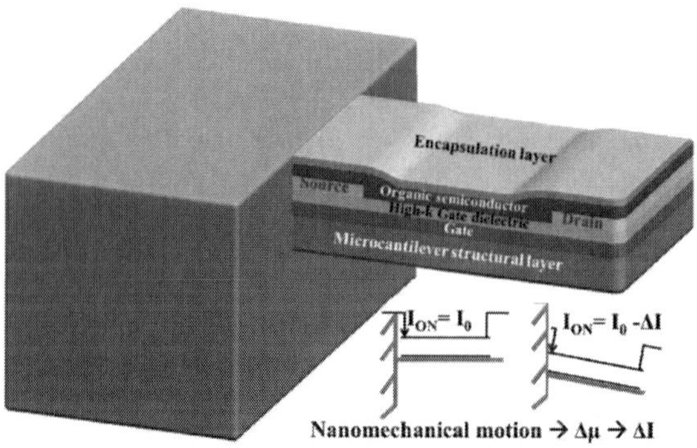

Figure 2.8 Concept of an "organic CantiFET" device.[23]

2.2.2.1 Fabrication process integration of organic CantiFET

The CantiFET device chip consists of two CantiFET devices for differential current measurement scheme, in which one of the devices was considered as the measurement cantilever and other being the reference cantilever.[23] Arrays of such CantiFET device chips were fabricated. The process integration of pentacene OFET with solution processed polymer materials like SU-8 is a real challenge as pentacene is known to get degraded by organic solvents. For this, a novel process sequence was developed, in which, one develops all the required layers except the pentacene layer for OFET on the SU-8 cantilever and then deposits pentacene layer after the releasing the device chips from the substrate to form the final organic CantiFET. This novel polymer MEMS process has three inherent advantages: (1) It avoids the exposure of pentacene layer to organic solvents used in SU-8 lithography, (2) it obviates the need for patterning of strain-sensitive layer, which is also known to be a non-trivial problem, and (3) this CantiFET platform could be used in future for realizing many polymer MEMS devices with integrated FETS with other strain-sensitive channel materials. The schematic of this fabrication sequence is given in Fig. 2.9 and process steps are detailed as below.

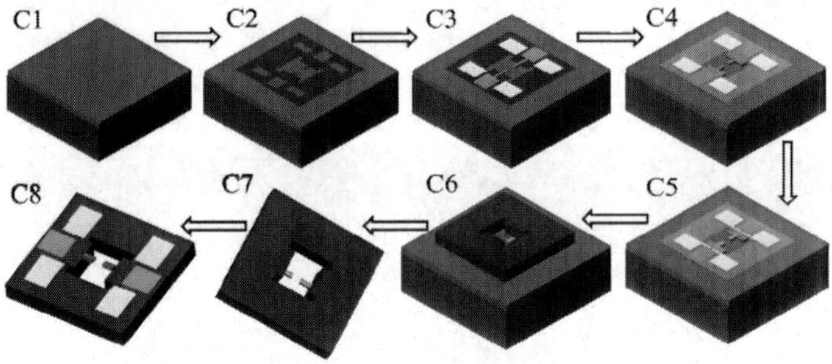

Figure. 2.9 Schematic of fabrication process for polymer CantiFET: (C1) Sacrificial layer. (C2) First layer of SU-8 defining the cantilever and contact vias. (C3) Au electrode patterning for gate of transistor. (C4) SU-8 as gate dielectric. (C5) Au electrodes defining the source and drain of OFET. (C6) Thick SU-8 layer defined for anchor or chip of the CantiFET device. (C7) Release of the CantiFET device from substrate and pentacene and encapsulation layer deposition. (C8) Final device structure showing cantilevers, source drain, and gate contacts through contact vias.[23]

Silicon wafer is used as the dummy substrate in this process, which was oxidized to form the silicon dioxide layer that served as the sacrificial layer for release of the final polymer structure from the dummy silicon substrate. The first layer for the cantilever die having the cantilever structural layer and contact vias was formed using 1.1 μm of SU-8 (Fig. 2.9 (C2)). Gate electrode and contact pads of the OFET were formed using gold (Au) with a thin layer of chrome (Cr) as the adhesion layer (Fig. 2.9 (C3)). The gate dielectric of the OFET was defined using SU-8 layer with a thickness of 900 nm (Fig. 2.9 (C4)). This process was followed by the formation of source and drain electrodes and the contacts using Cr/Au (Fig. 2.9 (C5)). A thick (more than 100 μm) SU-8 layer was patterned to form the polymer device chip and the frames for holding arrays of CantiFET device chips (Fig. 2.9 (C6)). Subsequently, the polymer device chips were released from the silicon substrate using isotropic etching of silicon dioxide in a buffered hydrofluoric acid. The final step in the fabrication of organic CantiFET was the vacuum thermal deposition of pentacene on the released SU-8 nanomechanical device to complete the integration of an OFET. The photograph of an array of released organic CantiFET chips along with the scanning electron micrographs (SEM) of one of the CantiFET devices is shown in Fig. 2.10. The pentacene OFET in the SU-8 cantilever has been encapsulated using a very thin (~15–20 nm) silicon nitride layer deposited using a hot wire chemical vapor deposition (HWCVD) method. The silicon nitride encapsulation layer was kept thin enough so as not to increase the stiffness of the cantilever structure.

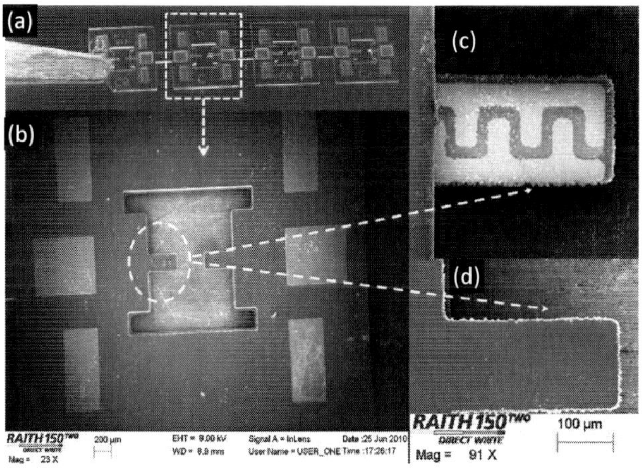

Figure 2.10 (a) Photographs of released arrays of organic CantiFETs. (b) SEM micrograph of the fabricated CantiFET device. (c) Bottom and enlarged view of cantilever portion of the CantiFET from SEM showing the interdigitated source drain electrode configuration (Type 1 and 3 CantiFET). (d) Top and enlarged view of cantilever portion of the CantiFET from SEM.[23]

2.2.2.2 Characterization of organic CantiFET

The organic CantiFETs were electrically characterized inside a shielded, vibration isolated probe station. The electrical measurements were performed at room temperature under ambient atmospheric conditions using semiconductor parametric analyzer. Output and transfer characteristics for these devices were recorded. These integrated OTFTs exhibited low gate leakage and good switching characteristics (Gate current density = 1.2 × 10^{-7} A/cm^2 @ V_{GS} = −40 V and I_{ON}/I_{OFF} = 2.2 × 10^3). The output characteristics of encapsulated Organic CantiFET exhibited a linear increase in the drain current at low drain bias (Fig. 2.11b). This is a clear indication of the existence of good ohmic contact at the interface of source/drain electrodes and the organic semiconductor. At high drain biases, proper saturation of drain current was also observed.

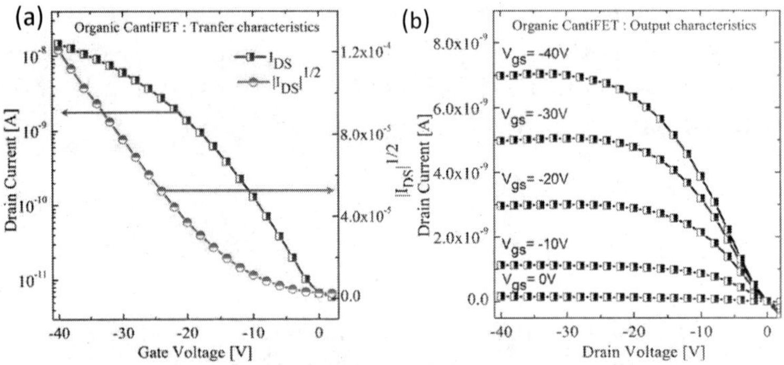

Figure 2.11 *I–V* characteristics of organic CantiFETs after silicon nitride encapsulation. (a) Transfer characteristics. (b) Output characteristics.[23]

The mechanical characterization of CantiFET devices were performed by beam bending technique using a nanoindenter tool. The spring constant obtained using this method is 0.4 N/m.[23]

In order to verify the suitability of CantiFET devices for biochemical sensing applications, the fabricated devices were electromechanically characterized to extract the surface stress sensitivity. The CantiFET device was subjected to different levels of compressive strain using calibrated micromanipulator and the *I–V* characteristics of the device was recorded after each level of strain, ε. It was observed that these organic CantiFET devices exhibit a good strain-sensitive behavior with the drain current increasing with compressive strain.[23] The percentage change in drain current, field effect mobility and threshold voltage were plotted as a function of change in the strain for the bias condition of $V_{GS} = V_{DS} = -40$ V (Fig. 2.12). From this, it can be observed that the transistor parameter that is strongly influenced by the strain is the field effect mobility and not the threshold voltage. This agrees well with the present understanding of hopping transport in pentacene.[30] The devices exhibited very high strain sensitivity ($\Delta I/I$ per unit strain) of the order of 103. The surface stress sensitivity for organic CantiFET devices ($\Delta I/I$ in ppm per unit surface stress in mN/m) extracted from the electromechanical characterization using Eq. 2.4 was 401 ppm $[\text{mN/m}]^{-1}$.

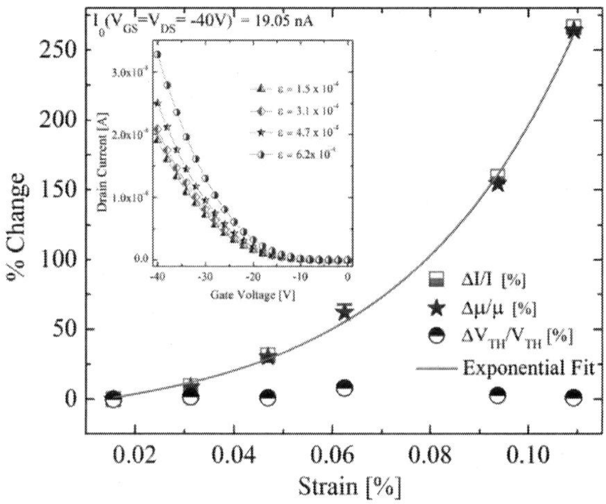

Figure 2.12 Percentage change in drain current, saturation field effect mobility and threshold voltage are plotted as a function of percentage strain. Inset: Transfer characteristics of organic CantiFETs under different strain conditions. Inset shows the I_{DS}–V_{DS} characteristics (@$V_{DS} = -40$ V) under identical strain conditions.[23]

The minimum detectable surface stress for CantiFET sensors was predicted by measuring the noise levels in these devices. A battery operated low-noise trans-impedance preamplifier (Stanford Research 570) with gain varying from 10^{-3} to 10^{-12} A/V was used to provide the gate bias to the CantiFETs probed inside a shielded probe station. The output or the drain terminal was connected to another SR570 in order to provide drain bias and to measure and amplify noise levels in the drain current. A spectrum analyzer (SR 750) was used to record the noise power spectrum in frequencies ranging from 1 Hz to a few kHz. The noise current level calculated for $1/f$ noise frequency range was 1.46 pA.[23]

The performance of CantiFET was compared with the MOSFET-embedded silicon cantilevers.[22] The deflection sensitivity of organic CantiFET sensors was 15 times higher compared to that of MOSFET-embedded silicon cantilevers. The extracted surface stress sensitivity value for organic CantiFET is more than three orders of magnitude higher in comparison to that of SU-8 microcantilevers with integrated Au strain gauge (0.3 ppm $[mN/m]^{-1}$) and 50 times higher than that of SU-8 nanocomposite cantilevers. With this high surface stress sensitivity and low noise levels, Organic CantiFETs should be able to detect surface stress values down to 0.2 mN/m. This capability of detecting very low surface stress makes the organic CantiFET a suitable candidate for many applications in bio/chemical sensor developments. CantiFET platform therefore opens up a new direction in realizing ultra-sensitive polymer MEMS sensors with integrated transistors with various strain-sensitive channel materials.

2.2.3 Al-Doped ZnO Thin-Film Transistor–Embedded Polymeric Micro-Cantilever as a Piezoresistive Sensor

Another polymeric cantilever based sensor is based on an aluminum-doped zinc oxide (AZO) thin-film transistor embedded inside the cantilever beam for nanomechanical sensing applications.[39] This device senses the surface stress due to a change in the carrier mobility of the semi-conducting layer. The low Young's modulus and high strain sensitivity of the AZO layer helps these micro-cantilevers exhibit a higher deflection sensitivity.

2.2.3.1 Material characterization

The semiconducting material that has been used here for thin-film transistor fabrication is Al-doped ZnO. The most explored deposition technique for this Al-doped ZnO (AZO) deposition is Radiofrequency (RF) magnetron sputtering system. The same technique also has been used for the deposition in this work. In this deposition method, the parameter that can be varied is RF power and Argon (Ar) pressure. Ar is the process gas for RF sputtering of AZO. The RF power has been varied from 60 W to 100 W. Al-doped ZnO film has shown a strong (002) diffraction pattern (34.3°). Since (002) plane is the most closely packed plane of ZnO, as can be seen from Fig. 2.13, this strong diffraction peak shows the preferred *c*-axis orientation to surface. The small difference in diffraction angle between Al-doped ZnO film and pure ZnO is because of the presence of Al.

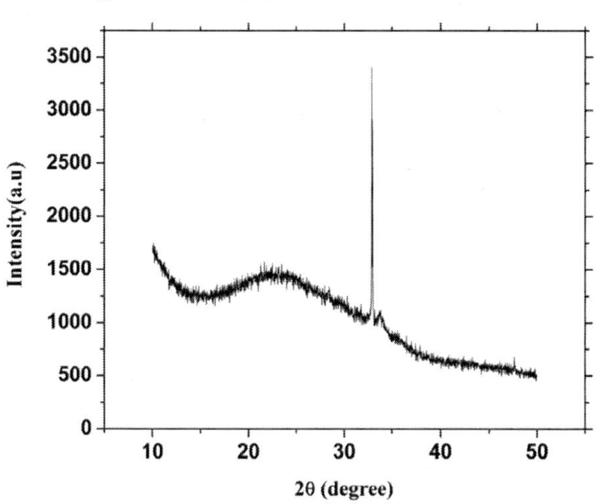

Figure 2.13 XRD pattern of Al-doped ZnO film by deposited by RF sputtering.[39]

To evaluate the effect of varying RF power on the surface morphology of the deposited ZnO films, the micrographs of these films obtained from Atomic force microscope (AFM) were analyzed, shown in Fig. 2.14. The surface morphology was found to be changing with RF power as can be observed in two dimensional AFM image. The surface roughness was also found to increase with RF power. With increasing RF power, there is an increase in the grain size of ZnO as shown in the AFM images (Fig. 2.14). However, the film porosity and film quality seem to degrade with increasing RF power.

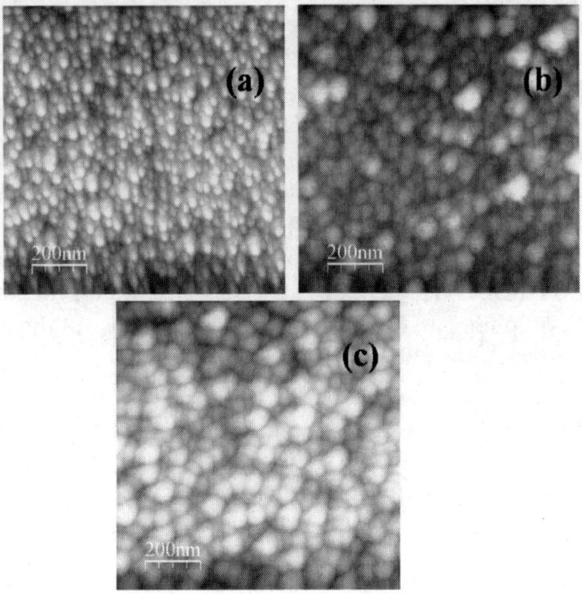

Figure 2.14 Atomic force microscope (AFM) topography images of Al-doped ZnO film by RF sputtering for different RF power values: (a) 60 W, (b) 75 W, and (c) 100 W.

2.2.3.2 Device fabrication

Fabrication process started with a RCA cleaned silicon wafer. A sacrificial layer of silicon dioxide with thickness 500 nm has been grown through a thermal oxidation method. A cantilever beam layer of SU-8 2002 was spin coated to get a thickness of 900 nm. This layer of SU-8 will act as a structural layer for the device. This SU-8 layer was patterned by UV-exposure at a dose of 72 mJ/cm^2 and developed with SU-8 developer. In the next process step the contact pad and gate have been fabricated by depositing Cr/Au layer of thickness 15 nm/80 nm and patterned by standard photo lithography step. In these devices the SU-8 was used as a gate dielectric because of its photo-patternability, low-temperature process compatibility, atmospheric stability and its low Young's modulus value. A 900 nm-thick SU-8 2002 was spin coated and patterned as a gate dielectric with the corresponding mask. Next, the source and drain electrodes are needed to be fabricated. For this purpose, a 15 nm/150 nm-thick Cr/Au layer was deposited using RF sputtering and then patterned using standard photolithography followed by wet etching of the metal layers. Now the device is ready for anchor layer fabrication, which will hold the cantilever beams. For this purpose, a viscous layer of SU-8-2100 was spin coated at a spin speed of 2200 rpm to achieve a thickness of 150 μm. It was patterned with the corresponding mask of anchor layer with the UV exposure dose of 231 mJ/cm^2. Now the devices need to be released from the silicon wafer. 5:1 BHF was used to etch the sacrificial layer of silicon dioxide. The final step in this process is to deposit the semi-conducting layer of Al-doped ZnO thin film on the back side of the cantilever over the source/drain metal electrode. For this purpose, sputtering was used to deposit AZO layer at a RF power of 60 W with the film having a thickness of 50 nm. Full fabrication process is graphically described in Fig. 2.15.[39]

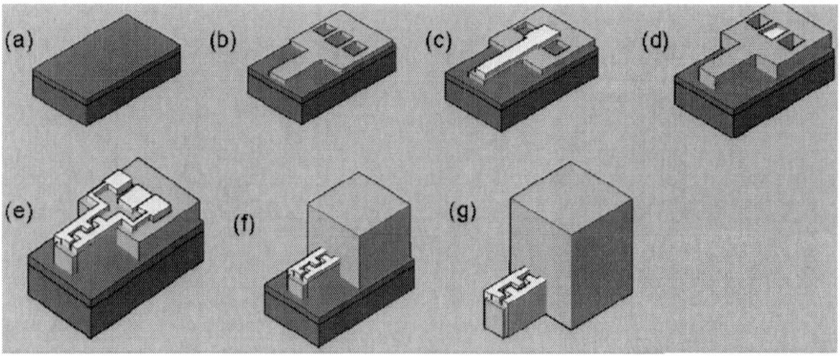

Figure 2.15 (a) Sacrificial layer, (b) SU-8 2002 structural layer, (c) contact pad and gate electrode of Cr/Au, (d) gate dielectric, (e) source drain contact of Cr/Au, (f) thick SU-8 patterning as anchor layer, (g) release of the device from the silicon wafer.[39]

2.2.3.3 Electrical and electromechanical characterization

The optical images of the released devices are shown in Fig. 2.16.

Keithley 2602 semiconductor source-measure unit with a shielded probe station has been used for electrical characterization of these aluminum-doped ZnO TFT–embedded cantilever devices. The drain current versus drain voltage at different gate voltages ranging from 0 to 40 V (output characteristic, Fig. 2.17a) and drain current versus gate voltage at a drain voltage of 40 V (transfer characteristic, Fig. 2.17b) were recorded.[39]

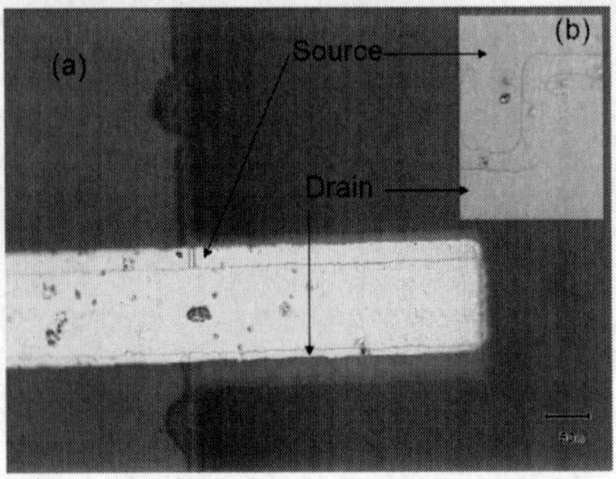

Figure 2.16 Optical images of the cantilever devices: (a) top view (b) bottom view.[39]

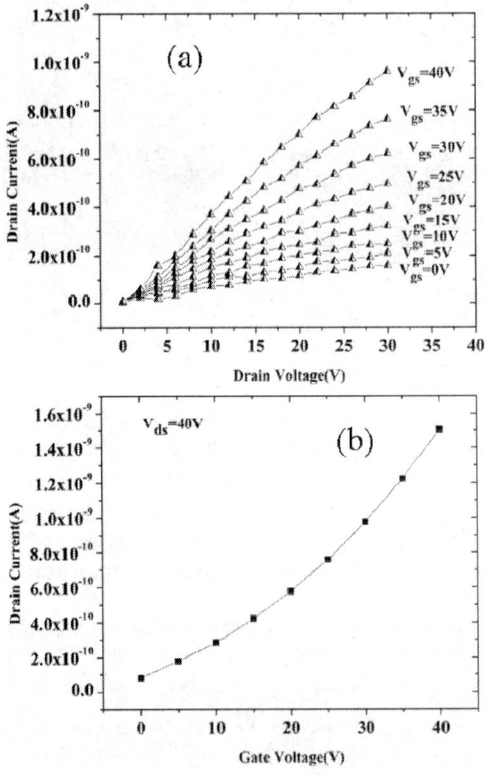

Figure 2.17 (a) Drain current versus drain voltage for different gate voltages. (b) Drain current versus gate voltage plot of AZO thin-film transistor embedded on a cantilever, with SU-8 as a gate dielectric.[39]

Next, these devices were tested to demonstrate their electromechanical response by applying force with the help of a micromanipulator and measuring the change in current using a probe station. The cantilever was subjected to different force levels. The applied force was calculated from the measure spring constant and deflection of the cantilever. For a different force, output characteristics (Fig. 2.18a) and transfer characteristics (Fig. 2.18b) were measured.[39]

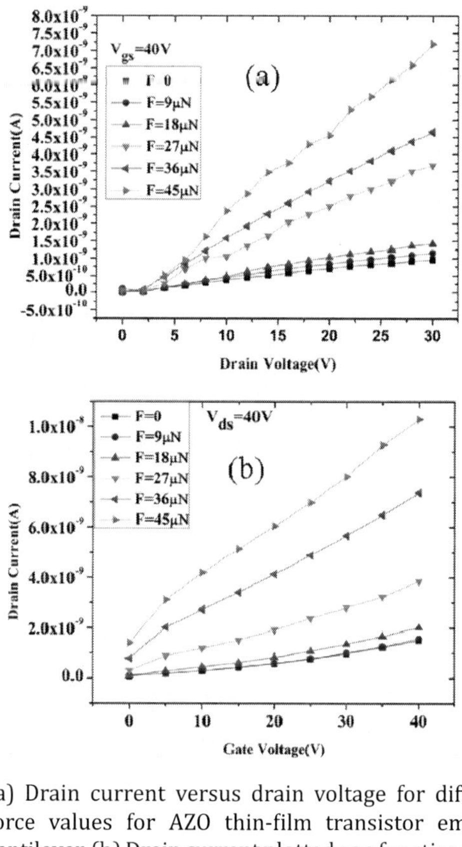

Figure 2.18 (a) Drain current versus drain voltage for different applied force values for AZO thin-film transistor embedded on a cantilever. (b) Drain current plotted as a function of gate voltage for different applied forces for a AZO thin-film transistor embedded in a cantilever.[39]

With the bending of the beam, the generated strain in the semiconducting film will decrease the energy spacing, which in turn helps to decrease the effective mass of the charge carriers. Effective mass of the charge carriers is inversely proportional to mobility. Higher mobility therefore will give rise to a change in the current with mechanical deflection. Also, the piezoelectric property of AZO can affect the force sensitivity. When the cantilever bends due to the applied force, a charge separation occurs in the channel layer owing to the piezoelectric property creating an additional potential. Additional carrier injection will happen in the channel from source and also drain current because of this reduced surface potential. These cantilever devices have shown high deflection sensitivity as result of this coupling effect of piezoresistive and piezoelectric effects shown in Fig. 2.19.[39]

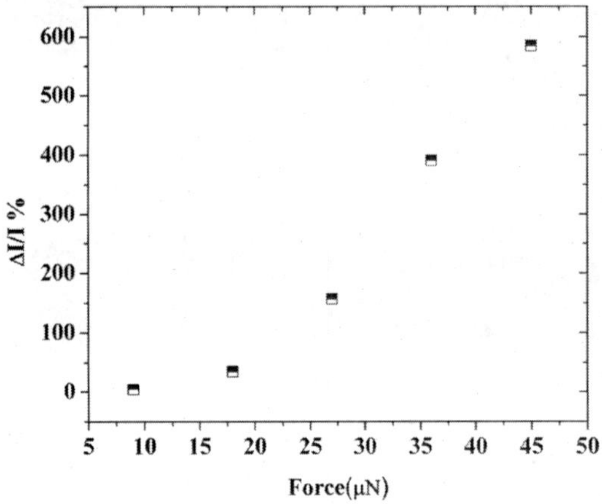

Figure 2.19 Percentage change in current as a function of applied force on an AZO thin-film transistor embedded on a cantilever.[39]

2.2.3.4 Mechanical characterization

The frequency characteristic of these devices is characterized using a laser Doppler vibrometer (LDV) as shown in Fig. 2.20. LDV detects the Doppler shift, i.e., the frequency shift of laser that is reflected from the moving object to the source. The object that is moving away from the laser source reflects light at a larger wave length as compared to the emitted light. Reverse effect can also be observed when the object is moving towards the laser source. The cantilever die has been mounted on the vibration source and laser was reflected from the cantilever beam. Now, using the Doppler shift of the laser beam, frequency and the amplitude of the cantilever vibrations can be calculated. The stiffness of this cantilever was also measured by Hysitron Triboscope nanoindentor. Nanoindentor probe was placed at the tip of the cantilever and the load versus displacement was plotted in Fig. 2.21. From the slope of this plot, spring constant was calculated to be 0.9 N/m. This high resonant frequency actually helps to retain the performance of these cantilevers in the presence of the environmental noise.[39]

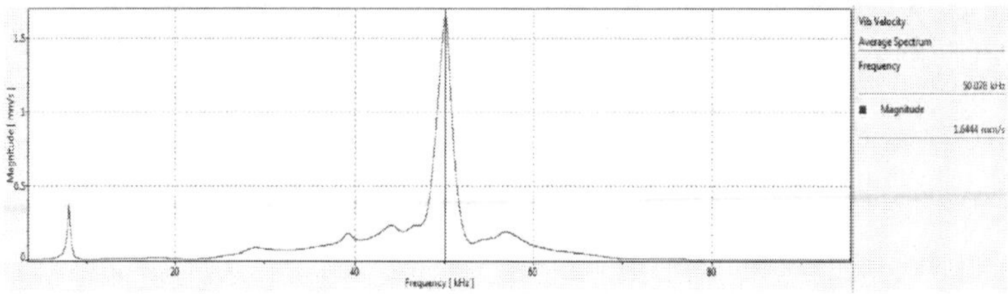

Figure 2.20 Frequency plot of cantilever with a AZO thin-film transistor embedded using laser Doppler vibrometer.[39]

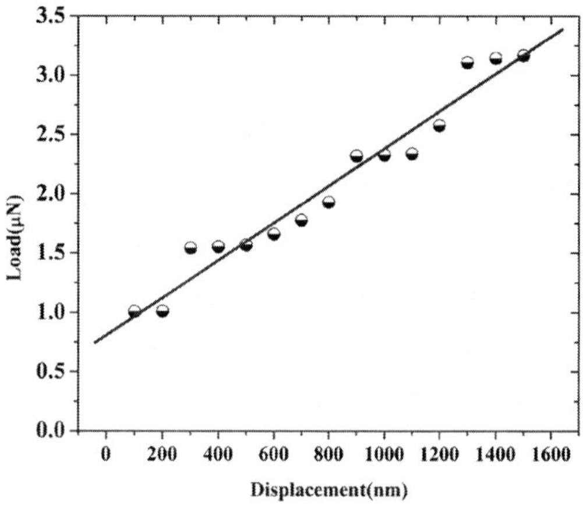

Figure 2.21 Load versus displacement plot from nanoindentation experiment for a AZO thin-film transistor–embedded cantilever.[39]

2.2.4 ZnO Nanowire–Embedded Strain Sensing Cantilever

In this section, a ZnO nanowire–embedded polymer cantilever structure is proposed as a strain sensor and realized using an innovative micro-fabrication process flow. A detailed electromechanical and mechanical characterization of this cantilever has been performed to understand its effectiveness as a physical sensor.[40]

2.2.4.1 Material characterization

The synthesized process to grow the ZnO nanowire that has been followed here is a low-temperature hydrothermal method.[41] A seed layer of ZnO thin film of 50 nm was deposited on Si by RF sputtering. For hydrothermal growth of ZnO NWs, a equimolar (30 mM) aqueous solution has been prepared with zinc nitrate hexahydrate ($Zn(NO_3)_2 \cdot 6H_2O$) and methenamine ($C_6H_{12}N_4$). The water that has been used is MilliQ with resistivity of 18.2 MΩ·cm. The sample was placed in the solution and kept in a hot air oven at 95°C for 6 h. After the growth, the sample was cleaned using MilliQ water to remove the residual salt and then dried. Figure 2.22a shows the surface images of grown ZnO NWs on silicon. AFM topography is shown in Fig. 2.22b. Crystalline structure of ZnO nanowire has been measured using X-ray Diffraction analysis (Fig. 2.23). The 2θ angle is measured to be around 34.4°, which represent vertically aligned ZnO NWs.

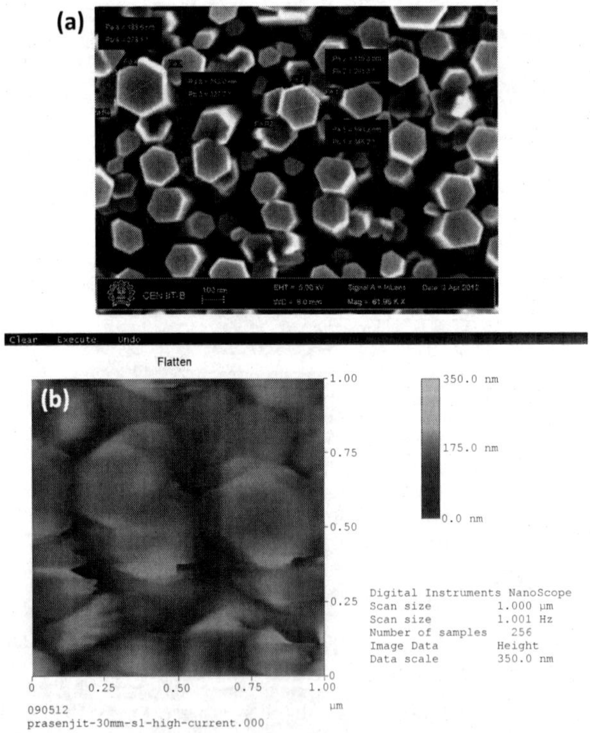

Figure 2.22 (a) SEM and (b) AFM images of the ZnO vertical nanowire.

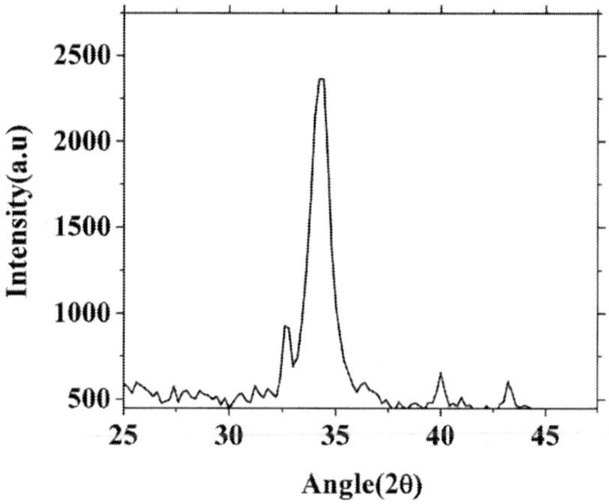

Figure 2.23 XRD pattern for ZnO nanowire film.[40]

A Hysitron Triboindentor (TI-900) has been used to indent the vertical ZnO NW. The force versus displacement was measured with a load resolution <1 nN and displacement resolution <0.0004 nm using the nanoindentor. Different forces have been employed and force–displacement plot has been recorded (Fig. 2.24a). Using Oliver-Phar method,[42] the reduced Young's modulus of the curve is described by Eq. 2.6;

$$E_{\mathrm{r}} = \frac{S}{2\beta}\sqrt{\frac{\pi}{A}}, \tag{2.6}$$

where, S is the unloading stiffness of the curve represented by dp/dh, i.e., the slope of the unloading curve, β is the constant whose value is 1 for Berkovich indenter and A is the contact area of indenter with sample. E_{r} taking into accounts the deformation for both the indenter and sample. So the actual Young's modulus of the NW is given by Eq. 2.7.

$$\frac{1}{E_{\mathrm{r}}} = \frac{1-v_{\mathrm{sample}}^2}{E_{\mathrm{sample}}} + \frac{1-v_{\mathrm{indenter}}^2}{E_{\mathrm{indenter}}}, \tag{2.7}$$

where E_{sample} and v_{sample} are Young's modulus and Poisson's ratio for NW while E_{indenter} and v_{indenter} are Young's modulus and Poisson's ratio for the indenter tip. For the indenter tip, the Poisson's ratio

($v_{indenter}$) and Young's modulus ($E_{indenter}$) are 0.07 and 1140 GPa, respectively. Poisson's ratio for ZnO is 0.358.

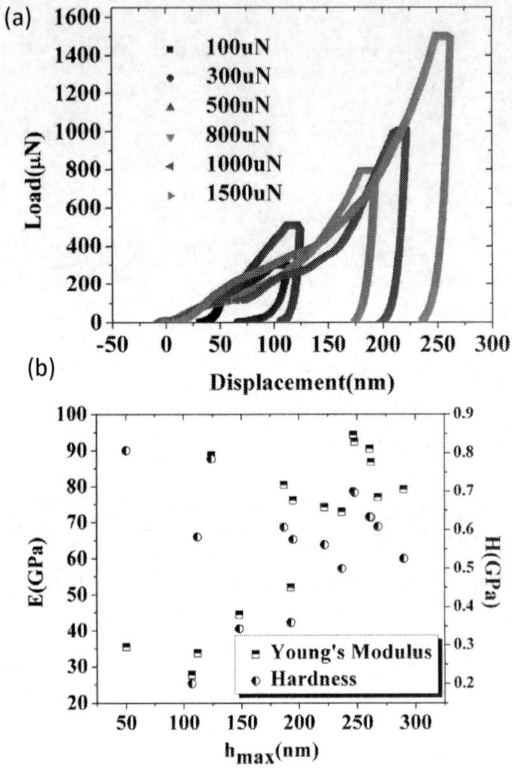

Figure 2.24 (a) Force–displacement curve for a varying force of 100 uN, 300 uN, 500 uN, 800 uN, 1000 uN, and 1500 uN on Vertical ZnO nanowire. (b) Young's modulus and hardness as function of indention depth.

The maximum indent depth has been observed to be 250 nm for 1500 uN force, which is around 12% of the NW height. This value is within the 20% limit, i.e., indenter depth should not cross 20% of the height of ZnO NW so that substrate effect on the measurement of elastic modulus can be ignored. The force–displacement curve at varying forces is shown in Fig. 2.24a. The measured Young's modulus value of ZnO nanowire is in the range of 70–93 GPa, whereas in the case of ZnO thin film, it is 101 GPa. Therefore, Young's modulus of ZnO NW is comparable to that of the ZnO thin film. The hardness of the ZnO vertical nanowire comes in the range of 0.6–0.69 GPa (Fig. 2.24b).

Next, this ZnO nanowire film has been embedded inside the SU-8 matrix and perform the indentation experiment on this SU-8/ZnO nanowire film. The load–displacement curve has been plotted for two different loads (Fig. 2.25a). Young's modulus and hardness have been calculated using the above method and plotted in (Fig. 2.25b).

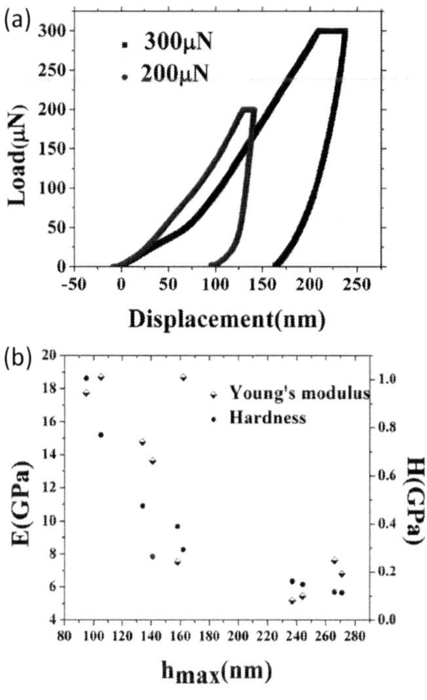

Figure 2.25 (a) Load–displacement curve using Berkovich indent. (b) Young's modulus and hardness as a function of indentation depth.

2.2.4.2 Device fabrication

The device fabrication starts with a RCA cleaned wafer. Since the device is eventually removed from the wafer by etching a sacrificial layer, the wafer can be reused in our process, which reduces the process costs. Here, a sacrificial layer is used for a final release of these devices from the wafer. For this purpose, either a 900 nm thick thermally grown oxide or 150–200 nm sputter deposited oxide can be used. The sputter deposited oxide is preferred than the thermally grown oxide due to its porosity, which enables a faster release. This porosity is not good for use of this oxide for gate dielectric, but most desirable as a sacrificial layer so that it will take less time to etch out the sacrificial layer (Fig. 2.26).

Figure 2.26 Process flow: (a) Silicon dioxide grown on a silicon wafer. (b) First encapsulation layer of SU-8. (c) Patterned Cr/Au layer for contact. (d) Patterning of ZnO seed layer. (e) Vertical growth of ZnO nanowire. (f) Bottom encapsulation layer of SU-8. (g) Patterning of SU-8 anchor layer. (h) Released device after sacrificial layer etching.[40]

After this step, SU-8 2002 has been spin coated at 6000 rpm to achieve a thickness of 900 nm. A pre-exposure bake had been given at 70 and 90°C with a slow ramp-up and ramp-down of temperature. This SU-8 layer needs to be patterned to act as an upper encapsulation layer of the vertical ZnO nanowire film. Standard photolithography process was used with the required UV exposure dose and patterned using an encapsulation layer mask. A post-exposure temperature bake was performed at the same parameters as the pre-exposure bake, and developed with SU-8 developer, to get the required pattern for the encapsulation layer.

Next the contact pad of the ZnO NW film was needed to be defined. For the contact pad, we used chromium as an adhesion layer and gold for the metallization. A 10 nm/150 nm Cr/Au layer has been deposited using RF sputtering system and patterned as a contact pad using photolithography. Then the Cr/Au layer was etched using wet etching method to get the required contacts.

The next layer to be patterned is the seed layer for ZnO NW growth. For that, a 100 nm ZnO layer was deposited using sputtering and patterned using a corresponding mask so that it stays inside the cantilever beam. Wet etching method is followed to pattern the seed layer of ZnO thin film. Vertical ZnO nanowire was grown using a low-temperature solution–based method. For the hydrothermal growth of ZnO NW, an equimolar (30 mM) aqueous solution has been prepared with zinc nitrate hexahydrate ($Zn(NO_3)_2 \cdot 6H_2O$) and methenamine ($C_6H_{12}N_4$). The water that has been used is MilliQ de-ionized water with a resistivity of 18.2 MΩ·cm. The sample has been placed in the solution and kept in a hot air oven at 95°C for 6 h. After the growth the sample was cleaned using MilliQ water to remove the residual salt and dried. The achieved height of the ZnO nanowire will be around 1.5–2.5 μm.

Now one SU-8 bottom encapsulation layer is required to protect this ZnO NW film. So SU-8 2005 was spin coated at 3000 rpm to get a thickness of 5 μm. Then it was post-baked at 70 and 90°C for an optimized time. With the help of standard optical lithography and using a corresponding encapsulation mask, this layer was cross-linked. Now it was developed to get the required pattern.

Sample is now ready for the patterning of an anchor layer of thickness 150 μm. The anchor layer, which will hold the array of cantilever beams, needs to be thick enough for easy handling. SU-8 2100 was spin coated at a spin speed of 2200 rpm. Now the anchor layer was exposed using a mask aligner to cross-link the anchor area and subsequently developed to get the anchor layer patterned.

To release the device from the silicon wafer, we dip the wafer in a 5:1 BHF solution and the sacrificial layer SiO_2 was etched out within 3 h duration. The optical images of the released devices are shown in Fig. 2.27a. The released cantilever device has a beam length of 250 μm and of width of each leg is 50 μm. The SEM images of the grown ZnO nanowires of height 2.5 μm are shown in Fig. 2.27b.[40]

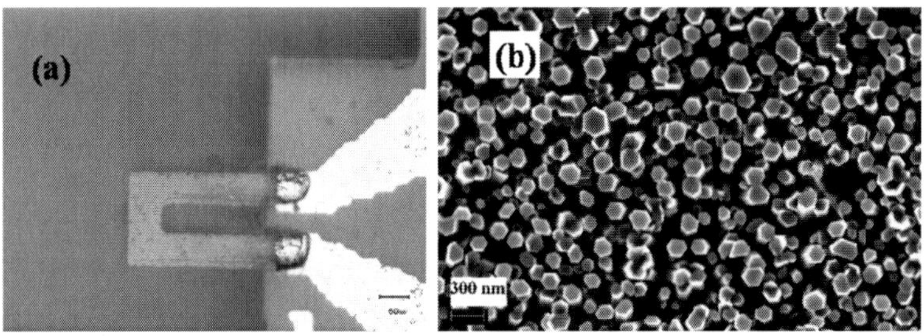

Figure 2.27 (a) Optical images of the fabricated cantilever. (b) SEM images of vertical ZnO nanowire.[40]

2.2.4.3 Electrical and mechanical characterization of the device

These micro-cantilever devices were explored to check the electrical and electromechanical performance using 2602 source-meter from Keithley. The strain on the cantilever beam has been generated by deflecting the cantilever beam using a calibrated micromanipulator from Suss Microtech. To make sure that only ZnO nanowire is conducting not the ZnO seed layer, we measured the current through the seed layer before growing the ZnO nanowires. The measured current–voltage characteristics with only the sputtered seed layer are shown in Fig. 2.28a. After the seed layer formation, nanowires were grown and encapsulated with a top SU-8 layer as discussed in the previous section. The fabricated cantilever beam was then deflected from 10 to 50 μm and the current is measured for varying voltages, which is plotted in Fig. 2.28b.[40]

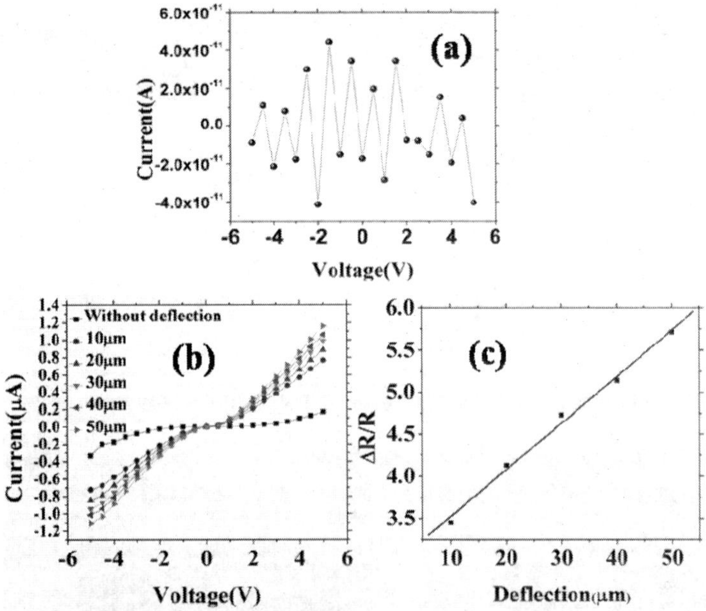

Figure 2.28 (a) Current–voltage characteristics through the ZnO seed layer, without the nanowires. (b) Current–voltage plot for different values of deflection for the cantilever beam with ZnO nanowires. (c) $\Delta R/R$ as a function of deflection for SU-8 cantilever with embedded ZnO nanowires.[40]

The conduction mechanism in these cantilevers is through a vertically grown ZnO nanowire film inside the beam. Current conduction between these ZnO nanowire random networks can be explained using thermo ionic diffusion theory.[45,46]

$$I = A_C A^* T^2 \exp\left(\frac{-q\Phi_B}{K_B T}\right)\left[\exp\left(\frac{qV}{K_B T}\right) - 1\right] \tag{2.8}$$

When any strain is generated on these cantilever beams, a strain is created inside the ZnO nanowire film, by which the contact area (A_C) and Schottky barrier height (Φ_B) are modified. The current through the film therefore changes because of these two parameters, which depends on the amount of strain in the film. A plot for $\Delta R/R$ as a function of deflection is shown in Fig. 2.28c.[40]

The mechanical characterization of these devices has also been done to get the frequency response and also to get an idea on the stiffness of these devices. Laser Doppler vibrometer of MSA 500 model from Polytec is used for measuring the frequency response. LDV detects the Doppler shift, i.e., the frequency shift of laser that is reflected from the moving object to the source. The device was mounted on the piezo-shaker to give the vibration and the laser beam was reflected back from the cantilever surface. The cantilever's amplitude and frequency were calculated from the Doppler shift of laser frequency. The natural vibration frequency for these cantilevers from LDV is found to be 30.4 kHz, which is for the first mode (Fig. 2.29a). Normally the stress sensitive cantilevers should have a resonant frequency higher than 5 kHz to get less effect from

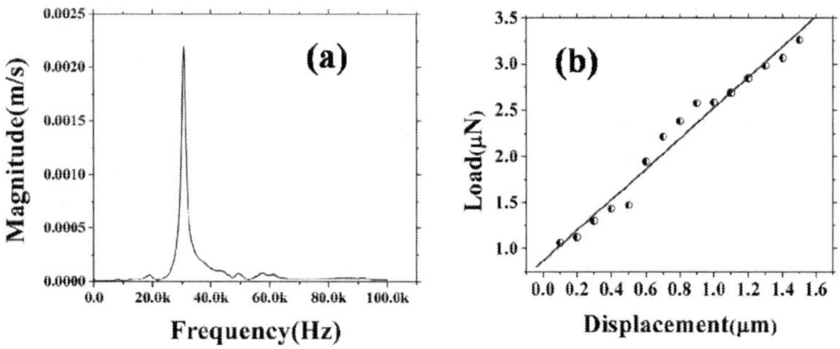

Figure 2.29 (a) Frequency plot of cantilever using a laser Doppler vibrometer. (b) Load versus displacement plot from nanoindentation experiment for a ZnO nanowire–embedded SU-8 cantilever.[40]

the environment. This high resonant frequency of our devices actually helps to retain the performance of these cantilevers in the presence of environmental noise. As part of the mechanical characterization, stiffness of this cantilever was also measured using a Hysitron Triboscope nanoindentor. Nanoindentor probe was placed at the tip of the cantilever and the load versus displacement obtained is plotted in Fig. 2.29b. From the slope of this plot, spring constant was estimated to be 1.16 N/m.[40]

2.3 Summary

In this chapter, we have summarized various approaches for electro-mechanical transduction of nanomechanical motion into an electrical readout using various piezoresistive approaches suitable for polymer cantilever technologies. Polymer composites utilizing a variety of nanoparticles such as carbon nanotubes, carbon black, ZnO, etc. have been shown to be highly effective in improving the deflection sensitivity of piezoresistive polymer micro-cantilevers. A novel polymer cantilever technology platform named CantiFET has been described, which integrates a strain-sensitive organic transistor on a polymeric micro-cantilever. The CantiFET technology has been further extended with the integration of an Al doped ZnO (AZO) Thin-Film Transistor (TFT) embedded on a polymeric cantilever. Deflection sensitivities for surface stress have been demonstrated with ΔI and ΔR values greater than 100s of parts-per-million (ppm) for a nanometer of deflection. Such novel transduction approaches embedded into polymer cantilever materials are expected to pave way for future low-cost and highly sensitive biochemical sensors suitable for a variety of applications.

References

1. Fatikow, S., and Rembold, U., *Microsystem Technology and Microrobotics*, 1997, Springer, Berlin.
2. Senturia, S., *Microsystem Design*, 1997, Kluwer Academic Press.
3. Beeby, S., Ensell, G., Kraft, M., and White, N., *MEMS Mechanical Sensor*, 2003, Artech house.
4. Engel, J., Chen, J., and Liu, C., Polymer-based MEMS multi-modal sensory array. Paper Presented at 226th National Meeting of the American Chemical Society (ACS). New York, NY, 2003.

5. Lei, J. F., and Will, H. A. (1998). *Sens. Actuators A.*, **65**(2–3), 187–193.
6. Klejwa, N., Harjee, N., Kwon, R., Coulthard, S. M., and Pruitt, B. L. (2007). Transparent SU-8 three-axis micro strain gauge force sensing pillar arrays for biological applications, *Solid-State Sensors, Actuators and Microsystems Conference, TRANSDUCERS*, pp. 2259–2262.
7. Huang, J. C. (2002). *Adv. Polym. Technol.*, **21**(4), 299–313.
8. Philip, B., Abraham, J. K., Chandrasekhar, A., and Varadan, V. K. (2003). *Smart Mater. Struct.*, **12**(6), 935–939.
9. Engel, J. M., Chen, N., Ryu, K., Pandya, S., Tucker, C., Yang, Y., and Liu, C. (2006). Multi-layer embedment of conductive and non-conductive PDMS for all-elastomer MEMS, *Solid State Sens. Actuators Microsyst. Workshop*, p. 3169.
10. Engel, J. M., Chen, J., Bullen, D., and Liu, C. (2005). Polyurethane rubber as a MEMS material: characterization and demonstration of an all-polymer two-axis artificial hair cell flow sensor, *18th IEEE MEMS Conference*, pp. 279–282.
11. Engel, J. M., Chen, J., Chen, N., Pandya, S., and Liu, C. (2005). Development and characterization of an artificial hair cell based on polyurethane elastomer and force sensitive resistors, *18th IEEE MEMS Conference*, pp. 1014–1017.
12. Boisen, A., Thaysen, J., Jensenius, H., and Hansen, O. (2000). *Ultramicroscopy,* **82**, 11–16.
13. Gammelgaard, L., Rasmussen, P. A., Calleja, M., Vettiger, P., and Boisen, A. (2006). *Appl. Phys. Lett.,* **88**, 113508.
14. Seena, V., Rajorya, A., Pant, P., Mukherji, S., and Rao, V. R. (2009). *Solid State Sci.,* **11**, 1606–1611.
15. Auld, B. A. (1973). *Acoustic Fields and Waves in Solids*, Wiley, p. 1.
16. Zhang, G., Xie, H., De Rosset, L. E., Fedder, and G. K. (1999). A lateral capacitive CMOS accelerometer with structural curl compensation, *12th IEEE MEMS Conference,* pp. 606–611.
17. Beliveau, A. Spancer, G. T., Thomas, K. A., and Roberson, S. L. (1999). *IEEE Des. Test Comput.,* **16**, 48–56.
18. Kenney, T. W., Waltman, S. B., Reynolds, J. K., and Kaiser., W. J. (1990). A micro-machined silicon electron tunneling sensor, in *Proc. IEEE Micro Electro Mechanical Systems Workshop (MEMS'90)*, Napa Valley, CA, pp. 192–196.
19. Kenny, T. W., Waltman, S. B., Reynolds, J. K., and Kaiser, W. J. (1991). *Appl. Phys. Lett.*, **58**, 100.

20. Waltman, S. B., and Kaiser, W. J. (1989). *Sens. Actuators A*, 19, 201.
21. Baski, A. A., Albrecht, T. R., and Quate, C. F. (1988). *J. Microsc.*, **152**, 73.
22. Shekhawat, G., Tark, S.-H., and Dravid, V. P. (2006). *Science*, **311**, 1592–1595.
23. Seena, V., Nigam, A., Pant, P., Mukherji, S., and Rao, V. R. (2012). *IEEE J. Microelectromech. Syst.*, **21**(2), 294–301.
24. Thaysen, J., Yalcinkaya, A. D., Vettiger, P., and Menon, A. (2002). *J. Phys. D: Appl. Phys.*, **35**, 2698–2703.
25. Kale, N. S., Nag, S., Pinto, R., and Rao, V. R. (2009). *J. Microelectromech. Syst.*, **18**, 79–87.
26. Seena, V., Fernandes, A., Pant, P., Mukherji, S., and Rao, V. R. (2008). *Nanotechnology*, **22**, 295501.
27. Bose, S., Bhattacharyya, A. R., Khare, R. A., Kulkarni, A. R., Patro, T. U., and Sivaraman, P. (2008). *Nanotechnology*, **19**, 335704.
28. Ray, P., Seena, V., Khare, R. A., Bhattacharyya, A. R., Apte, P. R., and Rao, R. (2011). SU-8/modified MWNT composite for piezoresistive sensor application. *MRS Proceedings*, 1299, mrsf10-1299-s04-11 doi:10.1557/opl.2011.57.
29. Mott, N. F. (1967). *Adv. Phys.* (*Philos. Mag. Suppl.*), **16**, 49.
30. Sekitani, T., Kato, Y., Iba, S., Shinaoka, H., and Someya, T. (2005). *Appl. Phys. Lett.*, **86**, 073511.
31. Ji, T., Jung, S., and Varadan, V. K. (2007). *IEEE Electron Device Lett.*, **28**(12), 1105–1107.
32. Kitamura, M., and Arakawa, Y. (2008). *J. Phys. Condens. Matter*, **20**, 184011.
33. Sekitani, T., Zschieschang, U., Klauk, H., and Someya, T. (2010). *Nat. Mater.*, **9**, 1015.
34. Jedaa, A., and Halik, M. (2009). *Appl. Phys. Lett.*, **95**, 103309.
35. Sekitani, T., Iba, S., Kato, Y., Noguchi, Y., and Someya, T. (2005). *Appl. Phys. Lett.*, **87**, 173502.
36. Yang, C., Yoon, J., Kim, S. H., Hong, K., Chung, D. S., Heo, K., Park, C. E., and Ree, M. (2008). *Appl. Phys Lett.*, **92**, 243305.
37. Kanari, M., Kunimoto, M., Wakamatsu, T., and Ihara, I. (2010). *Thin Solid Films*, **518**, 2764–2768.
38. Tark, S.-H., Srivastava, A., Chou, S., and Shekhawat, G. (2009). *Appl. Phys. Lett.*, **94**, 104101.
39. Ray, P., and Rao, V. R. (2013). *Appl. Phys. Lett.*, **102**, 064101.

40. Ray, P., and Rao, V. R. (2013). *J. Microelectromech. Syst.*, **22**(5), 995–997.
41. Vayssieres, L. (2013). *Adv. Mater.*, **201**(15), 464.
42. Oliver, W. C., and Pharr, G. M. (1992). *J. Mater. Res.*, **7**, 1564.
43. Kale, N. S., and Rao, V. R. (2006). (*IEEE/ASME*) *J. Microelectromechan. Syst.* (*J-MEMS*), 2006, **15**(6), 1789–1794.
44. Seena, V., Rajorya, A., Fernaundus, A., Dhale, K., Pant, P., Mukherji, S., and Rao, V. R. (2010). Fabrication and characterization of a novel polymer composite microcantilever sensors for explosive detection, *Proceedings of the 23rd IEEE International Conference on Micro Electro Mechanical Systems (MEMS 2010)*, pp. 851–854.
45. Donald, A. N., *Semiconductor Physics and Device*, McGraw-Hill, New York, p. 320.
46. Nagata, T., Ahmet, P., Yoo, Y. Z., Yamada, K., Tsutsui, K., Wada, Y., and Chikyow, T. (2006). *Appl. Surf. Sci.*, **252**, 2503–2506.
47. Seena, V., Rajorya, A., Pant, P., Mukherji, S., and Rao, V. R. (2009). *Solid State Science*, **11**(9), 1606–1611.

Part 2

Nonlinearity of Nanocantilever Beam Resonators

Chapter 3

Nonlinear Dynamics and Its Applications in Nanocantilevers

Najib Kacem

FEMTO-ST Institute—UMR 6174, Department of Applied Mechanics, University of Franche-Comté, 24 chemin de l'Épitaphe, 25000 Besançon, France

najib.kacem@femto-st.fr

3.1 Introduction

Nanoelectromechanical systems (NEMS) are drawing interest from the scientific community for a wide range of applications due to their unique properties. Nanocantilevers are among those of the possible NEMS realizations that offer access to fundamental resonant frequencies in the microwaves and active masses in the femtograms. Nanocantilever have been proposed for ultrafast sensors and actuators, signal processing components, quantum computing (Bose and Agarwal, 2006), and ultra sensitive force (Jiang et al., 2008) and mass (Li et al., 2007) detection.

Actually, it is a challenge to optimize NEMS mass sensors in order to achieve high resolutions. Although the linear design optimization and mechanical transduction gain of the devices have been thoroughly studied, the drive power has always been a priori limited by the onset of nonlinearities. Indeed, driving the

Nanocantilever Beams: Modeling, Fabrication, and Applications
Edited by Ioana Voiculescu and Mona Zaghloul

ISBN 978-981-4613-23-1 (Hardcover), 978-981-4613-24-8 (eBook)
www.panstanford.com

cantilever at large oscillation amplitude leads to better signal to noise ratio (SNR) and, thus, simplifies the design of the electronic feedback loop. However, doing so in the nonlinear regime reduces the sensor performances since the frequency instability of a nonlinear resonator is proportional to its oscillation amplitude. Moreover, even when NEMS resonators are used as oscillators in closed-loop, a large part of noise mixing (Kaajakari et al., 2005b) due to nonlinearities drastically reduces their dynamic range and alters their detection limit. NEMS cantilevers are promising candidates for the new generation of physical, chemical and biological sensing. One reason for this is that they are commonly said to have a very large linear dynamic range compared to clamped-clamped nanoresonators, without any formal proof, quantitative comparison, or thorough study. Models for doubly clamped beams (Kacem et al., 2009, 2011b) cannot be easily adapted to cantilevers: indeed, their real specificity comes from their complex nonlinear dynamics including geometric and inertial nonlinearities. This partly explains why so little has been done about nonlinear dynamics of electrostatically actuated cantilevers.

The nonlinear dynamics of cantilevers have received considerable attention because of their importance in many engineering applications. Crespo da Silva and Glynn (1978a,c) derived a set of integro-partial-differential equations governing flexural-flexural-torsional motions of inextensional beams, including geometric and inertia nonlinearities. They used these equations and the method of multiple scales to ascertain the importance of the geometric terms for the lower modes, especially the first mode. These equations have been also used for several investigations such as the non-planar responses of cantilevers to principal parametric and primary resonant excitations (Nayfeh and Pai, 1989; Pai and Nayfeh, 1990), the nonlinear response of an inextensional beam to a primary resonant excitation of one of its flexural modes when the first torsional frequency is of the same order as the lower flexural frequencies (da Silva and Zaretzky, 1994; Zaretzky and da Silva, 1994), as well as the nonlinear non-planar response of cantilever inextensional metallic beams to a parametric excitation of order two its flexural modes (Arafat et al., 1998). Chowdhury et al. (2005) provided a close-form model for the static pull-in voltage of electrostatically actuated cantilevers without including the geometric nonlinearities. Ahmadian et al. (2009) employed a finite element formulation for the dynamic analysis of nonlinear Euler cantilevers electrostatically actuated including main sources of nonlinearities, but the resonant case has not been considered. Liu et al. (2004) simulated an electrostatically controlled cantilever microbeam and qualitatively showed period-doubling bifurcation, chaos, Hopf bifurcation, and strange attractors using the Poincaré map method, which are hardly exploitable by MEMS and NEMS designers.

In this chapter, the physical limitations in NEMS mass sensors when operating beyond their critical amplitudes are investigated. Based on the nonlinear dynamics of nanomechanical cantilevers, the main idea is to provide simple analytical tools for NEMS designers in order to optimize mass resonant sensors designs and enhance their performances for precision measurement applications such as mass spectrometry.

3.2 Resonant Sensor Specifications

3.2.1 Mechanical Analysis

The mechanics of a resonant cantilever is theoretically investigated as an important step towards the determination of the resonance frequency and the sensitivity expressions. Then, based on the resonator dynamic response, the transfer function is deduced which will be used in an advanced step for noise analysis.

3.2.1.1 Resonance frequency

Assuming that the nonlinear terms are negligible, the equation of motion of a beam in bending subjected to an axial tensile force (F) can be written as

$$EI\frac{\partial^4 \tilde{w}}{\partial \tilde{x}^4} + \xi\frac{\partial \tilde{w}}{\partial \tilde{t}} - F\frac{\partial^2 \tilde{w}}{\partial \tilde{x}^2} + \rho A\frac{\partial^2 \tilde{w}}{\partial \tilde{t}^2} = 0, \tag{3.1}$$

where E is Young's modulus, I is the bending moment of inertia, ξ is the viscous damping coefficient, ρ is the material density and A is the beam section. The solution $\tilde{w}(\tilde{x},\tilde{t})$ can be solved by the method of separation of variables where

$$\tilde{w}(\tilde{x},\tilde{t}) = \phi(\tilde{x}) \cdot a(\tilde{t}) \tag{3.2}$$

The equation of the beam mode shapes with respect to the position coordinate $\tilde{x}$ can be written as

$$\phi(\tilde{x}) = C_1 \cos\left(\frac{\lambda\tilde{x}}{L}\right) + C_2 \sin\left(\frac{\lambda\tilde{x}}{L}\right) + C_3 \cosh\left(\frac{\lambda\tilde{x}}{L}\right) + C_4 \sinh\left(\frac{\lambda\tilde{x}}{L}\right) \tag{3.3}$$

Constants C_1–C_4 can be evaluated depending on the boundary conditions of the resonator; λ is a dimensionless parameter related to the wavelength, and it depends on the mode shape and the resonator boundary conditions. It can be evaluated numerically as listed in Table 3.1 for several bending modes of cantilevers as well as clamped-clamped beam resonators.

Table 3.1 Dimensionless parameter λ for the first four eigenfrequencies of cantilevers and c-c beams.

Mode/λ	Cantilever beam	Clamped-clamped beam
Mode 1	1.875	4.730
Mode 2	4.694	7.854
Mode 3	7.855	11.00
Mode 4	10.995	14.14

Using the Galerkin method, the time dependence can be cast in the form of a mass-spring-damper equation:

$$M_{\text{eff}}\ddot{a} + B_{\text{eff}}\dot{a} + K_{\text{eff}}a = 0, \tag{3.4}$$

where expressions for M_{eff} and K_{eff} can be written as (Roessig, 1998)

$$FM_{\text{eff}} = \int_0^L \rho A \phi^2(\tilde{x}) d\tilde{x} \tag{3.5}$$

$$K_{\text{eff}} = \frac{EI}{L^3}\int_0^L \left(\frac{\partial^2\phi}{\partial\tilde{x}^2}\right)^2 d\tilde{x} + \frac{F}{L}\int_0^L \left(\frac{\partial\phi}{\partial\tilde{x}}\right)^2 d\tilde{x} \tag{3.6}$$

Thus, for a null axial force, the resonator natural frequency is

$$f_0 = \frac{1}{2\pi}\sqrt{\frac{K_{\text{eff}}(F=0)}{M_{\text{eff}}}} \tag{3.7}$$

3.2.1.2 Sensitivity

For a resonator vibrating in its fundamental mode, the natural frequency for a constant axial force (F) along the length of the beam can be written in terms of the nominal resonant frequency by evaluating the integrals in Eqs. (3.5) and (3.6).

$$f = f_0\sqrt{1+F\cdot S} \tag{3.8}$$

$$S = \frac{\rho A L^2 \int_0^L \phi^2(\tilde{x})d\tilde{x} \int_0^L \left(\frac{\partial\phi}{\partial\tilde{x}}\right)^2 d\tilde{x}}{EI\int_0^L \left(\frac{\partial^2\phi}{\partial\tilde{x}^2}\right)^2 d\tilde{x}} \tag{3.9}$$

Thus, the force sensitivity of the resonator (mechanical scale factor) can be written as

$$SF_{\text{f}} = \frac{S}{2} f_0 \tag{3.10}$$

In the same way, the mass sensitivity of the resonator can be deduced as

$$SF_{\text{m}} = \frac{1}{2M_{\text{eff}}} f_0 \tag{3.11}$$

3.2.1.3 Dynamic response

The starting point for the simplest dynamic analysis is the characteristic differential equation describing the evolution of the displacement of the resonating element subjected to a linear time-varying drive force F_{d}.

$$M_{\text{eff}}\ddot{a} + B_{\text{eff}}\dot{a} + K_{\text{eff}}a = F_{\text{d}}\cos(\Omega\tilde{t}), \tag{3.12}$$

where Ω is the drive frequency. Equation (3.12) can be written in its canonical form as

$$\ddot{a} + \frac{\omega_0}{Q}\dot{a} + \omega_0^2 a = \frac{F_{\text{d}}}{M_{\text{eff}}}\cos(\Omega\tilde{t}), \tag{3.13}$$

where $\omega_0 = 2\pi f_0$ and Q is the resonator quality factor that can be estimated by evaluating the different system losses. Using the Fourier transform, the resonator transfer function can be deduced as follows

$$H(\Omega) = \frac{\frac{1}{M_{\text{eff}}}}{\Omega^2 - \omega_0^2 + j\frac{\Omega\omega_0}{Q}} \tag{3.14}$$

For MEMS and NEMS designers, the quality factor is an important parameter, since it defines the sensor bandwidth. Moreover, as shown in Eq. (3.14), the resonator transfer function depends on Q. Hence, it is important to estimate correctly the quality factor.

3.2.2 Quality Factor

The mechanical quality factor Q is a measure for the energy losses of a resonator or in other words, a measure for the mechanical damping. The Q-factor is defined as the ratio between the total energy stored in the vibration and the energy loss per cycle:

$$Q \approx 2\pi \frac{\text{total energy stored in vibration}}{\text{dissipated energy per period}} \tag{3.15}$$

Low energy losses imply a high Q-factor. The Q-factor cannot be determined directly, but instead can be deduced from the response characteristics of the resonator. One common method of determining Q is from the steady-state frequency plot of a resonator excited by a harmonic force with constant amplitude:

$$Q \approx \frac{\omega_{\text{res}}}{\Delta\omega_{-3\,\text{dB}}}, \tag{3.16}$$

where ω_{res} is the frequency with maximum frequency response and $\Delta\omega_{-3\,\text{dB}}$ is the half-power bandwidth of the frequency response.

In nanomechanical resonators, there are numerous possible sources of dissipation, which may broadly be classified as either intrinsic or extrinsic. Intrinsic sources of dissipation, such as phonon–phonon and phonon–electron interactions, result from properties of the resonating material, whereas extrinsic sources, such as gas friction, clamping loss, and surface loss, result from interactions with the environment. Obviously, little can be done to control dissipation from intrinsic sources other than careful choice of resonator material. Theoretical calculations have shown that these intrinsic sources of dissipation are small compared to the dissipation currently exhibited by nanomechanical resonators.

There are many extrinsic mechanisms of dissipation in nanomechanical resonators. They can be listed by their origin as follows:

3.2.2.1 Gas friction

At pressures above approximately 1 torr, viscous damping of a resonator by the surrounding gas is the dominant form of dissipation (Ekinci and Roukes, 2005). Here, the energy is radiated as sound. Fortunately, it is easy to achieve lower pressures where viscous damping no longer dominates. At these lower pressures, where the mean free path of the gas molecules is much larger than the relevant sound wavelength, energy may still be dissipated through momentum transfer to individual molecules. In this case, the dissipation is expressed as

$$Q_{\mathrm{gas}}^{-1} = \frac{pA}{M_{\mathrm{eff}}\omega_{\mathrm{r}}v}, \tag{3.17}$$

where p is the pressure, A is the surface area, M_{eff} is the effective mass of the resonator, ω_{r} is the resonator angular frequency, and v is the thermal velocity of the gas. According to Eq. (3.17) and multiple experiments, gas friction is not a significant source of dissipation below 10 mTorr.

3.2.2.2 Surface losses

Surface losses are caused by adsorbed molecules, dangling or broken bonds, an amorphous oxide layer, or other metastable systems that occur at a resonator's surface. These systems absorb energy from the fundamental resonant mode and irreversibly transfer it other modes and thermal energy. For resonating beams, the energy of a resonator is stored in the elastic strain throughout its volume and thus is proportional to its volume, V. If we assume that energy is predominately dissipated at the surface, then we would expect that the energy lost per cycle would be proportional to the surface area S, and thus,

$$Q_{\text{surface}}^{-1} \propto \frac{S}{V} \propto L^{-1} \tag{3.18}$$

Fortunately, it may be possible to control surface losses (Jensen et al., 2006) in NEMS resonators, through careful experimental techniques and the proper choice of resonator material.

3.2.2.3 Clamping loss

Clamping loss refers to mechanical energy dissipated through the supports of a resonator. Typically, this is theoretically modeled as elastic radiation of energy through the supports. There is still some contention as to the appropriate description of elastic clamping loss, though the most recent theoretical calculations predict a loss for a rectangular beam of

$$Q_{\text{clamping}}^{-1} \propto \frac{wt^4}{l^5}, \tag{3.19}$$

where w is the beam width in the direction of vibration, t is the beam thickness, l is the beam length, and the proportionality constant is dependent upon material properties (Geller and Varley, 2005). Clearly, to reduce clamping loss, a beam with a high aspect ratio is desirable. However, according to Eq. (3.19), clamping loss should be negligible for current resonator designs, including nanotube resonators with their extremely high aspect ratio.

3.2.2.4 Thermoelastic loss

Thermoelastic damping is the result of the transformation of elastic energy into thermal energy via thermal currents flowing between compressed and expanded regions of a deformed resonator. Zener (1938) first studied the phenomenon for a beam in flexure, giving the damping as

$$Q_{\text{Zener}}^{-1} = \frac{E\alpha^2 T}{C_P} \frac{\omega\tau_0}{1+\omega^2\tau_0^2}, \tag{3.20}$$

where E is Young's modulus, α is the thermal expansion coefficient, C_P is the constant-stress heat capacity, ω is the angular frequency of vibration, T is temperature, and $\tau_0 = \frac{h^2 C_P}{\pi^2 K}$ is the thermal relaxation time, with K the thermal conductivity and h the beam thickness.

3.2.2.5 Ohmic loss

Another type of dissipation associated with electrostatic actuation is ohmic losses from the electrons moving on and off the resonator due to capacitive coupling to a nearby gate. Following Sazonova (2006), the system can be represented as a variable capacitor in series with a resistor to which a voltage V is applied. If the time scales for the electrons to flow on the resonator and the time for one oscillation are matched perfectly, all of the charge flows through a resistor, dissipating energy through Joule heating. Thus, the ohmic losses are given by

$$Q_{\text{Ohmic}}^{-1} = \frac{1}{\pi\omega}\frac{R(C'V)^2}{M_{\text{eff}}}, \tag{3.21}$$

where ω is the angular frequency of vibration, C' is the gradient of the capacitance, R is the output resistor and M_{eff} is the effective mass of the considered mode. The smaller the resonator, the smaller the mass, the higher this contribution, hence NEMS are very sensitive to this effect.

In nanomechanical resonators, there are other extrinsic loss mechanisms such as anharmonic mode coupling and extrinsic noise (Mohanty et al., 2002), which can be neglected compared to the other sources of dissipation already cited.

3.2.3 Thermomechanical Noise

Thanks to the fluctuation-dissipation theorem, it can be written that the force noise spectral density due to thermomechanical fluctuations of the mass is (Postma et al., 2005)

$$S_{\text{f}}(\omega) = \frac{2}{\pi}K_{\text{B}}T\frac{M_{\text{eff}}\omega_0}{Q}, \tag{3.22}$$

where M_{eff} is the effective mass of the resonator, ω_0 is the angular frequency of vibration, Q is the quality factor, K_{B} is Boltzmann's constant, and T is the resonator temperature.

It may be assumed without loss of generality that the bandwidth BW used by the phase locked loop (PLL) readout is very narrow compared with the −3dB bandwidth of the resonator. Then, following Eq. (3.14), the transfer function of the resonator at resonance giving the displacement versus a constant force per unit length is

$$H_{\mathrm{fx}}(\omega)=\frac{Q}{K_{\mathrm{eff}}}=\frac{Q}{M_{\mathrm{eff}}\omega_0^2} \tag{3.23}$$

The displacement spectral density is then

$$S_{\mathrm{x}}(\omega_0)=S\left\|H_{\mathrm{fx}}(\omega)\right\|^2 S_{\mathrm{f}}(\omega)=\frac{2}{\pi}K_{\mathrm{B}}T\frac{Q}{M_{\mathrm{eff}}\omega_0^3} \tag{3.24}$$

Following Robins (1984), for a PLL-based readout technique, the frequency noise spectral density is

$$S_{\omega}(\omega)=\left(\frac{\omega_0}{2Q}\right)^2\frac{S_{\mathrm{x}}(\omega_0)}{P_0}, \tag{3.25}$$

where P_0 is the displacement carrier power, i.e., the RMS drive amplitude of the resonator $P_0 = \frac{1}{2}a_{\mathrm{d}}^2$. The latter should be driven below the hysteretic limit due to non-linearities. Even though this one will be higher when using a PLL-based technique, the open loop value may be used to stay on the safe side.

3.2.4 Resolution

The resolution is the lower limit of the dynamic range. It is set by the incoherent sum of all stochastic processes driving the resonator (Cleland and Roukes, 2002), such as thermomechanical fluctuations, quantum noise, noise from adsorption and desorption of gaseous species (Ekinci et al., 2004), and extrinsic sources such as vibrational and instrumental (read-out) noise. For simplicity, we solely consider thermomechanical noise in the case of a cantilever driven at its critical amplitude (open-loop stability limit). The resonator mass sensitivity (Eq. (3.11)) is

$$\mathrm{SF_m}=\frac{1}{2M_{\mathrm{eff}}}f_0\propto l^{-3}\cdot t^{-1} \tag{3.26}$$

Thus, the mass sensor resolution is given by

$$\delta m=\frac{\sigma_{\omega}}{\mathrm{SF_m}}\propto w^{-2}\cdot l^{7/2}\cdot t^{1/2} \tag{3.27}$$

At equilibrium, the concentration resolution for gas sensors is given by

$$\delta C = \frac{\delta m}{K_P V_P \rho_g} \propto w^{-3} \cdot l^{5/2} \cdot t^{1/2}, \tag{3.28}$$

where K_P is the partition coefficient depending on the couple polymer-gas combination, V_P is the polymer layer volume and ρ_g is the density of the gas analyte.

Equations (3.27) and (3.28) demand some important comments:

- In the case of resonant mass sensors, if the resonator dimensions are proportionally scaled down with respect to a given scale factor $N_{sf} << 1$, the resolution is proportionally improved with respect to N_{sf}^2. It proves that nanomechanical resonators, with their high fundamental resonance frequencies, diminished active masses and tolerable force constants, are extremely sensitive to mass changes. Therefore, for resonant mass sensors, NEMS resonators are a great alternative to improve the performances of such detectors.

$$\frac{\delta m(\text{NEMS})}{\delta m(\text{MEMS})} = N_{sf}^2 \tag{3.29}$$

- For resonant gas sensors, for a constant resonator width, if the resonator length and thickness are proportionally scaled down with respect to a given scale factor $N_{sf} << 1$ while keeping an acceptable slenderness ratio for Euler-Bernoulli beam model validity, the concentration resolution is proportionally improved with respect to N_{sf}^3. It proves that nanomechanical resonators are a great alternative for ultimate gas measurements.

3.2.5 Physical Nonlinearities

The validity of Eq. (3.1) is limited by the resonator physical nonlinearities. The mechanical nonlinearities are considered as a fundamental limit of the linear lower bound of the resonant sensor dynamic range (Cleland and Roukes, 2002). Furthermore, the actuation force can bring additional nonlinearities into the resonator dynamics. Electrostatic actuation is a good example for spring softening nonlinearities (Kacem et al., 2009, 2010).

Nanoscale mechanical resonant sensors offer a greatly enhanced performance that is unattainable with microscale devices. However, scaling down resonators from MEMS to NEMS makes nonlinearities quickly reachable (Cleland and Roukes, 2002) and drastically alters the sensor resolution.

To underline this fact, let us write Eqs. (3.27) and (3.28) without restrictions on the displacement carrier power.

$$\delta m = \frac{\delta_\omega}{\mathrm{SF_m}} \propto w^{-1} \cdot l^{7/2} \cdot t^{1/2} \cdot a_\mathrm{d}^{-1} \cdot Q^{-1/2} \tag{3.30}$$

$$\delta C = \frac{\delta m}{K_\mathrm{P} V_\mathrm{P} \rho_\mathrm{g}} \propto w^{-2} . l^{5/2} . t^{1/2} . a_\mathrm{d}^{-1} . Q^{-1/2} \tag{3.31}$$

To simplify the analysis of Eqs. (3.30) and (3.31) when the resonator dimensions are scaled down, we suppose that the quality factor Q is constant. Then, it is clear that the resolution of nanomechanical resonators depends on the drive oscillation. The latter is limited by the mechanical nonlinearity for thin clamped-clamped resonators.

Under this limit, Eqs. (3.30) and (3.31) show that proportionally scaling down the resonator improves the sensor resolution. It is even more convenient to reduce only the beam length l and thickness t while keeping an acceptable slenderness ratio for the validity of the Euler-Bernoulli model. Ultimate optimization depends on the drive amplitude of the resonator. Ideally, the resonator should be actuated to oscillate at the highest possible amplitude (below the pull-in for an electrostatic actuation). In open loop, the resonator is classically driven below its critical amplitude in order to ensure the stability of its dynamic response. This fundamental limit is set by the nonlinear dynamics of the resonator (details are given in Sections 3.3 and 3.4).

However, when used as a practical sensor, the resonator is most of the time used as an oscillator, embedded in a feedback loop, or a PLL. In such closed-loop operation, the phase is the control parameter of the system (the frequency is now an output) and hence stabilizes its dynamics: even in the non-linear regime, the frequency is a single valued function of the phase (Yurke et al., 1995; Juillard et al., 2008). In other words, the steady-state solution in the closed-loop case is always stable.

Now, the question is: What is the most important issue when the resonator is driven beyond its critical amplitude in either open or closed loop? The answer is detailed below.

3.2.6 Nonlinearities and Noise Mixing

In a capacitive resonator, mechanical and electrostatic nonlinearities are analytically combined to show that low-frequency voltage drift in the sustaining amplifier is directly converted into a frequency shift in the oscillator output. Experimental evidence of this effect is presented in Roessig et al. (1997), and it is shown that this is the dominant source of near-carrier frequency instability in tuning fork oscillators.

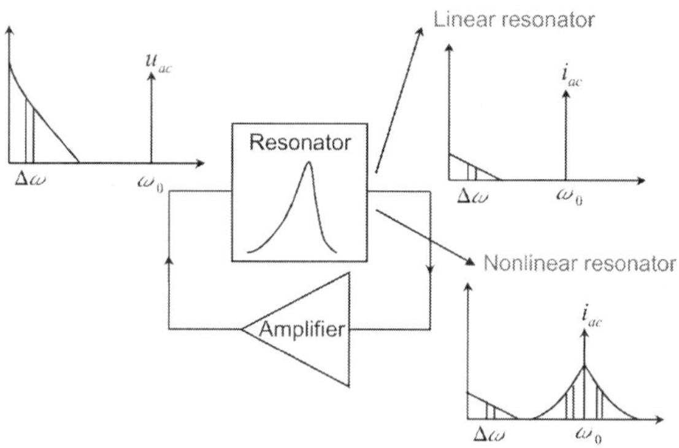

Figure 3.1 Schematic representation of noise aliasing in micro-oscillator. A linear resonator would filter out the amplifier low-frequency $\frac{1}{f}$-noise present at the resonator input, but nonlinear filtering element will result in noise aliasing (Kaajakari et al., 2005a).

Besides, a significant near-carrier noise source is the aliasing of $\frac{1}{f}$-noise to carrier side-bands due to the mixing of low-frequency noise and carrier signal in the active circuit elements. Kaajakari et al. (2005a) showed that, in addition to amplifier nonlinearities, the electrostatic transduction commonly used for coupling to silicon resonators is inherently nonlinear and leads to aliasing of noise. This process is illustrated in Fig. 3.1, which shows a schematic representation of an oscillator comprising a resonator and sustaining amplifier. In addition to amplifying oscillation signal u_{ac}, the amplifier output may present a significant amount of low-frequency $\frac{1}{f}$-noise to the resonator input. A linear resonator element would effectively filter out this low-frequency noise, but nonlinearities in the resonator will lead to unwanted aliasing of the low-frequency noise to carrier side-bands. Thus, the capacitive coupling is expected to be intrinsically more prone to noise aliasing. A detailed analysis of the noise-mixing mechanisms can be found in Kaajakari et al., (2005a), where the capacitive force nonlinearity was found to be the dominant up-mixing mechanism in electrostatic transduction.

Since the capacitive transduction is commonly used for resonant nanocantilevers, any source of frequency instability in the oscillator is detrimental to the noise behavior of the transducer, so these sources must be understood and minimized or canceled.

Practically, in order to avoid most of noise, which reduces the resonant sensor performances, the resonator should be driven linearly beyond its fundamental critical amplitude. Therefore, for ultimate optimizations, one should investigate the open loop nonlinear dynamics of nanocantilevers.

3.3 Nanocantilever Based on Electrostatic Detection

In order to develop a model for micro/nanocantilever beams, a slender uniform flexible beam is considered as shown in Fig. 3.2. The beam is initially straight, and it is clamped at one end and free at the other end, subject to viscous damping with a coefficient $\tilde{c}$ per unit length and actuated by an electric load $v(t) = V_{dc} + V_{ac} \cos(\tilde{\Omega}\tilde{t})$, where V_{dc} is the DC polarization voltage, V_{ac} is the amplitude of the applied AC voltage, $\tilde{t}$ is time, and $\tilde{\Omega}$ is the excitation frequency. In addition, the beam follows the Euler–Bernoulli beam theory, where shear deformation and rotary inertia terms are negligible.

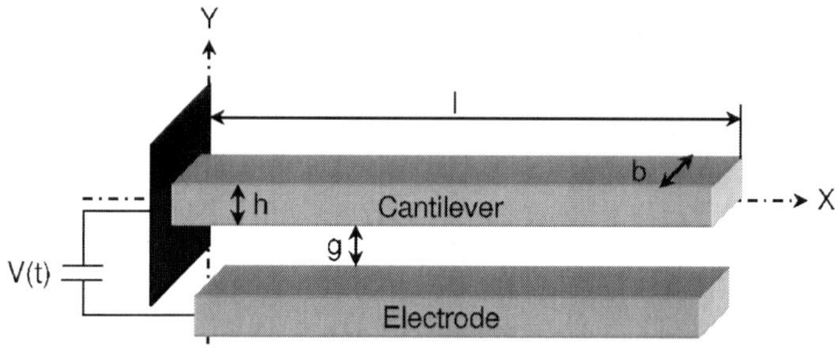

Figure 3.2 Schema of an electrostatically actuated nanocantilever.

3.3.1 Equation of Motion

Following a variational approach based on the extended Hamilton principle and used by Crespo da Silva and Glynn (1978b,d) and Crespo da Silva (1988a,b), the nonlinear equation of motion describing the flexural vibration of a cantilever beam electrostatically actuated can be written as

$$EI\{\tilde{w}'''' + [\tilde{w}'(\tilde{w}'\tilde{w}'')']'\} + \rho bh\ddot{\tilde{w}} + \tilde{c}\dot{\tilde{w}}$$
$$= -\frac{1}{2}\rho bh\left\{\tilde{w}'\int_l^s\left[\frac{\partial^2}{\partial\tilde{t}^2}\int_0^{s_1}(\tilde{w}')^2 ds_2\right]ds_1\right\}'$$
$$+\frac{1}{2}\varepsilon\frac{C_n b[V_{dc} + V_{ac}\cos(\tilde{\Omega}\tilde{t})]^2}{(g-\tilde{w})^2}, \qquad (3.32)$$

where primes and dots denote, respectively, the partial differentiation with respect to the arclength s and to the time $\tilde{t}$; $\tilde{w}$ is the beam bending deflection, E and I are Young's modulus and geometrical moment of inertia of the cross section, respectively; l and b are the length and width of the nanobeam, respectively; ρ is the material density; h is the nanobeam thickness in the direction of vibration; g is the capacitor gap width; and ε is the dielectric constant of the gap medium.

The first term in the left-hand side of Eq. (3.32) is due to the nonlinear expression for the curvature of the beam, while the first term in the right-hand side, which involves a double time derivative, is the nonlinear inertial term. The last term in Eq. (3.32) represents an approximation of the electrostatic force assuming a complete overlap of the area of the nanobeam and the stationary electrode where C_n is the fringing field coefficient computed using an existing analytical model (Nishiyama and Nakamura, 1990). The boundary conditions are

$$\tilde{w}(0,\tilde{t}) = \tilde{w}'(0,\tilde{t}) = \tilde{w}''(l,\tilde{t}) = \tilde{w}'''(l,\tilde{t}) = 0 \tag{3.33}$$

3.3.2 Normalization

For convenience and equation simplicity, we introduce the nondimensional variables:

$$w = \frac{\tilde{w}}{g}, \quad x = \frac{s}{l}, \quad t = \frac{\tilde{t}}{\tau}, \tag{3.34}$$

where $\tau = \frac{2l^2}{h}\sqrt{\frac{3\rho}{E}}$. Substituting Eq. (3.34) into Eqs. (3.32) and (3.33), we obtain

$$\begin{aligned} & w^{iv} + \ddot{w} + c\dot{w} + \delta_1[w'(w'w'')']' \\ & = -\delta^2\left\{w'\int_1^x\left[\frac{\partial^2}{\partial t^2}\int_1^{x_1}(w')^2 dx_2\right]dx_1\right\}' \\ & + d_3\frac{V_{dc}}{V_{ac}}\frac{[1+\frac{V_{ac}}{V_{dc}}\cos(\Omega t)]^2}{(1-w)^2} \end{aligned} \tag{3.35}$$

$$w(0,t) = w'(0,t) = w''(1,t) = w'''(1,t) = 0 \tag{3.36}$$

where primes and dots denote, respectively, the partial differentiation with respect to the dimensionless arclength x and to the dimensionless time t. The parameters appearing in Eq. (3.35) are

$$\begin{aligned} & c = \frac{\tilde{c}l^4}{EI\tau}, \quad \delta_1 = \left[\frac{g}{I}\right]^2, \quad \delta_2 = \frac{1}{2}\left[\frac{g}{l}\right]^2 \\ & \delta_3 = 6V_{ac}V_{dc}\frac{\varepsilon l^4}{Eh^3g^3}, \quad \Omega = \tilde{\Omega}\tau \end{aligned} \tag{3.37}$$

3.3.3 Solving

The beam total displacement $w(x, t)$ can be written as a sum of a static dc displacement $w_s(x)$ and a time-varying ac displacement

$w_d(x, t)$. However, for devices under low pressure, i.e., quality factors Q are in the range of 10^3 to 10^4, the static deflection is negligible with respect to the dynamic deflection.

A reduced-order model is generated by modal decomposition transforming Eq. (3.35) into a multi-degree-of-freedom system consisting of ordinary differential equations in time. We use the undamped linear mode shapes of the cantilever as basis functions in the Galerkin procedure. To this end, we express the deflection as

$$w(x,t) = \sum_{k=1}^{n} a_k(t)\phi_k(x), \tag{3.38}$$

where $a_k(t)$ is the kth generalized coordinate and $\phi_k(x)$ is the kth linear undamped mode shape of the straight microbeam, normalized such that $\int_0^1 \phi_k \phi_j = \delta_{kj}$ where $\delta_{kj} = 0$ if $k \neq j$ and $\delta_{kj} = 1$ if $k = j$. The linear undamped mode shapes $\phi_k(x)$ are governed by

$$\frac{d^4\phi_k(x)}{dx^4} = \lambda_k^2 \phi_k(x) \tag{3.39}$$

$$\phi_k(0) = \phi_k'(0) = \phi_k''(0) = \phi_k'''(1), \tag{3.40}$$

Here, λ_k is the kth natural frequency of the cantilever. The electrostatic force in Eq. (3.35) is expanded in a fifth order Taylor series in order to capture 5 possible amplitudes for a given frequency in the mixed behavior (Kacem and Hentz, 2009). Then, Eq. (3.38) is substituted into the resulting equation, Eq. (3.39) is used to eliminate $\frac{d^4\phi_k(x)}{dx^4}$, and the outcome is multiplied by ϕ_k and integrated from $x = 0$ to 1 for $k \in [1, n] \cap \mathbb{N}$. Thus, a system of coupled ordinary differential equations in time is obtained.

The DC voltage, which is generally at least ten times higher than the AC voltage, makes the second harmonic $\cos(2\Omega t)$ negligible with respect to the first harmonic $\cos(\Omega t)$. Also, assuming that the first mode is the dominant mode of the system, the study can be restricted to the case $n = 1$. Then, we obtain

$$40.44\delta_1 a_1^3 + 0.39\frac{V_{ac}}{V_{dc}}\delta_3 + 0.78\frac{V_{dc}}{V_{ac}}\delta_3 - \frac{V_{ac}}{V_{dc}}\delta_3 a_1$$
$$-2\frac{V_{dc}}{V_{ac}}\delta_3 a_1 + 2.22\frac{V_{ac}}{V_{dc}}\delta_3 a_1^2 + 4.43\frac{V_{dc}}{V_{ac}}\delta_3 a_1^2$$

$$
\begin{aligned}
&-4.7\frac{V_{\mathrm{ac}}}{V_{\mathrm{dc}}}\delta_3 a_1^3-9.4\frac{V_{\mathrm{dc}}}{V_{\mathrm{ac}}}2\delta_3 a_1^3+9.75\frac{V_{\mathrm{ac}}}{V_{\mathrm{dc}}}\delta_3 a_1^4\\
&+19.5\frac{V_{\mathrm{dc}}}{V_{\mathrm{ac}}}\delta_3 a_1^4-20\frac{V_{\mathrm{ac}}}{V_{\mathrm{dc}}}\delta_3 a_1^5-40\frac{V_{\mathrm{dc}}}{V_{\mathrm{ac}}}\delta_3 a_1^5\\
&+a_1\lambda_1^2+c\dot{a}_1+9.2\delta_2 a_1\dot{a}_1^2+\ddot{a}_1+9.2a_1^2\delta_2\ddot{a}_1\\
&+1.56\delta_3\cos(\Omega t)-4\delta_3 a_1^3\cos(\Omega t)\\
&+9\delta_3 a_1^2\cos(\Omega t)-19\delta_3 a_1^3\cos(\Omega t)
\end{aligned}
\tag{3.41}
$$

We recognize in the Eq. (3.41) some canonical nonlinear terms such as the Duffing nonlinearity as well as the parametric excitation (Mathieu term). However, the presence of other high order nonlinearities makes the described system in Fig. 3.2 as a forced nonlinear cantilever under multifrequency parametric excitation. This kind of equation is not so frequently treated in the literature: It includes terms coming from mechanical and electrostatic nonlinearities.

To analyze the equation of motion (3.41), it is convenient to invoke perturbation techniques, which work well with the assumptions of "small" excitation and damping, typically valid in NEMS resonators. To simplify the perturbation approach, in this case the averaging method, a standard constrained coordinate transformation is introduced, as given by

$$a_1 = A(t)\cos[\Omega t+\beta(t)] \tag{3.42}$$

$$\dot{a}_1 = -A(t)\Omega\sin[\Omega t+\beta(t)] \tag{3.43}$$

$$\ddot{a}_1 = -A(t)\Omega^2\cos[\Omega t+\beta(t)] \tag{3.44}$$

$A(t)$ and $\beta(t)$ are slowly time-varying functions. In addition, since near-resonant behavior is the principal operating regime of the proposed system, a detuning parameter σ is introduced, as given by

$$\Omega = \omega_1+\xi\sigma, \tag{3.45}$$

where $\omega_1=\sqrt{\lambda_1^2-\frac{V_{\mathrm{ac}}}{V_{\mathrm{dc}}}\delta_3-2\frac{V_{\mathrm{dc}}}{V_{\mathrm{ac}}}\delta_3}$ and ξ is the small non-dimensional bookkeeping parameter. Separating the resulting equations and averaging them over the period $\frac{2\pi}{\Omega}$ in the t-domain results in the system's averaged equations in terms of amplitude A and phase β given by

$$\dot{A} = \frac{\xi\delta_3 \sin\beta}{\omega_1}(0.78 + 1.11A^2 + 2.44A^4) - \frac{\xi c}{2}A + O(\xi^2) \tag{3.46}$$

$$\begin{aligned}\dot{\beta} = \sigma\xi - \xi\frac{\delta_3 \cos\beta}{\omega_1}\left(\frac{0.78}{A} + 3.32A + 12.19A^3\right) \\ + \xi\frac{A^4\delta_3}{\omega_1}\left(12.51\frac{V_{dc}}{V_{ac}} - 6.25\frac{V_{ac}}{V_{dc}}\right) - \xi\frac{15.16A^2\delta_1}{\omega_1} \\ + \frac{\xi\delta_3 A^2}{\omega_1}\left(1.76\frac{V_{ac}}{V_{dc}} + 3.52\frac{V_{dc}}{V_{ac}}\right) + 2.3\xi A^2\delta_2\omega_1 + O(\xi^2)\end{aligned} \tag{3.47}$$

The steady-state motions occur when $\dot{A} = \dot{\beta} = 0$, which corresponds to the singular points of Eqs. (3.46) and (3.47). Thus, the frequency response equation can be written in its parametric form $A = K_1(\beta)$, $\Omega = K_2(\beta)$ as a function of the phase β. This set of two equations is easily implementable in Matlab or Mathematica. This ability makes the model suitable for NEMS designers as a quick tool to optimize the resonant sensors performance.

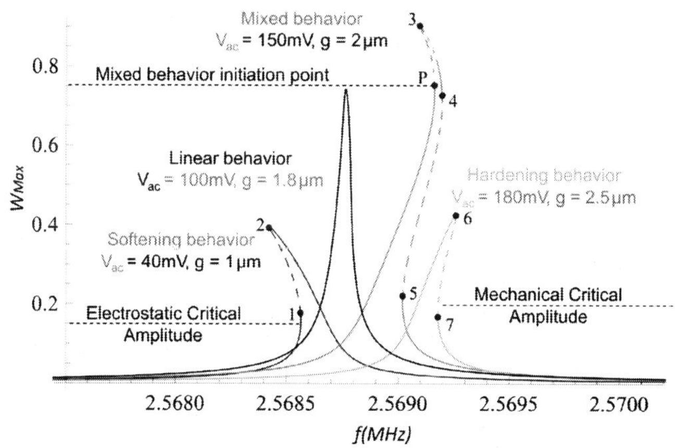

Figure 3.3 Analytical forced frequency responses for $Q = 10^4$ and several values of g and V_{ac}; W_{max} is the beam displacement at its free end normalized by the gap g, A_c is the critical amplitude above which bistability occurs, the different bifurcation points are {1, 2, 3, 4, 5, 6, 7, P}, the P point characterizes the initiation of the mixed behavior.

The plots of Fig. 3.3 were carried out with the following set of parameters: l = 12.5 μm, h = 300 nm, b = 500 nm, and $V_{dc} = 50V_{ac}$; g and V_{ac} were used for parametric studies. This analytical model enables the capture of all the nonlinear regimes in the resonator dynamics and describes the competition between the mechanical hardening and the electrostatic softening behaviors. In addition, the model permits the optimization of the resonator design by tuning the geometrical parameters in order to cancel out nonlinearities as shown in Fig. 3.3. For g = 1.8 μm, V_{dc} = 5 V and $Q = \rho bh \frac{\omega_1}{\tau\tilde{c}} = 10^4$ (black curve). The obtained linear behavior enhances the detection limit of NEMS resonant sensors.

3.3.4 Critical Amplitude

The critical amplitude is the oscillation amplitude A_c above which bistability occurs. Thus, A_c is the transition amplitude from the linear to the nonlinear behavior.

3.3.4.1 The mechanical critical amplitude

By using Eqs. (3.46) and (3.47) when the mechanical nonlinearities (terms proportional to δ_1 or δ_2) are dominating the cantilever dynamics, the parametric form of the frequency response can be written as

$$\Omega = 1 + \zeta_0 \cot\beta + 0.077\delta_1\zeta_1^2 \sin^2\beta \tag{3.48}$$

$$A = 2\zeta_1 \sin\beta, \tag{3.49}$$

where $\zeta_0 = 0.142206c$ and $\zeta_1 = \frac{0.445\delta_3}{c}$.

Mathematically, A_{cm} is defined as the oscillation amplitude for which the equation $\frac{d\Omega}{d\beta} = 0$ (infinite slope) has a unique solution $\beta_{cm} = \frac{\pi}{3}$. Thus, the critical electrostatic force is deduced as $\zeta_{1c} = \frac{0.456\sqrt{\zeta_0}}{\delta_1}$.

The critical amplitude A_{cm} is obtained by substituting the critical electrostatic force into Eq. (3.49) at the point $\beta = \frac{\pi}{2}$ and multiplying by the gap g and the value of the first linear undamped mode shape function ϕ_1 at the free point of the beam. Finally, we obtain the following close-form solution:

$$A_{\text{cm}} = 6.3\frac{1}{\sqrt{Q}} \tag{3.50}$$

Remarkably, the mechanical critical amplitude of a resonant cantilever depends only on its length and its quality factor. However, for a clamped-clamped resonator, the mechanical critical amplitude is $A_{\text{cm}} = 1.68\frac{h}{\sqrt{Q}}$ (Kacem et al., 2009); it depends only on the resonator width in the direction of vibration and its quality factor. For a given quality factor, the ratio between both critical amplitudes is $R_c = \frac{A_{\text{cm}}^{\text{c-f}}}{A_{\text{cm}}^{\text{c-c}}} = 3.75\frac{l}{h} = 3.75\lambda_e$, where λ_e is the slenderness ratio of the beam. While the resonator is sufficiently slender to validate the Euler–Bernoulli theory, the dynamic range ratio between cantilevers and clamped-clamped beams is very high (>20), which makes nanocantilevers an advantageous candidate for NEMS resonators-based applications.

3.3.4.2 The electrostatic critical amplitude

In this case, the electrostatic nonlinearities in Eqs. (3.46) and (3.47) are supposed to be dominating the cantilever dynamics. Considering only nonlinear terms up to the third order, while neglecting the parametric terms and the terms proportional to V_{ac}^2, the form of the frequency response can be written as

$$\Omega = \frac{c}{2}\cot\beta - \frac{8.64\frac{V_{\text{dc}}}{V_{\text{ac}}}\delta_3^3}{c^2\omega_1^3}\sin\beta^2 + \frac{6.2}{\omega_1} - \frac{V_{\text{dc}}^2\delta_3}{\omega_1} + \frac{\omega_1}{2} \tag{3.51}$$

$$A = \frac{3.13\delta_3}{c\omega_1}\sin\beta \tag{3.52}$$

Mathematically, A_{ce} is defined as the oscillation amplitude for which the equation $\frac{d\Omega}{d\beta} = 0$ has a unique solution $\beta_{\text{ce}} = \frac{\pi}{3}$. Thus, the critical electrostatic AC voltage is deduced as

$$V_{\text{ac}_c} = 11.2V_{\text{dc}}\frac{\delta_3^3}{c^3\omega_1^3} \tag{3.53}$$

The electrostatic critical amplitude A_{ce} is obtained by sub-stituting Eq. (3.53) into Eq. (3.52) at the pointt $\beta = \frac{\pi}{2}$ and multiplying by the gap g and the value of the first linear undamped mode shape function ϕ_1 at the free point of the beam.

$$A_{ce} = 2 \times 10^9 g^{\frac{5}{2}} \frac{h}{l\sqrt{QV_{dc}}} \left(\frac{7.5 \times 10^7 h^2}{l^4} - \frac{3.8 \times 10^{-15} V_{dc}^2}{g^3 h} \right)^{\frac{1}{4}} \tag{3.54}$$

Unlike the mechanical critical amplitude (Eq. (3.50)), the electrostatic critical amplitude of a resonant cantilever depends on its length *l*, its width in the direction of vibration *h*, the gap *g*, the DC voltage as well as the quality factor *Q*.

3.3.4.3 Engineering optimization

The hysteresis suppression is based on the counterbalance between hardening mechanical nonlinearities and softening electrostatic nonlinearities. The mixed behavior, captured by including the fifth order nonlinear electrostatic terms, is less pronounced than in electrostatically driven clamped-clamped beams. Therefore, while neglecting the fifth order terms, the nonlinearity cancellation based on the critical amplitude expressions can be written as

$$A_{c_m} = A_{c_e} \tag{3.55}$$

Thus, assuming a constant quality factor *Q*, the optimal *DC* drive voltage is

$$V_{dc_{OP}} = \sqrt{\frac{1}{2}\sqrt{\frac{1.65 \times 10^{39} g^{14} h^6}{l^{16}} + \frac{3.2 \times 10^{42} g^{10} h^6}{l^{12}}} - \frac{8.1 \times 10^{19} g^7 h^3}{l^8}} \tag{3.56}$$

In reality, the overall quality factor may decrease when V_{dc} increases because of the ohmic losses from the electrons moving on and off the resonator due to capacitive coupling to a nearby electrode (Sazonova, 2006). This ohmic contribution adds up to the other sources of dissipation (thermomechanical, anchor losses, adsorption/desorption ...) like $Q_{total}^{-1} = Q_{thermo}^{-1} + Q_{anchor}^{-1} + \cdots + Q_{ohmic}^{-1}$ and may be expressed as $\left(Q_{ohmic}^{-1} = \frac{1}{\pi\omega} \frac{R(C'V_{dc})^2}{M_{eff}} \right)$, (Sazonova, 2006) where C' is the gradient of the capacitance, *R* is the output resistor and M_{eff} is the effective mass of the considered mode.

Then, the quality factor should be changed to $\frac{1}{Q^{-1} + Q_{ohmic}^{-1}}$, for the electrostatic critical amplitude and thus, the optimal DC drive voltage can be deduced using the same Eq. (3.55).

In the particular case of Fig. 3.3, the mechanical critical amplitude is $A_{c_m} = 0.2$ g. When $g = 1.8$ μm and for a quality factor $Q = 10^4$, the optimal DC drive voltage taking into account the ohmic losses, is around 5 V. At this voltage, as shown by the black curve of Fig. 3.3, the peak amplitude is linear and beyond the critical amplitude ($A_{peak} = 0.8$ g). Therefore, the enhancement $\left(\frac{A_{peak}}{A_c}\right)$ rate of the sensor performance is around 4.

Remarkably, the electrostatic critical amplitude is independent on the AC voltage. This is due to the use of a low AC voltage compared to the DC voltage for the cantilever actuation, which makes the contribution of V_{ac} in the electrostatic Duffing term negligible. Hence, in this configuration, the compensation of the nonlinearities is independent on the AC voltage. This interesting result makes possible the enhancement of the nanocantilever performances up to very high displacements comparable to the gap in the case of an electrostatic actuation by increasing the AC voltage, and limited by an upper bound instability such as the pull-in (Nayfeh et al., 2007).

The aim of this model is to provide practical analytical rules for MEMS and NEMS designers in order to optimize resonant sensor performances. Hence, it is important to check its validity experimentally on nanocantilever electrostatically actuated.

3.3.5 Fabrication

Practical applications of nanocantilever beams benefit much from on-chip signal processing, whereby optimal performance is achieved in case of monolithic integration with CMOS (Verd et al., 2006, 2008; Arcamone et al., 2008, 2007). Such NEMS/CMOS made of silicon combine unique sensing attributes, thanks to the high resonance frequency mobile mechanical part, with the possibility to electrically detect the output signal in enhanced conditions. For those reasons, the nanobeams that are experimentally studied in this work have been monolithically integrated with a dedicated CMOS circuit to enhance the capacitive readout of their resonance "motional" current. This heterogeneous integration has required a very specific fabrication process described hereafter. More details about the process and the functionality of this NEMS/CMOS system can be found in Arcamone et al. (2007, 2008).

The cantilevers fabrication process used here has been performed at CNM-IMB (CSIC) in Barcelona (Spain) and at EPFL in Lausanne (Switzerland). It is based on a post-processing approach in which CMOS circuits are first fabricated according to a standard CMOS technology, then NEMS resonators are subsequently patterned by nanostencil lithography (nSL) (van den Boogaart et al., 2004) and fully fabricated. nSL is a shadow-mask based, full-wafer and parallel nanopatterning technique providing a resolution down to 200 nm and a high fabrication throughput.

After concluding the fabrication of the CMOS circuits, dedicated areas (located close to each circuit, 1 per circuit) are selectively patterned with an 80 nm thick aluminum layer by nSL. The following step is a reactive ion etching (RIE) of silicon that transfers the aluminum patterns to the polysilicon structural layer of the resonators. The last step consists in releasing the resonators and removing the Al mask by a local wet etching based on HF acid. The circuit is robustly protected during this etching by an adequately annealed photoresist layer. This entire process is described in Arcamone et al. (2008).

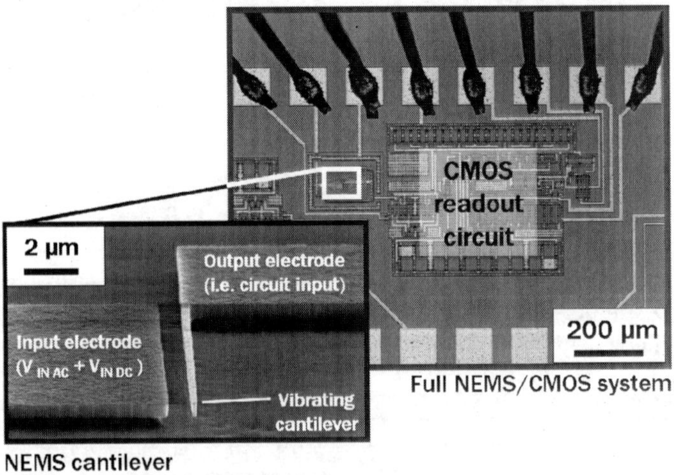

Figure 3.4 Optical picture of the NEMS resonator/CMOS readout circuit system. The scanning electron micrograph zooms the cantilever beam itself and its driving electrode.

CMOS wafers containing each 2000 fully fabricated nanomechanical devices of diverse types have been obtained with that process; all connected to dedicated CMOS circuits for signal interfacing and amplification. Figure 3.4 depicts a fully fabricated nanocantilever beam (optimized for in-plane motion) which is monolithically integrated with its dedicated CMOS readout circuit.

3.3.6 Electrical Characterization

Electrostatic actuation and capacitive detection are used for detecting in-plane oscillations (in the MHz range) of those Si nanocantilever resonators. When the resonator is electrostatically driven by a DC + AC voltage, the readout electrode (i.e. the cantilever itself in our case, see Fig. 3.4), electrically connected to the closely located CMOS circuit input, collects a capacitive current in enhanced conditions since parasitic capacitances at the NEMS output are drastically reduced to the few fF range. The motional current I_{M}, which is a fraction of the total NEMS output current I_{MEMS}, is specifically generated by the variation of electrode/resonator capacitance due to the mechanical motion itself. The other part, the background current I_{BG}, is related to the capacitive feed through between the NEMS input and output electrodes, one part of it being the "static" capacitance between the cantilever and the in-front electrode, the other part being the parasitic fringing capacitance between both.

The dedicated CMOS readout circuit ensures a constant voltage biasing at the resonator output, while it amplifies the readout current by a factor 100 and converts it to an output voltage according to an external load resistor R_{LOAD} (see Fig. 3.5). Arcamone et al. (2007, 2008) give more details on the circuitry and the readout scheme.

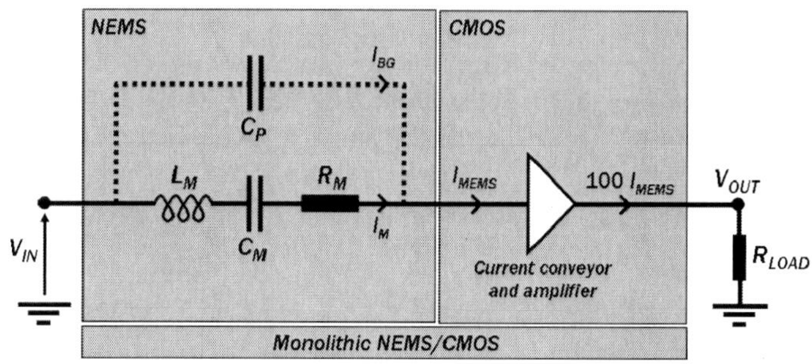

Figure 3.5 Electrical scheme of the monolithic NEMS/CMOS system.

The mechanical frequency response of those nanocantilevers has been electrically characterized with a network analyzer (AGILENT E5100A) in air (with a probe station) and vacuum (with wire-bonded samples). In both configurations, the capacitive detection scheme including the CMOS circuit successfully transduced into an electrical signal the mechanical motion corresponding to the fundamental in-plane flexural resonance mode of vibrating cantilevers. The data collected by the network analyzer, in terms of magnitude and phase, are under the following form:

$$R_{\mathrm{NA}}(f) = 20\log\frac{V_{\mathrm{OUT\,AC}}}{V_{\mathrm{IN\,AC}}} \tag{3.57}$$

The NEMS resonator total current (at the cantilever output, not at the circuit output) being

$$I_{\mathrm{BG}} + I_{\mathrm{M}} = I_{\mathrm{MEMS}} = \frac{V_{\mathrm{IN\,AC}}}{100R_{\mathrm{LOAD}}}10^{\frac{R_{\mathrm{NA}}}{20}} \tag{3.58}$$

The modulus and argument of the NEMS resonator electrical admittance Y can be calculated based on the circuit characteristics and the network analyzer response as follows:

$$|Y| \approx \frac{I_{\mathrm{MEMS}}}{V_{\mathrm{IN\,AC}}} = \frac{10^{\frac{R_{\mathrm{NA}}}{20}}}{100R_{\mathrm{LOAD}}} \tag{3.59}$$

$$\arg(Y) = P - \Delta\phi, \tag{3.60}$$

where P is the phase signal as measured by the network analyzer and $\Delta\phi$ is the phase delay introduced by the circuit; in our case $\Delta\phi \approx \frac{\pi}{4}$. In the 2-port configuration of these measurements, a NEMS cantilever can be modeled as two parallel branches (see Fig. 3.5):

- a capacitive branch (of capacitance C_{P}), of admittance Y_{BG}, corresponding to the background signal, in which I_{BG} flows. Y_{BG} is given by $Y_{\mathrm{BG}} = sC_{\mathrm{P}}$ with $s = j\omega$.
- an RLC branch of admittance Y_{M}, corresponding to the resonating part, in which I_{M} flows. Y_{M} is given by $Y_{\mathrm{M}} = \frac{Cs}{LCs^2 + RCs + 1}$

The total admittance is then $Y = Y_{BG} + Y_M$. The theoretical model described in the previous section is expressed in terms of Y_M only, i.e., the term describing the oscillations. Therefore, it is required to apply a simple data treatment to extract the motional admittance Y_M from the measured admittance Y, Y_M being given by $Y_M = Y - Y_{BG} = Y - sC_P$. Hence, its modulus is given by

$$|Y_{\mathrm{M}}| = \sqrt{[|Y|\cos(\arg(Y))]^2 + [|Y|\sin(\arg(Y)) - \omega C_{\mathrm{P}}]^2} \tag{3.61}$$

C_P can be estimated by considering the electrical response $R_{NA\text{-}OOR}$ at a given frequency f_{OOR} that is out of resonance, then C_P is given by

$$C_{\mathrm{P}} = \frac{1}{2\pi f_{\mathrm{OOR}}} \frac{10^{\frac{R_{\mathrm{NA}}\text{-OOR}}{20}}}{100R_{\mathrm{LOAD}}} \tag{3.62}$$

Figure 3.6a shows the raw electrical response R_{NA} around the mechanical resonance of cantilever A (l = 14.5 μm, b = 460 nm, h = 400 nm, g = 600 nm) as directly measured by the network analyzer. Figure 3.6b shows the motional admittance Y_M extracted from the data of Fig. 3.6a according to Eq. (3.61). Using the developed model, the motional admittance can be computed as follows:

$$I_{\mathrm{M}} = Y_{\mathrm{M}} V_{\mathrm{ac}} = \frac{dC_{\mathrm{res}}}{dt} V_{\mathrm{dc}} \tag{3.63}$$

$$\frac{dC_{\mathrm{res}}}{dt} = \int_0^1 \frac{bC_{\mathrm{n}}\varepsilon_0\phi_1(x)\dot{a}_1(t)}{(1-a_1)(t)\phi_1(x))^2} dx \tag{3.64}$$

The derivative of the resonator capacitance with respect to the dimensionless time t has been expanded in a fifth order Taylor series, which enables the analytical computation of the integral in Eq. (3.64). Then, Eqs. (3.42) and (3.43) are substituted into the outcome equation and the trigonometric functions are linearized. Since the electrical measurement filters out all frequency components of the readout signal except which of the drive frequency, the first harmonic of the Fourier transform of Eq. (3.63) gives the motional current frequency response including the coupling between the dynamics of the resonator and the read-out voltage Eq. (3.64). Although this coupling brings extra nonlinear terms, their contribution happens to be negligible and the read-out voltage is proportional to the dynamic deflection.

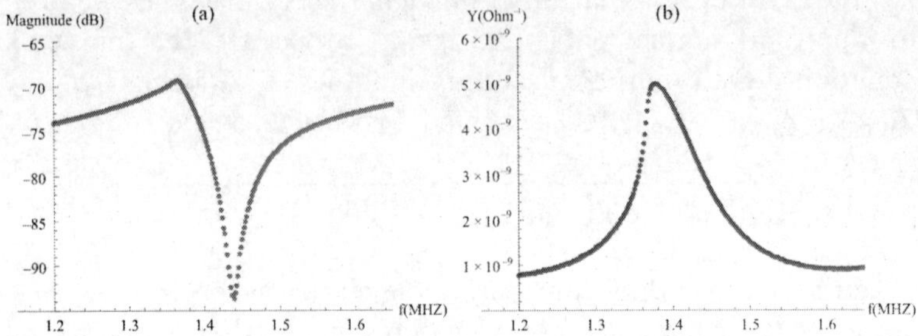

Figure 3.6 (a) Raw electrical response R_{NA} around the mechanical resonance of a nanocantilever. This is the response as measured by the NA of the full NEMS-CMOS system. (b) Motional admittance frequency response extracted from the data of Fig. 3.6a according to Eq. (3.61).

3.3.6.1 Measurements in air

It is important to underline that all the inputs of the model are known physical parameters including the fringing field coefficients computed using the analytical formulae (Nishiyama and Nakamura, 1990), except the measured quality factor Q and the parasitic capacitance C_P. Therefore, to evaluate the model, Q has been fitted using linear curves. For a fully analytical prediction, the quality factor may be computed using existing models taking into account the thermoelastic damping (Lifshitz and Roukes, 2000), the support loss (Hao et al., 2003), and the surface loss (Yang et al., 2002). Such a computation gives results in good agreement with experimental measurements.

Figure 3.7 shows the characteristic responses analytically computed and electrically measured of cantilever A (l = 14.5 μm, b = 460 nm, h = 400 nm, g = 600 nm) operated in air for several DC voltages. Due to some mass that has been deposited at the free end of the cantilever, its resonance frequency was shifted from 2 MHz down to 1.5 MHz. The first linear curve (V_{dc} = 16V) is fitted by adjusting the parasitic capacitance C_P and the quality factor Q. Then, the same value of C_P has been used for the three other curves while adjusting the Q factor for each one.

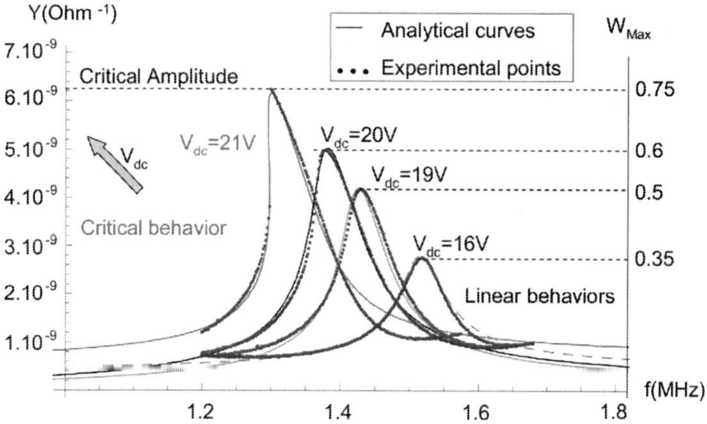

Figure 3.7 Analytical and measured motional admittance frequency curves (in air) of cantilever A. $W_{\max}$ is the cantilever displacement at its free end normalized by the gap.

Figure 3.7 shows that the resonance frequency of the cantilever can be tuned by varying the applied DC voltage and a clear spring-softening effect is seen with increasing V_{dc}. The analytical curves are in good agreement with experimental results and the critical amplitude predicted analytically for a 21 V DC voltage has been experimentally confirmed as shown in Fig. 3.7. which demonstrates the accuracy and performance of the model. Out of resonance, the experimental curves slightly increase which is due to the variation of $\Delta\phi$ during the measurements.

When experiments are performed in air, in order to reach detectable signals, the applied DC voltage has to be high ($V_{dc} > 10$ V). As the quality factor is low ($15 < Q < 20$), the dynamic behavior of the nanocantilever is dominated by the nonlinear softening electrostatic forces which significantly increases the critical amplitude Eq. (3.54).

3.3.6.2 Measurements in vacuum

The chip containing cantilever B ($l = 14.5$ µm, $b = 570$ nm, $h = 260$ nm, $g = 820$ nm) has been wire bonded and it has been measured in a vacuum chamber under a pressure of 10^{-2} mBar. The fits have been made in the same way as for the experimental curves measured in air. Figure 3.8 shows two measured frequency responses and their predicted analytical curves.

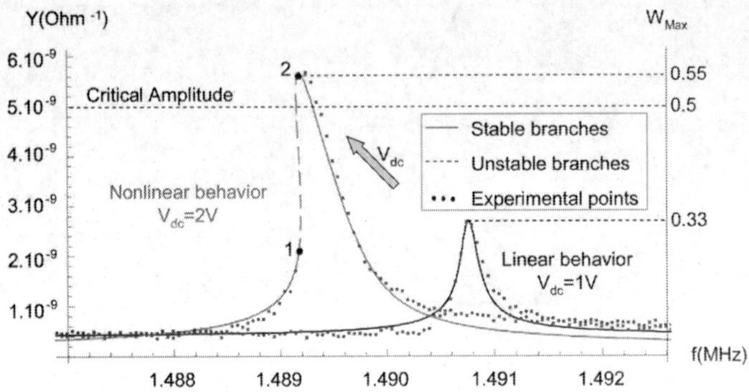

Figure 3.8 Analytical and measured motional admittance frequency curves (in vacuum) of cantilever B. $W_{\max}$ is the cantilever displacement at its free end normalized by the gap.

Due to the high quality factors obtained in vacuum, low drive voltages are enough to polarize the device: 1 or 2V DC added to a 0.092V AC. A 2V DC polarization is sufficient to provoke a nonlinear behavior yielding almost vertical slopes both in magnitude and in phase. The extracted Q factor decreases from 9150 down to 6650 for 1 and 2V DC polarization due to the ohmic losses (Sazonova, 2006).

The high quality factors reduce the critical amplitude A_c as shown in Fig. 3.8, which confirms the accuracy of the model beyond the critical amplitude (second curve with 2V DC). The effect of the spring softening is also present due to the electrostatic negative stiffness. Even with high quality factors, the cantilever B displays a softening behavior. This can be explained by the fact that the nonlinearities coming from the electrostatic force are stronger than the mechanical hardening nonlinearities due to the length of the beam and the small gap thickness.

In air, NEMS cantilever have a large critical amplitude (75% of the gap as shown in Fig. 3.7 for cantilever A) which implies a large dynamic range compared to clamped-clamped NEMS resonators (their critical amplitude could be lower than 1% of the gap). In addition, even in vacuum with high quality factors, the critical amplitude of nanocantilevers still interestingly large (50% of the gap as shown in Fig. 3.8 for cantilever B). This important property makes cantilevers the best candidates for resonant mass and gas sensing devices.

The electric characterization of the NEMS/CMOS devices earlier described was an important step towards the validation of the nonlinear model for NEMS cantilevers. As previously shown, the competition between both behaviors (softening and hardening) is controlled by the design parameters, the quality factor as well as the DC voltage. The model shows that for appropriate values of those parameters, it is possible to suppress the hysteresis and to obtain an optimal design for which the mechanical and the electrostatic nonlinearities would be equilibrated. The next step is the complete validation of the model through the compensation of the nonlinearities, which is the subject of the next section.

3.4 Nanocantilever Based on Piezoresistive Detection

3.4.1 Device Description

The device is composed of a fixed-free lever beam and two piezoresistive gauges connected to the cantilever at a distance $d = 0.15l$ from its fixed end. This value was chosen to maximize the stress inside the gauges due to the cantilever motion (Fig. 3.9). The gauges have been etched along the <110> direction in order to benefit from the high gauge factor associated with p^{++} doped silicon. A driving electrode was patterned along one side of the resonant cantilever for electrostatic actuation. The NEMS cantilever is actuated electrostatically at the primary resonance of its first linear undamped mode shape. The cantilever oscillation induces stress inside the piezoresistive gauges and the collected strain is transduced into a resistance variation due to the piezoresistive effect. Thus, the sensor frequency response is obtained via a piezoresistive read-out perfectly decoupled from the capacitive actuation of the resonator. The amount of molecules absorbed by the functionalized surface of the cantilever changes its effective mass, which lowers its frequency resonance. By evaluating the frequency shift, the mass of the added species can be estimated.

Thus, the considered device can be used as either a mass or a gas sensor.

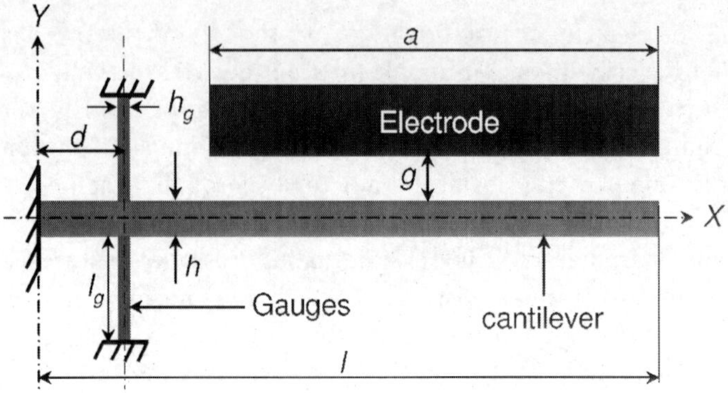

Figure 3.9 Resonant nanocantilever based on piezoresistive detection.

3.4.2 Transduction

The cantilever motion is detected by using the piezoresistive transduction principle (Mo Li and Roukes, 2007). The piezoresistive transducers consist on semiconducting silicon p++ (boron) doped suspended nanogauges. The rationale for using piezoresistive doped silicon nanogauges is related to the giant piezoresistance effect of these materials appearing for sub-100 nm dimensions (He and Yang, 2006). They are suitable for integrated transducers and for self-sensing devices (Arlett et al., 2006). The figure of merit characterizing these materials is the piezoresistive gauge factor γ defined as

$$\gamma = (1+\nu) + \frac{1}{\varepsilon_l}\frac{\Delta\rho}{\rho}, \tag{3.65}$$

where ρ is the resistivity, ε_l the gauge elongation and ν the Poisson ratio. The gauge factor relates the mechanical strain applied on the gauges to its relative resistance change. The resistance change depends on two effects. The first term in Eq. (3.65) is a geometrical consequence and is associated with elastic deformation, while the latter is related to the modification of the energy bands inside a crystal and thus altering its resistivity. In metals only the first term participates which ranges from 1–2 and is the way chosen by Roukes et al. (Yang, 2006). In semiconductors, the second term provides a significant contribution, which was shown to be more than three orders of magnitude higher (He and Yang, 2006). The force applied to the lever is amplified by the appropriate design and transferred to the gauges. This design makes possible to exploit a first order piezoresistance effect with the suspended gauges acting as strain collectors instead of second order one (He et al., 2008). We should notice that there is no need for further metallization layers, which lead to additional damping and energy dissipation. The strain collected by the gauges is transduced into a resistance variation due to the piezoresistance effect proportional to

$$\frac{\Delta R(\Omega)}{R} = \gamma\varepsilon_1(\Omega) \tag{3.66}$$

3.4.3 Equation of Motion

The nonlinear equations of motion describing the flexural vibration of the device described in Fig. 3.9 can be derived using the extended Hamilton principle (Silva and Glynn, 1978b,d; Silva, 1988a,b). The cantilever bending deflection $\tilde{w}_j$ is decomposed into $\tilde{w}_0$ for $\tilde{s} \in [0, d]$ and $\tilde{w}_1$ for $\tilde{s} \in [d, l]$.

$$\begin{aligned} E_c I_c &= \{\tilde{w}_j'''' + [\tilde{w}_j'(\tilde{w}_j'\tilde{w}_j'')']'\} + \rho bh\ddot{\tilde{w}}_j + \tilde{c}\dot{\tilde{w}}_j \\ &= -\frac{1}{2}\rho bh\left\{\tilde{w}_j'\int_{(1-j)d+jl}^{s}\left[\frac{\partial^2}{\partial\tilde{t}^2}\int_{jd}^{s_1}(\tilde{w}_j')^2 ds_2\right]ds_1\right\}' \\ &\quad + \frac{j}{2}\varepsilon\frac{C_n b[V_{dc} + V_{ac}\cos(\tilde{\Omega}\tilde{t})]^2}{(g-\tilde{w}_j)^2}H(S+a-1) \end{aligned} \tag{3.67}$$

where s is the arclength; E_c and I_c are Young's modulus and moment of inertia of the nanocantilever cross section, respectively; l and h are the length and width of the nanobeam, respectively; b is the device thickness; ρ is the material density; g is the capacitor gap width; and ε is the dielectric constant of the gap medium. The last term in Eq. (3.67) represents an approximation of the electrostatic force assuming a partial distribution along the nanobeam length. H is a Heaviside function and C_n is the fringing field coefficient. The boundary conditions are

$$\tilde{w}_0 = (0, \tilde{t}) = \tilde{w}_0'(0, \tilde{t}) = \tilde{w}_1''(l, \tilde{t}) = \tilde{w}_1'''(l, \tilde{t}) = 0 \tag{3.68}$$

$$\tilde{w}_0(d, t) - \tilde{w}_1(d, \tilde{t}) = \tilde{w}_0'(d, \tilde{t}) - \tilde{w}_1'(d, \tilde{t}) = 0 \tag{3.69}$$

$$\tilde{w}_0''(d,\tilde{t}) - \tilde{w}_1''(d,\tilde{t}) = -\frac{I_g}{I_c l_g}\tilde{w}_0'(d,\tilde{t}) \tag{3.70}$$

$$\tilde{w}_0'''(d,\tilde{t}) - \tilde{w}_1'''(d,\tilde{t}) = -\frac{2h_g b}{I_c l_g}\tilde{w}_0'(d,\tilde{t}), \tag{3.71}$$

where $\tilde{t}$ is time, h_g and I_g are the width and the moment of inertia of the gauge cross section. Equations (3.70) and (3.71) are obtained by writing the force and torque moment equilibrium equations at the point $s = d$.

$$T_{sc}|_d + 2T_{ag}|_d = EI_c[\tilde{w}_0'''(d,\tilde{t}) - \tilde{w}_1'''(d,\tilde{t})] + 2Ebh_g\frac{\tilde{w}_0(d, \tilde{t})}{l_g} = 0 \tag{3.72}$$

$$M_{bc}|_d + M_g|_d = EI_c[\tilde{w}_0''(d, \tilde{t}) - \tilde{w}_1''(d, \tilde{t})] - EI_g\frac{\tilde{w}_0'(d, \tilde{t})}{l_g} = 0, \tag{3.73}$$

where T_{sc} is the shear force applied to the cantilever, M_{bc} is its bending moment, T_{ag} is the axial force applied to the gauges, and M_g is its corresponding torque moment.

3.4.4 Normalization and Solving

Similarly to the cantilever model presented in Section 3.3, we introduce the same nondimensional variables. Then, a reduced-order model is generated by modal decomposition (Eq. (3.38)) transforming the normalized equation of motion into a multidegree-of-freedom system consisting in ordinary differential equations in time. The undamped linear mode shapes, which are defined as piecewise functions for the device described in Fig. 3.10, are used as basis functions in the Galerkin procedure.

The electrostatic force is expanded in a fifth order Taylor series, and Eq. (3.38) is substituted into the resulting equation. Then,

Eq. (3.39) is used to eliminate $\frac{d^4\phi_k(x)}{dx^4}$, and the outcome is multiplied by ϕ_k and integrated from $x = 0$ to 1 for $k \in [1, n] \cap \mathbb{N}$. Thus, a system of coupled ordinary differential equations in time is obtained.

Figure 3.10 shows that between the clamped end of the cantilever and the gauges, the vibrations of the sensor are relatively negligible with respect to its dynamics between the gauges and the free end of the cantilever for the first four modes. Consequently, the nonlinear coupling between the mode shapes is negligible for $x \in [0, d]$ and when the sensor is actuated on its first mode, its dynamics can be approximated by the dynamics of a cantilever of length $l - d$.

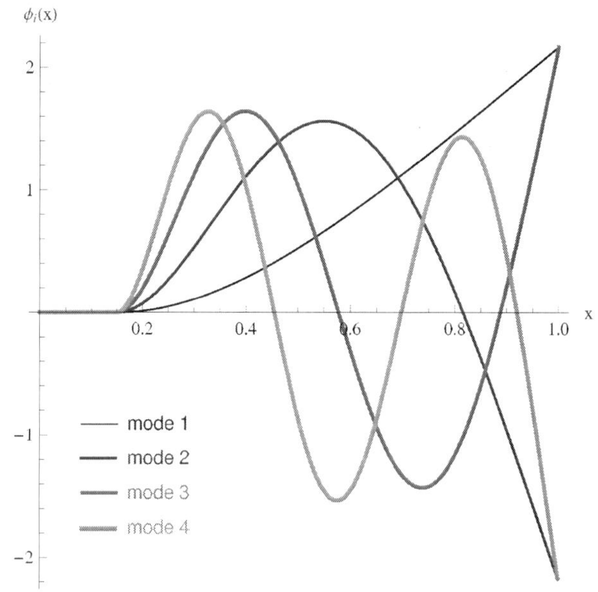

Figure 3.10 The first four linear undamped mode shapes of the device described in Fig. 3.9.

Assuming that the first mode is the dominant mode of the system, the study can be restricted to the case $n = 1$, which results in a nonlinear Mathieu-Duffing Equation. It can be solved using a perturbation technique which permits the transformation of the nonlinear second order equation into two first order nonlinear ordinary differential equation that describe the amplitude and phase modulation of the system frequency response. For

$V_{ac} << V_{dc}$, the second harmonic terms are neglected. The resulting phase and amplitude averaged equations over the period $\frac{2\pi}{\Omega}$ and around the primary resonance ($\Omega = \lambda_1 + \xi\sigma$) are

$$\dot{A} = f_1(A, \beta) \tag{3.74}$$

$$\dot{\beta} = f_2(A, \beta) \tag{3.75}$$

The steady-state motions occur when $\dot{A} = \dot{\beta} = 0$, which corresponds to the singular points of Eqs. (3.74) and (3.75). The normalized displacement W_{max} with respect to the gap at the free end of the cantilever and the drive frequency Ω can be expressed in function of the phase β. Thus, the frequency response curve can be plotted parametrically as shown in Fig. 3.11 for the following parameters: $l = 5$ µm, $b = 160$ nm, $h = 300$ nm, $l_g = 500$ nm, $h_g = 80$ nm, $a = 350$ nm and $V_{ac} = 0.1V_{dc}$. The gap g and the DC voltage V_{dc} were used for parametric studies.

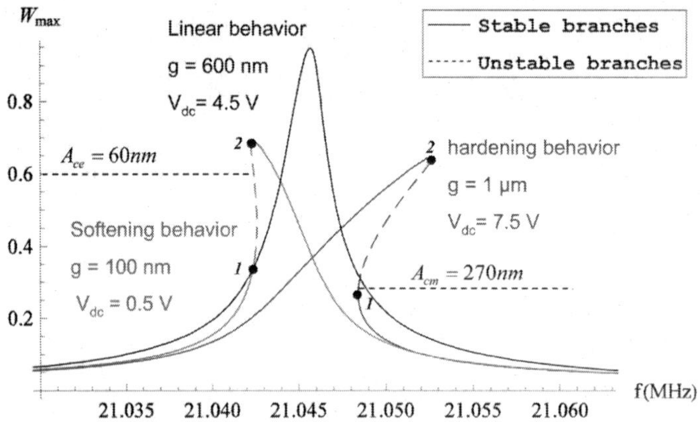

Figure 3.11 Analytical forced frequency responses of the resonant piezoresistive device presented in Figs. 3.9 and 3.12 for a quality factor $Q = 10^4$. W_{max} is the displacement of the beam normalized by the gap g at its free end.

3.4.5 The Critical Amplitude

3.4.5.1 The critical mechanical amplitude

The nonlinearities acting between the fixed end of the cantilever and the nanogauges are negligible since the vibration amplitudes of this part are close to zero for the first linear undamped mode shape as shown in Fig. 3.10. Therefore, the NEMS behaves dynamically as a resonant nanocantilever of length $l - d$. Hence, using Eq. (3.50), the critical mechanical amplitude can be written as

$$A_{\mathrm{cm}} = 6.3\frac{l-d}{\sqrt{Q}} \tag{3.76}$$

3.4.5.2 The critical electrostatic amplitude

In this case, the mechanical nonlinearities are neglected. Also, the electrostatic nonlinearities are acting only on the sensor part comprised between the gauges and the free end of the cantilever. By considering only nonlinear terms up to the third order, while neglecting the parametric terms and the terms proportional to V_{ac}^2 and following Eq. (3.54), the critical electrostatic amplitude can be expressed as

$$A_{\mathrm{ce}} = \frac{2\times 10^9 hg^{5/2}}{(l-d)\sqrt{Q}V_{\mathrm{dc}}}\left(\frac{7.5\times 10^7 h^2}{(l-d)^4} - \frac{3.8\times 10^{-15} V_{\mathrm{dc}}^2}{g^3 h}\right)^{1/4} \tag{3.77}$$

3.4.5.3 Engineering optimization

As shown in Fig. 3.11, when $g \ll h$, the mechanical nonlinearities are negligible with respect to the electrostatic nonlinearities. Then, the NEMS forced frequency curve displays a softening behavior (red curve of Fig. 3.11) and the critical amplitude is given by Eq. (3.77), which depends on the quality factor Q, the cantilever width h, the gap g, the DC voltage V_{dc} and the distance $l - d$ between the piezoresistive nanogauges and free end of the cantilever. In this case, the open-loop stability of the NEMS resonant sensor is limited by an oscillation amplitude around 60 nm.

If $g \gg h$, the electrostatic nonlinearities are negligible with respect to the mechanical nonlinearities. Then, the NEMS forced frequency curve displays a hardening behavior (blue curve of Fig. 3.11) and the critical amplitude is given by Eq. (3.76) which only depends on the quality factor Q and the distance $l - d$. In this case, the open-loop stability of the NEMS resonant sensor is limited by an oscillation amplitude around 270 nm: more than four times higher than the previous case. Thus, the mass resolution is enhanced by a factor $\Pi_{\mathrm{enh}} = 4$ compared to the first case.

Hence, designing NEMS cantilevers displaying a softening behavior is disadvantageous and can significantly alter the sensor resolution especially when this supposes that we are able to fabricate structures with very small gaps, which is rather awkward. Indeed, assuming that the upper bound limit (the pull-in) occurs at an amplitude of the gap order, even if the cantilever can vibrate linearly up to very high amplitudes comparable to the gap, the sensor performances can be altered due to its small dimensional amplitude limited by the gap. In other words, enhancing the dimensionless critical amplitude (red curve of Fig. 3.11) is not important when the gap is significantly reduced.

The optimal gap is g_p = 600 nm for which the mechanical and the electrostatic nonlinearities are balanced which permits the linearization of the frequency response as shown in Fig. 3.11 (black curve). For this design, which is technologically feasible, the mass resolution is enhanced by a factor $\Pi_{enh} = 9$, compared to first case.

3.4.6 Fabrication

The NEMS device presented in Fig. 3.12 was fabricated in CEALETI clean rooms using CMOS compatible materials with nano-electronics state-of-the-art lithography and etching techniques. A 200 mm silicon-on-insulator (SOI) wafer of <100> orientation was used. It comprises a 160 nm-thick top silicon structural layer (resistivity ≈ 10 Ωcm) and a 400 nm-thick sacrificial oxide layer. The top silicon layer was implanted with boron ions (p-type) through a thin layer of thermal oxide. Homogenous doping (3×10^{19} cm^{-3}) in the whole thickness of the top silicon was obtained through specific annealing step (for material reconstruction and doping activation), resulting in top layer resistivity of approximately 6 mΩcm. A hybrid e-beam/DUV lithography technique (Colinet et al., 2009) was used to define the nanoresonators and electrode pads, respectively. Top silicon layer was etched by anisotropic RIE. In order to decrease the lead resistances, the interconnecting leads have been thickened with a 650 nm thick AlSi layer, a typical metal for CMOS interconnection process. Finally, the nanoresonators have been released using a vapor HF isotropic etching to remove the sacrificial layer oxide beneath the structures. The main process steps are summarized in Fig. 3.13.

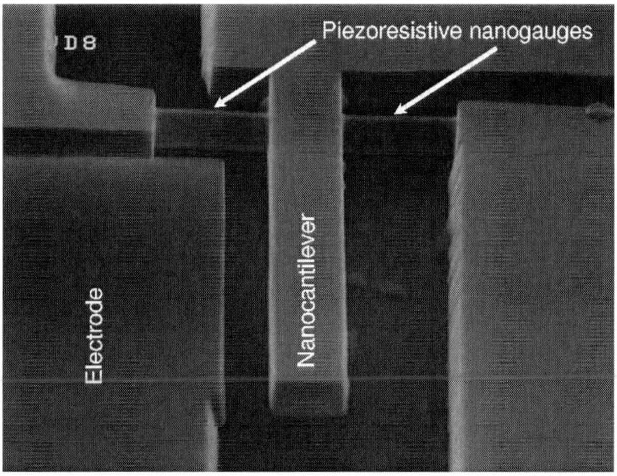

Figure 3.12 SEM image of the in-plane piezoresistive structure.

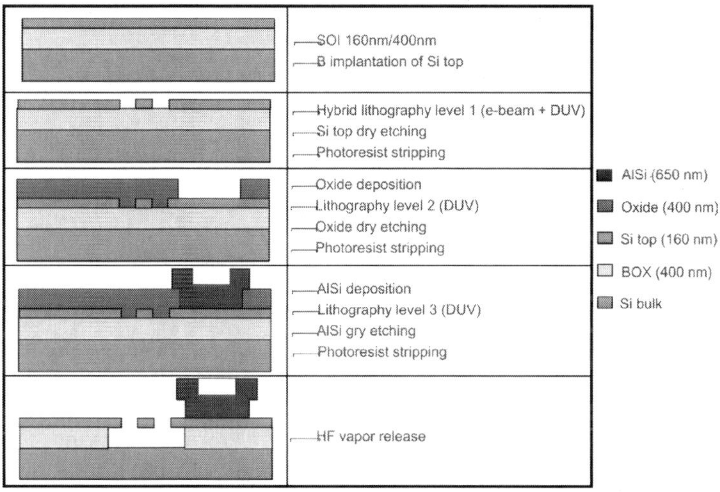

Figure 3.13 In-plane piezoresistive structure process flow.

3.4.7 Electrical Characterization

The strain collected by the gauges is transduced into a resistance variation due to the piezoresistance effect proportional to

$$\frac{\Delta R(\Omega)}{R} = \gamma\varepsilon_1(\Omega) = \gamma\frac{1}{24}l^3h_g^2 gE[\phi_{11}'''(d) - \phi_{01}'''(d)]a_1(\Omega) \tag{3.78}$$

where R is the gauge resistance and E Young's modulus. The displacement frequency response $a_1(\Omega)$ can be written in its parametric form in order to plot parametrically the resistance frequency response with respect to the phase β.

3.4.7.1 ω Down-mixing technique

The devices under test were connected to a radio frequency (RF) circuit board through wire bonding and loaded to a RF vacuum chamber for room temperature measurements. The beam is actuated electrostatically through capacitive coupling and detected through piezoresistive displacement transduction. The electrical read-out at high frequency is complicated by parasitic capacitances, which change the expected behavior of the electrical circuit. In order to avoid parasitic impedances and to easily reach the nonlinear regime, an ω down-mixing technique has been used to read-out the resistance variation at a lower frequency $\Delta\omega$ (Bargatin et al., 2005) (a schematic of the setup is shown in Fig. 3.14). The change in resistance $\Delta R(\omega)$ is read by applying a proper bias $I_b(\omega - \Delta\omega)$ to the gauges, which are acting as signal mixers and measuring the potential at the bridge center. The output voltage at low frequency is proportional to

$$V_{\text{out}}(\Delta\omega)=\frac{1}{2}I_{\text{b}}\cdot\Delta R\cdot\cos(\Delta\omega t) \tag{3.79}$$

The two gauges situated on opposite sides of the lever work in tensile and compressive strain alternatively offering a double advantage. Firstly they allow making a differential measurement at the centre, working both at the same time, thus contributing twice on the output signal. Second this flexible design constitutes a balanced bridge configuration which permits suppression of the background at the middle point by applying two 180 degrees out of phase voltage signal to the gauges extremities. The adequate decoupling of actuation and detection by using orthogonal principles as well as separating them in frequency has a direct consequence on the background reduction. As shown in Fig. 3.15, a huge signal of the order of 2–3 mV at resonance and a very low background was obtained with these devices giving rise to a signal to background ratio of more than 60 dB.

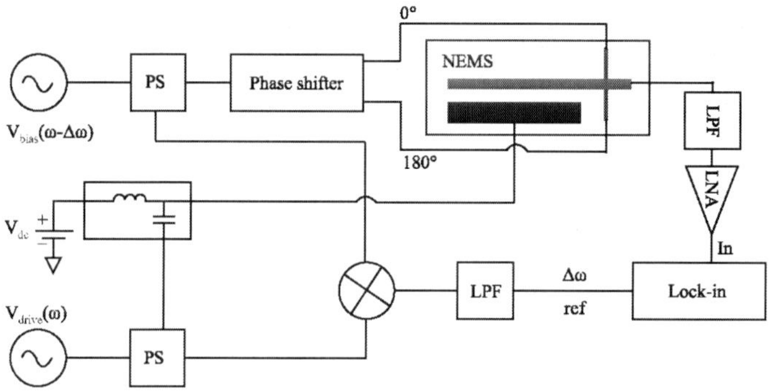

Figure 3.14 Test-bench for motion detection of piezoresistive resonant NEMS based on an ω down-mixing technique. PS, LPF are power splitter and phase shifter, respectively.

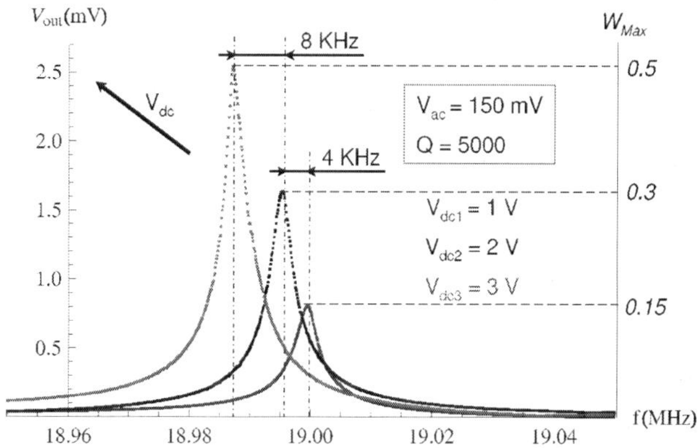

Figure 3.15 Linear resonance frequency responses measured using an ω zown-mixing technique. The effect of the DC voltage on the resonance frequency is presented.

Several measurements were performed on the device for a fixed bias voltage (V_{bias} = 1.56 V peak-peak). The cantilever displacement depends on the applied electrostatic force which is proportional to $F = \frac{1}{2}C\ (V_{\text{dc}} + V_{\text{ac}}\cos(\omega t))^2$. This force will have an AC ($F_{\text{ac}}(\omega)$) and a static (F_{dc}) component proportional to V_{dc}^2 for $V_{\text{dc}} >> V_{\text{ac}}$. The first will have a direct consequence on the displacement amplitude while the second affects the lever stiffness thus changing the resonance frequency. This is confirmed by the experimental results showing that the resonance frequency curve shifts to the low frequencies due to the electrostatic negative stiffness (Fig. 3.15).

Figure 3.15 shows three linear resonance peaks obtained for V_{ac} = 150 mV and DC voltages going from 1 V up to 3 V. The analytical resonance frequency is around 21 MHz, while the measured one is around 19 MHz. This can be due to many factors such as silicon residual stress, size effect on Young's modulus as well as micro and nanofabrication tolerances. The measured quality factor of the first linear curve (V_{dc} = 1 V) is around 5000. Remarkably, increasing the DC voltage did not degrade the quality factor. In fact, in these devices, since the detection is piezoresistive, there is no correlation between the electrical resistance of the device and the measured quality factor. The expected dissipation from this mechanism (ohmic losses) is thus negligible. The last resonance curve of Fig. 3.15 (in red) is close to the critical amplitude that has been analytically computed using the developed model (Eq. (3.77) for a softening behavior), which results in $A_c \approx 90$ nm.

Then, in order to reach the nonlinear regime, the cantilever has been actuated using high DC voltages. Moreover, the frequency response has been tracked experimentally using a lock-in amplifier in frequency sweep-up and -down in order to obtain a full characterization of the resonator bifurcation topology. No extramechanism loss has been observed due to the nonlinear dynamics of the cantilever and therefore the same quality factor has been conserved (Q = 5000). Figure 3.16 shows two nonlinear resonance peaks:

- The first resonance curve (in dashed line) was obtained for V_{ac} = 150 mV and V_{dc} = 5 V. It displays a softening behavior characterized by a jump-up frequency at the bifurcation point B_2 and a jump-down frequency at the bifurcation point B_3 for which the cantilever oscillation amplitude is around 75% of the gap (150 nm).
- The second resonance curve was obtained for V_{ac} = 75 mV and V_{dc} = 8 V. Remarkably, in frequency sweep down, two jumps have been observed: a jump-up at the bifurcation point B_1 and a jump down at the highest bifurcation point in the softening domain B_3 for which the cantilever oscillation amplitude is around 150 nm. This characterizes a particular

mixed hardening-softening behavior (Kacem and Hentz, 2009), which is not the logical expected result since the increase in the DC voltage, amplifies the nonlinear negative stiffness due to the electrostatic forces. Combined with an oscillation amplitude below the first softening curve, this should ensure negligible mechanical nonlinearities with respect to the electrostatic nonlinearities. Hence, the dynamic behavior should be purely softening.

Nevertheless, in frequency sweep up, only a jump-up has been identified at the bifurcation point B_2. Then, the resonance response follows a softening branch. In this configuration, the nonlinear dynamic behavior of the cantilever is complex (between a softening and a mixed behavior) which leads to a sensitivity to the initial conditions.

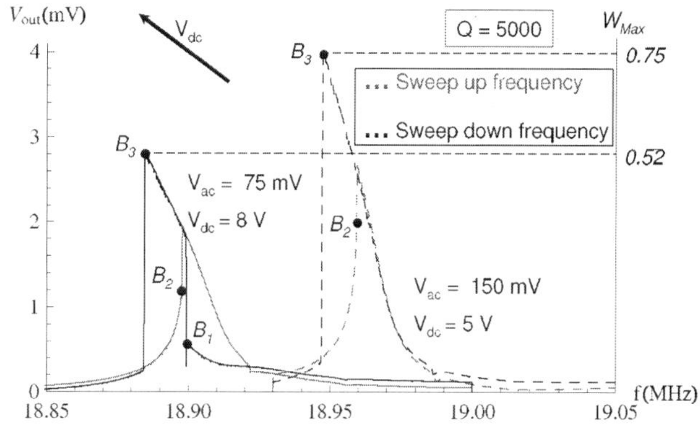

Figure 3.16 Nonlinear resonance frequency responses measured using an ω down mixing technique and showing the location of the different bifurcation points $\{B_1, B_2$ and $B_3\}$. $W_{\max}$ is the cantilever displacement at its free end normalized by the gap.

3.4.7.2 Optimal DC voltage

For this piezoresistive resonant NEMS, the quality factor Q is constant with respect to the DC and AC voltages. Then, using Eqs. (3.76) and (3.77), the optimal DC drive voltage is

$$V_{\mathrm{dcOP}} = \sqrt{\frac{h^3 g^7}{2}\sqrt{\frac{1.65 \times 10^{39}}{(l-d)^{16}} + \frac{3.2 \times 10^{42}}{g^4 (l-d)^{12}}} - \frac{8.1 \times 10^{19}}{(l-d)^8}} \quad (3.80)$$

The computed DC voltage that permits the hysteresis suppression is then computed using Eq. (3.80) resulting in $V_{dc_{OP}} = 1$ V. Hence, a high AC voltage is needed in order to validate the nonlinearity cancellation. In practice and as shown in Fig. 3.15, the AC voltage should be higher than 0.5 V which makes the assumption of neglected second harmonic terms invalid and the used model must be corrected by including additional linear and nonlinear terms.

Nevertheless, one can use a 2ω configuration which enables the third order nonlinearity cancellation under primary resonance combined with a dynamic stabilization due to the superharmonic resonance (Kacem et al., 2011a).

3.4.7.3 2ω Down-mixing technique

In order to actuate the cantilever at its primary and super harmonic resonances simultaneously, a 2ω down-mixing technique has been used enabling a read-out of the resistance variation at a lower frequency $\Delta\omega$ (a schematic of the setup is shown in Fig. 3.17).

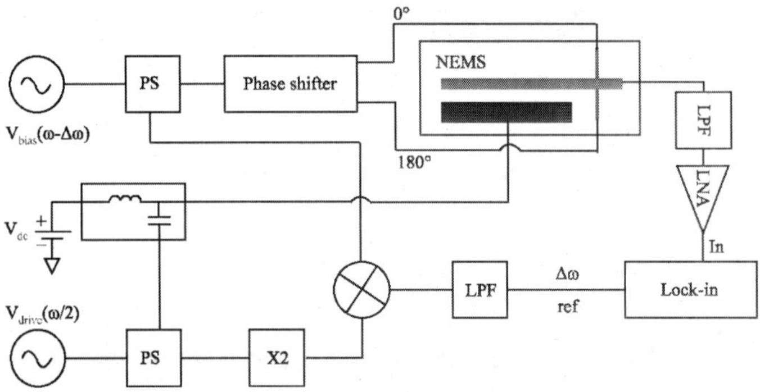

Figure 3.17 Test-bench for motion detection of piezoresistive resonant NEMS based on a 2ω down-mixing technique. PS, LPF are power splitter and phase shifter, respectively.

Several measurements were performed on the device for a fixed bias voltage ($V_{bias} = 1.56$ V peak-peak). The cantilever displacement depends on the applied electrostatic force which is proportional to $F = \frac{1}{2}C\left(V_{dc} + V_{ac}\cos\left(\frac{\omega}{2}t\right)\right)^2$. This force will have an AC ($F_{ac}(\omega)$) and a static (F_{dc}) component proportional to $V_{dc}^2 + \frac{1}{2}V_{ac}^2$.

Figure 3.18 shows two linear peaks obtained using a 2ω down-mixing technique for V_{ac} = 2 V. The measured quality factor is about 5000, which confirms the independence of Q on the DC and AC voltages. When, the DC voltage is increased from 0.2 V up to 0.3 V, the variation of the negative stiffness is negligible and consequently, no remarkable frequency shift has been observed. Interestingly, the measured output signal of the second linear peak is close to 3 mV for a low DC voltage, which was not reachable linearly for the ω down-mixing configuration.

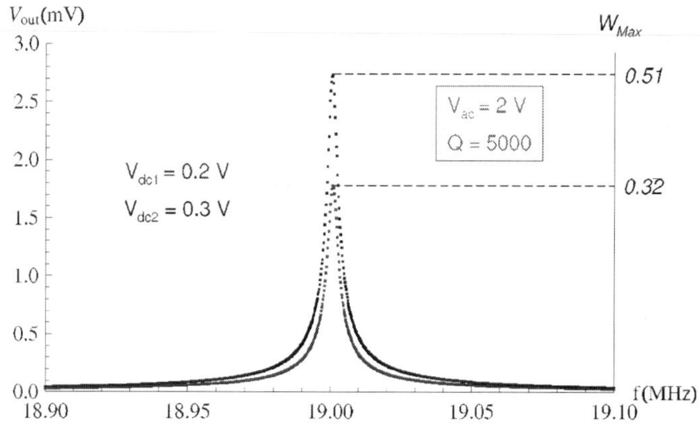

Figure 3.18 Linear resonance frequency responses measured using a 2ω down-mixing technique. The effect of the DC voltage on the resonance frequency is negligible. W_{max} is the cantilever displacement at its free end normalized by the gap.

Since the superharmonic resonance has no effect on the bistability limit of the resonator (Kacem et al., 2012), the mechanical critical amplitude of a cantilever under simultaneous resonance is then $A_{cm} = 6.3\frac{l}{\sqrt{Q}}$. However, in the electrostatic critical amplitude, one must add the contribution of the AC voltage in the nonlinear electrostatic stiffness, which changes substantially the close form solution of the optimal drive DC voltage. The latter has been estimated using the model for V_{ac} = 2 V resulting in V_{dc} = 0.5 V.

The resonance peak of Fig. 3.19 displays a slightly softening behavior close to the critical amplitude. The measured peak has been obtained using a 2ω down-mixing configuration for V_{ac} = 2 V and V_{dc} = 0.5 V. Analytically, for this set of parameters the nonlinear electrostatic and mechanical stiffness's are balanced and the oscillation amplitude of the cantilever is close to 200 nm at its free end. Indeed, the maximum of induced stress into the piezoresistive gauges is reached, as the free end of cantilever touched the electrode without a damageable pull-in for which the cantilever becomes unstable and collapses. In order to verify that the pull-in amplitude has been reached, the DC voltage has been increased successively from 0.5 V up to 2 V. Consequently, the cantilever nonlinearity becomes potentially softening which should increase the oscillation amplitude of the NEMS sensor.

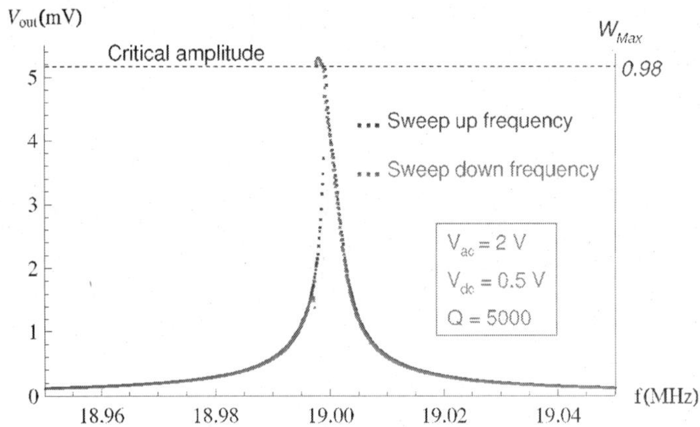

Figure 3.19 Slightly softening resonance frequency response measured using a 2ω down-mixing technique at the optimal DC voltage. The peak is close to the critical amplitude. Wmax is the cantilever displacement at its free end normalized by the gap.

Figure 3.20 shows a softening resonance curve obtained for V_{dc} = 2 V. The increase of the electrostatic softening nonlinear stiffness is displayed clearly by the distance between the two bifurcation points (softening domain) significantly enlarged in comparison with the frequency response in Fig. 3.18. Remarkably, the output signal at the peak is around 5.4 V which is the same value of V_{out} at V_{dc} = 0.5 V. Moreover, the slope of the softening branch between the two bifurcation points is close to zero, which confirms that the pull-in amplitude is reached giving the maximum of stress variation into the piezoresistive gauges.

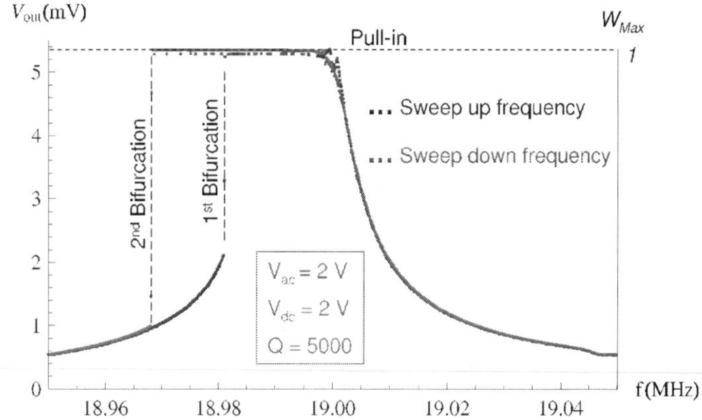

Figure 3.20 Softening frequency response measured using a 2ω down-mixing technique at V_{dc} = 2 V. The maximal stress on the piezoresistive gauges is reached for the pull-in amplitude. W_{max} is the cantilever displacement at its free end normalized by the gap.

3.4.8 Mass Resolution Enhancement

NEMS are usually embedded in a phase locked loop (PLL) or a self-excited loop in order to monitor time evolution of their resonant frequency. The frequency stability of the overall system (e.g., of the NEMS and the supporting electronics) is characterized by the Allan deviation, defined as (Mo Li and Roukes, 2007)

$$\frac{\delta\omega_0}{\omega_0} = \sqrt{\frac{1}{N-1}\sum_{1}^{N}\left(\frac{\overline{\omega}_{i+1}-\overline{\omega}_i}{\omega_0}\right)^2} \tag{3.81}$$

where ω_i is the average angular frequency in the ith time interval, N is the number of independent frequency measurements, which is assumed to be a sufficiently large number. The mass resolution δm is then $\sqrt{2}M_{\text{eff}}\frac{\delta\omega_0}{\omega_0}$ for 1 second integration time. At the linear regime and for a cantilever oscillation amplitude around 65 nm, the dynamic range (DR) experimentally measured was about 100 dB (Mile et al., 2010). This would lead to a theoretical ultimate Allan deviation $\left.\frac{\delta\omega_0}{\omega_0}\right|_{\text{th}} = \frac{10^{-\frac{DR}{20}}}{\sqrt{2Q}}$ of around 10^{-9} (Ekinci et al., 2004). For an effective mass of 200 fg and a Q-

factor of 6500, this would result in a potential mass resolution of $\delta m = \frac{M_{\text{eff}}}{Q} 10^{-\frac{DR}{20}} \approx 0.3 \text{ zg}$ at room temperature and at relatively low frequency (20 MHz).

Using the drive conditions of Fig. 3.19, at an extremely enhanced critical amplitude of the gap order, the mass sensor dynamic range can be potentially enhanced to reach the level of 110 dB. Consequently, a resolution around 100 Da (0.1 zg) is achievable. However, to reach this performance, the temperature fluctuation should be controlled at least below 10^{-2} K (Giessibl, 2003). Actually, the experimental Allan deviation leads to a mass resolution of approximately 105 zg at room temperature (Mile et al., 2010).

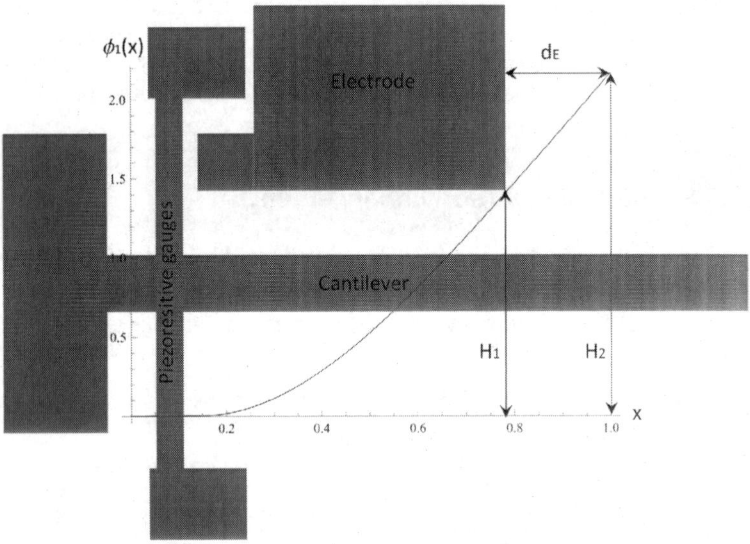

Figure 3.21 The next generation of NEMS resonant mass/gas sensor currently in fabrication in the clean rooms of LETI.

Once the noise contributions from the actuation voltage and the thermal bath issue are solved at low temperature, the ultimate resolution is then 100 Da. At this level, the cantilever probably touches the electrode as explained in Fig. 3.20. Consequently, no further optimizations are possible and one should think about a next generation of the studied device where the gap is quite larger than 200 nm. Nevertheless, the more we enlarge the gap, the more the applied drive voltage must be significantly increased to achieve very high oscillations. Practically, one of the best solutions for the next generation of the NEMS resonant mass sensor consists in moving the actuation electrode closer to the piezoresistive gauges so that the free end of the cantilever is allowed to oscillate at amplitudes larger than the gap g. As shown in Fig. 3.21, the cantilever can potentially undergo oscillations of the order $\frac{H_2}{H_1}g$ at its free end. Combined with the use of an advanced top-down nanowire fabrication techniques (Ernst et al., 2008) with expected giant gauge factors, as well as a possible nonlinearity cancellation, this may greatly decrease the resolution down to one single Dalton.

3.5 Conclusions

In this chapter, the development of an analytical model and its validation to quantitatively assess the nonlinear dynamics of nano-cantilever have been presented. This model includes the main sources of nonlinearities (mechanical and electrostatic) and is based on the modal decomposition using the Galerkin procedure combined with a perturbation technique (the averaging method).

As a first step, the experimental validation of the model has been performed on NEMS cantilevers fabricated using wafer-scale nanostencil lithography (nSL) enabling the definition of very low critical dimension devices. These cantilevers were monolithically integrated with CMOS circuits, which made possible the electrical characterization of their frequency responses. The NEMS devices have been driven in different conditions (in air and in vacuum). All parameters of the model, except the quality factor and the parasitic capacitance, are set prior to the comparison, which shows an excellent agreement in resonance frequency, peak shape and amplitude. Hence, it proves the efficiency of the model as a predictive tool.

The effects of some design parameters on the nonlinear behavior of nanocantilevers have been analytically investigated and close-form solutions of the critical amplitude under dominating mechanical nonlinearities and electrostatic nonlinearities, respectively, have been provided which demonstrates the large dynamic range of NEMS cantilevers compared to doubly clamped nanobeams. The mechanical critical amplitude of a cantilever is then $A_{cm} = 6.3\frac{l}{\sqrt{Q}}$ (Kacem et al., 2010). More specifically, the analytical expression of the optimal DC drive voltage has been extracted which is an interesting tool for resonant sensors designers. Theoretically, it allows for the cancellation of the nonlinearities in order to drive the NEMS cantilever linearly beyond its critical amplitude. Consequently, this may be a great gain in sensors' sensitivity, as the resonator's carrier power is largely increased while keeping a linear behavior; this may prevent most of noise mixing (Kaajakari et al., 2005a).

In a second step the model has been validated on a high frequency NEMS device electrostatically actuated based on piezoresistive detection (160 nm thick) fabricated using a hybrid e-beam/DUV lithography technique. The nanomechanical sensor has been characterized using a down-mixing technique. The ω configuration is first used in order to easily reach the nonlinear regime. Then, the optimal DC voltage being very low, a 2ω down-mixing configuration has been used in order to enable the compensation of the nonlinearities as predicted using the model.

The experimental results show an excellent agreement with the predicted dynamic behaviors. Particularly, the compensation of the nonlinearities has been validated for cantilever displacements up to the gap. Consequently and in a stable linear fashion, the optimal stress variation into the piezoresistive gauges has been reached using the 2ω down-mixing technique. Moreover, in this configuration the mixed behavior has not been observed up to the pull-in amplitude due to the effect of the superharmonic resonance in retarding and suppressing undesirable behaviors. An impressive ultimate resolution about 100 Da is achievable at low temperature and linearly at an oscillation amplitude comparable to the gap for which the maximum of strain collected by the piezoresistive gauges is reached. In order to overcome the gap limitation for the cantilever oscillations, the next generation of the studied device involves an actuation electrode shifted to the gauges side.

Very Large Scale Integration (VLSI) of such devices (Fig. 3.21) will potentially enable a wide range of new sensors, such as massive arrays of oscillating NEMS and sensitive multigas sensors. Indeed, the analytical rules provided in this chapter are applicable for resonant chemical and biological nanosensors in order to ensure the optimal mass resolution. Hence, these nonlinear analyses could be very interesting for many nanotechnology challenges such as sub-single-atom resolution in NEMS mass spectrometry (Boisen, 2009).

Acknowledgments

I gratefully acknowledge financial support from the CARNOT institute (Carnot NEMS Project), from the ANR (MNTEurop Project) and from the European Commission within the project NaPa (contract no NMP4-CT-2003-500120). I am also pleased to acknowledge LETI and CNM clean room staff, LETI NEMS teams and Prof. Jürgen Brugger and his team (LMIS1) at EPFL for their contribution to sample processing.

References

Ahmadian, M., Borhan, H., and Esmailzadeh, E. (2009). Dynamic analysis of geometrically nonlinear and electrostatically actuated micro-beams. *Commun. Nonlinear Sci. Numerical Simulation*, **14**(4), 1627–1645.

Arafat, H. N., Nayfeh, A. H., and Chin, C.-M. (1998). Nonlinear nonplanar dynamics of parametrically excited cantilever beams. *Nonlinear Dynamics,* **15**, 31–61.

Arcamone, J., Misischi, B., Serra-Graells, F., van den Boogaart, M. A. F., Brugger, J., Torres, F., Abadal, G., Barniol, N., and Perez-Murano, F. (2007). A compact and low-power CMOS circuit for fully integrated NEMS resonators. *IEEE Trans. Circ. Syst. II: Expr. Briefs,* **54**(5), 377–381.

Arcamone, J., van den Boogaart, M. A. F., Serra-Graells, F., Fraxedas, J., Brugger, J., and Prez-Murano, F. (2008). Full-wafer fabrication by nanostencil lithography of micro/nanomechanical mass sensors monolithically integrated with CMOS. *Nanotechnology*, **19**(30), 305302.

Arlett, J. L., Maloney, J. R. B., Gudlewski, M. M., and Roukes, M. L. (2006). Self-sensing micro-and nanocantilevers with attonewton-scale force resolution. *Nano Lett.*, **6**(5), 1000–1006.

Bargatin, I., Myers, E. B., Arlett, J., Gudlewski, B., and Roukes, M. L. (2005). Sensitive detection of nanomechanical motion using piezoresistive signal downmixing. *App. Phys. Lett.*, **86**(13), 133109.

Boisen, A. (2009). Nanoelectromechanical systems: Mass spec goes nanomechanical. *Nat. Nanotechnol.*, **4**, 404–405.

Bose, S., and Agarwal, G. S. (2006). Entangling pairs of nano-cantilevers, cooper-pair boxes and mesoscopic teleportation. *N. J. Phys.*, **8**(3), 34.

Chowdhury, S., Ahmadi, M., and Miller, W. C. (2005). A closed-form model for the pull-in voltage of electrostatically actuated cantilever beams. *J. Micromechan. Microeng.*, **15**(4), 756.

Cleland, A. N., and Roukes, M. L. (2002). Noise processes in nanomechanical resonators. *J. App. Phys.*, **92**(5), 2758–2769.

Colinet, E., Durand, C., Duraffourg, L., Audebert, P., Dumas, G., Casset, F., Ollier, E., Ancey, P., Carpentier, J.-F., Buchaillot, L., and Ionescu, A. (2009). Ultra-sensitive capacitive detection based on SGMOSFET compatible with front-end CMOS process. *IEEE J. Solid-State Circ.*, **44**(1), 247–257.

da Silva, M. R. M. C., and Zaretzky, C. L. (1994). Nonlinear flexural-flexural-torsional interactions in beams including the effect of torsional dynamics. (i) Primary resonance. *Nonlinear Dyn.*, **5**, 3–23.

Ekinci, K. L., and Roukes, M. L. (2005). Nanoelectromechanical systems. *Rev. Sci. Inst.*, **76**(6), 061101.

Ekinci, K. L., Yang, Y. T., and Roukes, M. L. (2004). Ultimate limits to inertial mass sensing based upon nanoelectromechanical systems. *J. App. Phys.*, **95**(5), 2682–2689.

Ernst, T., Duraffourg, L., Dupré, C., Bernard, E., Andreucci, P., Becu, S., Ollier, E., Hubert, A., Halte, C., Buckley, J., Thomas, O., Delapierre, G., Deleonibus, S., De Salvo, B., Robert, P., and Faynot, O. (2008). Novel Si-based nanowire devices: Will they serve ultimate MOSFETs scaling or ultimate hybrid integration? in Electron Devices Meeting, 2008. IEDM 2008. IEEE International, pp. 1–4.

Geller, M. R., and Varley, J. B. (2005). Friction in nanoelectromechanical systems: Clamping loss in the GHz regime, http://arxiv.org/abs/cond-mat/0512710v1.

Giessibl, F. J. (2003). Advances in atomic force microscopy. *Rev. Mod. Phys.*, **75**(3), 949–983.

Hao, Z., Erbil, A., and Ayazi, F. (2003). An analytical model for support loss in micromachined beam resonators with in-plane flexural vibrations. *Sens. Actuators A Phys.*, **109**(1–2), 156–164.

He, R., Feng, X. L., Roukes, M. L., and Yang, P. (2008). Self-transducing silicon nanowire electromechanical systems at room temperature. *Nano Lett.*, **8**(6), 1756–1761.

He, R., and Yang, P. (2006). Giant piezoresistance effect in silicon nanowires. *Nat. Nanotechnol.*, **1**, 42–46.

Jensen, K. Girit, C., Mickelson, W., and Zettl, A. (2006). Tunable nanoresonators constructed from telescoping nanotubes. *Phys. Rev. Lett.*, **96**(21), 215503.

Jiang, Y. G., Ono, T., and Esashi, M. (2008). Fabrication of piezoresistive nanocantilevers for ultra-sensitive force detection. *Meas. Sci. Technol.*, **19**(8), 084011.

Juillard, J., Bonnoit, A., Avignon, E., Hentz, S., Kacem, N., and Colinet, E. (2008). From MEMS to NEMS: Closed-loop actuation of resonant beams beyond the critical Duffing amplitude. *in Sensors, 2008 IEEE*, pp. 510–513.

Kaajakari, V., Koskinen, J., and Mattila, T. (2005a). Phase noise in capacitively coupled micromechanical oscillators, *IEEE Trans. Ultrason. Ferroelectr. Freq. Control.*, **52**(12), 2322–2331.

Kaajakari, V., Koskinen, J. K., and Mattila, T. (2005b). Phase noise in capacitively coupled micromechanical oscillators. *IEEE Trans. Ultrason. Ferroelectr. Freq. Control.*, **52**(12), 2322–2331.

Kacem, N., Arcamone, J., Perez-Murano, F., and Hentz, S. (2010). Dynamic range enhancement of nonlinear nanomechanical resonant cantilevers for highly sensitive NEMS gas/mass sensor applications. *J. Micromechan. Microeng.*, **20**(4), 045023.

Kacem, N., Baguet, S., Dufour, R., and Hentz, S. (2011a). Stability control of nonlinear micromechanical resonators under simultaneous primary and superharmonic resonances. *App. Phys. Lett.*, **98**(19), 193507.

Kacem, N., Baguet, S., Hentz, S., and Dufour, R. (2011b). Computational and quasi-analytical models for nonlinear vibrations of resonant MEMS and NEMS sensors, *Int. J. Nonlinear Mechan.*, **46**(3), 532–542.

Kacem, N., and Hentz, S. (2009). Bifurcation topology tuning of a mixed behavior in nonlinear micromechanical resonators. *App. Phys. Lett.*, **95**(18), 183104.

Kacem, N., Hentz, S., Hentz, S., and Dufour, R. (2012). Pull-in retarding in nonlinear nanoelectromechanical resonators under super-harmonic excitation. *J. Computat. Nonlinear Dyn.*, **7**(2), 021011.

Kacem, N., Hentz, S., Pinto, D., Reig, B., and Nguyen, V. (2009). Nonlinear dynamics of nanomechanical beam resonators: Improving the performance of NEMS-based sensors. *Nanotechnology*, **20**(27), 275501.

Li, M., Tang, H. X., and Roukes, M. L. (2007). Ultra-sensitive NEMS-based cantilevers for sensing, scanned probe and very high-frequency applications. *Nat. Nanotechnol.*, **2**, 114–120.

Lifshitz, R., and Roukes, M. L. (2000). Thermoelastic damping in micro-and nanomechanical systems. *Phys. Rev. B*, **61**(8), 5600–5609.

Liu, S., Davidson, A., and Lin, Q. (2004). Simulation studies on nonlinear dynamics and chaos in a MEMS cantilever control system. *J. Micromechan. Microeng.*, **14**(7), 1064.

Mile, E., Jourdan, G., Bargatin, I., Labarthe, S., Marcoux, C., Andreucci, P., Hentz, S., Kharrat, C., Colinet, E., and Duraffourg, L. (2010). In-plane nanoelectromechanical resonators based on silicon nanowire piezoresistive detection. *Nanotechnology*, **21**(16), 165504.

Mo, Li, H. X. T., and Roukes, M. L. (2007). Ultra-sensitive NEMS-based cantilevers for sensing, scanned probe and very high-frequency applications. *Nat. Nanotechnol.*, **2**, 114–120.

Mohanty, P., Harrington, D. A., Ekinci, K. L., Yang, Y. T., Murphy, M. J., and Roukes, M. L. (2002). Intrinsic dissipation in high-frequency micromechanical resonators. *Phys. Rev. B*, **66**(8), 085416.

Nayfeh, A. H., and Pai, P. F. (1989). Nonlinear non-planar parametric responses of an inextensional beam. *Int. J. Nonlinear Mechan.*, **24**, 139–158.

Nayfeh, A. H., Younis, M. I., and Abdel-Rahman, E. M. (2007). Dynamic pull-in phenomenon in MEMS resonators. *Nonlinear Dyn.*, **48**, 153–163.

Nishiyama, H., and Nakamura, M. (1990). Capacitance of a strip capacitor. *IEEE Trans. Components Hybrids Manufacturing Technol.*, **13**(2), 417–423.

Pai, P. F., and Nayfeh, A. H. (1990). Nonlinear nonplanar oscillations of a cantilever beam under lateral base excitations. *Int. J. Nonlinear Mechan.*, **25**, 455–474.

Postma, H. W. C., Kozinsky, I., Husain, A., and Roukes, M. L. (2005). Dynamic range of nanotube-and nanowire-based electromechanical systems, *Appl. Phys. Lett.*, **86**(22), 223105.

Robins, W. P. (1982). Phase noise in signal sources: Theory and applications, IEE Telecommunications Series, Publish by Peregrinus, ISBN 978-090-6048-76-4, LCCN 82208863.

Roessig, T. (1998). Integrated MEMS tuning fork oscillators for sensor applications, Ph.D. thesis, University of California, Berkeley.

Roessig, T. A., Howe, R. T., and Pisano, A. P. (1997). Nonlinear mixing in surface-micromachined tuning fork oscillators. in *Frequency Control Symposium 1997. Proceedings of the 1997 IEEE International.*, pp. 778–782.

Sazonova, V. A. (2006). A tunable carbon nanotube resonator, Ph.D. thesis, Cornell University.

Silva, M. R. M. C. D. (1988a). Nonlinear flexural-flexural-torsional-extensional dynamics of beams. i: Formulation. *Int. J. Solids Struct.,* **24**, 1225–1234.

Silva, M. R. M. C. D. (1988b). Nonlinear flexural-flexural-torsional extensional dynamics of beams. ii: Response analysis. *Int. J. Solids Struct.,* **24**, 1235–1242.

Silva, M. R. M. C. D., and Glynn, C. C. (1978a). Nonlinear flexural-flexural-torsional dynamics of inextensional beams. i: Equations of motion. *J. Struct. Mechan.,* **6**, 437–448.

Silva, M. R. M. C. D., and Glynn, C. C. (1978b). Nonlinear flexural-flexural-torsional dynamics of inextensional beams. i: Equations of motion. *J. Struct. Mechan.,* **6**, 437–448.

Silva, M. R. M. C. D., and Glynn, C. C. (1978c). Nonlinear flexural-flexural-torsional dynamics of inextensional beams. ii: Forced motions. *J. Struct. Mechan.,* **6**, 449–461.

Silva, M. R. M. C. D., and Glynn, C. C. (1978d). Nonlinear flexural-flexural-torsional dynamics of inextensional beams. ii: Forced motions. *J. Struct. Mechan.,* **6**, 449–461.

Van den Boogaart, M., Kim, G., Pellens, R., Van den Heuvel, J. P., Brugger, J. (2004). Deep-ultraviolet microelectromechanical systems stencils for high-throughput resistless patterning of mesoscopic structures. *J. Vac. Sci. Technol.*, B, **22**, 3174–3178.

Verd, J., Uranga, A., Abadal, G., Teva, J., Torres, F., Lopez, J., Perez-Murano, E., Esteve, J., and Barniol, N. (2008). Monolithic CMOS MEMS oscillator circuit for sensing in the attogram range. *Electron Device Lett. IEEE,* **29**(2), 146–148.

Verd, J., Uranga, A., Teva, J., Lopez, J., Torres, F., Esteve, J., Abadal, G., Perez-Murano, F., and Barniol, N. (2006). Integrated CMOS MEMS with on-chip readout electronics for high-frequency applications. *IEEE Electron Device Lett.*, **27**(6), 495–497.

Yang, Y. T., Callegari, C., Feng, X. L., Ekinci, K. L., Roukes, M. L. (2006). Zeptogram-scale nanomechanical mass sensing. *Nano Lett.,* **6**(4), 583–586.

Yang, J., Ono, T., and Esashi, M. (2002). Energy dissipation in submicrometer thick single-crystal silicon cantilevers. *J. Microelectromechan. Syst.*, **11**(6), 775–783.

Yurke, B., Greywall, D. S., Pargellis, A. N., and Busch, P. A. (1995). Theory of amplifier-noise evasion in an oscillator employing a nonlinear resonator. *Phys. Rev. A*, **51**(5), 4211–4229.

Zaretzky, C. L., and da Silva, M. R. M. C. (1994). Nonlinear flexural-flexural-torsional interactions in beams including the effect of torsional dynamics. ii: Combination resonance. *Nonlinear Dyn.*, **5**, 161–180.

Zener, C. (1938). Internal friction in solids ii. general theory of thermoelastic internal friction. *Phys. Rev.*, **53**(1), 90–99.

Chapter 4

Intentional Nonlinearity for Design of Micro/Nanomechanical Resonators

Hanna Cho,[a] Lawrence A. Bergman,[b] Min-Feng Yu,[c] and Alexander F. Vakakis[d]

[a]*Department of Mechanical Engineering, The Ohio State University, Columbus, OH, USA*
[b]*Department of Aerospace Engineering, University of Illinois at Urbana-Champaign, Urbana, IL, USA*
[c]*School of Aerospace Engineering, Georgia Institute of Technology, Atlanta, GA, USA*
[d]*Department of Mechanical Science and Engineering, University of Illinois at Urbana-Champaign, Urbana, IL, USA*

cho.867@osu.edu

Dynamic micro/nanomechanical systems are devices employing mechanical elements (e.g., a beam, a membrane, etc.) at micro/nanometer scales, designed to exhibit mechanical motions usually at or near their resonance frequencies; these are called "micro/nanomechanical resonators" or just "micro/nanoresonators." The mechanical elements are often accompanied by electrical or optical functionality to achieve the ultimate design objectives originating from their unique mechanical motion. The attributes that the tiny size provides are high resonant frequency and low damping, which make micro/nanomechanical systems appealing for many

Nanocantilever Beams: Modeling, Fabrication, and Applications
Edited by Ioana Voiculescu and Mona Zaghloul

ISBN 978-981-4613-23-1 (Hardcover), 978-981-4613-24-8 (eBook)
www.panstanford.com

applications including extremely sensitive sensors and high frequency RF electronic components.

One of the major research directions with the goal of achieving higher performance has been toward decreasing the device size in order to reach higher frequencies and lower damping. Building such tiny devices relies on remarkable development in micro- and nanofabrication techniques. The main difficulties in decreasing the device size do not originate from fabricating a small device but from realizing the required operation combined with actuation and detection techniques. Extensive research over the last two decades has overcome such difficulties by developing effective schemes to actuate these tiny devices and detect the resulting small motion with extreme sensitivity. However, the operation of these devices is still limited by the intrinsic traits of readily realizable nonlinear characteristics in micro/nanomechanical systems.

A mechanical resonator at micro/nanoscale can easily transit from linear to nonlinear resonance operation, because of its remarkable properties, namely small size and low damping. There are many sources of system nonlinearity such as nonlinear dissipative mechanisms, geometric/kinematic nonlinearities, nonlinear potential force-laws, etc. Moreover, as the device size decreases and, accordingly, the corresponding damping becomes very small, the system becomes vulnerable to such sources of nonlinearity and may exhibit strongly nonlinear behavior whether this is by design and wanted or not.

Early studies on micro/nanomechanical systems, however, were concerned with their operation in the linear dynamical range. As the systems were designed to operate within the linear regime from the beginning, the operation naturally strayed from the original design specifications when nonlinear effects appeared, leading to performance degradation. In fact, in order to restrict the dynamics within the linear regime, the resonant amplitude was restricted to be smaller than the critical amplitude for the onset of nonlinearity, which is often comparable to the amplitude of thermal vibration for nanoscale devices. As a result, the dynamic operational range was narrowly confined and, thus, performance and applications were also limited.

Recently, efforts by our research group were undertaken to intentionally implement strong nonlinearities in micro/

nanomechanical systems. The main focus of our approach is (i) to develop predictive capacity for the passive dynamics of micro/nanomechanical systems with intentional strong nonlinear elements, and (ii) based on this knowledge and capacity to design systems with unprecedented properties subject to a broad range of operational and excitation conditions. Hence, our approach can be considered as contrarian, since instead of treating system nonlinearities as unwanted and detrimental to the design objectives, we intentionally implement them with the goal of improving the operational performance of devices, which would be otherwise unattainable in linear settings. Indeed, far from regarding nonlinearities as detrimental to the design objectives, we intend to harness unique nonlinear behaviors in order to enhance the performance of micro/nanomechanical systems and develop new paradigm-shifting applications.

In synopsis, the main theme of this section is the intentional integration of nonlinear dynamics in the design of micro/nanomechanical resonators, whose behavior is otherwise linear or almost linear. The section is organized as follows: First we provide an introduction to nonlinear micro/nanomechanical resonating systems and a brief discussion of their dynamics. Then, we introduce three different designs of nano/micromechanical resonators that incorporate intentional nonlinearities for different design objectives and discuss their unique nonlinear performance and potential applications.

4.1 Nonlinear Dynamics of Micro/Nanomechanical Resonators

A thorough review of nonlinear dynamics in micro/nanoresonators can be found in review papers.[1,2] Previous studies show that most common nonlinear effects entering into the equation of motion are in the form of a force proportional to the cube of the displacement, which results in Duffing- or Duffing-like characteristics. Here we discuss various sources of nonlinearity arising in micro/nanomechanical resonators, followed by an analytical study of the most representative nonlinear equation, namely, the Duffing oscillator.

4.1.1 Origin of Nonlinearity

4.1.1.1 Geometric nonlinearity: Nonlinear elastic restoring force

One of the most common sources of nonlinearity encountered in micro/nanoresonators is geometric nonlinearity.[3-6] Considering a beam in the x–y plane with the axial coordinate x and transverse coordinate y, the transverse vibration can be expressed by the well-known Euler–Bernoulli beam equation:

$$m\frac{\partial^2 y(x,t)}{\partial^2 t}+EI\frac{\partial^4 y(x,t)}{\partial^4 x}=0, \tag{4.1}$$

where m denotes mass per unit length, E Young's modulus, and I area moment of inertia. When a mechanical structure with clamped boundary conditions, i.e., both ends fixed, undergoes an oscillatory motion in the transverse direction, the extension in length (midplane stretching) induces axial tension in the element. In the presence of tension, Eq. 4.1 needs to be modified as

$$m\frac{\partial^2 y}{\partial t^2}+EI\frac{\partial^4 y}{\partial x^4}=T\frac{\partial^2 y}{\partial x^2} \tag{4.2}$$

where T is the tension induced in the beam. Assuming no pretension, the additional tension in the beam of length L is given by the strain, $\Delta L/L$, multiplied by the axial stiffness of the beam EA, where A is the cross-sectional area of the beam. The strain and axial tension in the beam are then calculated by

$$\frac{\Delta L}{L}=\frac{\int_0^L\sqrt{1+\left(\frac{\partial y}{\partial x}\right)^2}dx-L}{L}\simeq\frac{\int_0^L\left(\frac{\partial y}{\partial x}\right)^2 dx}{2L}$$

$$T=EA\frac{\Delta L}{L}=\frac{EA}{2L}\int_0^L\left(\frac{\partial y}{\partial x}\right)^2 dx \tag{4.3}$$

where we assume a small gradient, i.e., $(\partial y/\partial x)^2 << 1$. Substituting Eq. 4.3 into Eq. 4.2 results in the following nonlinear equation for the beam:

$$m\frac{\partial^2 y}{\partial t^2}+EI\frac{\partial^4 y}{\partial x^4}-\frac{EA}{2L}\int_0^L\left(\frac{\partial y}{\partial x}\right)^2 dx=0 \tag{4.4}$$

in which the nonlinearity is created by the geometric effect of midplane stretching.

This system is further studied analytically in Section 4.2.1 by considering a nonlinear carbon nanotube (CNT) resonator with geometric nonlinearity. Employing modal reduction techniques, it is shown that the leading modal amplitude is approximately governed by a Duffing equation with a hardening effect. Note that the induced axial tension makes the beam stiffer as the oscillating amplitude increases causing the frequency response curve to bend towards high frequencies. Even though this kind of geometric nonlinearity is also present in macroscale mechanical systems, the effect is generally not significant enough to prevail over the dominant linear restoring force. In small-sized devices at micro/nanoscales, however, one often encounters nonlinear behavior due to this geometric effect, producing a hardening effect and, eventually, nonlinear hysteresis. Now, the discussion proceeds to address answers to the following important questions: Why is nonlinearity more conveniently realizable at smaller scales, and what is the critical amplitude that separates the linear and nonlinear regimes?

Driven by even a small force, micro/nanoresonators undergo large amplitude of oscillations due to their intrinsic properties engendered by small size, e.g., high quality factor (i.e., low damping), small mass, and low bending stiffness, which is proportional to the fourth power of the device width. As a result, the relative magnitude of the involved resonance amplitude readily becomes significant when compared to the device size.

For very small oscillation amplitudes, the response of a forced Duffing oscillator is nearly Lorentzian, like a linear oscillator. As the amplitude increases, however, the curve starts to bend and eventually has a multi-valued region with two stable branches and one unstable branch. The point where this bistability emerges gives the condition for the onset of nonlinearity. This can be obtained by setting

$$d\omega/d(a^2) = 0 \tag{4.5}$$

in the equation relating frequency ω to amplitude squared, a^2, dependence (see Eq. 4.12 below), which will be derived in the following section. Then, the critical amplitude to determine whether the system is in the regime of linearity or nonlinearity can be approximated by

$$a_c = \omega_0 L^2 \sqrt{\frac{\sigma\sqrt{3}}{EQ}}. \tag{4.6}$$

4.1.1.2 Nonlinear external potential

Another source of nonlinearity originates from the applied external potential to actuate or detect resonance, which is intrinsically nonlinear. The actuation technique can be subdivided into two different approaches: Direct excitation and parametric excitation. For parametric excitation, one (or more) of the physical parameters in a mechanical resonator is modulated as a function of time, which in turn modulates its frequency. Parametric excitation is readily realized in small-scale devices through careful design of the system[7] and more than one instability region is realizable, unlike in macroscopic systems, due to small damping. Thus, parametric resonance systems have been an interesting research subject for various applications,[7-9] but are not of particular interest in this section.

Direct excitation employs various kinds of forces for interrogating a mechanical structure such as electrostatic, electromagnetic, and piezoelectric forces, which often exhibit nonlinearity. For example, for a system driven by electrostatic force, the offset in the equilibrium position due to the presence of charge resulting from dc voltage gives rise to a nonlinear term.[1] The effect of this nonlinear force generally makes the resonator softer and, accordingly, decreases its resonant frequency. For nanoscale devices, geometric nonlinearity due to tension appears much earlier and is more dominant, and the effect of the nonlinear potential field is not often observed. For microscale resonators, however, relatively large potential is used to actuate the system, and one often encounters both the hardening effect of geometric nonlinearity and the softening effect of external forces, to produce very unique and complex dynamics[10-14] such as transitions between hardening and softening resonances.

4.1.1.3 Nonlinear interactions

When a cantilever system in micro/nanometer scale is used as a force sensor (e.g., probe-based microscopy such as atomic force microscopy), geometric nonlinearity does not play a role because of its boundary condition of one free end, which releases the tension in the structure. However, while the mechanical element itself is governed by linear dynamics, the target force interacting between the cantilever and the nearby sample is highly nonlinear and, thus, the system exhibits nonlinear behavior. Numerous experimental and theoretical research efforts are devoted to improving the understanding of mechanical motion and dynamics of the cantilever probe, especially for atomic force microscopy (AFM) applications.[15] There are various types of interactions, including long-range interactions such as van der Waals, electrostatic, capillary, and magnetic forces, as well as short-range interactions such as repulsive van der Waals forces. For example, one of common models to describe the tip–sample interaction during AFM operations is the combined van der Waals force and Derjaguin–Muller–Toporov contact model,[16] which is further discussed in Section 4.3.3.

4.1.1.4 Nonlinear damping

As emphasized earlier, damping plays an important role in the dynamics of micro/nanomechanical systems. Mostly, it is possible to describe damping by a linear damping force, proportional to the velocity of the resonators. However, recent work on nanomechanical resonators employing boron nitride nanotubes (BNNTs),[17] CNTs, and graphene[18] shows that the simple linear damping scenario breaks down for resonators with atomic scale transverse dimensions. Instead, a nonlinear damping force, proportional to the product of the square of displacement and velocity, needs to be included to explain the amplitude-dependent damping effect. A parametric study of nonlinear damping is further discussed in Section 4.3.2.

4.1.2 Linear and Nonlinear Response

In this section, we study the dynamic response and stability of a Duffing resonator with cubic nonlinearity driven by an external harmonic force, expressed by Eq. 4.7:

$$\ddot{x} + 2\mu\dot{x} + \omega_0 x + \alpha x^3 = q_0 \cos\Omega t \tag{4.7}$$

where μ, ω_o, α, q_o, and Ω are the damping constant, linearized resonant frequency, the nonlinear constant, amplitude of exciting force, and excitation frequency, respectively.

There are several ways to analytically study this system, e.g., harmonic balance, multiple scales, and averaging. Since the detailed procedure for each of these methods is introduced in the following sections, here we discuss the dynamics and stability of the Duffing equation for primary resonance based on the solution based on the multiple scales method.

To start with the multiple scales method, we introduce the small parameter ε for bookkeeping purposes and the rescalings, $y \to \varepsilon y, \mu \to \varepsilon\mu, q_o \to \varepsilon q_o$; the proximity of the excitation frequency to the linearized resonant frequency is represented by introducing the detuning parameter σ:

$$\Omega = \omega_o + \varepsilon\sigma \tag{4.8}$$

Then, we define multiple time scales explicitly as $T_n = \varepsilon^n \tau$, and the periodic (steady state) solution of Eq. 4.7 is sought in the form

$$y(\tau, \varepsilon) = y_o(T_o, T_1) + \varepsilon y_1(T_o, T_1) + \cdots \tag{4.9}$$

The resulting first approximation to the steady-state solution is given by

$$y(t) = a\cos(\Omega t - \gamma) + O(\varepsilon) \tag{4.10}$$

where α and γ are defined by the autonomous set of ordinary equations

$$a' = -\mu a + \frac{1}{2}\frac{q_o}{\omega_o}\sin\gamma \tag{4.11}$$

$$-a\gamma' = \sigma a - \frac{3}{8}\frac{\alpha}{\omega_0}a^3 + \frac{1}{2}\frac{q_o}{\omega_o}\cos\gamma$$

By setting $a' = \gamma' = 0$ in Eq. 4.11, the frequency response equation for the amplitude of the steady-state response as a function of the frequency detuning parameter σ and the excitation amplitude q_o is obtained as

$$\left[\mu^2 + \left(\sigma - \frac{3}{8}\frac{\alpha}{\omega_o} a_{ss}^2\right)^2\right] a_{ss}^2 = \frac{q_o^2}{4\omega_o} \tag{4.12}$$

In Fig. 4.1, we compare the frequency response curves of the linear (α = 0) and nonlinear (α > 0) oscillators. The linear curve in Fig. 4.1a is symmetric with respect to the resonant frequency within a narrow bandwidth, which is determined by damping. The effect of nonlinearity is to bend the curve toward higher frequencies (Fig. 4.1b), creating a multivalued region. Note that when the nonlinear constant α is negative, the response curve bends in the opposite direction.

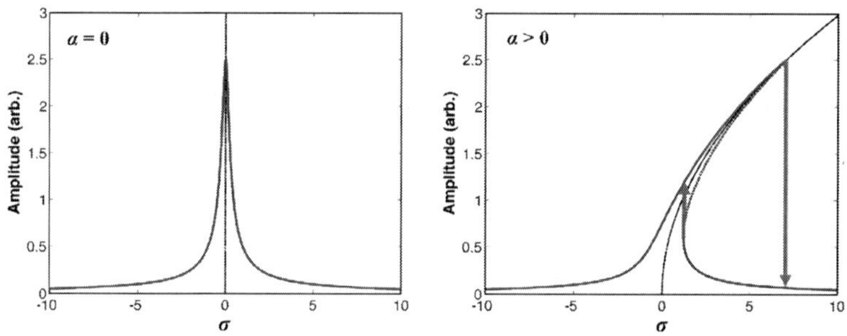

Figure 4.1 Comparison of linear and nonlinear response curves.

The multivalued characteristic of the response curves due to nonlinearity leads to jump phenomena, i.e., to abrupt changes in amplitude and phase for a very small change in frequency. There are two branches of stable resonances (solid line) that are connected by an unstable branch (dashed line). The stability of each branch can be examined by applying stability analysis, considering small perturbation of the equilibrium position. As the frequency is swept up through low frequencies, the resonance amplitude in the upper branch of stable resonances increases to the maximum possible amplitude and then drops abruptly to a lower value as the forced motion makes a transition to the lower stable branch. If the frequency is swept down through high frequencies, the amplitude follows the lower stable branch up to the point where it merges with the unstable branch. Then, a further decrease in frequency causes a sudden increase in amplitude, to follow the upper stable branch. The difference between the frequency points of drop-jump and up-jump phenomena leads to nonlinear hysteresis loops.

From this simplest form of nonlinear system, we can discuss the characteristics of nonlinear behavior. First, unlike a linear system having a fixed resonant frequency determined by the mechanical structure's geometry and material properties, the resonant frequency of a nonlinear system depends on the energy engaged with the system. Therefore, the resonant bandwidth can be tuned by simply controlling the force exerted to the system. Moreover, one can easily exploit large resonant amplification over a broad frequency bandwidth. Also, such energy-frequency dependency can be used for various design purposes including energy absorbing and harvesting systems. Second, the characteristic sharp transition of nonlinear behavior can improve the sensing capability tremendously, because the instability point is extremely sensitive to various external perturbations. Most importantly, rich dynamics that nonlinear design brings into the system have a potential to develop new paradigm-shifting designs that a linear system can never realize.

The goal of research introduced in this section is to harness these valuable nonlinear characteristics. Next, we discuss how we intentionally implement nonlinearity to otherwise linear micro/nanomechanical systems and how this enhances their performance.

4.2 Implementation of Intentional Nonlinearity

As discussed above, there are various sources that induce nonlinearity in micro/nanomechanical resonators. Among them, our particular aim is implementing geometrically nonlinear effects. These are inherent in the dynamics of these systems so they can be readily incorporated into practical device development, no matter what kind of operation or detection methods are used.

4.2.1 Intrinsically Nonlinear Nanoresonator

4.2.1.1 Design strategy

Figure 4.2 shows the scanning electron microscope (SEM) images of a nonlinear nanoresonator using a doubly clamped CNT.[19] The intrinsic nonlinearity is simply introduced into the nanoscale resonance system through a geometric design as described in what follows.

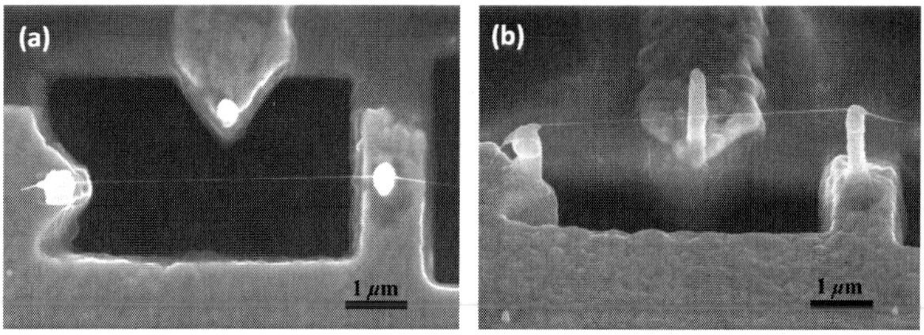

Figure 4.2 SEM images in top view (left) and tilted view (right) of a representative nanoresonator employing a CNT suspended between and fixed at both ends on the fabricated platinum electrode posts.[19]

Consider a fixed-fixed mechanical resonator employing a linear elastic wire with negligible bending stiffness and no initial axial pretension, as shown in Fig. 4.3. When driven transversely by a periodic excitation force applied locally at the middle of the wire, it will exhibit strong geometric nonlinearity and become an intrinsically (purely) nonlinear resonator. In such a resonator, the force–displacement dependence is described by the relation $F = kx\ [1 - L(L^2 + x^2)^{-1/2}] \approx (k/2L^2)x^3 + O(x^5)$,[20] where F is the transverse point force applied to the middle of the wire, x is the transverse displacement at the middle of the wire, and L and k are the half-length and the effective axial spring constant of the wire, respectively. The total absence of a linear force–displacement dependence term (i.e., a term of the form kx) results in the realization of a geometrically nonlinear force–displacement dependence of pure cubic order. This resonator has no preferential resonant frequency, and its resonant response is broadband,[20] conceptually different from typical linear mechanical resonators. Moreover, the apparent resonant frequency is completely tunable through the instantaneous energy of the system. If the bending effects are non-negligible, or if an initial pretension exists in the wire, a nonzero linear term in the previous force–displacement relation is included, giving rise to a preferential resonant frequency. However, as long as this preferential frequency is sufficiently small compared to the frequency range of the nonlinear resonance dynamics, the previous conclusions still apply.

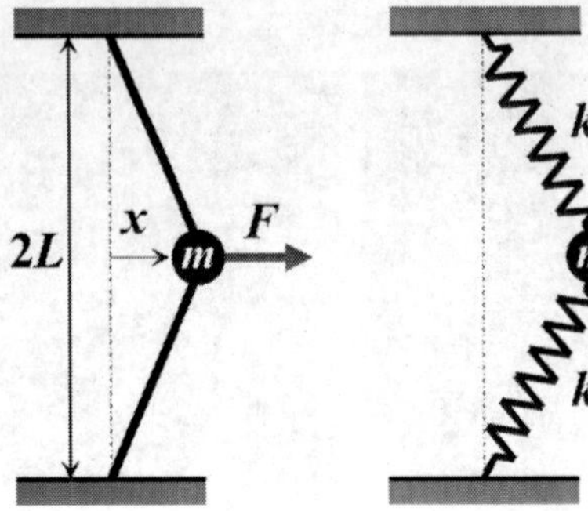

Figure 4.3 Schematic showing a simple doubly clamped mechanical beam and its equivalent spring mass model having intrinsic geometric nonlinearity.

The geometric layout of the device is schematically depicted in Fig. 4.4a. The platinum post on the counter electrode to apply an electric field to the CNT is modeled as a sphere and the CNT beam as a cylinder. When the radius of the sphere (R) is much smaller than the distance (d) between sphere and cylinder ($R << d$), the total induced charge on the sphere is given by $Q_S = (4\pi\varepsilon_o)RV$, where ε_o is electric permittivity and V is the potential difference between sphere and cylinder. The charge distributed on a specific location on the cylinder is inversely proportional to the distance r, so the charge at position x is described by $q(x) = k/r$, where k is a proportional constant. Practically assuming that the total amount of induced charge on the cylinder is the same as the charge on the sphere, k can be obtained from

$$Q_s = \int_{-L}^{+L} q(x)dx = \int_{-L}^{+L} \frac{k}{\sqrt{x^2+d^2}}dx = k\ln\frac{L+\sqrt{L^2+x^2}}{-L+\sqrt{L^2+x^2}} \quad (4.13)$$

The electrostatic force per unit length at x is then

$$F^*(x) = \frac{1}{4\pi\varepsilon_o}\frac{Q_s q(x)}{r^2} = \frac{Q_s k}{4\pi\varepsilon_o}\frac{1}{(x^2+L^2)^{3/2}}, \quad (4.14)$$

and the force components in the transverse and longitudinal directions are $F_y^* = f(x)\cos\theta$ and $F_x^* = f(x)\sin\theta$, respectively. The distribution of the transverse force per unit length applied on the CNT is thus calculated based on the experimental parameters (R = 100 nm, d = 1.5 μm, $2L$ = 6 μm, and V = 10 V) and is shown in Fig. 4.4b. The force at the middle of the beam is over an order of magnitude higher than at the ends, approximating a center-concentrated drive force necessary for realizing the geometric nonlinear resonance. The total force is obtained by integrating Eq. 4.14 over the whole beam length and is calculated to be ~26 pN, which is larger than the force, ~7 pN, estimated from experiments. Note that the electrostatic calculation described here overestimates the induced charge on the CNT and thus the interaction force, as the distribution of induced charge on the surrounding objects, such as conductive leads, was not considered in the model.

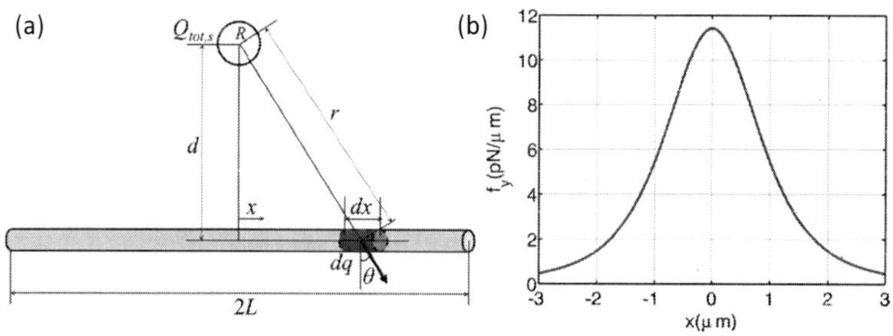

Figure 4.4 Theoretical model: (a) Schematic of a device used in the experiment, (b) transverse force distribution per unit length along the CNT.[19]

4.2.1.2 Theoretical model

We now proceed to analyze a doubly clamped Euler–Bernoulli beam having an added mass (m_c) attached at its middle and excited transversely by an alternating center-concentrated force. Considering the geometric nonlinearity induced by axial tension during oscillation, the vibration of the beam is described by

$$[\rho A + m_c \delta(x-L)]w_{tt} + (m\omega_o/Q)w_t + EIw_{xxxx} - (EA/4L)w_{xx}\int_0^{2L} w_x^2 dx$$
$$= F\cos\omega t\delta(x-L), \qquad (4.15)$$

where $w(x, t)$ is the transverse displacement of the beam with x and t denoting the spatial and temporal independent variables,

E and ρ are Young's modulus and mass density, A and L are the cross-sectional area and half-length of the beam, I is the area moment of inertia of the beam, Q is the quality factor of the resonator in the linear dynamic regime, F is the excitation force applied at the middle of the beam, $\omega(=2\pi f)$ is the driving frequency, and $\omega_o(=2\pi f_o)$ is the linearized natural resonant frequency of the beam. It is assumed that no initial axial tension exists when the beam is at rest, and subscript notation for partial differentiation is used.

The transverse displacement of the beam can be approximated as

$$w(x,t)=\sum_{i=1}^{N} W_i(x)\gamma_i(t) \quad (4.16)$$

by discretizing the continuous system using an eigenfunction expansion, where $W_i(x)$ is the i-th mode shape of the linearized beam, $\phi_i(t)$ the corresponding i-th modal amplitude, and N the number of beam modes considered in the approximation. Here, the i-th linearized mode shape of the beam is given by

$$W_i(x) = k_i[\sin\lambda_i x - \sinh\lambda_i x] + [\cos\lambda_i x - \cosh\lambda_i x], \quad (4.17)$$

where $k_i = (\cos 2\lambda_i L - \cosh 2\lambda_i L)/(\sin 2\lambda_i L - \sinh 2\lambda_i L)$ and the eigenvalues λ_i are the positive roots of the eigenvalue equation, $\cos\lambda_i \cosh\lambda_i = 1$. The displacement of the first mode at the middle of the beam is $W_1(L)\phi_1(t)$ with $W_1(L) = 1.59$ for a doubly clamped beam.

The leading modal amplitude, $\phi_1(t)$, is then approximately governed by a Duffing equation obtained by discretizing Eq. 4.13 through a standard one-mode Galerkin approach[21]

$$(1+M)\ddot{\phi}_1 + \frac{\omega_o}{Q}\dot{\phi}_1 + \omega_o^2\phi_1 + \alpha\phi_1^3 = q\cos(\omega t). \quad (4.18)$$

Here, $M = [m_c/2\rho AL]W_1^2(L) = (m_c/m_o)W_1^2(L)$ is the ratio of the added mass to the overall mass of the beam multiplied by a factor due to the center-concentrated geometry of the added mass distribution (when the added mass is distributed evenly on the beam, $M = (m_c/m_o)$; the amplitude of the drive force per unit mass in Eq. 4.16 is defined by $q = W_1(L)F/m_o$, and the nonlinear coefficient is defined by $\alpha = -E/(32\rho L^4)\int_0^{2L}(W_1')^2dx$.

When there is damping, the steady-state vibration will have a phase lag, Δ, and we assume that

$$\phi_1 = c_1 \cos(\omega t - \Delta) \tag{4.19}$$

Then, by applying the Ritz method, the relation among the drive frequency, the amplitude c_1 and the phase Δ are given by

$$\left(\frac{3}{4}\frac{\alpha}{\omega_o^2}\right)^2 c_1^3 = \left((1+M)\frac{\omega^2}{\omega_o^2} - 1\right)c_1 - \frac{q}{\omega_o^2}\sqrt{1 - \frac{(\omega\omega_o c_1/Q)^2}{q^2}} \tag{4.20}$$

$$\Delta = \tan^{-1}\left(\frac{\omega\omega_o/Q}{-(1+M)\omega^2 + \omega_o^2 + \frac{3}{4}\alpha c_1^2}\right) \tag{4.21}$$

The "backbone" curve, corresponding to the response of the nonlinear free vibration,[21] is obtained by setting q equal to zero in Eq. 4.20.

$$\left(\frac{3}{4}\frac{\alpha}{\omega_o^2}\right)^2 c_1^2 = \left((1+M)\frac{\omega^2}{\omega_o^2} - 1\right) \tag{4.22}$$

Substituting the equation of the point loci where the spectrum intersects with the backbone curve, $c_1 = \frac{q}{\omega\omega_o/Q}$, into Eq. 4.20 yields

$$(1+M)r_{\text{drop}}^4 - r_{\text{drop}}^4 - \frac{3}{4}\frac{\beta q^2 Q^2}{\xi_0^6} = 0 \tag{4.23}$$

The positive root of Eq. 4.23 is the so-called "drop-frequency," which is given by

$$r_{\text{drop}} = \frac{f_{\text{drop}}}{f_o} = \left(\frac{1+\sqrt{1+(1+M)\Gamma}}{(1+M)}\right)^{1/2} \tag{4.24}$$

where

$$\Gamma = \frac{3\alpha q^2 Q^2}{\omega_0^6} = \gamma \cdot \left(\frac{FQ}{E}\right)^2 \left(\frac{2L}{D}\right)^6 \left(\frac{1}{D^4}\right) \tag{4.25}$$

and γ = 0.0303. The response spectrum of the Duffing oscillator described by Eq. 4.18 forms a multi-valued region when the oscillation amplitude exceeds a critical value. Specifically, there are two branches of stable resonances that are connected by a branch of unstable resonances. As the frequency is swept up, the resonance amplitude in the upper branch of stable resonances increases up to the maximum possible amplitude and then drops abruptly to a lower value as the forced motion makes a transition to the lower stable branch. The drop-frequency, f_{drop}, where this jump phenomenon occurs, is determined by Eq. 4.24. From this equation, it is clear that the drop-frequency of this nonlinear resonator strongly depends on the attached center mass and damping, besides the geometry of the beam and the applied excitation force. A similar computation can be performed for the reverse jump-up frequency during a decreasing frequency sweep; in that case, the dynamics follows a transition from the lower stable resonance branch to the upper.

4.2.1.3 Device fabrication and experiment

The device was fabricated through micromachining and nanomanipulation. A silicon (100) wafer was coated with a 500 nm-thick silicon nitride layer followed by 1.5 μm-thick silicon dioxide. A thin Cr/Au layer was then sputter-coated onto the silicon wafer and subsequently patterned through photolithography to form a three-electrode layout. This silicon wafer was back etched in KOH to make a thin membrane of silicon dioxide under the electrodes. The window was then milled with a focused ion beam (FIB) to create three suspended electrodes. Three vertical platinum posts were fabricated onto these three electrodes through electron beam induced deposition. A high quality multiwall CNT produced with arc-discharge was then selected and manipulated inside an electron microscope and suspended between two of the platinum posts with both ends fixed with electron beam induced deposition of a small amount of platinum. The remaining platinum post was used as the driving electrode for applying the localized oscillating electric field to drive the oscillation of the CNT. The overall design of the device maximized the localization of the excitation force applied to the CNT beam. According to the previous discussion, the localization of the applied force is necessary for creating the strong geometric nonlinearity in the resonance system.

To acquire the response spectrum of the nanoresonator, the frequency of the applied AC driving voltage (V_{ac}) was swept up and then down, while the oscillation amplitude at the middle of the CNT was measured from the acquired images in a SEM at room temperature and at a vacuum pressure of $\sim 10^{-6}$ Torr. To evaluate the effect of an added mass on the dynamic behavior of the nanoresonator, a small amount of platinum was deposited at the middle of the CNT with electron beam induced deposition, and its mass was estimated from the measured dimension.

4.2.1.4 Nonlinear response

Figure 4.5 shows the acquired response spectrum for a nonlinear nanoresonator incorporating a CNT of $2L = \sim 6.2$ μm and $D = \sim 33$ nm driven with an AC signal of 10 V in amplitude. The initiation of the oscillation started around 4 MHz, near the natural resonant frequency of this doubly clamped CNT. The amplitude of the resonance oscillation increased continuously during the increasing frequency sweep up to 14.95 MHz, at which point the amplitude suddenly dropped to zero. This response resembled closely what was modeled previously for an intrinsically nonlinear nanoresonator, and corresponded to a resonance bandwidth of over 10 MHz. During the ensuing decreasing frequency sweep, the resonator stayed mostly in a non-resonance state until the neighborhood of the natural resonance frequencies of the CNT, where transitions back to resonant oscillations occurred. By fitting the obtained drop-jump and up-jump frequencies with the model prediction, the drive force was estimated to be ~7 pN and the Q-factor of the system ~260, which were in agreement with the estimate from an electrostatic analysis based on the experimental setup and the reported Q-factor values for typical CNT-based resonators,[22] respectively.

The occurrence of multiple up-jump transitions during the decreasing frequency sweep appears to be due to the existence of multiple natural resonance frequencies in the multiwall CNT and thus multiple modes of resonance. In theory,[23] there is the same number of fundamental frequencies and resonance modes as the number of cylinders in a multiwall CNT. Moreover, in a recent computational study[24] it was shown that in the strongly nonlinear regime there can be coupling between multiple radial and axial modes of a double-walled CNT, with van der Waals forces provoking dynamical transitions between the modes of the inner and outer walls. Such strongly nonlinear modal interactions can be studied using asymptotic techniques in the context of coupled nonlinear oscillators.[25]

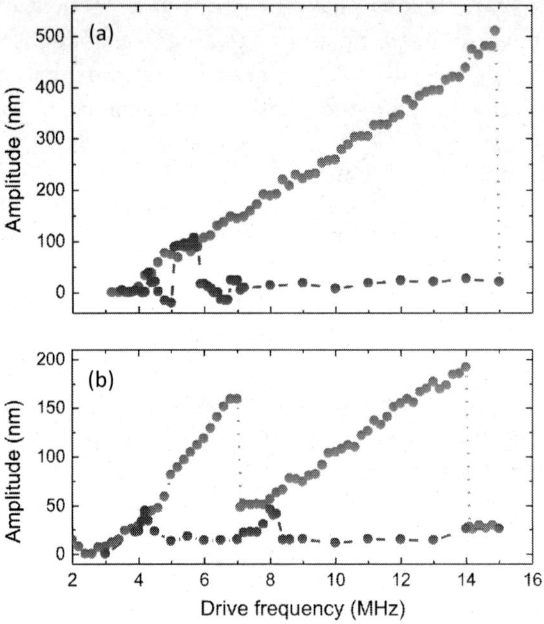

Figure 4.5 The acquired response spectra of a CNT ($2L$ = ~6.2 μm, D = ~33 nm) nonlinear nanoresonator driven with AC voltage signals of 10 V (a) and 5 V (b) in amplitude.[19]

The existence of multiple natural modes in this multiwall CNT-based nonlinear resonator can also be revealed in an increasing frequency sweep when the drive force is reduced. Figure 4.5b shows the response spectrum acquired from the same resonator when the applied AC amplitude was reduced to 5 V. Two distinct resonance modes were excited in this case. The first mode appeared around 4 MHz and its drop-jump occurred at 7.05 MHz. The second mode then initiated right after the drop-jump of the first mode, and jumped down at 14.15 MHz. As shown previously, when the drive force was increased, it appeared that the first mode resonance became dominant and suppressed the initiation of the second mode in the increasing frequency sweep; while in the decreasing frequency sweep, since there was no dominant mode, those modes were excited in the neighborhoods of their linearized resonance frequencies. Similar observations have been reported in coupled nonlinear resonators[26] but not, until now, for a multiwall CNT intentionally operated in a highly nonlinear regime.

4.2.2 Nonlinear Microcantilever-Nanotube System I[27]

4.2.2.1 Design strategy

We now introduce another nonlinear system composed of two micromechanical cantilevers coupled by a nanotube, in which nonlinear dynamics was, again, intentionally introduced. The SEM images of the representative system are presented in Fig. 4.6.[27] This system consists of two micromechanical cantilevers bridged between two free ends by a multi-walled BNNT. Titled views at 52° in the SEM images on the right reveal that the two cantilevers are nearly coplanar for Fig. 4.6a and non-coplanar in Fig. 4.6b. The two microcantilevers are separately anchored to a solid base, so the only mechanical coupling is through the attached BNNT. The end segments of the BNNT are rigidly fixed onto the surfaces of the free ends of the microcantilevers through localized e-beam induced deposition of platinum. In the absence of the nanotube, there is no mechanism to introduce stiffness nonlinearity into these two cantilevers, and their dynamics is linear within the range of their operation. The two microcantilevers are designed with their first eigenfrequencies sufficiently apart so that when one of the microcantilevers oscillates near its fundamental resonant frequency, the other is stationary and acts as a nearly fixed anchor for the nanotube. The microcantilevers are coupled through the attached nanotube and the main (and dominant) source of nonlinearity in our system is the geometrical stiffening or softening nonlinear effect induced by the nanotube when stretched or compressed axially at every period of oscillation of the microcantilever assembly. Note that the nanotube is elastically soft in the lateral direction but comparatively rigid in the axial direction. As shown in the zoomed-in image in Fig. 4.6b, the gap between the free ends of the cantilevers is designed to be narrow to accommodate a short nanotube attachment at the scale of several micrometers in order to maximize the induced geometric nonlinearity.

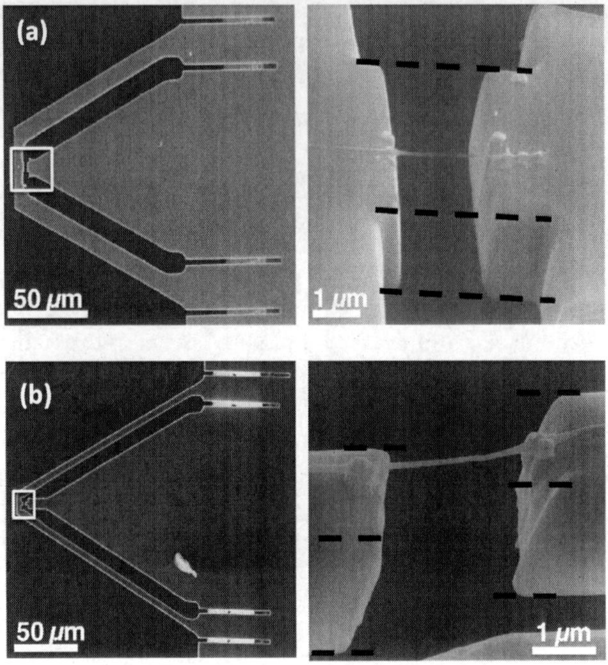

Figure 4.6 SEM images of two nonlinear microcantilever systems with integrated geometric nonlinearity through a nanotube attachment. The images on the right is a magnified view of the marked region in square in the images on the left at a sample tilt angle of 52°: (a) Two cantilevers in (a) are almost coplanar, while in (b) non-coplanar.[27]

4.2.2.2 Theoretical model

Using a simple lumped parameter model, we describe the forced resonance of this intentionally nonlinear system, i.e., when the microcantilevers oscillate with a dominant frequency equal to that of the applied harmonic excitation. Figure 4.7 shows the schematic model of the system. The microcantilever oscillating in its fundamental mode is modeled by the vertical linear harmonic oscillator with mass (m), vertical spring (k_1), and viscous damper (c_1), excited by a harmonic force ($F_o \cos\omega t$). The microcantilever in a non-resonant state is approximately modeled as a rigid ground. The nanotube is modeled as a massless linear spring (k_2), which introduces geometric nonlinearity into the system as the mass m transverses in the vertical direction. To be more general, considering that in a fabricated device there might exist a non-coplanar offset in the attached nanotube between the free ends of the microcantilevers, we ascribe an initial offset (tilt) angle φ in the model for the attachment of the nanotube. It will be shown later that this angle plays an important role in the geometric nonlinearity induced in our system: Depending on its value, the system can produce either hardening or softening responses. Referring to the notation of Fig. 4.2, the length of the nanotube spring due to the vertical displacement of the mass (m) is $L_2' = \sqrt{y^2 + L_2^2 - 2yL\sin\varphi}$. The vertical component of the tension T in the nanotube caused by the oscillation of the microcantilever is then $T\cos\theta = k_2(L_2' - L_2)(y - L_2\sin\varphi)/L_2'$. Performing the balance of vertical forces applied on the mass (m) provides the following equation of motion of the microcantilever system with the coupling nanotube

$$m\ddot{y} + c_1\dot{y} + k_1 y + T\cos\theta = F_o\cos\omega t \tag{4.26}$$

where overdot denotes differentiation with respect to time. Expanding the expression for the tension in the nanotube in Taylor series for small vertical displacements yields

$$T\cos\theta = (k_2\sin^2\varphi)y - \left(\frac{3k_2L_2\sin\varphi}{2}\right)\left(\frac{y}{L_2}\right)^2 + \left(\frac{k_2L_2}{2}\right)\left(\frac{y}{L_2}\right)^3 + \cdots, \left|\frac{y}{L_2}\right| \ll 1 \tag{4.27}$$

Here, we assumed small motions and small offset. Substituting Eq. 4.27 into Eq. 4.26 yields

$$m\ddot{y} + c_1\dot{y} + \tilde{k}_1 y + \tilde{k}_2 y^2 + \tilde{k}_3 y^3 = F_o\cos\omega t \tag{4.28}$$

where

$$\tilde{k}_1 = k_1 + k_2\sin^2(\varphi), \quad \tilde{k}_2 = -\frac{3k_2}{2L_2}\sin\varphi, \quad \tilde{k}_3 = \frac{k_2}{2L_2^2}. \tag{4.29}$$

As expected, the nanotube spring introduces the cubic nonlinearity into the system, regardless of the existence of the offset angle. However, the offset angle φ contributes to both the quadratic and linear spring constants of the system. This means that depending on the value of this angle either the hardening effect due to the cubic stiffness term or the softening effect due to the quadratic stiffness term can potentially dominate the nonlinear dynamics of the system. Note that the system with $\varphi = 0$ is a typical Duffing oscillator with a hardening spring as the sign of k_3 is always positive.

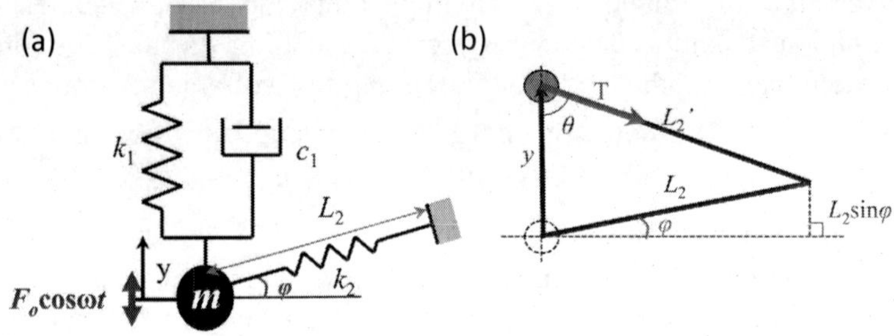

Figure 4.7 Model of the system of Fig. 4.6: (a) Schematic representation of the microcantilever system integrated with nanotube coupling; (b) geometry describing the stretch of the attached nanotube.[27]

To facilitate the analysis, Eq. 4.28 is normalized and expressed in nondimensional form as

$$\bar{y}'' + 2\zeta\bar{y}' + \bar{y} + a_2\bar{y}^2 + a_3\bar{y}^3 = q_o \cos \Omega\tau \tag{4.30}$$

by introducing the nondimensional variables, $\bar{y} = \frac{y}{L_2}$, $\tau = \omega_0 \tau$ (note that overbars have been dropped for convenience). Primes denote differentiation with respect to dimensionless time τ, whereas the parameters appearing in Eq. 4.30 are given by

$$\tilde{\omega}_o = \sqrt{\frac{\tilde{k}_1}{m}} \approx \sqrt{\frac{k_1}{m} + \frac{k_2}{2m}\sin^2\varphi}.$$

$$2\zeta = \frac{c_1}{m\tilde{\omega}_o}, \quad q_o = \frac{F_o}{\tilde{k}_1 L_2}, \quad \Omega = \frac{\omega}{\tilde{\omega}_o},$$

$$a_2 = \frac{\tilde{k}_2}{\tilde{k}_1} L_2 = -\frac{3}{2}\frac{k_2 \sin\varphi}{k_1 + k_2\sin^2\varphi}, \quad a_3 = \frac{\tilde{k}_3}{\tilde{k}_1} L_2^2 = \frac{k_2}{2(k_1 + k_2\sin^2\varphi)}. \tag{4.31}$$

The method of multiple scales[21] is employed to find the dynamic response of system at the fundamental mode resonance of the microcantilever, i.e., when $\Omega \approx 1$. Through procedures similar to those introduced in Section 4.1.2, we can approximate the relation between steady-state amplitude (α_{ss}) and frequency ($\Omega = 1 + \varepsilon^2\sigma$) as

$$\left[\zeta^2 + \left(\sigma - \frac{9a_3 - 10a_2^2}{24}\alpha_{ss}^2\right)^2\right]\alpha_{ss}^2 = \pm\frac{q_o^2}{4}, \quad \gamma = \frac{\pi}{2}, \tag{4.32}$$

Examining the frequency response of the system at steady state for the fundamental resonance described in Eq. 4.32, the effect of the offset angle becomes more consequential. First we derive the backbone curves for the corresponding Hamiltonian system, i.e., the undamped and unforced oscillator resulting from Eq. 4.32 by setting $\zeta = q_o = 0$. This produces the following approximate frequency-steady-state amplitude curve

$$\alpha_{ss} = \pm\left(\frac{24}{9a_3 - 10a_2^2}\sigma\right)^{1/2} \tag{4.33}$$

As shown in this equation, the sign of the coefficient of the frequency detuning parameter determines whether the dynamics of the system exhibits nonlinear softening or hardening resonance. Accordingly, we define a parameter $\mu = 9a_3 - 10a_2^2$ quantifying the type of nonlinearity in the system. Substituting the physical parameters into the expression for μ gives the critical offset angle φ_{cr} required to eliminate both hardening and softening nonlinear behavior by counterbalancing the quadratic and cubic terms

$$\sin^2\varphi_{cr} = \frac{k_1}{4k_2} \tag{4.34}$$

When the offset is smaller than this critical angle, the nonlinear dynamic response will be hardening otherwise it will be softening. Hence, by designing the offset angle we may induce hardening or softening geometric stiffness nonlinearities in the coupled assembly.

4.2.2.3 Device fabrication and experiment

The nonlinear cantilever system was fabricated with micro-machining and nanomanipulation. First, the two micromechanical cantilevers were fabricated with a silicon-on-insulator (SOI) wafer having a 2 μm-thick device layer (Si), 2 μm-thick buried insulator layer (SiO_2), and 400 μm-thick substrate layer (Si). The device layer forms a cantilever structure, whereas the substrate layer serves as a base of the suspended cantilevers. The general steps of the fabrication involved the positive photoresist (SPR-220) coating of both the top and bottom surfaces of the SOI wafer; the patterning of the photoresist with photolithography according to the design of the cantilevers and the anchoring base; the etching of the top Si layer to define the cantilevers; and the deep reactive ion etching (DRIE) of the wafer backside to define the anchoring base. In the last step, the wafer was immersed in hydrofluoride (HF) solution to etch away the buried oxide layer to release the cantilevers from the wafer to realize the freestanding formation of the cantilevers.

The micromachining process provided two micromechanical cantilevers connected to each other at the free ends. Focused ion beam cutting was applied to cut through a narrow gap between the ends of the two cantilevers to separate them. A high quality BNNT was then selected and manipulated inside a SEM with a nanomanipulator and placed across the narrow gap. Finally, both ends of the placed BNNT were fixed on the cantilever surface by electron beam-induced deposition of a small amount of platinum.

The cantilever system was driven by a piezoelectric stack placed underneath the cantilever-anchoring base in an air ambient environment. A sinusoidal AC driving signal supplied from a function generator actuated the stack. The frequency was first swept up and then down, and the oscillation amplitude at the end of the cantilever was measured using a laser Doppler Microvibrometer (Polytec UHF-120). Fast Fourier transform (FFT) of the data collected from the vibrometer at each driving frequency provided the spectral response of the system. It is noted that nonlinearity is not introduced in either the actuation with the piezoelectric stack or the measurement laser of the vibrometer operated at a constant power less than 5 mW.

4.2.2.4 Nonlinear response

In order to test the assumptions in our model, the dynamic responses of the inner and outer microcantilevers were measured before being coupled through the nanotube. The experimental results are plotted in Fig. 4.8 and are fitted with solid curves, which represent the predictions according to model of a forced linear harmonic oscillator. The responses of both cantilevers closely follow the fitting curves; hence, in the absence of the nanotube, the dynamics of each microcantilever is reasonably within the linear regime over its respective operational range. The fundamental frequencies, damping coefficients, and base excitation amplitudes of the piezoelectric stack are estimated to be 196.8 kHz, 0.0022, and 1.24 nm for the inner cantilever and 121.1 kHz, 0.0033, and 1.28 nm for the outer cantilever, respectively. It is important to note that the fundamental eigenfrequencies of the inner and outer microcantilevers are located enough apart, so there is no significant inherent (linear) coupling between them.

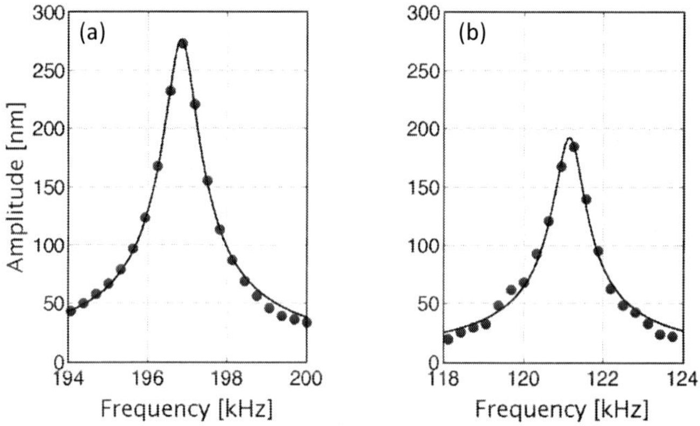

Figure 4.8 Experimentally acquired linear responses (solid circle) for the inner (a) and outer (b) microcantilevers shown in Fig. 4.6a, measured from the microcantilevers with no nanotube coupling; the solid lines are fitting curves based on model of a forced linear harmonic oscillator.[27]

Figure 4.9 depicts the experimentally acquired hardening resonances for the inner cantilever in the system configuration shown in Fig. 4.6a. The attached BNNT is ~2.6 μm in length and ~80 nm in diameter. The results from Figs. 4.9a–f correspond to a monotonic increase of the oscillation amplitude due to an increase of the driving amplitude. The circles/crosses in the plots denote the acquired steady-state experimental measurements made during upward/downward frequency sweeps, respectively. The measurements were performed on a device having an almost perfect coplanar alignment between the cantilevers as shown in Fig. 4.6a (i.e., the offset angle was approximately $\varphi = 0$). Hence, the quadratic term in the equation of motion can be considered to be nearly negligible, which leads to a series of hardening resonances. At small amplitudes of oscillation (e.g., Fig. 4.9a), the dynamic response remains nearly linear since the axial tension induced in the nanotube is insignificant. As the oscillating amplitude increases, however, the spectral curves bend toward higher frequencies due to the hardening effect introduced from axially stretching in the nanotube, leading to hysteresis loops including (well-documented) discontinuous jumps. The displacement of the outer beam within the frequency range of interest was also examined, per the inset of Fig. 4.6e, and shown to be comparable to the base excitation amplitude, confirming its non-resonance state.

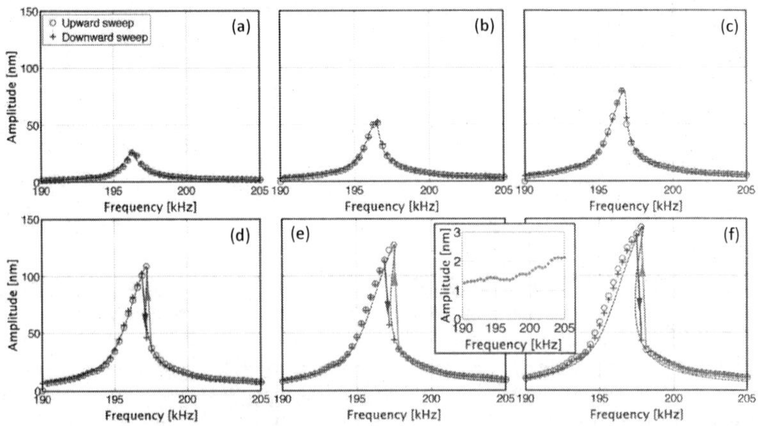

Figure 4.9 Experimentally acquired hardening resonances for the inner microcantilever shown in Fig. 4.6a with increasing driving amplitude, dashed lines are the fitting curves based on model (30). The inset in (e) represents the response of the outer cantilever within the frequency range of interest, whose amplitude is in the same order of magnitude as the base excitation amplitude.[27]

The experimental results were compared with the analytical model results. The linearized fundamental resonant frequency (f_0) and the nonlinear constant (μ) were estimated by fitting the maximum amplitude values of the response curves into the backbone curve described by Eq. 4.33. The fitting, as shown in Fig. 4.10, gives f_0 = 196.7 kHz and μ = 1.06 × 10^{-5}. The damping coefficient (ζ) of the cantilever was obtained from curve fitting to the response curves acquired at small oscillation amplitude and was found to be ~0.0023. Finally, the equivalent applied forces (q_0) were obtained by fitting the drop-jump instabilities in the backbone curve. Substituting those estimated parameters into Eq. 4.32, the corresponding frequency responses predicted by the model were plotted with dashed lines in Fig. 4.9, which showed good agreement with the experimental results. The discrepancy between model and experiment grows as the driving force increases, possibly due to the amplitude-dependent increase in damping of the cantilever itself. The increased energy dissipation due to the deformation of the nanotube,[28,29] which was not considered in our analysis, may also be partly responsible for the discrepancy between theory and experiment.

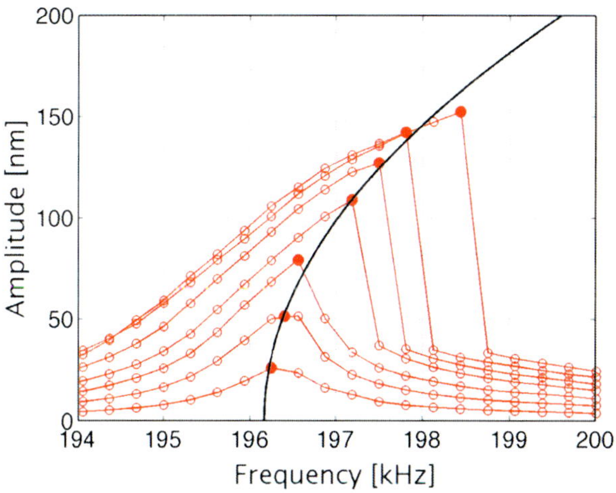

Figure 4.10 Backbone curve fit for the hardening resonances presented in Fig. 4.9. The maximum oscillation amplitude in each resonance response (red circle) under given driving amplitude is fitted into a modeled backbone curve (in black solid line) for the nonlinear system.[27]

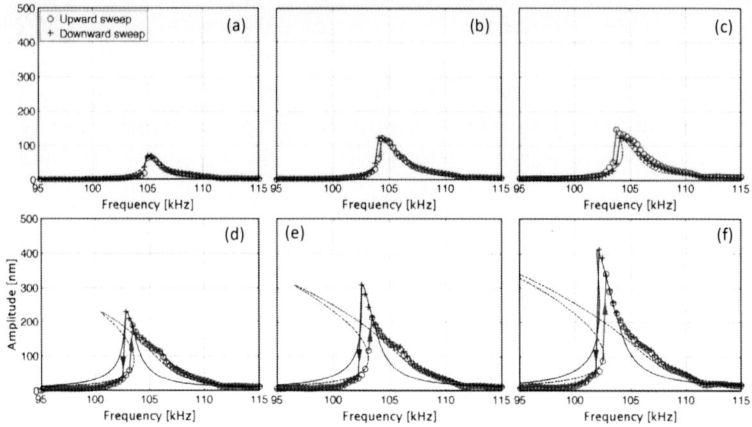

Figure 4.11 Experimentally acquired softening resonances for the inner microcantilever shown in Fig. 4.6b. Two fitting curves are plotted based on the parameters estimated from backbone curve fitting shown in Fig. 4.12. The dashed/solid curves represent the nonlinear responses of the system in the compression/tension-dominated regimes of the attached nanotube, respectively.[27]

In the microcantilever system having non-coplanar cantilever surfaces (shown in Fig. 4.6b) the attached nanotube is severely tilted off the reference ground plane. Such a microcantilever system exhibited a clear softening response as shown in Fig. 4.11. As explained previously, such a softening response is the direct result of the tilting in the coupling nanotube and the consequent appearance of a strong quadratic term in Eq. 4.34. The relevant parameters were estimated, enabling comparison of the experimental results with those predicted by the theoretical model. The backbone curve in Fig. 4.12, derived solely from data acquired at low oscillation amplitude, failed to predict the experimentally acquired response at high oscillation amplitude. Thus, two sets of parameters (f_0, μ), (105.4 kHz, -3.6×10^{-5}) and (103.1 kHz, -2.6×10^{-6}) for low and high amplitude oscillations, respectively, were used to fit the experimental data over the entire frequency range. Such qualitatively different responses at varying oscillation amplitudes are believed to originate from the different stiffness behaviors of the attached nanotube as it undergoes different magnitudes of deformation. Indeed, when there is an initial offset angle, the axial stress applied to the attached nanotube during dynamic oscillation of the cantilever is divided into two regimes; one regime that is compression-dominated at low oscillating amplitudes, and a second regime that is tension-dominated at high oscillating amplitudes. Note that in the absence of this offset angle the nanotube can only experience tensile strain due to axial stretching induced by the oscillatory motions of the cantilevers. Due to its high aspect ratio structure, a nanotube under compression can exhibit elastic Euler buckling (or rippling bending) at small compressive strain, resulting in a reduction in its effective stiffness.[30] Alternately, a nanotube in tension can sustain high tensile strain maintaining high axial stiffness until failure.[31] We conjecture that the difference in axial stiffness when in tension or in compression causes the change of sign of the nonlinear constant (μ) between the tension- and compression-dominated regimes.

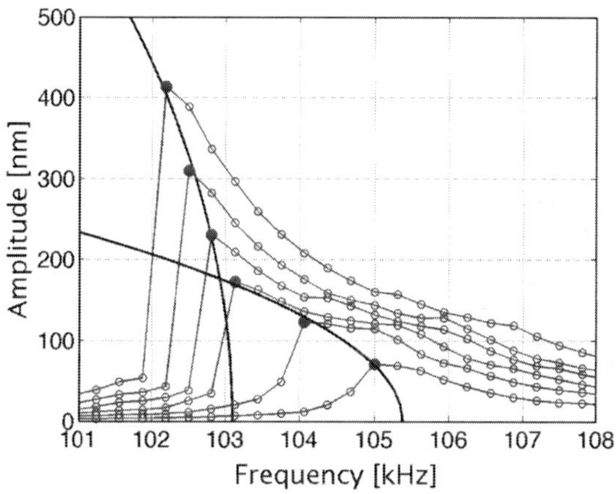

Figure 4.12 Backbone curve fitting for the softening resonances presented in Fig. 4.11; to describe the entire range of dynamic response, two backbone curves (in black) from two different sets of fitting parameters for f_0 and μ are used to fit the responses (blue cross), reflecting the different mechanical behaviors of the attached nanotube under compression at small oscillation amplitude and tension at large oscillation amplitude.[27]

Two sets of responses are thus determined from the model and plotted in Fig. 4.11; dashed and solid lines used fitting parameters obtained in compression-dominated and tension-dominated regions, respectively. When the amplitude of oscillation is small, the experimentally measured response follows the dashed line. Thereafter, as the oscillation amplitude increases, the slope of the upper stable branch increases to follow the solid line. It is interesting to note that the small "shoulders" experimentally observed in the upper branch near the intersection of the two fitted curves (cf. Figs. 4.11d–f) are indicative of the transition from the compression-dominated to the tension-dominated regime.

4.2.3 Nonlinear Microcantilever-Nanotube System II

4.2.3.1 Design strategy

The second design of an intentionally nonlinear microcantilever-nanotube system is composed of a commercial microcantilever system with dimensions and operating conditions similar to those widely used in scanning probe microscopy.[17] Figure 4.13 depicts the SEM images of the nonlinear microcantilever fabricated for incorporating significant intentional geometrical nonlinearity. An inner paddle was carved out in a rectangular cantilever structure to create an inner cantilever having its free end close to the fixed base of the overall structure. It is expected (and later demonstrated in our modeling result) that when the microcantilever system is driven to resonate near one of its bending modal frequencies,

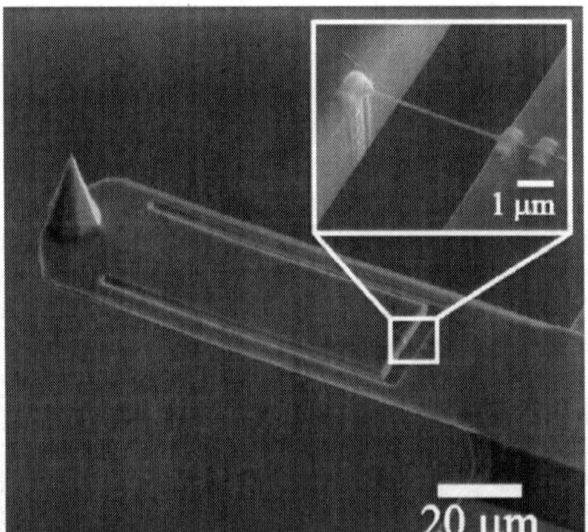

Figure 4.13 SEM images of the nonlinear microcantilever-nanotube system. The inset shows the attached BNNT, whose diameter and suspended length are 120 nm and 4 μm, respectively.[17]

the free end of the inner paddle can exhibit a large vertical (out-of-plane) displacement relative to the fixed base located just across a small gap. To introduce intentional strong nonlinearity, a single BN nanotube is attached across this gap as shown in the inset in Fig. 4.13. This local modification of the microcantilever structure alters the relation between the overall force acting on the structure and the resulting displacement, and introduces strong geometric and kinematic nonlinearity due to axial stretching of the nanotube in the dynamic mechanical resonance of the system.

4.2.3.2 Theoretical model

In Fig. 4.14, we present the simplified, lumped-parameter model of the nonlinear microcantilever system of Fig. 4.13. In Fig. 4.14, m denotes the effective mass of the mode, and k_1 and c_1 represent the linear effective stiffness and the damping coefficients of the microcantilever, respectively; k_2 and c_2 are parameters arising from the nanotube attachment, standing for the axial stiffness and the viscous damping coefficient of the BNNT, respectively. The schematic model is nearly identical to the one of the previous section except for the exclusion of the offset angle and the inclusion of the additional horizontal nonlinear damping. Since the commercial AFM cantilever we used has almost negligible pretension within the structure, the paddle and the cantilever are almost coplanar after patterning the inner paddle (this is why we may neglect the initial offset angle in the model of Fig. 4.14). Moreover, in that model, the vertically attached parallel spring and damper represent the microcantilever structure at its second (linear) bending mode in the absence of the nanotube attachment, since we will be exciting the system at that frequency range. The point near the base where the nanotube is anchored is modeled as a rigid ground since that point is nearly stationary. The nanotube is modeled as a massless system of a linear spring in parallel with a viscous damper, which is attached horizontally between the mass and the ground. The reason why we include the damper in the present model is that we operate the device under vacuum. Indeed, under ambient pressure (as was the case in the previous application), the damping of the nanotube is assumed negligible compared with the large overall damping of the system resulting from both structural and ambient pressure dissipation;

under vacuum pressure, however, the damping within the nanotube structure is expected to play a much more significant role in the form of possibly a nonlinear dissipation mechanism, a fact that will be confirmed later in this section.

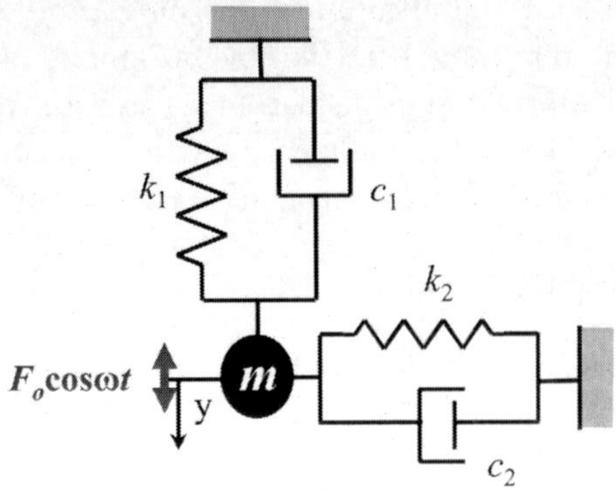

Figure 4.14 Schematic model of the system of Fig. 4.13.

Returning to the model of Fig. 4.14, the mass moves along the vertical direction, with y denoting the out-of-plane displacement of the free end of the inner paddle. Then the magnitude of the force T exerted on the microcantilever by the horizontally attached nanotube is $T = k_2\delta + c_2\dot{\delta}$ where δ is the axial deformation of the nanotube, k_2 is its linear stiffness and c_2 its linear viscous damper. The vertical component of the force T (which represents the transverse force applied to the microcantilever), can be approximately expressed as

$$T\sin\theta = \frac{k_2}{2L^2}y^3 + \frac{c_2}{L^2}y^2\dot{y} + O(y^4) \tag{4.35}$$

This includes a cubic term dependent on x (the amplitude) leading to "pure" geometric nonlinearity (i.e., there is an absence of a linear term, rendering the expression strongly nonlinear—in fact, non-linearizable), and a term associated with nonlinear dissipation originating from the viscoelastic behavior of the nanotube leading to damping nonlinearities as we will show later.

To facilitate the analysis, the normalization process is applied by introducing the following parameters:

$$z = \frac{y}{L_0},\ \tau = \omega_0 t,\ \omega_0 = \sqrt{\frac{k_1}{m}}, \tag{4.36}$$

where L_0, ω_0, τ and z are an arbitrary scaling length, the natural frequency, the normalized time and the normalized displacement, respectively. Then, the equation of motion of the lumped-parameter model can be expressed in the dimensionless form[17]

$$z'' + (2\xi_1 + \xi_3 z^2)z' + z + \alpha z^3 = q_0 \cos\Omega\tau \tag{4.37}$$

The dimensionless parameters in Eq. 4.37 are

$$\begin{aligned} &2\xi_1 = \frac{c_1}{m\omega_0}, \quad \xi_3 = \frac{c_3}{m\omega_0}L_0^2, \quad \alpha = \frac{k_3}{k_1}L_0^2 \\ &q_0 = \frac{F_0}{k_1 L_0}, \quad \Omega = \frac{\omega}{\omega_0}, \quad (\ )' = d/d\tau \end{aligned} \tag{4.38}$$

The approximate analytical expression for the frequency-amplitude dependence of the dynamics of Eq. 4.37 can be derived by a perturbation approach[21] and can be written as[17]

$$\left[\xi_{\text{eff}}^2 + \left(\sigma - \frac{3}{8}\alpha A_{\text{ss}}^2\right)^2\right] A_{\text{ss}}^2 = \left(\frac{q_0}{2}\right)^2 \tag{4.39}$$

where σ is the frequency detuning parameter (i.e., $\Omega = 1 + \varepsilon\sigma$) and A_{ss} is the steady-state amplitude. Here we define an effective damping coefficient ξ_{eff} through

$$\xi_{\text{eff}} = \xi_1 + \frac{\xi_3}{8}A_{\text{ss}}^2 = \frac{1}{2}\left(\frac{1}{Q_0} + \frac{1}{Q_{\text{nl}}}\right) = \frac{1}{2Q_{\text{eff}}} \tag{4.40}$$

where Q_0 is the Q-factor for the linear damping term, $Q_{\text{nl}}^{-1}/2$ is the damping effect contributed by the nonlinear damping effect, and $Q_{\text{eff}}^{-1}/2$ is the representative overall damping of the system. It should be noted that Q_{nl} and Q_{eff} represent the nonlinear and overall effective Q-factors, respectively, adopted from the general linear concept of Q-factor in order to obtain a better quantitative measure of the contribution of the nonlinear damping term to the overall dissipation compared to the linear damping term (which is expected to persist in the dynamics independent of the energy and ambient pressure level). In our system, Q_0 arises from the viscous effect exerted on the microcantilever by the surrounding gaseous environment and the intrinsic damping of the beam itself, whereas Q_{nl} arises due to geometric and kinematic nonlinearities; that is, it originates from the intrinsic nonlinear damping characteristic of the attached nanotube during the oscillation.

4.2.3.3 Device fabrication and experiment

The nonlinear microcantilever system was fabricated by modifying a commercially available AFM cantilever (Mikro-Masch, NSC-14), and the inner paddle was carved out with a FIB based machining process. A multi-walled BNNT was then manipulated and deposited across the gap by nanomanipulation inside an SEM, and was subsequently fixed on both ends through an electron beam induced platinum deposition process. The same experimental setup as in the previous section is used. To investigate the effective damping of the overall system (which, as discussed above, includes both linear and nonlinear components), the system was placed in a vacuum chamber where the vacuum pressure was varied from 750 Torr down to 4.5 mTorr.

4.2.3.4 Nonlinear response

Figure 4.15 shows the experimentally acquired response spectrum for a microcantilever system driven with an AC excitation amplitude of 1.5 V at a vacuum pressure P of 4.5 mTorr. Red circles denote data points acquired when the frequency was swept upward and blue triangles downward. Note that our initial measurement of the first two bending modes of the system revealed that the free end of the inner paddle produced greater displacement relative to the neighboring point across the gap in the second bending mode than in the first when driven with the same excitation. We thus chose the second bending mode in our study in order to involve the stronger geometric nonlinearity due to the tensile stretching of the nanotube. As seen from the acquired response spectrum, abrupt transitions are observed between the upper and lower stable branches as a result of nonlinear hysteresis, indicating the existence of strong geometric stiffness nonlinearity induced by the nanotube attachment.

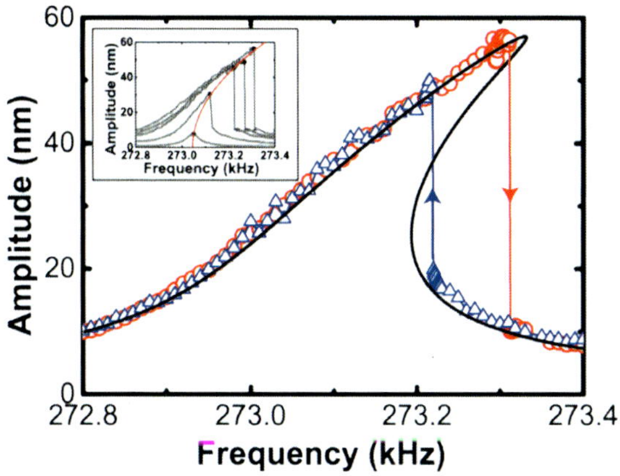

Figure 4.15 Nonlinear dynamical steady state response of the microcantilever with the attached BNNT for V_{pp} = 1.5 V at a vacuum pressure of 4.5 mTorr; black line is the analytical solution based on the extracted α and Q_{eff}. Inset shows the nonlinear resonance curves at the same pressure level as the excitation amplitude was varied (0.1, 0.5, 1.0, 1.25 and 1.5 V). The amplitudes at the drop frequencies are fitted by the red line (the backbone curve), resulting in $\alpha = 8.52 \times 10^{-7}$.[17]

The experimentally obtained results were compared with the results from the model developed previously. The inset of Fig. 4.15 shows the measured response curve at increasing excitation amplitude V_{pp}. With the increase of V_{pp}, the frequency band of resonant operation widens, and the drop frequency increases. The amplitude values at the drop frequency in these response curves fit precisely into the so-called "backbone" curve derived from Eq. 4.39 when all dissipative terms are set to zero and the dynamic system is Hamiltonian. The nonlinear stiffness constant a and linearized resonant frequency $f_o = \omega_0/2\pi$ in Eq. 4.37 were obtained through this fitting, which gave $\alpha = 8.52 \times 10^{-7}$ and f_0 = 273.046 kHz. The dimensionless force constant, q_0, was extracted from fitting the response curve acquired at an ambient pressure environment (P = 750 Torr), where the nonlinear damping effect was negligible, because the amplitude of oscillation of the microcantilever was small due to strong (linear) viscous damping caused by thc surrounding gaseous environment and intrinsic damping; in this case q_0 is measured to be 1.88 × 10^{-2} at a drive amplitude of V_{pp} = 1.5 V. We then fit the whole response curve as shown in Fig. 4.15 with our model result (i.e., Eq. 4.39) and the effective damping coefficient was found to be ξ_{eff} = 1.65 × 10^{-4} (Q_{eff} = 3036). The fitting produced response curves that matched the experimental result quite well as shown in Fig. 4.15.

4.3 Applications

Linear micro/nanomechanical resonators have exceptional properties, such as high resonant frequencies ranging 10^4–10^9 Hz and low damping with quality factors in the range of 10^2–10^5. Based on these properties, micro/nanoresonators are ideal systems for sensing applications because the sharp resonance peak at high resonant frequency is sensitive to extremely small perturbations in physical quantities, such as mass and force. For nonlinear micro/nanoresonators such as the one considered in this section, however, the resonance band broadens due to the softening or hardening nonlinear effect, and, thus, we can no longer apply this traditional methodology. Notwithstanding that, the nonlinear system exhibits superb performance as a sensor by utilizing its characteristic jump phenomena. A nonlinear resonance system can exploit the instabilities intrinsically existing within the system that are highly sensitive to external perturbation. The sharp transitions at those bifurcation points are all highly favorable characteristics from the precision measurement point of view, which can potentially enhance the measurement sensitivity of a practical sensing system. In the following exposition, we will present sensing capabilities and applications of the nonlinear micro/nanomechanical systems introduced above.

4.3.1 Mass Sensing

As we theoretically studied in an earlier section, the jump frequency of nonlinear resonance depends not only on the geometry and material properties of the beam but also on mass, damping, and excitation force. The first application as a mass sensor is introduced in this section.

First, we estimate the mass responsivity (R_m) from Eq. 4.24, defined as the shift in drop-frequency with respect to the change in the added center mass as

$$R_m = \lim_{\Delta m_c \to 0} \frac{\Delta f_{drop}}{\Delta m_c} = -\frac{f_o}{2m_o} r_{drop} \left(1 - \frac{r_{drop}^2 - 1}{2r_{drop}^2 - 1}\right) W_1^2(L). \qquad (4.41)$$

Compared with a mass sensor based on a linear resonator, for which the responsivity is $-f_o/2m_o$, the nonlinear resonator utilizing the drop-frequency as an indicator has a better responsivity by a factor of $r_{drop}[1 - (r_{drop}^2 - 1)/(2r_{drop}^2 - 1)]$, when ignoring the term $W_1^2(L)$ and if $r_{drop} \geq 1.618$. The mass responsivities for three representative doubly clamped beams with E = 100 GPa and ρ = 2600 kg/m^3, and a single wall CNT beam with E = 1 TPa, for which parameters are listed in the inset table, are plotted in Fig. 4.16 as a function of the normalized frequency f_{drop}/f_o. The value at $f_{drop}/f_o = 1$ indicates the responsivity of a linear resonator. It is apparent that the responsivity is enhanced not only by considering a nonlinear resonator with smaller intrinsic mass and higher resonant frequency, but also by increasing the ratio of the drop-frequency over the natural resonant frequency. This means that the performance of a mass sensor based on a nonlinear nanoresonator can be considerably raised by increasing its resonance bandwidth which, as we will show later, is practically tunable by the nonlinearity.

In order for a nonlinear resonator to have such an intrinsically nonlinear behavior and a highly broadband resonance response, several parameters, including the quality factor, the size of the mechanical beam, and the driving force, should to be optimized to provide a larger value of Γ according to Eq. 4.25. Here, it is noted that the resonance bandwidth can be extended by simply increasing the excitation force, while keeping all other parameters of the resonator fixed. In fact, an indication of the tunability of the bandwidth is that it can be increased up to two orders of magnitude by simply changing the excitation force applied to a nonlinear mechanical nanoresonator.[19]

The mass sensing capability of the nonlinear nanoresonator is evaluated by adding a small platinum (Pt) deposit at the middle of a suspended CNT, as shown in Fig. 4.17a. In this case, the CNT is ~6.0 μm long and ~26 nm in diameter. The Pt deposit at the middle is measured to approximate an ellipsoid from the acquired SEM image and has a size of 190 nm × 90 nm × 50 nm and a volume of 4.5 × 10^5 nm^3. The volume of CNT inside the ellipsoid is subtracted to get the volume of the actual Pt deposit, 3.5 × 10^5 nm^3. Taken the mass density of the bulk platinum, 21g/cm^3, the added mass is estimated to be ~7 fgr. In Fig. 4.17b, the acquired response spectra of this CNT nonlinear nanoresonator before (blue triangle) and after (red circle) depositing the Pt center mass are shown. The added mass caused both a 2.0 MHz shift of the linearized natural frequency, approximately defined as the frequency where the resonance oscillation initiated, and a more significant 7.4 MHz shift of the drop-frequency. The corresponding mass responsivity calculated from the shift in the drop-frequency ($R_{\text{m,nonlinear}}$ = 1.06 Hz/zg) was 3.7 times the one calculated from the linearized natural frequency ($R_{\text{m,linear}}$ = 0.29 Hz/zg). These mass responsivity values compare favorably with our model prediction from which $R_{\text{m,nonlinear}}$ = 2.18 Hz/zg and $R_{\text{m,linear}}$ = 0.60 Hz/zg.

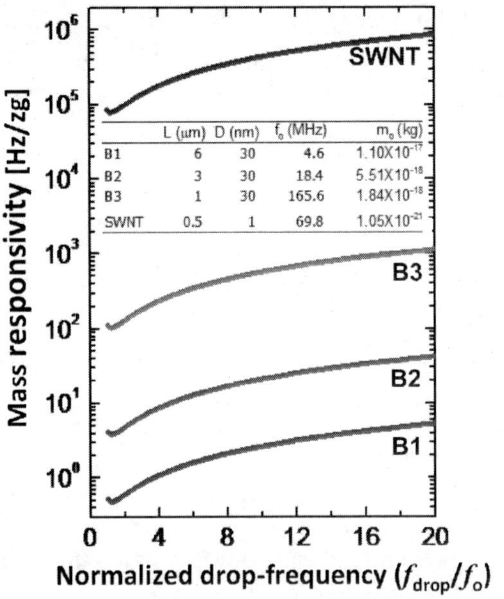

Figure 4.16 Sensing performance of a nonlinear nanoresonator to mass; mass responsivities of four different doubly clamped beams (listed in the inset table) are presented as functions of normalized drop-frequency.[19]

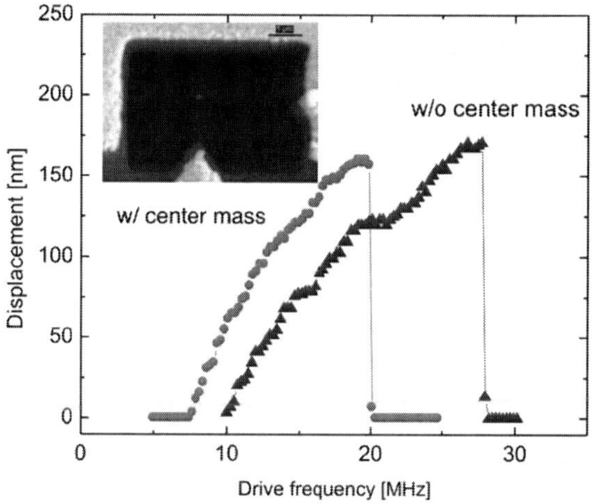

Figure 4.17 Mass sensing with a nonlinear CNT nanoresonator: inset presents SEM image showing the Pt deposit at the middle of a suspended CNT ($2L$ = ~6.0 μm, D = ~26 nm). The acquired response spectrum of this CNT nonlinear nanoresonator before (▲) and after (●) depositing a center mass with the electron beam-induced deposition.[19]

This demonstration of a relatively simple nonlinear nanoresonator incorporating intrinsic geometric nonlinearity offers a model system for expanded studies of the nonlinear resonance behavior, which has been shown to be rich in physics, providing opportunities for practical applications from the macroscale down to the micro/nanoscale. In this study, we show a prototype as a mass sensor, which can be applied to other types of high sensitivity sensing applications. Comparing with the sensing principles applied in nanoscale linear resonance systems, a nonlinear resonance system can exploit the instabilities intrinsically existing within the system that are highly sensitive to external perturbation. The large oscillation amplitude and the sharp transition at those bifurcation points are all highly favorable characteristics from the precision measurement point of view, which can potentially enhance the measurement sensitivity of a practical sensing system; the large oscillation amplitude implies less susceptibility of the resonance system to thermal noise and a sharp transition allows for a narrow measurement bandwidth. However, the use of such a nonlinear system for high sensitivity sensing applications may ultimately rely on our more detailed understanding of the robustness of such transitions related to the instability and the effect of external noises (thermal noises and stochastic perturbations).[32] As shown in previous other studies,[5,26,32,33] the complex dynamical phenomena exhibited by nonlinear systems and the instabilities associated with them are theoretically predictable, and are robust enough for implementation in practical sensing applications.

4.3.2 Damping Sensing

Since the jump phenomena are also closely related with the effective damping in the resonator, the nonlinear system should be able to sense the energy dissipation during the dynamical operation. Using the system introduced in the Section 4.2.3, we studied the change in damping while controlling the vacuum pressure level. We found that in order to account for the experimentally observed mechanical damping behavior in the system, we needed to include geometrically nonlinear dissipation effects, beside the typical linear (viscous) damping present in the nanotube. Indeed, the geometric nonlinearity of the dynamic response amplifies the nonlinear damping associated with the tensile stretching of the nanotube. Here we show that the existence of this nonlinear damping behavior, which is amplitude-dependent, results in the narrowing of the broadband resonance of the overall microcantilever system. The designed microcantilever system can thus both serve as a practical nonlinear resonance system for sensing-related applications, and also work as a testing platform for directly studying the nonlinear damping behavior of individual attached nanostructures.

We acquired a series of response curves at different vacuum pressures and at different excitation amplitudes, and performed fitting procedures similar to those described in Section 4.2.3 in order to extract the corresponding effective Q-factors. The results are summarized in Fig. 4.18a where we plot Q_{eff} as function of the vacuum pressure for four different drive amplitudes V_{pp}. Q_0 is defined in Eq. 4.40 and represents the contribution from linear damping (such as the hydrodynamic viscous and intrinsic thermoelastic damping). Q_{nl}, as defined in Eq. 4.40 represents the contribution from the nonlinear damping and is an amplitude-dependent quantity. As expected, $Q_{\mathrm{nl}}^{-1} << Q_0^{-1}$ when the device is resonating at small amplitudes and the dynamic response is almost purely linear. For example, at an excitation of $V_{\mathrm{pp}} = 0.1$ V, the maximum oscillation amplitude is of the order of a few nanometers even at high vacuum pressure, so Q_{nl}^{-1} is negligible, and $Q_{\mathrm{eff}} \approx Q_0 \cdot Q_0$ as a function of the vacuum pressure acquired at small excitation amplitudes (i.e., $V_{\mathrm{pp}} = 0.1$ V) is plotted in Fig. 4.18a. It is note worthy that the variation of Q_0 reveals the existence of three distinct linear or nonlinear dissipation regimes for varying level of vacuum pressure: Starting from the viscously damped regime at the level of high pressure ($K_{\mathrm{n}} < 0.01$, $P > \sim 100$ Torr) where the linear dissipative effects are dominant, to the intermediate regime ($0.01 \leq K_{\mathrm{n}} \leq 10$), and then transitioning to the free molecular flow regime as the vacuum pressure decreases ($K_{\mathrm{n}} > 10$, $P < \sim 0.1$ Torr).[34,35] At large excitation amplitudes (where the oscillation amplitude of the inner cantilever can reach up to around 60 nm), the effective Q-factor Q_{eff} decreases and saturates toward higher vacuum pressures. This deviation of Q_{eff} from Q_0 is more clearly seen in Fig. 4.18b, where Q_{eff} is plotted against the corresponding Q_0. For reference, we performed the same measurement on the same microcantilever without the nanotube attachment driven with the same set of excitation amplitudes and under the same set of vacuum pressures. The extracted Q_{eff} is plotted against Q_0 as shown in the inset in Fig. 4.18b, exhibiting a completely linear damping response across the full range of the pressure and excitation amplitude variations, and showing no dependence on the oscillation amplitude as expected for a linear resonator.

The results demonstrate the critical importance of incorporating the nonlinear damping contribution into our nonlinear microcantilever resonance system model. We proceed to extract the nonlinear damping coefficient ξ_3 (i.e., Q_{nl}^{-1}) based on Eq. 4.40. Fig. 4.18c shows the dependence of the extracted Q_{nl}^{-1} as a function of the amplitude at the drop frequency (A_{drop}) measured from the acquired response curves at an excitation amplitude of $V_{\mathrm{pp}} = 1.5$ V and at varying vacuum pressures. Recalling that $Q_{\mathrm{nl}}^{-1} = \xi_3/A_{\mathrm{ss}}^2/4$ according to Eq. 4.40, ξ_3 was estimated to be 2.1×10^{-7} by fitting the measured data in Fig. 4.18c. As ξ_3 is directly linked to c_2, which is associated with the intrinsic viscoelastic property of the nanotube, it is reasonably expected that ξ_3 is also an intrinsic property and independent of the dynamics of our microcantilever system. This is confirmed by the excellent fit in Fig. 4.18a–b between the experimental data and the solid fitting curves obtained by assuming the same estimated value of ξ_3, even though the experimental data are acquired at varying vacuum pressures and excitation amplitudes.[4,19,36–47]

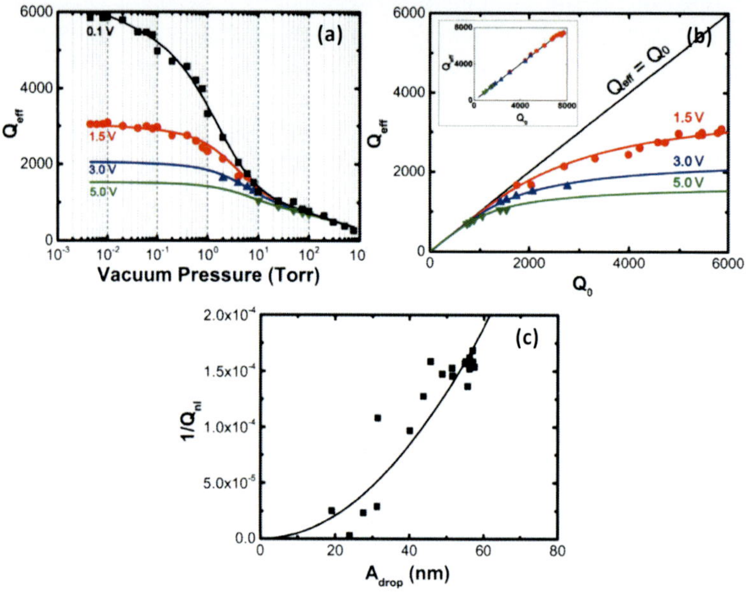

Figure 4.18 Nonlinear damping effect in the microcantilever with the attached BNNT: (a) Q_{eff} as function of vacuum pressure, where at low excitation amplitudes (V_{pp} = 0.1 V) the linear damping effect dominates (denoted by the black line), whereas at higher amplitudes the effective Q-factor shows saturation towards higher vacuum pressures (depicted by the red, blue and green solid lines for V_{pp} = 1.5 V, 3.0 V and 5.0 V, respectively) reflecting the contribution of the nonlinear damping Q_{nl}; (b) Q_{eff} plotted against the corresponding Q_0 at each vacuum pressure; (c) Q_{nl}^{-1} plotted as function of the corresponding oscillation amplitude at the drop frequency (A_{drop}) when V_{pp} = 1.5 V and the vacuum pressure is varied from 8 Torr to 4.5 mTorr.[17]

The saturation of Q_{eff} towards high vacuum pressures can thus be easily understood from the consideration of the linear and nonlinear damping co-existing in our system: Towards high vacuum pressure operation, the linear damping decreases (or equivalently, the energy dissipation by linear damping becomes smaller), which leads to an increase of the oscillation amplitude. However, the increase of the oscillation amplitude aggravates the nonlinear damping effect through its amplitude-dependence. This counterbalance leads to the eventual saturation of Q_{eff} towards high vacuum pressures. This Q-factor saturation is also expected to occur in some recent studies on clamped-clamped nanoresonators, although it was not directly reported in those studies.

4.3.3 Atomic Force Microscopy

The design of the microcantilever-nanotube systems discussed in Sections 4.2.2 and 4.2.3 is based on structural modifications of typical AFM cantilever probes. In this section, we discuss how these nonlinear designs help to improve the performance of current AFM systems.[48]

The dynamic motion of either design when in interaction with a sample as in a typical dynamic AFM experiment can be modeled, as per previous sections, as a harmonic oscillator with a mass (m), a vertical spring (k_1) parallel to a viscous damper (c_1), and a horizontal spring (k_2) giving rise to geometrically nonlinear effects. The equation of motion in the non-dimensional form is then

$$\tilde{z}'' + \frac{1}{Q}\tilde{z}' + \tilde{z}' + \alpha\tilde{z}^3 = q_{\text{o}}\cos\Omega\tau + q_{\text{ts}}. \tag{4.42}$$

with the normalizations $\tilde{z} = z/Z_{\text{c}}$ and $\tau = \sqrt{k_1/mt} = \omega_{\text{o}}t$, where Z_{c} is the static equilibrium separation between the tip and the sample, $Q = (m\omega_{\text{o}})/c_1$, $\alpha = 1/2(k_2/k_1)(Z_{\text{c}}/L_2)$, $\Omega = \omega/\omega_{\text{o}}$, $q_{\text{o}} = u_{\text{o}}/Z_{\text{c}}$, and $q_{\text{ts}} = F_{\text{ts}}/(k_1 Z_{\text{c}})$, where F_{ts} represents the tip–sample interaction. The tip–sample interaction is modeled according to the Derjaguin–Müller–Toporov (DMT) contact model[16] and normalized as

$$q_{\text{ts}}[\tilde{z}] = \begin{cases} -\dfrac{HR}{6k_1 Z_{\text{c}}^3(1+\tilde{z})^2} & (1+\tilde{z} > a_{\text{o}}/Z_{\text{c}}) \\ -\dfrac{HR}{6k_1 a_0^2 Z_{\text{c}}} + \dfrac{4}{3k_1}E^*\sqrt{RZ_{\text{c}}}(a_{\text{o}}/Z_{\text{c}} - 1 - \tilde{z})^{3/2} & (1+\tilde{z} \le a_{\text{o}}/Z_{\text{c}}) \end{cases} \tag{4.43}$$

Where H is the Hamaker constant, R is the AFM tip radius of curvature, E^* is the effective elastic modulus of the tip–sample interaction, and a_0 is the intermolecular distance.

We applied the method of averaging[1] to find an analytical solution of this problem. Correct to first-order approximation, we consider a solution in the form $z(\tau) = A(\tau)\cos(\Omega\tau + \phi(\tau)) = A\cos\theta$. Following the method of averaging we derive two (slow flow) modulation equations governing the modulations of the amplitude A and phase ϕ in this solution.[1] The steady-state solutions are then determined by setting $A' = \theta' = 0$, leading to the steady-state values

$$A = \frac{q_o}{\sqrt{(\Omega_e^2 - \Omega^2) + (\Omega/Q)^2}} \tag{4.44}$$

$$\tan\phi = \frac{\Omega}{Q(\Omega^2 - \Omega_e^2)} \tag{4.45}$$

with amplitude-dependent effective resonant frequency

$$\Omega_e^2(A) = 1 + \frac{3\alpha}{4}\Omega A^2 - \frac{1}{\pi A}\int_0^{2\pi} q_{ts}[A\cos\theta]\cos\theta d\theta. \tag{4.46}$$

For the case without the horizontal spring, i.e., for a conventional linear cantilever, the solutions are obtained by simply setting α to be zero. When more than one solution exists, the eigenvalues of the variational matrices associated with the modulation equations for $A(\tau)$ and $\theta(\tau)$ close to these steady-state equilibrium states. A solution is stable, namely physically realizable, only if both eigenvalues of the variational matrices have non-positive real parts.

The frequency spectra of the conventional linear cantilever and the nonlinear cantilever when in free oscillation (dashed curves) or in interaction with a sample surface (solid curves) are compared in Fig. 4.19 using the parameters listed in Table 4.1. Note that the analytically obtained spectra applying the averaging method show good agreement with numerically derived steady-state solutions using direct numerical integrations based on the Runge-Kutta method (marked with ■ in the plots). As seen in Fig. 4.19a–b, in the absence of the tip–sample interaction (i.e., during free oscillation), the oscillation of the linear cantilever is theoretically stable on both sides of the resonant frequency with the resonance having a narrow bandwidth determined by the quality factor (typically around several hundred) for a typical cantilever probe used in an AFM system. As the cantilever is driven to higher amplitudes and interacts with the sample surface during the dynamic AFM imaging process, the attractive (due to van der Waals or meniscus interactions) and repulsive (due to tip–sample contact) tip–sample interactions bend the stable branches, producing more than one stable solution for the oscillation of the system within a certain frequency range as shown in Fig. 4.19a–b. On the left-hand side of the resonant frequency, the multi-valued region is relatively small compared with that on the right-hand side. Within each region, there exist two stable branches and one unstable branch (marked with crosses). The existence of bi-stable solutions can lead the dynamic motion of the cantilever to switch unexpectedly between these two distinct branches. The bi-stable region on the right-hand side is typically extended out over a significant frequency range overlapping the originally stable branch due to the substantial repulsive interaction, which is one of the reasons for selecting an operating frequency that is slightly lower than the resonant frequency in the practical operation of dynamic mode AFM.

Table 4.1 Parameters used in this study

Dimensionless parameters		Tip-sample interaction parameters	
Q	300	H	3e-19 J
α	0.3	R	30 nm
q_0	0.005	k_1	10 N/m
		Z_c	100 nm
		a_0	0.2 nm
		E^*	1 GPa

Introducing the geometric nonlinearity into this linear cantilever system, the corresponding frequency responses in Fig. 4.19c–d show totally different dynamics. Without the tip–sample interaction (dashed line), the so-called backbone curve is bent toward the right-hand side due to the hardening effect provided by the constraint of the added nanowire, and the

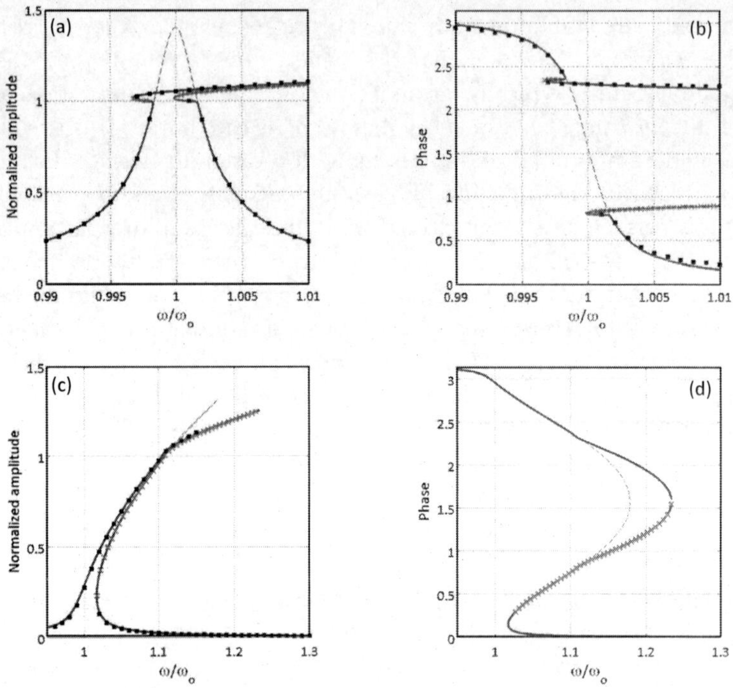

Figure 4.19 Analytically obtained amplitude and phase spectra of the linear (a-b), and nonlinear (c-d) cantilevers using the averaging method for parameters listed in Table 4.1; numerically calculated steady state solutions are marked with squares and show good agreement with theoretical predictions; dashed/solid curves indicate dynamics without/with tip–sample interactions, respectively, and unstable solution branches are marked with crosses.[48]

frequency spectrum has a broad multi-valued region. The response of this nonlinear system follows the upper stable branch (the unstable branch is again marked with crosses) and remains in resonance up to the maximum possible amplitude as the frequency is swept upward, followed by a sudden jump to the lower, almost non-resonance stable branch at the drop-jump frequency. This jump phenomenon is well-known in many nonlinear systems, in which the drop-jump frequency is dependent on the driving force (the driving amplitude of the piezoelectric element in a typical AFM cantilever system) and damping coefficient of the dynamic system.[1] The near non-resonance lower stable branch, though coexisting, is practically impossible to excite once the upper branch is triggered except at the drop-jump frequency (discussed further in the next section). When the tip–sample interaction is applied to this nonlinear system, the influence of attractive interaction, forcing an upper stable branch to bend toward the left, is counterbalanced by the hardening effect in the nonlinear system as shown in Fig. 4.19c. Only if the tip–sample interaction involves a substantial attraction can instability be introduced, and even then the frequency range of this instability is expected to be much narrower than that in the linear resonance case, and is certainly significantly narrower compared to the overall broad resonance bandwidth of this nonlinear system. Equally important, the wide bi-stable region on the right-hand side of the resonant frequency in the linear system disappears in this nonlinear system because the branch on the right-hand side of the backbone curve is intrinsically unstable and physically unattainable, in contrast with the linear system. Note that the discrepancy of the predicted drop-jump frequency between analytical and numerical (marked with squares) responses is due to the first-order approximation of our solution.

To more clearly visualize the stability exhibited by the non-linear cantilever system in the dynamic mode AFM, we investigate the domains of attraction for both the linear cantilever system and the nonlinear cantilever system in interaction with a sample surface in Fig. 4.20. The global stability of the response at various initial conditions was determined by computing the time-varying trajectories numerically and superimposing the solutions for several representative frequencies at which multiple solutions exist. Fig. 4.20a–b describe the phase planes of the linear cantilever system with the tip–sample interaction at 0.998 and 1.001 of normalized frequency (Ω), respectively. There are two domains of attraction manifested by two stable foci, P1 and P3, separated by two separatrix branches emerging from an unstable saddle point, P2. Note that the domain of attraction of P3 is surrounded by the domain of attraction of P1, overlapping on P2 in phase throughout the amplitude trajectories. At the controlled steady-state amplitude (the set-point amplitude in AFM dynamic mode imaging), a sudden change in phase due to external perturbation can easily lead to an unexpected jump to another stable solution, which introduces the well-known bi-stability in dynamic imaging with the AFM. However, in the phase plane of the nonlinear cantilever system illustrated in Fig. 4.20c, two stable foci, P1 and P3, still exist, but values of which in phase and amplitude are far apart from each other. Moreover, the domains of attraction of P3 occupy the whole phase range at its steady-state amplitude; consequently, instability is not initiated by a change in phase in contrast to the linear system, giving a significant advantage when implementing the nonlinear cantilever system in dynamic mode AFM of global stability of the dynamic response of the imaging system.

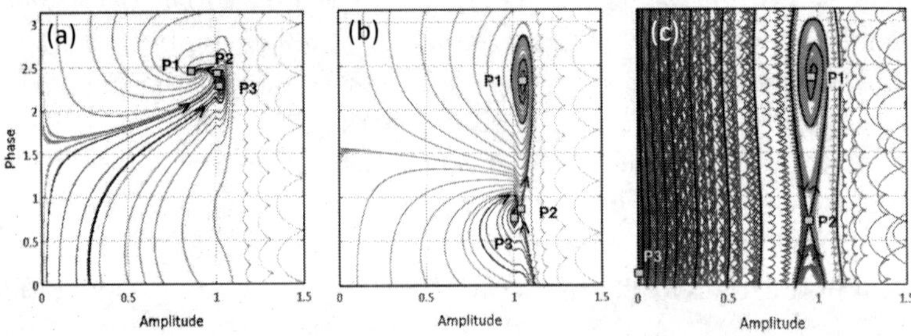

Figure 4.20 Sets of numerically computed time varying trajectories (in basins of attraction of the stable steady state solutions) using various initial conditions plotted in the phase plane: (a) Linear system, ω/ω_0 = 0.998; (b) linear system, ω/ω_0 = 1.001; and (c) nonlinear system, ω/ω_0 = 1.1.[48]

One more significant advantage in adopting a nonlinear cantilever system lies in the greatly broadened resonance bandwidth operationally useful for AFM imaging. We define an operational frequency bandwidth as the range of frequency at which the free oscillating amplitude is large enough for the AFM tip to make contact with the sample (i.e., the normalized amplitude is greater than unity), and the amplitude–frequency slope is positive as required by the built-in negative feedback control used in typical dynamic mode AFM. For the linear cantilever system, this bandwidth (normalized by the resonant frequency), defined by the frequency at which the normalized free oscillation amplitude is unity (marked by the arrowed line in the inset of Fig. 4.21a), is plotted in Fig. 4.21a. The bandwidth is mainly a function of the normalized driving force (q_0) exerted on the system and for a lesser extent the quality factor (Q); here, the normalized driving force $q_o = u_o/Z_c$, u_o is the oscillating amplitude of the excitation piezo element driving the AFM cantilever probe and Z_c is the tip–sample separation as defined before. The operational bandwidth of the linear cantilever system is not able to exceed 20% of the fundamental resonant frequency of the cantilever even at a practically large normalized driving force of 0.3, i.e., the amplitude of the excitation piezo element is 30% of the tip–sample separation. For comparison, the operational bandwidth of the nonlinear cantilever system, defined from the frequency at the unity amplitude to the drop-jump frequency, is calculated analytically and plotted in Fig. 4.21b. Different from the linear cantilever system, the bandwidth is critically dependent on both Q and q_o and, in addition, on the nonlinear constant, α, as the drop-jump frequency in a nonlinear dynamic system is closely related to the damping in the system. The operational bandwidth is found to reach several times the linearized natural frequency even at a relatively small normalized driving force of less than 0.1 for a cantilever having a typical Q-factor of ~300.

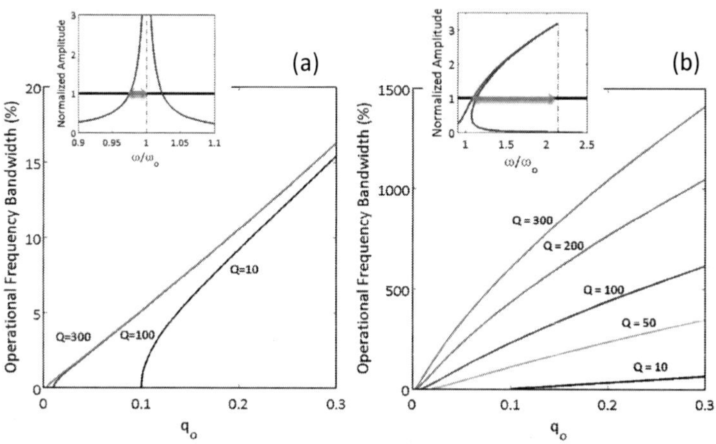

Figure 4.21 Operational frequency bandwidth, i.e., frequency range at which the free oscillation amplitude is large enough for the AFM tip to make contact with the sample, with the amplitude-frequency slope being positive, for the linear cantilever system (a), and for the nonlinear cantilever system (b).[48]

The inherent broadband resonance and the localization-governed stability of such a nonlinear cantilever system incorporating essential geometric nonlinearity provide new practical strategies for enabling the broad use of AFM in the field of nanospectroscopy, where the high force or force gradient sensitivity of the micro-cantilever is used to discern specific types of tip–sample interactions related to the mechanical properties and chemical composition of the sample surface. For example, the frequency-dependent rate of energy dissipation in the tip–sample interaction during the dynamic AFM can be sensitively acquired by operating the AFM at various frequencies in the broad resonance band afforded by the nonlinear AFM cantilever system, which, in turn, can be used to resolve the mechanical heterogeneities in the sample surface beyond the elastic property differences. The fact that the nonlinear AFM cantilever system still operates in the resonance mode with high-Q ensures the amplified sensitivity of sensing across a broad frequency spectrum, which is different from the linear AFM system where only within a narrow frequency band such resonant sensing is possible, and is also different from the force modulation AFM-based technique where the AFM is operated in the off-resonance state.

4.4 Concluding Remarks

The research work reviewed herein was based on the concept of implementing intentional strong nonlinearity for design of micro/nanoscale resonators. We have designed nonlinear micro/nanomechanical resonators and studied their dynamics and applications both in analytical and experimental ways. In the applications discussed in this section we were mainly concerned with the implementation of geometric/kinematic nonlinearities induced by midplane stretching of elastic elements, leading to Duffing or Duffing-like reduced order oscillators with cubic nonlinearities; however, other ways of implementing strong nonlinearities are also possible. Such nonlinear systems present unique dynamics, which cannot be realized in linear systems; e.g., broad resonance bandwidths, tunability, sharp transition between stable co-existing solutions, rich dynamics, etc. We showed that it is possible to carefully harness those nonlinear characteristics in practical applications; e.g., to improve the sensing capacity and enhance AFM performance. Further understanding of the nonlinear dynamics of micro/nanoresonators will have a significant payoff, providing designers the opportunity to embrace nonlinear behavior in the development of novel systems in ways that are not realizable in current linear designs. Below, we briefly discuss several specific topics having great potentials to extend this new field of study.

Nonlinear Atomic Force Microscopy

As shown in our previous work, controlled integration of nonlinearity in AFM systems can improve their performance. Another design of nonlinear AFM cantilever system is also possible through application of the concepts of nonlinear targeted energy transfer,[20] a new field of study developed during the last decade. This can be achieved by modifying conventional AFM cantilevers by attaching essentially nonlinear components, with the aim of drastically enhancing the generation of higher frequency harmonics in the response, which, in turn, could be harnessed for higher sensitivity in characterizing material properties and imaging topological features.

Broadband Energy Harvesting

The integration of individual nanodevices into a fully functioning nanosystem is an emerging trend, which drives demand for energy harvesting in order to attain independent and sustainable operation. Scavenging energy from various environmental sources such as solar, magnetic, chemical, thermal, wind, and vibration is broadly investigated. Here harvesting mechanical energy from vibration, which can achieve maintenance-free characteristics, is of particular interest. The most desirable operational condition is through motion amplification gained by mechanical resonance. This, however, is technically challenging due to the high resonant frequency and narrow bandwidth of current micro/nanodevices, originating from the intrinsic traits of small mass and damping. Thus, it is difficult to accommodate environmental sources with time-varying low frequencies. Properly designed nonlinear systems employing essentially nonlinear components, having no preferential resonance frequencies, and with capacity for broadband resonance, can successfully resolve such difficulties. Such work holds promise for designing efficient nonlinear harvesting systems with enhanced performance.

Fundamental Study of Nonlinear Dynamics

Micro/nanoresonators are effective tools for studying fundamental nonlinear dynamics experimentally due to the ease of realization of nonlinear behavior. Taking advantage of exceptional properties of nanomaterials and/or readily manageable fabrication of resonating systems, we can find extensive examples to be investigated. One such example is the study of multi-walled CNT, whose unique structure (i.e., assemblies of concentric cylindrical shells coupled by van der Waals forces) provides inherently rich nonlinear dynamical behavior. Also, multi-coupled resonating systems, easily fabricated with controllability of frequencies (i.e., identical, closely packed, and commensurable) and/or nonlinearities, will bring unprecedented opportunity to study nontrivial dynamics, including internal resonances, chaos, and directional energy transfers. These types of fundamental studies can be the basis for many practical applications of intentionally nonlinear resonators at the micro/nanoscales.

Acknowledgment

This work was supported in part by National Science Foundation (NSF) Grant CMMI-100615. This support is gratefully acknowledged.

References

1. Lifshitz R., Cross M. C. *Nonlinear Dynamics of Nanomechanical and Micromechanical Resonators*. Wiley-VCH Verlag GmbH & Co. KGaA: 1–52.
2. Rhoads J., Shaw S., Turner K. L. (2010). Nonlinear Dynamics and Its Applications in Micro- and Nanoresonators. *J. Dyn. Syst. Meas. Control.*, **132**(3), 034001.
3. Husain A., Hone J., Postma H. W. C., et al. (2003). Nanowire-based very-high-frequency electromechanical resonator. *Appl. Phys. Lett.*, **83**(6), 1240–1242.

4. Sazonova V., Yaish Y., Ustunel H., Roundy D., Arias T. A., McEuen P. L. (2004). A tunable carbon nanotube electromechanical oscillator. *Nature*, **431**(7006), 284–287.

5. Kozinsky I., Postma H. W. C., Bargatin I., Roukes M. L. (2006). Tuning nonlinearity, dynamic range, and frequency of nanomechanical resonators. *Appl. Phys. Lett.*, **88**(25), 253101.

6. He R., Feng X. L., Roukes M. L., Yang P. (2008). Self-transducing silicon nanowire electromechanical systems at room temperature. *Nano Lett.*, **8**(6), 1756–1761.

7. Karabalin R. B., Feng X. L., Roukes M. L. (2009). Parametric nanomechanical amplification at very high frequency. *Nano Lett.*, **9**(9), 3116–3123.

8. Patil S., Dharmadhikari C. V. (2003). Small signal amplification using parametric resonance in NcAFM imaging. *Appl. Surf. Sci.*, **217**(1–4), 7–15.

9. Rhoads J., Shaw S., Turner K., Baskaran R. (2005). Tunable microelectromechanical filters that exploit parametric resonance. *J. Vib. Acoust.*, **127**, 423–430.

10. Kacem N., Hentz S. (2009). Bifurcation topology tuning of a mixed behavior in nonlinear micromechanical resonators. *Appl. Phys. Lett.*, **95**(18), 183104.

11. Kacem N., Arcamone J., Perez-Murano F., Hentz S. (2010). Dynamic range enhancement of nonlinear nanomechanical resonant cantilevers for highly sensitive NEMS gas/mass sensor applications. *J. Micromech. Microeng.*, **20**(4), 045023.

12. Mestrom R., Fey R., Phan K. (2009). Experimental validation of hardening and softening resonances in a clamped-clamped beam MEMS resonator. *Procedia. Chem.*, **1**, 812–815.

13. Elshurafa A. M., Khirallah K., Tawfik H. H., Emira A., Abdel Aziz A. K. S., Sedky S. M. (2011). Nonlinear dynamics of spring softening and hardening in folded-MEMS comb drive resonators. *J. Microelectromech. Syst.*, **20**(4), 943–958.

14. Mestrom R. M. C., Fey R. H. B., Phan K. L., (2010). Nijmeijer H. Simulations and experiments of hardening and softening resonances in a clamped–clamped beam MEMS resonator. *Sens Actuator A-Phys*, **162**(2), 225–234.

15. Raman A., Reifenberger R., Melcher J., Tung R. (2009). *Noncontact Atomic Force Microscopy*. Berlin, Heidelberg: Springer Berlin Heidelberg, 361–395.

16. Derjaguin B. V., Muller V. M., P T. Y. (1975). Effect of contact deformations on the adhesion of particles. *J. Colloid Interface Sci.*, **53**(2), 314–326.
17. Jeong B., Cho H., Yu M.-F., Vakakis A. F., McFarland D. M., Bergman L. A. (2013). Modeling and measurement of geometrically nonlinear damping in a microcantilever–nanotube system. *ACS Nano.*, **7**(10), 8547–8553.
18. Eichler A., Moser J., Chaste J., Zdrojek M., Wilson-Rae I., Bachtold A. (2011). Nonlinear damping in mechanical resonators made from carbon nanotubes and graphene. *Nat. Nanotech.*, **6**(6), 339–342.
19. Cho H., Yu M.-F., Vakakis A. F., Bergman L. A., McFarland D. M. (2010). Tunable, Broadband nonlinear nanomechanical resonator. *Nano Lett.*, **10**(5), 1793–1798.
20. Vakakis A. F., Gendelman O. V., Bergman L. A., McFarland D. M., Kerschen G., Lee Y. S. *Nonlinear Targeted Energy Transfer in Mechanical and Structural Systems.* Springer Netherlands, 2009.
21. Nayfeh A. H., Mook D. T. *Nonlinear Oscillations.* Wiley, 1995.
22. Léonard F. *Physics of Carbon Nanotube Devices.* William Andrew Publishing, 2009.
23. Yoon J. (2013). Vibration of an embedded multiwall carbon nanotube. *Compos. Sci. Technol.*, **63**(11), 1533–1542.
24. Yan Y., Zhang L. X., Wang W. Q. (2008). Dynamical mode transitions of simply supported double-walled carbon nanotubes based on an elastic shell model. *J. Appl. Phys.*, **103**(11), 113523.
25. Vakakis A. F. (1992). Fundamental and subharmonic resonances in a system with a "1–1" internal resonance. *Nonlinear Dyn.*, **3**, 123–143.
26. Karabalin R. B., Cross M. C., Roukes M. L. (2009). Nonlinear dynamics and chaos in two coupled nanomechanical resonators. *Phys. Rev. B*, **79**(16), 165309.
27. Cho H., Jeong B., Yu M.-F., Vakakis A. F., McFarland D. M., Bergman L. A. (2012). Nonlinear hardening and softening resonances in micromechanical cantilever-nanotube systems originated from nanoscale geometric nonlinearities. *Int. J. Solids Struct.*, **49**(15–16), 1–7.
28. Yum K., Wang Z., Suryavanshi A. P., Yu M.-F. (2004). Experimental measurement and model analysis of damping effect in nanoscale mechanical beam resonators in air. *J. Appl. Phys.*, **96**(7), 3933–3938.
29. Jiang H., Yu M.-F., Liu B., Huang Y. (2004). Intrinsic energy loss mechanisms in a cantilevered carbon nanotube beam oscillator. *Phys. Rev. Lett.*, **93**(18), 185501.

30. Yap H. W., Lakes R. S., Carpick R. W. (2007). Mechanical instabilities of individual multiwalled carbon nanotubes under cyclic axial compression. *Nano Lett.,* **7**(5), 1149–1154.

31. Qian D., Wagner G. J., Liu W. K., Yu M.-F., Ruoff R. S. (2002). Mechanics of carbon nanotubes. *Appl. Mech. Rev.*, **55**(6), 495–533.

32. Aldridge J. S., Cleland A. N. (2005). Noise-enabled precision measurements of a duffing nanomechanical resonator. *Phys. Rev. Lett.*, **94**(15), 156403.

33. Kozinsky I., Postma H. W. C., Kogan O., Husain A., Roukes M. L. (2007). Basins of attraction of a nonlinear nanomechanical resonator. *Phys. Rev. Lett.*, **99**(20), 207201.

34. Bhiladvala R. B., Wang Z. J. (2004). Effect of fluids on the Q factor and resonance frequency of oscillating micrometer and nanometer scale beams. *Phys. Rev. E*, **69**(3), 1–5.

35. Verbridge S. S., Ilic B., Craighead H. G., Parpia J. M. (2008). Size and frequency dependent gas damping of nanomechanical resonators. *Appl. Phys. Lett.*, **93**(1), 013101.

36. Wang Z., Wei J., Morse P., Dash J. G., Vilches O. E., Cobden D. H. (2010). Phase transitions of adsorbed atoms on the surface of a carbon nanotube. *Science*, **327**(5965), 552–555.

37. Jensen K., Weldon J., Garcia H., Zettl A. (2007). Nanotube radio. *Nano Lett.,* **7**(11), 3508–3511.

38. Gouttenoire V., Barois T., Perisanu S., et al. (2010). Digital and FM Demodulation of a doubly clamped single-walled carbon-nanotube oscillator: Towards a nanotube cell phone. *Small*, **6**(9), 1060–1065.

39. Lassagne B., Tarakanov Y., Kinaret J., Garcia-Sanchez D., Bachtold A. (2009). Coupling mechanics to charge transport in carbon nanotube mechanical resonators. *Science*, **325**(5944), 1107–1110.

40. Steele G. A., Huttel A. K., Witkamp B., et al. (2009). Strong coupling between single-electron tunneling and nanomechanical motion. *Science*, **325**(5944), 1103–1107.

41. Laird E. A., Pei F., Tang W., Steele G. A., (2011). Kouwenhoven LP. A high quality factor carbon nanotube mechanical resonator at 39 GHz. *Nano Lett.,* **12**(1), 193–197.

42. Witkamp B., Poot M., van der Zant H. S. J. (2006). Bending-mode vibration of a suspended nanotube resonator. *Nano Lett.,* **6**(12), 2904–2908.

43. Hüttel A. K., Steele G. A., Witkamp B., et al. (2013). Carbon nanotubes as ultrahigh quality factor mechanical resonators. *Nano Lett.,* **9**(7), 2547–2552.

44. Garcia-Sanchez D., van der Zande A. M., Paulo A. S., Lassagne B., McEuen P. L., Bachtold A. (2008). Imaging mechanical vibrations in suspended graphene sheets. *Nano Lett.*, **8**(5), 1399–1403.

45. Shivaraman S., Barton R. A., Yu X., et al. (2010). Free-standing epitaxial graphene. *Nano Lett.*, **9**(9), 3100–3105.

46. Chen C., Rosenblatt S., Bolotin K. I., et al. (2009). Performance of monolayer graphene nanomechanical resonators with electrical readout. *Nat. Nanotech.*, **4**(12), 861–867.

47. Singh V., Sengupta S., Solanki H. S., et al. (2010). Probing thermal expansion of graphene and modal dispersion at low-temperature using graphene nanoelectromechanical systems resonators. *Nanotechnology*, **21**(16), 165204.

48. Cho H., Yu M.-F., Vakakis A. F., Bergman L. A., McFarland D. M. (2012). Dynamics of microcantilever integrated with geometric nonlinearity for stable and broadband nonlinear atomic force microscopy. *Surf. Sci.*, **606**(17–18), L74–L78.

Part 3

Applications of Nanocantilever Beams

Chapter 5

Electromechanical Properties and Applications of Carbon Nanotube Nanocantilevers

Changhong Ke

Department of Mechanical Engineering,
State University of New York at Binghamton,
Binghamton, NY 13902, USA

cke@binghamton.edu

5.1 Introduction

Nanoelectromechanical systems (NEMS) are made of electromechanical components with characteristic dimensions within the range of a few to hundreds of nanometers. NEMS offer unprecedented and intriguing properties in the fields of electronic computing and sensing where technological demands require both substantially reduced component size and cost and dramatically increasing device performance and functionality. For instance, NEMS devices can have fundamental frequencies in microwave range (~100 GH)[1]; mechanical quality factors in the tens of thousands,[2,3] meaning low energy dissipation; active mass in the femtogram range[4]; force sensitivity at the attonewton

Nanocantilever Beams: Modeling, Fabrication, and Applications
Edited by Ioana Voiculescu and Mona Zaghloul

ISBN 978-981-4613-23-1 (Hardcover), 978-981-4613-24-8 (eBook)
www.panstanford.com

level[5]; mass sensitivity up to yoctogram (yg, 10^{-24} g) level,[6,7] which essentially enables mass spectroscopy with single-atom resolution; heat capacities far below a "yoctocalorie"[8]; power consumption in the order of 10 attowatts[9]; and extreme high integration level, approaching 10^{12} elements per square centimeter.[1] All these distinguished properties of NEMS devices originate from the exploration of the nanoscale size effect and deviate greatly from their predecessor microelectromechanical systems (MEMS), paving the way to a variety of applications, such as memory elements, force/mass sensors, chemical sensors, and biological sensors.

The interesting properties of NEMS devices typically arise from the behaviors of their active components, which, in most cases, are in the forms of cantilevers or doubly clamped beams with dimensions at the nanometer scale. Carbon nanotubes (CNT), a cylindrical type of nearly one-dimensional nanostructures with high aspect ratio and perfectly terminated surfaces, have been considered ideal building blocks for NEMS not only because of their distinguished electrical and mechanical properties and chemical stability, but also due to the significant advances achieved during the past decades with respect to the growth, manipulation, fabrication, and characterization of nanotube structures. A large number of CNT-based NEMS devices have been proposed and demonstrated so far. The family of cantilevered CNT-based NEMS includes nanoswitches,[10,11] tweezers,[12,13] relays,[14] radios,[15] resonators,[16,17] just to name a few.

The unique structural architectures and properties of NEMS devices pose significant challenges in the device design, manufacturing, and characterization, which need to be solved before mass commercialization of NEMS can occur. The optimal design of NEMS devices requires simple and reliable electromechanical models. Atomic simulations of NEMS typically require prohibitively expensive computational resources or exceed the current computational power. Simplified models based on continuum theories need to be verified experimentally before they can be confidently used for the design of NEMS. On the other hand, experimental characterization at the nanoscale presents many challenges. For instance, manipulation of nano-objects is difficult and lacks reproducibility, real time observation of the nanoscale mechanical response requires the use of in situ electron microscopy techniques, and electrical measurements require excellent control of noise and parasitic effects.

The purpose of this chapter is to provide a review of the recent advances in the electromechanical modeling and characterization of CNT nanocantilevers and their electromechanical applications. This chapter is organized as follows: in Section 5.2, we briefly review the basic structural and physical properties of CNTs; in Section 5.3, we present an overview of the electromechanical modeling and characterization of CNT nanocantilevers; in Section 5.4, we present two representative types of cantilevered CNT-based NEMS devices.

5.2 Carbon Nanotubes

Carbon nanotubes, first discovered by Sumio Iijima in 1991,[18] are a subset of the family of fullerene structures and are the most exciting and researched nanotube structures due to their exceptional physical properties and application prospects. The physical properties of CNTs depend on the atomic architecture, tube diameter, and length. From the structural perspective, CNTs are composed of seamless covalent C–C hexagonal networks and are formed through rolling graphene sheets into cylinders, as shown in Fig. 5.1a. The rolling direction of the graphene nanosheet is denoted by the chirality vector (n, m). The indexes n and m of the chirality vector denote the number of unit vectors along the two directions in the honeycomb lattice of graphene nanosheet. The nanotubes are called "*zig-zag*" if $m = 0$, or "*arm-chair*" if $m = n$. CNTs can crystallize in both single-walled and multi-walled nanophases. Multi-walled carbon nanotubes (MWCNTs) are simply composed of multiple concentric single-walled carbon nanotubes (SWCNTs), as illustrated in Fig. 5.1b. The spacing between the neighboring graphite layers in MWCNTs is ~0.335 nm. Two adjacent layers interact with each other via van der Waals interactions. Figure 5.1c shows two representative high-resolution transmission electron microscopy (HRTEM) images of SWCNTs (left) and MWCNTs (right), respectively.[19] The methods to synthesize CNTs include electric arc-discharge,[20,21] laser ablation,[22] and catalytic chemical vapor deposition (CVD) methods.[23] It is noted that the CVD method is capable of producing directionally aligned CNTs by applying external electric fields.[24]

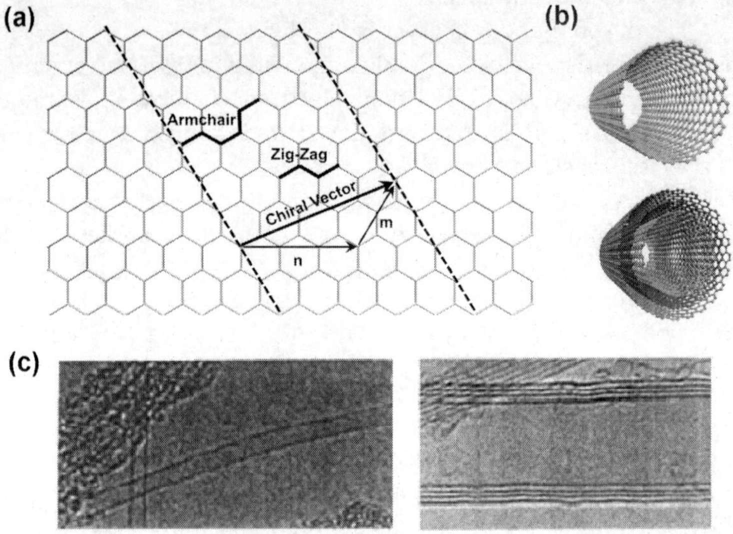

Figure 5.1 (a) Graphene sheet rolling direction and nanotube chirality; (b) Schematic drawings of single- and multi-walled carbon nanotubes (SWCNTs and MWCNTs); (c) High-resolution transmission electron microscopy images of typical SWCNTs and MWCNTs. Reprinted with permission from [19], P. M. Ajayan, *Chem. Rev.*, **99**, 1787 (1999). © 1999, American Chemical Society.

CNTs possess many extraordinary material properties. For instance, CNTs have excellent mechanical and electrical properties, thermal conductivity, and chemical stability. The strength of CNTs has been found to be of the order of 10–100 GPa with Young's modulus of about 1–1.2 TPa and tensile failure strain of up to 30%.[25] The thermal conductivity of CNTs is nearly 3500 $W \cdot m^{-1} \cdot K^{-1}$.[26] SWCNTs are either metallic or semi-conductive depending on the tube chirality, while MWCNTs typically show metallic behaviors. It is reported that CNTs have the capability to carry huge electrical currents and the current densities as high as 10^9 A/cm^2 can be sustained.[27] More detailed mechanical and electrical properties of CNTs can be found in the review papers contributed by Qian et al.[25] and McEuen et al.,[28] respectively. Besides, studies of the coupled electromechanical properties of

CNTs show that their electrical properties are sensitive to the structure variation and can be changed dramatically due to the change of the atomic bonds induced by mechanical deformations. Transitions of the electrical properties of CNTs from semiconductor-to-metal or metal-to-semiconductor when subjected to mechanical deformations have been documented.[29–37]

5.3 Electromechanical Properties of Carbon Nanotube Cantilevers: Modeling and Experiments

5.3.1 Electromechanical Modeling

A generalized configuration of cantilevered nanotube-based electromechanical devices is illustrated in Fig. 5.2. The mechanical motion of the nanostructure is actuated by its electrostatic and van der Waals interactions with the substrate. In the following sections, we present an overview of the continuum theories for the involved energy domains in the equilibrium of the nanostructure and the governing equations for the deflection of the nanotube nanocantilever in both small and finite deformation regimes.

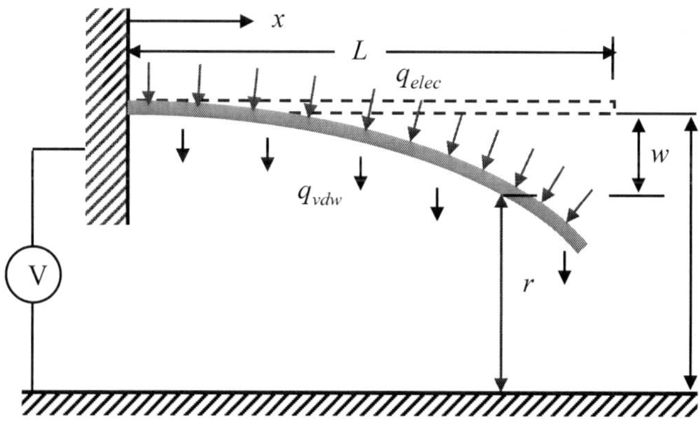

Figure 5.2 Schematic of the deformed configuration of a nanotube cantilever device subjected to electrostatic and van der Waals forces. Reprinted with permission from [50], C.-H. Ke et al., *J. Appl. Mech.*, **72**, 726 (2005). © 2005, American Society of Mechanical Engineers.

5.3.1.1 Van der Waals interactions

The van der Waals (vdw) energy originates from the interaction between atoms. The Lennard-Jones (L-J) potential is one of the most commonly used models to describe the vdw interaction.[38] The L-J potential includes both the attractive and repulsive terms. The L-J potential between two atoms i and j is given by $\phi_{ij} = \frac{C_{12}}{r_{ij}^{12}} - \frac{C_6}{r_{ij}^6}$, in which r_{ij} is the distance between atoms i and j and C_6 and C_{12} are attractive and repulsive constants, respectively. For the carbon-carbon interaction, C_6 = 15.2 eVÅ^6 and C_{12} = 24.1 keVÅ^{12} and the equilibrium spacing r_0 = 3.414 Å.[39] The repulsive component of the L-J potential decays extremely fast and plays an important role only when the distance is close to or smaller than r_0. The total vdw energy can be computed by means of a pair-wise summation over all the atoms. Because the computational cost is proportional to the square of the number of atoms in the system, for nanodevices with millions of atoms, this technique is prohibitively expensive. Instead, a continuum model was established to calculate the vdw energy through a double volume integral of the L-J potential,[40,41]

$$E_{\text{vdw}} = \int_{v_1}\int_{v_2} n_1 n_2 \left(\frac{C_{12}}{r^{12}(v_1, v_2)} - \frac{C_6}{r^6(v_1, v_2)} \right) dv_1 dv_2, \tag{5.1}$$

where v_1 and v_2 represent the two domains of integration, and n_1 and n_2 are the densities of atoms in the domains v_1 and v_2, respectively; $r(v_1, v_2)$ is the distance between any two points on v_1 and v_2.

For a horizontally placed SWCNT above a ground plane consisting of N-layer graphene sheets with interlayer distance d = 3.35 Å, as illustrated in Fig. 5.3a, the vdw energy per unit length between the nanotube and the substrate is given by[41]

$$\frac{E_{\text{vdw}}}{L} = 2\pi\sigma^2 R \sum_{n=1}^{N} \int_{-\pi}^{\pi} \left(\frac{C_{12}}{10[(n-1)d + r_{\text{init}} + R + R\sin\theta]^{10}} - \frac{C_6}{4[(n-1)d + r_{\text{init}} + R + R\sin\theta]^4} \right) d\theta, \tag{5.2}$$

where L and R are the nanotube length and radius, respectively; r_{init} is the distance between the bottom of the nanotube and the top graphene sheet; $\sigma \cong 38$ nm^{-2} is the graphene surface density.

When $r_{\text{init}} \gg r_0$, the repulsive component can be ignored and Eq. 5.2 can be simplified as

$$\frac{E_{\text{vdw}}}{L} = C_6\sigma^2\pi^2 R \sum_{r=r_{\text{init}}}^{(N-1)d+r_{\text{init}}} \frac{(R+r)(3R^2+2(r+R)^2)}{2((r+R)^2-R^2)^{7/2}}. \tag{5.3}$$

For an MWCNT with an inner radius R_{int} and an outer radius R_{ext}, as illustrated in Fig. 5.3b, the vdw energy per unit length can be readily obtained by summing up the interactions between all the separated nanotube shells and substrate layers and is given by

$$\frac{E_{\text{vdw}}}{L} = C_6\sigma^2\pi^2 \sum_{R=R_{\text{int}}}^{R_{\text{ext}}} \sum_{r=r_{\text{init}}}^{(N-1)d+r_{\text{init}}} \frac{R(R+r)(3R^2+2(r+R)^2)}{2((r+R)^2-R^2)^{7/2}}. \tag{5.4}$$

By differentiating the vdw energy with respect to r, the per-unit-length vdw force applied on the nanotube is obtained as

$$q_{\text{vdw}} = -C_6\sigma^2\pi^2 \sum_{R=R_{\text{int}}}^{R_{\text{ext}}} \sum_{r=r_{\text{init}}}^{(N-1)d+r_{\text{init}}} \frac{R\sqrt{r(r+2R)}(8r^4+32r^3R+72r^2R^2+80rR^3+35R^4)}{2r^5(r+2R)^5}. \tag{5.5}$$

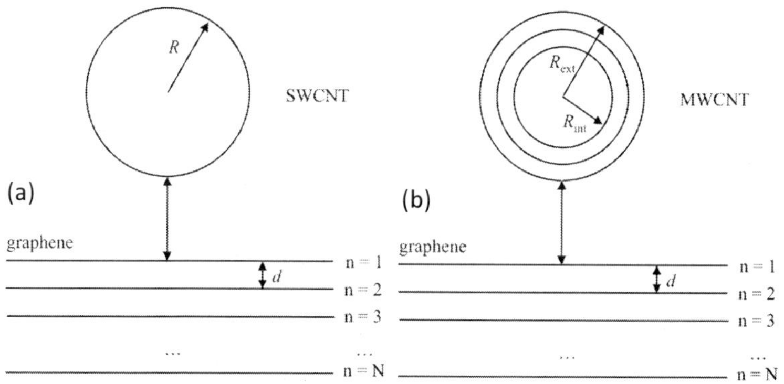

Figure 5.3 Van der Waals integration of an SWCNT (a) and an MWCNT (b) over a graphite ground plane. Reprinted with permission from [41], M. Dequesnes et al., *Nanotechnology*, **13**, 120 (2002). © 2002, Institute of Physics.

5.3.1.2 Electrostatic interactions

When a biased conductive nanotube is placed above a conductive substrate, induced electrostatic charges on both the tube and the substrate occur. The electrostatic force acting on the tube can be calculated using classical capacitance models. In spite of having hollow structures, CNTs with capped ends are more electro-chemically stable than those with open ends.[42] Thus, nanotubes with finite length can be geometrically approximated as conductive nanocylinders. For small-scale nanocylinders, the density of states on their surfaces is finite. The screening length, defined as the distance that surface charges actually penetrate into the cylinder interior, is found to be a nanometer-scale quantity.[43] For SWCNTs of diameters approaching the screening length, the quantum effect as a result of the finite size and density of states has to be considered thoroughly in the calculation of the surface/volume charge distribution.[44,45] For MWCNTs of large outer diameters that well exceed the screening length, this quantum effect can be considered negligible. Thus, the charge distribution on large-diameter CNTs can be approximated through applying classical electrostatic analysis on perfectly conductive cylinders with the same geometry.

For infinitely long metallic cylinders, the capacitance per unit length along their longitude directions is given by[46]

$$C_{\mathrm{d}}(r)=\frac{\pi\varepsilon}{\cosh^{-1}\left(1+\frac{r}{R}\right)}, \tag{5.6}$$

where r is the distance between the lower fiber of the cylinder and the substrate, R is the radius of the cylinder, and ε is the permittivity of the medium. For vacuum permittivity, $\varepsilon_0 = 8.854 \times 10^{-12}$ $C^2N^{-1}m^{-2}$. Equation 5.6 can be applied for infinitely long MWCNTs with large outer diameters ($R = R_{\mathrm{ext}}$).

For the charge distribution on infinitely long SWCNTs, Bulashevich et al.[45] proposed a quantum correction, rendering the capacitance per unit length as

$$C=\frac{C_{\mathrm{d}}}{1+\frac{C_{\mathrm{d}}}{C_{\mathrm{Q}}}}\approx C_{\mathrm{d}}\left(1-\frac{C_{\mathrm{d}}}{C_{\mathrm{Q}}}\right), \tag{5.7}$$

where $C_{\mathrm{Q}} = e^2\nu_{\mathrm{m}}$; ν_{m} is the constant density of the states near the electro-neutral level measured from the Fermi level.

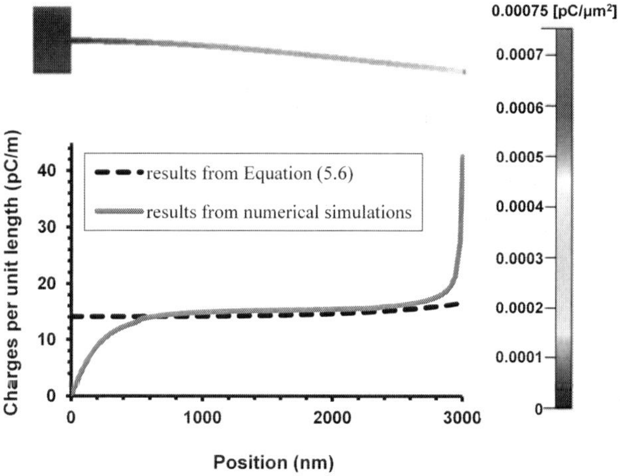

Figure 5.4 (Top) Side view of the charge distribution on a deflected nanotube cantilever; (Bottom) charge distribution per unit length along the deflected nanotube cantilever. The solid line represents the numerical simulation results and the dashed line is plotted based on Eq. 5.6. Reprinted with permission from [49], C.-H. Ke et al., *J. Appl. Mech.*, **72**, 721 (2005). © 2005, American Society of Mechanical Engineers.

For cylinders of finite length, there are two types of boundary surfaces: the cylindrical side surface and the planar end surface. Significant charge concentration at the cylinder end surface has been observed in the distribution of the charge density on the cylinder surface based on classical electrostatic simulations.[44,47,48] Figure 5.4 shows the charge distribution along a deflected nanotube cantilever with a tube radius R of 20 nm, a tube length L of 3 μm and an initial gap distance to the substrate H of 500 nm. The nanotube deflection profile was assumed to follow a second order polynomial and the gap between the free end and the substrate was set to be 236 nm. The contour plot shows the charge density (side view), while the solid curve represents the amount of electrical charges per unit length along the nanotube. The numerical simulation was performed using CFD-ACE$^+$, a commercial package from CFD Research Corporation based on finite and boundary element methods. There is significant charge concentration on the free end and relatively uniform charge distribution in the central portion of the cantilever, which nearly follows Eq. 5.6. It is noticed that the clamped end has a significant influence on the charge distribution in the region close to it.[43] However, this effect can be considered negligible because its contribution to the deflection of the nanotube is quite limited. An empirical formula about the charge distribution on deflected nanotube cantilevers, including the charge concentration effect, was derived from a parametric analysis and is given by[49]

$$C(r(x)) = C_{\mathrm{d}}(r(x))\{1 + 0.85[(H+R)^2 R]^{\frac{1}{3}} \delta(x - x_{\mathrm{tip}})\} = C_{\mathrm{d}}(r(x))\{1 + f_{\mathrm{c}}\} \tag{5.8}$$

where the first term in the bracket accounts for the uniform charges along the side surface of the tube and the second term, f_c, accounts for the concentrated charges at the end of the tube. $\delta(x)$ is the Dirac function and $r(x) = H - w(x)$ with w being the tube deflection. $x = x_{\mathrm{tip}} = L$ for the small displacement regime, while $x = x_{\mathrm{tip}} \neq L$ for finite kinematics, i.e., the large displacement regime. The per-unit-length electrostatic force acting on the nanotube is given by means of differentiation of the electrostatic energy as follows[50]:

$$q_{\mathrm{elec}} = \frac{1}{2} V^2 \frac{\mathrm{d}C}{\mathrm{d}r} = -\frac{\pi \varepsilon_0 V^2}{\sqrt{r(r+2R)}\left(\cosh^{-1}\left(1 + \frac{r}{R}\right)\right)^2}(1 + f_{\mathrm{c}}). \tag{5.9}$$

5.3.1.3 Governing equations

The electromechanical characteristics of electrically actuated nanotube cantilevers, as illustrated in Fig. 5.2, can be determined by coupling the vdw and electrostatic forces applied on the nanotube surface with the elastic (i.e., spring) forces from their mechanical deformations. Considering only the bending of the nanotube cantilever, the governing equation under the small deformation assumption is given by[41]

$$EI \frac{\mathrm{d}^4 r}{\mathrm{d}x^4} = q_{\mathrm{elec}} + q_{\mathrm{vdw}}, \tag{5.10}$$

where $r = r(x)$ is the gap between the nanotube and the ground plane, x is the position along the tube, E is Young's modulus of CNTs, and

I is the moment of inertia, given by $I = \frac{\pi}{4}(R_{\text{ext}}^4 - R_{\text{int}}^4)$ for MWCNTs; q_{elec} and q_{vdw} are given by Eqs. 5.5 and 5.9, respectively.

For cantilevers exhibiting large displacements, the curvature of the deflection should be considered and the governing equation changes to[50]

$$EI\frac{\mathrm{d}^2}{\mathrm{d}x^2}\left(\frac{\dfrac{\mathrm{d}^2 r}{\mathrm{d}x^2}}{\left(1+\left(\dfrac{\mathrm{d}r}{\mathrm{d}x}\right)^2\right)^{\frac{3}{2}}}\right) = (q_{\text{vdw}} + q_{\text{elec}})\sqrt{1+\left(\frac{\mathrm{d}r}{\mathrm{d}x}\right)^2}. \tag{5.11}$$

The aforementioned governing equations can be numerically solved by either direct integration methods or finite difference methods.

5.3.1.4 Analytical solutions of the pull-in voltage

In this section, we briefly present the analytical solutions of the pull-in voltage, a key parameter in the design and optimization of MEMS and NEMS devices based on energy methods.[51]

The deflection of the cantilever nanotube can be approximated by the following quadratic function, i.e., $w(x) \approx \frac{x^2}{L^2}c$, in which c is a constant that represents the displacement at the free end of the cantilever. The total energy of the system E_{total} is expressed as

$$E_{\text{total}}(c) = E_{\text{elas}}(c) + E_{\text{elec}}(c) + E_{\text{vdw}}(c), \tag{5.12}$$

where the elastic energy $E_{\text{elas}}(c)$, the electrostatic energy $E_{\text{elec}}(c)$ and the vdw energy $E_{\text{vdw}}(c)$ are given by the following integrations, respectively:

$$E_{\text{elas}}(c) = \frac{EI}{2}\int_0^L \left(\frac{\mathrm{d}^2 w}{\mathrm{d}x^2}\right)^2 \mathrm{d}x;\quad E_{\text{elec}}(c) = \int_0^L \frac{\mathrm{d}E_{\text{elec}}}{\mathrm{d}x}\mathrm{d}x;\quad E_{\text{vdw}}(c) = \int_0^L \frac{\mathrm{d}E_{vdw}}{\mathrm{d}x}\mathrm{d}x, \tag{5.13}$$

The equilibrium condition for the deflection of the nanotube cantilever is reached when the total energy reaches a minimum value, i.e., $\frac{\mathrm{d}E_{\text{total}}}{\mathrm{d}c} = 0$. Similarly, the instability of the deflected nanotube cantilever, i.e., pull-in, happens when the second order derivative of the total energy equals zero, i.e., $\frac{d^2E_{total}}{dc^2} = 0$.

By assuming that the nanotube's external diameter R is much smaller than the distance r between the nanotube and ground plane, the pull-in voltage is given by

$$V_{PI} \approx 0.85\sqrt{\frac{1+K^{FK}}{1+K^{TIP}}\frac{H}{L^2}\ln\left(\frac{2H}{R}\right)}\sqrt{\frac{EI}{\varepsilon_0}}, \tag{5.14}$$

where $K^{FK} \approx \frac{8H^2}{9L^2}$ and $K^{TIP} \approx \frac{2.55[R(H+R)^2]^{1/3}}{L}$ represent the effects of nonlinear finite kinematics and concentrated charges, respectively.

5.3.2 Electromechanical Characterization

5.3.2.1 Small deformation regime

The experimental data of the nanotweezers reported by Akita et al.,[13] as shown in Fig. 5.5a, are used to assess the model accuracy on the deflection of electrically actuated CNT cantilevers in the small deformation regime. The nanotweezers were composed of two parallel MWCNTs, both of which were mounted as cantilevers to an atomic force microscopic probe coated with a Ti/Pt film using a three-stage manipulator. The Ti/Pt film was separated into two by focused ion beam (FIB) etching so that the two nanotube arms could be operated by applying DC voltages to the separated Ti/Pt films. The nanotweezers shown in Fig. 5.5a have a cantilever length of about 2.5 μm and an original tip separation of about 780 nm. Various voltages were applied between the two nanotube arms so that they were bent and pulled closer by the attractive electrostatic force. The scanning electron microscopy (SEM) images in Fig. 5.5a clearly show the bending of the nanotube arms and the reducing tip separation with the increase of the applied voltage V. The tip separation became 500 nm at $V = 4$ V and zero at $V > 4.5$, indicating the occurrence of pull-in.

From the modeling point of view, the nanotweezers shown in Fig. 5.5a are equivalent to a nanotube cantilever with a length of 2.5 μm freestanding above an electrode with an initial gap of 390 nm. A comparison between the analytically predicted nanotube cantilever deflection using the models presented in Section 5.3.1.4

and the experimentally measured data is presented in Fig. 5.5b.[51] The analytical predictions take into account the effect of the charge concentration at the free end of the nanotube cantilever.

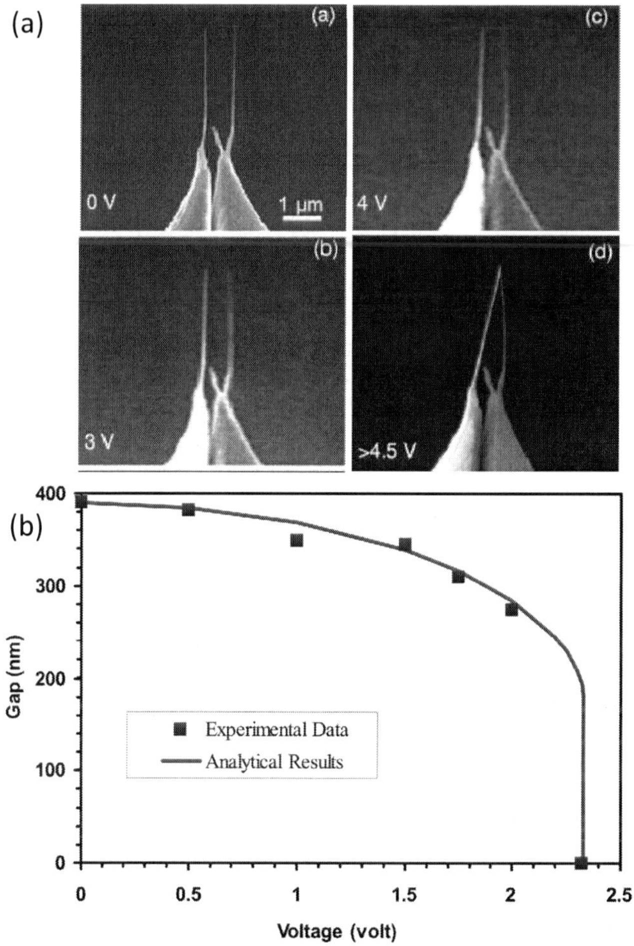

Figure 5.5 (a) SEM images of the motion of nanotube arms in a pair of nanotweezers as a function of the applied voltage. Reprinted with permission from [13], S. Akita et al., *Appl. Phys. Lett.*, **79**, 1691 (2001). © 2001, American Institute of Physics; (b) Comparison between the experimental data and the analytical predictions for the deflection of the nanotube arms shown in (a). Reprinted with permission from [51], C.-H. Ke et al., *J. Mech. Phys. Solids*, **53**, 1314 (2005). © 2005, Elsevier, Ltd.

The parameters of the nanotube used in the analytical prediction include Young's modulus E = 1 TPa, external radius $R = R_{ext}$ = 5.8 nm, and inner radius R_{int} = 0. The predicted pull-in voltage based on the analytical solution (Eq. 5.14) is 2.34 V, while the experimentally measured pull-in value was 2.33 V. It is clear that the analytical predictions of the deflection of the nanotube cantilever as a function of the applied voltage are in very good agreement with experimental measurements.

5.3.2.2 Finite kinematics regime

Experimental measurements on the deflection of electrically actuated CNT cantilevers in the finite kinematics regime essentially require a low aspect ratio between the nanotube length and the original gap distance, and were demonstrated by in situ SEM electromechanical measurements.[51]

The configuration of the in situ measurement is schematically shown in drawing (a) in Fig. 5.6a. The electrode was made of a small piece of silicon wafer coated with a thin Au film. This Si chip was attached onto a side surface of a Teflon block and mounted to the SEM sample holder at an angle of 93° with respect to the holder plane. A nanotube was first mounted to the tip of a conductive manipulator probe as a cantilever structure of 6.8 μm in length by means of nanomanipulation and electron beam-induced deposition of Pt, and was then placed horizontal and parallel to the electrode surface. By focusing on the electrode surface and adjusting the working distance of the electron beam, features on the electrode surface staying in the same horizontal plane with the nanotube were identified and schematically marked as a straight line in drawing (a) as well as images (b) to (d) in Fig. 5.6a. SEM image (b) in Fig. 5.6a shows that the nanotube stayed at a horizontal gap distance of 3 μm from the electrode surface, indicating a ratio of 2.3 between the nanotube length and the gap distance. Therefore, the deflection of the nanotube cantilever was expected to be in the finite kinematics regime.

SEM images (c) and (d) in Fig. 5.6b show two selected deflection conformations of the nanotube cantilever as it was subjected to increasing applied voltages. These images clearly reveal changes in nanotube deflection and local curvature as a function of the applied voltage. Pull-in occurred at an applied voltage of 48 V. The experimentally measured nanotube cantilever deflections, in the finite kinematics regime, are plotted in Fig. 5.6b.

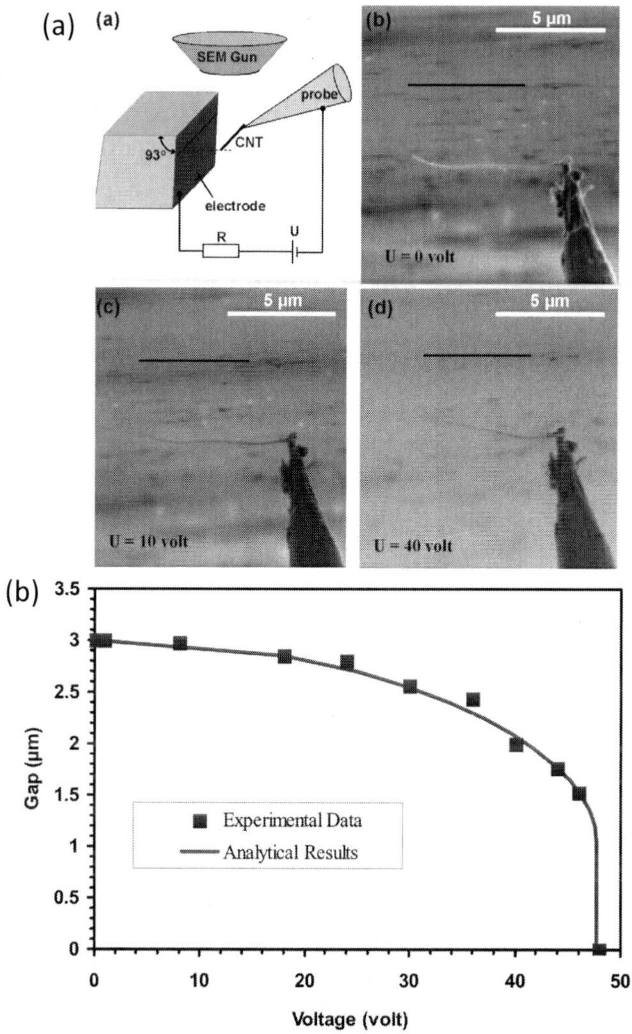

Figure 5.6 (a) Schematic of in situ SEM electromechanical testing configuration and three selected SEM images of a deformed CNT cantilever under various applied voltages; (b) Comparison between the experimental data and the analytical predictions for the deflection of the nanotube shown in (a). Reprinted with permission from [51], C.-H. Ke et al., *J. Mech. Phys. Solids,* **53**, 1314 (2005). © 2005, Elsevier, Ltd.

The figure also shows a comparison between analytical predictions and experimental measurements. The analytical predictions take into account both the effects of finite kinematics and charge concentration at the free end of the nanotube cantilever. The following parameters were employed in the analytical predictions: $E = 1$ TPa, $R = R_{ext} = 23.5$ nm, and $R_{int} = 0$. The pull-in voltage given by the analytical solution based on Eq. 5.4 is 47.8 V and is in very good agreement with the experimental value (48 V).

5.4 Carbon Nanotube Cantilever-Based Electromechanical Devices

5.4.1 Feedback-Controlled Nanocantilever Switches

A cantilevered CNT-based NEMS device with feedback control features is schematically illustrated in Fig. 5.7a.[10] The device is made of a conductive CNT cantilever, in series with a resistor R and a power supply. When the applied voltage U is below the pull-in voltage, the electrostatic force together with the vdw force applied on the nanotube cantilever is balanced by the elastic force from its deflection and the nanotube cantilever remains in the "upper" equilibrium position. When the applied voltage reaches or exceeds the pull-in voltage, the electrostatic force becomes larger than the elastic force and the nanotube cannot maintain its equilibrium, deflecting downward in an accelerating fashion. When the nanotube tip is in very close proximity to the bottom electrode, a junction is formed. A substantial tunneling current passes through the junction when its gap distance falls into the sub-nanometer range. As a result of the potential drop on the resistor R in the circuit, the voltage applied to the nanotube drops, weakening the electric field between the nanotube and the bottom electrode. Because of the kinetic energy of the nanotube, it continues to deflect downward, resulting in an increase of the tunneling current and thus a further decrease of the electric field strength. When the elastic force becomes larger than the electrostatic force, the nanotube decelerates and eventually reverses the direction of motion. The increase of the gap distance, in return, results in a decrease of the tunneling current, recovering the electric field. The cycle of the gap increasing and decreasing processes repeats and eventually stops when the kinetic energy of the nanotube is dissipated due to the damping in the system. The nanotube stays at a so-called "lower" equilibrium position, at which the electrostatic force equals the elastic force and a stable tunneling current is established in the device circuit. A gradual decrease of the applied voltage U results in an increase of the gap distance and thus a decrease of the current in the circuit, while the cantilever still maintains its stable equilibrium. When U decreases to a critical value, called pull-out voltage, the cantilever is released from its lower equilibrium position and returns back to its upper equilibrium position, essentially zeroing the current in the device. Therefore, the pull-in and pull-out processes together with the "upper" and "lower" equilibrium positions form a hysteretic loop for the applied voltage and the current in the device. The upper and lower equilibrium positions can be considered as the "OFF" and "ON" states of a switch, respectively. It is noted that the existence of the feedback resistor and the tunneling current in the device circuit make the "lower" equilibrium state very robust, which is key to many applications of interest. The potential applications of the device include NEMS switches, memory elements, logic devices, gap sensors, and ultrasonic wave detectors for monitoring the health of materials and structures.

The electromechanical behavior of the feedback controlled cantilevered nanotube switch has been demonstrated experimentally using in situ electron microscopy electromechanical testing techniques.[52] The employed test configuration is illustrated in Fig. 5.7b. An MWCNT was welded to the tip of a conductive manipulator probe that acts as one electrode. A second electrode employed in the testing configuration is a piece of silicon chip coated with a thin Au film. The device circuit was incorporated with a 1 GΩ resistor in series connection with both electrodes. The nanotube cantilever welded to the manipulator probe was displaced to approach the Au electrode until a desired gap distance was reached, which was typically 0.5~3 μm depending on the length and diameter of the nanotube. The testing configuration shown in Fig. 5.7b is considered electromechanically equivalent to the device configuration illustrated in Fig. 5.7a. Figure 5.7c shows an experimental current–voltage curve recorded on a nanotube of 9 μm in length during its pull-in/pull-out and equilibrium processes, which is also contrasted with theoretical predictions.

(a)
(b)
(c)

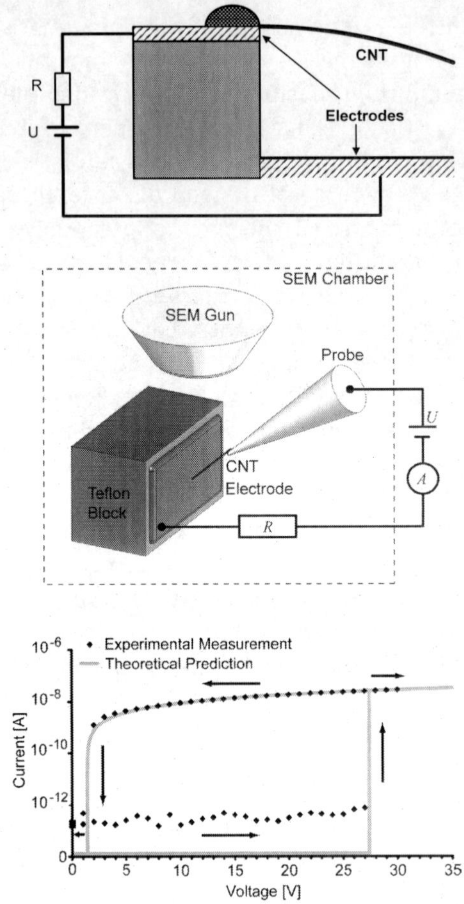

Figure 5.7 (a) Schematic of a feedback controlled cantilevered nanotube-based electromechanical nanoswitch. Reprinted with permission from [10], C.-H. Ke et al., *Appl. Phys. Lett.*, **85**, 681 (2004). © 2004, American Institute of Physics; (b) Schematic of the in situ electromechanical testing configuration of a nanotube cantilever inside a high resolution scanning electron microscope; (c) Measured voltage–current characteristic curve on a nanotube cantilever of 9 μm in length during its pull-in/pull-out and equilibrium processes and a comparison with theoretical predictions. The arrows show the directions in which the hysteretic loop evolves during the increase and decrease of the driving voltage *U*. Reprinted with permission from [52], C.-H. Ke et al., *Small*, **2**, 1484–1489 (2006). © 2006, Wiley-VCH Verlag GMBH.

The measured current–voltage curve exhibited the theoretically predicted bi-stability and hysteretic loop. The arrows in the plot show the directions in which the hysteretic loop evolves during the increase and decrease of the driving voltage.

5.4.2 Dual-Side Actuated Nanocantilever Resonators

Cantilevered nanotubes have been also employed to build NEMS resonators, in which the nanotube structure is excited by alternating driving signals. Studies on nanocantilever-based nano-resonators reported in the literature have focused on single-side actuated devices. Figures 5.8a,b illustrate a double-side driven nano-resonator based on a nanocantilever.[17] The nano-resonator comprises a nanotube cantilever with length L and outer diameter R, which is actuated by two parallel-plate electrodes (I and II). Both electrodes have an equal separation with the nanotube, H. The driving signals applied on the two electrodes have the same DC bias V_{dc}, AC amplitude V_{ac}, and period T, while having opposite phases. The nanotube cantilever is actuated in a *"pull and push"* manner and its vibrating motion is symmetric with respect to its original un-actuated position, resulting in zero static deflection for the nanotube beam.

The unique characteristic of the dual-side actuated resonator compared to its single-side counterpart is its symmetric tunable oscillation with larger stable motion range. It is well recognized that resonant frequency tuning is a key aspect in the functioning and performance of nano-resonators. This is because challenges associated with nanodevice fabrication and unique nanoscale phenomena, such as van der Waals interactions, quantum effects and molecular contamination, have much more significant impacts on the dynamic behaviors of nanodevices compared with their micro- or larger-scale counterparts. The approach of tuning the resonant frequency in such resonating devices is typically coupled with their actuation schemes. For single-side driven tunable resonators, the nanostructure is actuated in a *"pull only"* manner by attractive electrostatic and van der Waals forces. The constant force component in the electrostatic loading induces a static deflection, thus pulling down the nanostructure and reducing its gap distance to the driving electrode. The decrease of the gap distance results in a decrease of the maximum stable oscillation amplitude of the nanostructure because the unstable dynamic pull-in phenomenon occurs at smaller oscillation amplitudes. Therefore, the maximum stable oscillation range of the nanostructure is adversely limited as a result of the resonant frequency tuning. Furthermore, the smaller stable oscillation range of the nanostructure makes the sensing and detection of its nanoscale motion an even more challenging task. These limitations do not arise in resonators employing the dual-side actuation scheme.

The resonant motion of the dual-side driven resonator was investigated using energy methods.[17] From a system point of view, the added energy to the resonator is the electrostatic energy provided by the parallel-plate electrodes or the power supply, while the dissipated energy can be approximated by the energy loss due to the mechanical damping in the system. For each oscillation period, if the added energy equals the dissipated energy, the oscillation of the nanotube cantilever reaches its equilibrium or steady state. If the added energy cannot be balanced by the dissipated energy, dynamic pull-in, or resonant pull-in when the nanotube cantilever is in resonant oscillations, takes place and the cantilever beam subsequently snaps onto one of the electrodes. For the case of $V_{dc} >> V_{ac}$, the tuned resonant frequency of the nanotube cantilever f_{tun} is determined by the bias voltage V_{dc} and can be approximated by

$$\frac{f_{tun}}{f_n} \approx \sqrt{1 - 0.364\left(\frac{V_{dc}}{V_{SPI}}\right)^2\left(\frac{2}{\ln(2H/R+2)}+1\right)}, \tag{5.15}$$

where V_{SPI} is the quasi-static pull-in voltage of cantilevered nanotube devices with single-side electrostatic actuation and is given by Eq. 5.14. The steady-state resonance amplitude of the nanotube cantilever D can be obtained from the following relationship:

$$\frac{S(D)}{[D/(H+R)]^2} = \frac{0.549V_{SPI}^2 L}{QV_{dc}V_{ac}[\ln(2H/R+2)]^2}\sqrt{1 - 0.364\left(\frac{V_{dc}}{V_{SPI}}\right)^2\left(\frac{2}{\ln(2H/R+2)}+1\right)}, \tag{5.16}$$

where Q is the quality factor. It is noted that the $2\pi\varepsilon_0 S(D)$ represents the capacitance difference between the two nanotube–electrode

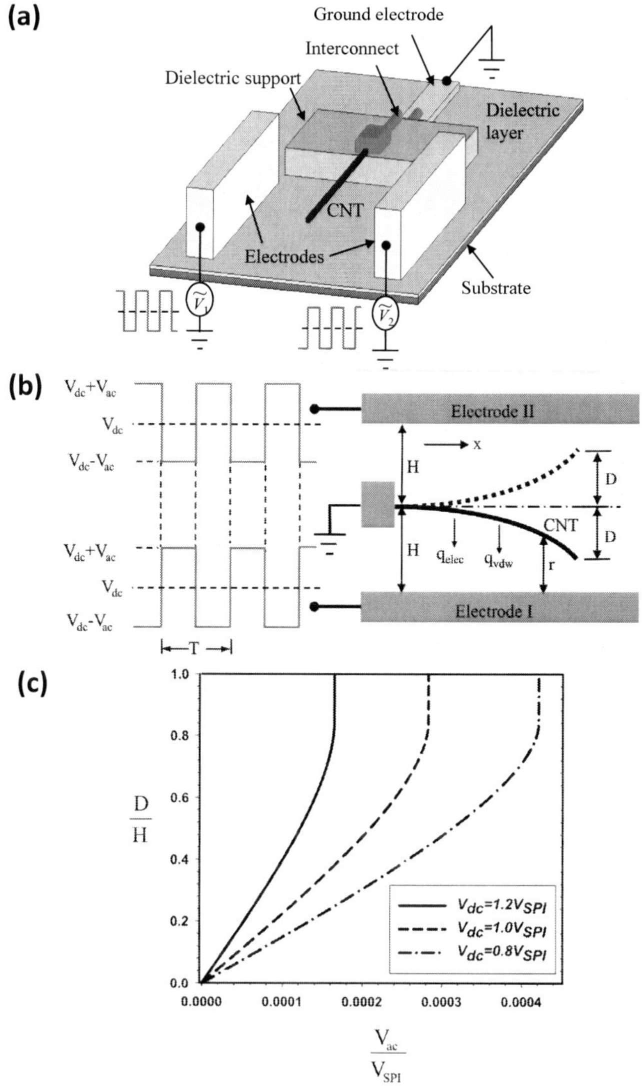

Figure 5.8 (a) 3D blueprint of a double-side actuated cantilevered nanotube-based tunable electromechanical resonator; (b) Schematic illustrations of the double-side actuated resonator (right) and driving signals (left); (c) The steady-state resonant characteristics of the double-side driven nanotube resonator for three tuned resonant frequencies. Reprinted with permission from [17], C.-H. Ke, *J. Appl. Phys.*, **15**, 024301 (2009) © 2009 American Institute of Physics.

combinations. Equations 5.15 and 5.16 clearly show that tuning the resonant frequency and controlling the steady-state resonance amplitude of the nanotube resonator are decoupled and can be controlled separately by V_{dc} and V_{ac}, respectively. Figure 5.8c shows the steady-state oscillation amplitude of a dual-side driven nanotube as a function of V_{ac} for three different resonant frequencies tuned by V_{dc}. It is shown that the maximum stable steady-state oscillation range of the dual-side driven nanotube resonator is independent of the electrostatic force-based resonant frequency tuning and can reach up to 90% of the gap between the driving electrodes.

Acknowledgments

A substantial portion of the research work presented in this chapter is part of the author's doctoral dissertation research under the supervision of Professor Horacio D. Espinosa at Northwestern University, which was financially supported by National Science Foundation and Federal Aviation Administration. Special appreciation is expressed to Meng Zheng and Xiaoming Chen for their assistance with the manuscript preparation.

References

1. Rueckes, T., Kim, K., Joselevich, E., Tseng, G. Y., Cheung, C. L., and Lieber, C. M. (2000). Carbon nanotube-based nonvolatile random access memory for molecular computing, *Science*, **289**(5476), 94–97.
2. Solanki, H. S., Sengupta, S., Dubey, S., Singh, V., Dhara, S., Kumar, A., Bhattacharya, A., Ramakrishnan, S., Clerk, A. A., and Deshmukh, M. M. (2011). High-Q electromechanics with InAs nanowire quantum dots, *Applied Physics Letters*, **99**(21), 213104.
3. Huttel, A. K., Steele, G. A., Witkamp, B., Poot, M., Kouwenhoven, L. P., and van der Zant, H. S. J. (2009). Carbon nanotubes as ultrahigh quality factor mechanical resonators, *Nano Letters*, **9**(7), 2547–2552.
4. Ekinci, K. L., and Roukes, M. L. (2005). Nanoelectromechanical systems, *Review of Scientific Instruments*, **76**(6), 061101.
5. Sazonova, V., Yaish, Y., Ustunel, H., Roundy, D., Arias, T. A., and McEuen, P. L. (2004). A tunable carbon nanotube electromechanical oscillator, *Nature*, **431**(7006), 284–287.

6. Jensen, K., Kim, K., and Zettl, A. (2008). An atomic-resolution nanomechanical mass sensor, *Nature Nanotechnology*, **3**(9), 533–537.

7. Lassagne, B., Garcia-Sanchez, D., Aguasca, A., and Bachtold, A. (2008). Ultrasensitive mass sensing with a nanotube electromechanical resonator, *Nano Letters*, **8**(11), 3735–3738.

8. Roukes, M. L. (1999). Yoctocalorimetry: phonon counting in nanostructures, *Physica B-Condensed Matter*, **263**, 1–15.

9. Roukes, M. L. (2000). Nanoelectromechanical systems. *Technical Digest of the 2000 Solid-State Sensor and Actuator Workshop*. Hilton Head Island, SC, June 4-8, 2000 (ISBN 0-9640024-3-4).

10. Ke, C. H., and Espinosa, H. D. (2004). Feedback controlled nanocantilever device, *Applied Physics Letters*, **85**(4), 681–683.

11. Jang, J. E., Cha, S. N., Choi, Y., Amaratunga, G. A. J., Kang, D. J., Hasko, D. G., Jung, J. E., and Kim, J. M. (2005). Nanoelectromechanical switches with vertically aligned carbon nanotubes, *Applied Physics Letters*, **87**(16), 163114.

12. Kim, P., and Lieber, C. M. (1999). Nanotube nanotweezers, *Science*, **286**(5447), 2148–2150.

13. Akita, S., Nakayama, Y., Mizooka, S., Takano, Y., Okawa, T., Miyatake, Y., Yamanaka, S., Tsuji, M., and Nosaka, T. (2001). Nanotweezers consisting of carbon nanotubes operating in an atomic force microscope, *Applied Physics Letters*, **79**(11), 1691–1693.

14. Kinaret, J. M., Nord, T., and Viefers, S. (2003). A carbon-nanotube-based nanorelay, *Applied Physics Letters*, **82**(8), 1287–1289.

15. Jensen, K., Weldon, J., Garcia, H., and Zettl, A. (2007). Nanotube radio, *Nano Letters*, **7**(11), 3508–3511.

6. Garcia-Sanchez, D., Paulo, A. S., Esplandiu, M. J., Perez-Murano, F., Forro, L., Aguasca, A., and Bachtold, A. (2007). Mechanical detection of carbon nanotube resonator vibrations, *Physical Review Letters*, **99**(8), 085501.

17. Ke, C.-H. (2009). Resonant pull-in of a double-sided driven nanotube-based electromechanical resonator, *Journal of Applied Physics*, **15**, 024301.

18. Iijima, S. (1991). Helical microtubules of graphitic carbon, *Nature*, **354**(6348), 56–58.

19. Ajayan, P. M. (1999). Nanotubes from carbon, *Chemical Reviews*, **99**(7), 1787–1799.

20. Journet, C., Maser, W. K., Bernier, P., Loiseau, A., delaChapelle, M. L., Lefrant, S., Deniard, P., Lee, R., and Fischer, J. E. (1997). Large-scale

production of single-walled carbon nanotubes by the electric-arc technique, *Nature*, **388**(6644), 756–758.

21. Ebbesen, T. W., and Ajayan, P. M. (1992). Large-scale synthesis of carbon nanotubes, *Nature*, **358**(6383), 220–222.
22. Thess, A., Lee, R., Nikolaev, P., Dai, H. J., Petit, P., Robert, J., Xu, C. H., Lee, Y. H., Kim, S. G., Rinzler, A. G., Colbert, D. T., Scuseria, G. E., Tomanek, D., Fischer, J. E., and Smalley, R. E. (1996). Crystalline ropes of metallic carbon nanotubes, *Science*, **273**(5274), 483–487.
23. Li, W. Z., Xie, S. S., Qian, L. X., Chang, B. H., Zou, B. S., Zhou, W. Y., Zhao, R. A., and Wang, G. (1996). Large-scale synthesis of aligned carbon nanotubes, *Science*, **274**(5293), 1701–1703.
24. Zhang, Y. G., Chang, A. L., Cao, J., Wang, Q., Kim, W., Li, Y. M., Morris, N., Yenilmez, E., Kong, J., and Dai, H. J. (2001). Electric-field-directed growth of aligned single-walled carbon nanotubes, *Applied Physics Letters*, **79**(19), 3155–3157.
25. Qian, D., Wagner, G. J., Liu, W. K., Yu, M. F., and Ruoff, R. S. (2002). Mechanics of carbon nanotubes, *Applied Mechanics Review*, **55**, 495–533.
26. Pop, E., Mann, D., Wang, Q., Goodson, K., and Dai, H. J. (2006). Thermal conductance of an individual single-wall carbon nanotube above room temperature, *Nano Letters*, **6**(1), 96–100.
27. Mintmire, J. W., Dunlap, B. I., and White, C. T. (1992). Are fullerene tubules metallic, *Physical Review Letters*, **68**(5), 631–634.
28. McEuen, P. L., Fuhrer, M. S., and Park, H. K. (2002). Single-walled carbon nanotube electronics, *IEEE Transactions on Nanotechnology*, **1**(1), 78–85.
29. Tombler, T. W., Zhou, C. W., Alexseyev, L., Kong, J., Dai, H. J., Lei, L., Jayanthi, C. S., Tang, M. J., and Wu, S. Y. (2000). Reversible electromechanical characteristics of carbon nanotubes under local-probe manipulation, *Nature*, **405**(6788), 769–772.
30. Liu, B., Jiang, H., Johnson, H. T., and Huang, Y. (2004). The influence of mechanical deformation on the electrical properties of single wall carbon nanotubes, *Journal of the Mechanics and Physics of Solids*, **52**(1), 1–26.
31. Kuzumaki, T., and Mitsuda, Y. (2004). Dynamic measurement of electrical conductivity of carbon nanotubes during mechanical deformation by nanoprobe manipulation in transmission electron microscopy, *Applied Physics Letters*, **85**(7), 1250–1252.

32. Gulseren, O., Yildirim, T., Ciraci, S., and Kilic, C. (2002). Reversible band-gap engineering in carbon nanotubes by radial deformation, *Physical Review B*, **65**(15), 155410.

33. Kinoshita, Y., and Ohno, N. (2010). Electronic structures of boron nitride nanotubes subjected to tension, torsion, and flattening: A first-principles DFT study, *Physical Review B*, **82**(8), 085433.

34. Kim, Y. H., Chang, K. J., and Louie, S. G. (2001). Electronic structure of radially deformed BN and BC3 nanotubes, *Physical Review B*, **63**(20), 205408.

35. Shan, B., Lakatos, G. W., Peng, S., and Cho, K. J. (2005). First-principles study of band-gap change in deformed nanotubes, *Applied Physics Letters*, **87**(17), 173109.

36. Crespi, V. H., Cohen, M. L., and Rubio, A. (1997). In situ band gap engineering of carbon nanotubes, *Physical Review Letters*, **79**(11), 2093–2096.

37. Barboza, A. P. M., Chacham, H., and Neves, B. R. A. (2009). Universal response of single-wall carbon nanotubes to radial compression, *Physical Review Letters*, **102**(2), 025501.

38. Lennard-Jones, J. E. (1930). Perturbation problems in quantum mechanics, *Proceeding of Royal Society A*, **129**, 598–615.

39. Girifalco, L. A., Hodak, M., and Lee, R. S. (2000). Carbon nanotubes, buckyballs, ropes, and a universal graphitic potential, *Physical Review B*, **62**(19), 13104–13110.

40. Girifalco, L. A. (1992). Molecular-properties of C-60 in the gas and solid-phases, *Journal of Physical Chemistry*, **96**(2), 858–861.

41. Dequesnes, M., Rotkin, S. V., and Aluru, N. R. (2002). Calculation of pull-in voltages for carbon-nanotube-based nanoelectromechanical switches, *Nanotechnology*, **13**(1), 120–131.

42. Lou, L., Nordlander, P., and Smalley, R. E. (1995). Fullerene nanotubes in electric-fields, *Physical Review B*, **52**(3), 1429–1432.

43. Krcmar, M., Saslow, W. M., and Zangwill, A. (2003). Electrostatics of conducting nanocylinders, *Journal of Applied Physics*, **93**(6), 3495–3500.

44. Rotkin, S. V., Shrivastava, V., and Bulashevich, K. A., and Aluru, N. R. (2002). Atomistic capacitance of a nanotube electromechanical device, *International Journal of Nanoscience*, **1**, 337.

45. Bulashevich, K. A., and Rotkin, S. V. (2002). Nanotube devices: A microscopic model, *JETP Letters*, **75**(4), 205–209.

46. Hayt, W., and Buck, J. (2001). *Engineering Electromagnetics.* McGraw-Hill, New York.

47. Keblinski, P., Nayak, S. K., Zapol, P., and Ajayan, P. M. (2002). Charge distribution and stability of charged carbon nanotubes, *Physical Review Letters,* **89**(25), 255503.

48. Smythe, W. R. (1956). Charged right circular cylinder, *Journal of Applied Physics,* **27**(8), 917–920.

49. Ke, C. H., and Espinosa, H. D. (2005). Numerical analysis of nanotube-based NEMS devices—Part I: Electrostatic charge distribution on multiwalled nanotubes, *Journal of Applied Mechanics-Transactions of the ASME,* **72**(5), 721–725.

50. Ke, C. H., Espinosa, H. D., and Pugno, N. (2005). Numerical analysis of nanotube based NEMS devices—Part II: Role of finite kinematics, stretching and charge concentrations, *Journal of Applied Mechanics-Transactions of the ASME,* **72**(5), 726–731.

51. Ke, C. H., Pugno, N., Peng, B., and Espinosa, H. D. (2005). Experiments and modeling of carbon nanotube-based NEMS devices, *Journal of the Mechanics and Physics of Solids,* **53**(6), 1314–1333.

52. Ke, C. H., and Espinosa, H. D. (2006). In situ electron microscopy electromechanical characterization of a bistable NEMS device, *Small,* **2**(12), 1484–1489.

Chapter 6

Membrane-Type Surface Stress Sensor

Genki Yoshikawa

WPI Center for Materials Nanoarchitectonics (MANA),
National Institute for Materials Science (NIMS),
1-1, Namiki, Tsukuba, Ibaraki, 305-0044, Japan

YOSHIKAWA.Genki@nims.go.jp

The membrane-type surface stress Sensor (MSS) is an optimized nanomechanical sensor platform designed for the so-called "static mode" operation. MSS consists of an "adsorbate membrane" supported with four constricted "sensing beams," on which piezoresistors are embedded for electrical read-out of analyte-induced surface stress on the adsorbate membrane. With this configuration, each sensing beam experiences the cumulative deformation of the membrane, and thus, piezoresistors embedded at these sensing beams can efficiently detect the whole surface stress applied on the adsorbate membrane. The MSS design offers several advantages including high sensitivity (limit of detection ~0.1 mN/m), high stability (without free-ends), low drift (with self-compensation of full Wheatstone bridge), easy coating of receptors (with double-side coating method), low

Nanocantilever Beams: Modeling, Fabrication, and Applications
Edited by Ioana Voiculescu and Mona Zaghloul

ISBN 978-981-4613-23-1 (Hardcover), 978-981-4613-24-8 (eBook)
www.panstanford.com

cost, compact system, and the capability of mass production (with CMOS compatibility). In this chapter, the properties of MSS are discussed along with recent results.

6.1 Introduction

Nanomechanical sensors detect volume- or mass-induced mechanical changes of a sensing structure by means of so-called static or dynamic mode operations, respectively. Since all substances have these basic properties, nanomechanical sensors can be used for detecting almost any kind of samples as demonstrated in various examples.[1–6] Thus, nanomechanical sensors have been expected as a versatile sensor in various fields, including medicine, security, and environmental science. There is, however, still no nanomechanical sensor actually working in industry or in our daily life even after the decades of studies since their first reports.[7,8] It is important to develop a platform for practical applications in which nanomechanical sensors can contribute exclusively. For this purpose, each component of a nanomechanical sensor has to be optimized for fulfilling the practical requirements, such as small, simple, and low-cost, in addition to enhancing fundamental properties, including sensitivity, stability, and reproducibility.

In this chapter, the recent developments of a membrane-type surface stress sensor (MSS), which can be regarded as one of the most practical platforms of nanomechanical sensors, are summarized in terms of the sensing elements, coating schemes, and measurement systems. For sensing elements, the comprehensive structural optimization from a cantilever-based structure to a membrane-based structure will be reviewed in Section 6.2 with a summary of fine-tuning of each component and experimental verifications. In Section 6.3, the improvement in coating schemes and measurement systems based on the MSS structure will be summarized with a comprehensive comparison between each platform including typical cantilevers with the optical and piezoresistive read-outs. Here we focus on the static mode because it can be operated with a simpler system without an actuator and does not suffer from damping effects in liquid environment, providing wider opportunities for nanomechanical sensors.

6.2 Membrane-Type Surface Stress Sensor

6.2.1 Piezoresistive Read-Out

Among various techniques for transducing nanomechanical deformation into recognizable electrical signals, piezoresistive read-out is one of the most promising solutions for realizing small, easy-to-use, and cost-effective devices.[9–20] In contrast to the commonly used optical (laser) read-out, the lever-integrated piezoresistive sensing elements do not require bulky peripheries for read-out. This self-sensing platform makes it possible to detect target molecules even in opaque liquids such as blood, where optical methods cannot be applied. To simultaneously attain both high sensitivity and the capability for mass production, we focus on the standard p-type Si(100) as a piezoresistor. It has large piezoresistor coefficients [21–23] and compatibility with CMOS fabrication process. Since the chip fabrication technology and peripheral electronics have been well developed for this material in industry, the sensing elements based on p-type Si(100) are ready for further developments towards actual products once the optimized structure is established. In spite of these fascinating features, piezoresistive read-out had not been widely used for sensing purposes because of its low sensitivity. Although there were a lot of trials to increase the sensitivity of the piezoresistive nanomechanical cantilever sensors,[24–28] it had been difficult to compete with the commonly used optical read-out cantilever sensors.

6.2.2 Structural Optimization for Efficient Transduction of Induced Stress

To enhance the sensitivity of piezoresistive nanomechanical sensors, it is important to take account of the basic properties of the material. In the case of p-type Si(100), relative resistance change is described as follows[22,29]:

$$\frac{\Delta R}{R} \approx \frac{1}{2}\pi_{44}(\sigma_x - \sigma_y), \tag{6.1}$$

where π_{44} (~138.1 × 10^{-11} Pa^{-1}) is one of the fundamental piezoresistance coefficients of the silicon crystal. σ_x and σ_y are the stresses induced on the piezoresistor in [110] and [1–10] directions of the silicon crystal, respectively. According to this equation, both enhancement of $\sigma_x (\sigma_y)$ and suppression of $\sigma_y (\sigma_x)$ are required to yield a substantial amount of $\Delta R/R$. In the case of surface stress sensing, however, the stress is basically isotropic, i.e., $\sigma_x \sim \sigma_y$, resulting in $\Delta R/R \sim 0$. Therefore, the piezoresistive signal is virtually zero on the entire surface irrespective of simple modifications such as constrictions (Fig. 6.1a). Because of this intrinsic material property, it is difficult to achieve a high sensitivity with simple cantilever-type structures. It should be noted that the piezoresistive detection for force sensing applications, such as atomic force microscope (AFM), is very different from those for surface stress sensing. For point-force loading, the induced stress on the surface of a cantilever is uniaxial (i.e., $\sigma_x >> \sigma_y$), and the maximum stress at the clamped end is easily amplified by a simple constriction.[13] Thus, the piezoresistors embedded at the constricted parts can yield large signal for point-force loading, while it is not the case for the uniformly induced surface stress.

The strategy to overcome this intrinsic problem is summarized in Fig. 6.1. The first important step to enhance the sensitivity is the utilization of a double lever geometry (Fig. 6.1b).[30,31] In this case, the surface stress induced by the adsorbed analytes uniformly bends the longer lever (so-called "adsorbate lever"). The accumulated deflection at the free-end of the adsorbate lever is then transduced onto the rigidly connected short lever (so-called "sensing lever"). For the sensing lever, the stress distribution is similar with that for a point-force-loaded cantilever; the dominant stress applied on the sensing lever is in x-direction (i.e., $\sigma_x >> \sigma_y$) and it can be easily enhanced by a simple modification, such as a constriction. Although the double lever geometry can achieve significant improvement in sensitivity, this structure seems not so suitable for actual applications because of its asymmetric configuration, which also loses some amount of sensitivity at the connecting part. Alternatively, a simpler structure like that presented in Fig. 6.1c can achieve the similar or even better sensitivity. The adsorbate-covered part directly induces a uniaxial stress on the piezoresistive sensing part. It should be noted that at this stage, we have departed from the concept of a "cantilever (i.e., one-end-clamed beam)."

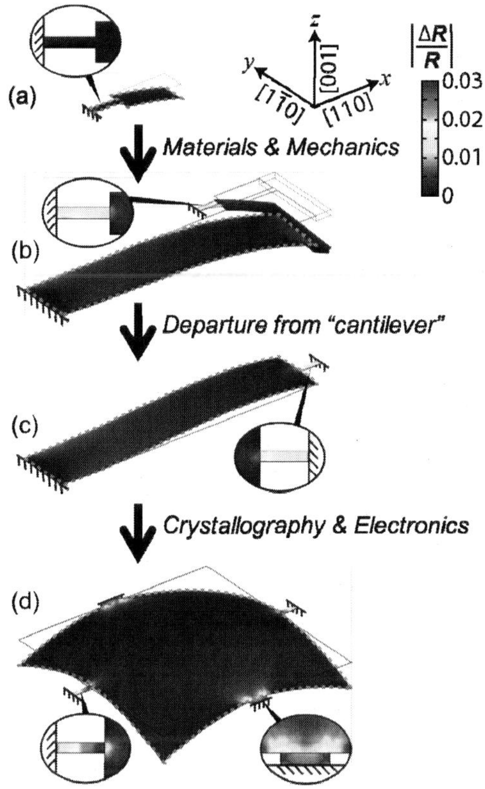

Figure 6.1 The Finite Element Analysis (FEA) results of various geometries, assuming that all of the geometries are made of p-type single-crystalline Si(100) and current flow in x-direction. Surface stress of 3.0 N/m is induced on the parts indicated by gray-dashed-lines. (a) Standard cantilever geometry, in which it is intrinsically difficult to gain signals even with some modifications, such as a constriction near to the clamped-end, because of the basic property of Si(100) (i.e., $\sigma_x \sim \sigma_y$, resulting in $\Delta R/R \sim 0$). (b) Double-lever geometry, in which the "adsorbate lever" and "sensing lever" are rigidly connected at their free-ends. Since the accumulated deflection caused by uniform surface stress on the adsorbate lever is transduced on the sensing lever in a similar form of point-force, the dominant stress induced on the sensing lever is along x-direction, $\sigma_x >> \sigma_y$, leading to $\Delta R/R >> 0$. (c) Double-

end-clamped geometry, in which both sensing and adsorbate levers are simply connected along x-direction. This structure is essentially same as that of the double-lever geometry, while it is rather simple with the symmetric structure. (d) Membrane-type surface stress sensor (MSS) geometry. An "adsorbate membrane" is supported with four constrictions, "sensing beams," on which piezoresistors are embedded. The isotropic deformation, which is accumulated at the periphery of the membrane, is effectively converted into a force at the connection between the membrane and the small sensing beams. With this configuration, each sensing beam experiences the cumulative deformation of the membrane, and thus, piezoresistors embedded at these sensing beams can efficiently detect the whole surface stress applied on the adsorbate membrane.

For further improvement, we have to take into account the actual read-out procedure of the obtained relative resistance change of the piezoresistors. To measure the resistance change, we can use a Wheatstone bridge as a simple and reliable method. In case that all four resistors (R_1–R_4) in a Wheatstone bridge are practically equal and that the relative resistance changes are small with $\Delta R_i/R_i << 1$ (I = 1 – 4), the total output signal V_{out} of the Wheatstone bridge can be approximated by Eq. 6.2:

$$V_{out} = \frac{V_B}{4}\left(\frac{\Delta R_1}{R_1} - \frac{\Delta R_2}{R_2} + \frac{\Delta R_3}{R_3} - \frac{\Delta R_4}{R_4}\right), \tag{6.2}$$

where V_B is a bias voltage of the Wheatstone bridge. Thus, if the sign of the resistance changes ΔR_1 and ΔR_3 are opposite to that of ΔR_2 and ΔR_4, the full Wheatstone bridge yields an amplification of another factor of 4. Here, we can again utilize the basic property of single crystalline silicon. For p-type Si(100), the relative resistance change with a current flow in x-direction is given by Eq. 6.1. Thus, if we configure the structure like Fig. 6.1d with an electrical configuration shown in Fig. 6.2, the dominant stresses induced by surface stress on the adsorbate membrane are σ_x in R_1, R_3 and σ_y in R_2, R_4, resulting in opposite signs for the relative resistance changes in each set of resistors, efficiently detecting the whole induced surface stress. We call this structure the membrane-type surface stress sensor (MSS, Fig. 6.2).[32] It realizes fascinating properties for practical applications such as high sensitivity, self-compensating low-drift operation with full Wheatstone bridge, and stable operation without free-end. The MSS structure is also beneficial for coating of a receptor layer as shown later in Section 6.3.2.

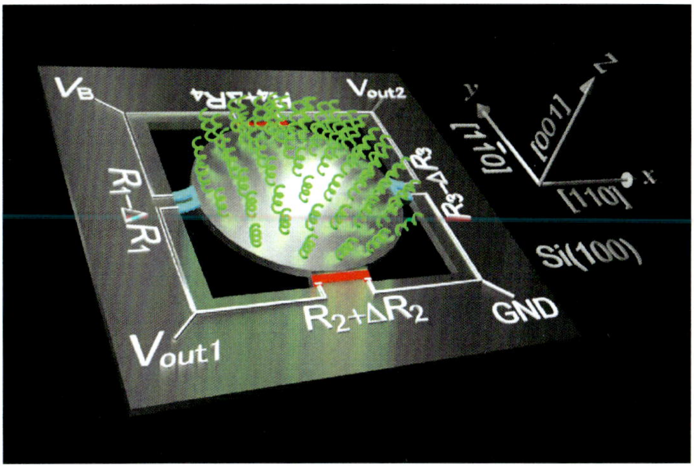

Figure 6.2 (a) Schematic illustration of MSS with the configuration of typical electrical wirings. The whole surface stress induced on the round center membrane is efficiently detected by piezoresistors (red and blue colored parts) embedded in the constricted beams, composing a full Wheatstone bridge.

6.2.3 Fine-Tuning of Each Part

Based on the MSS concept, we fabricated a prototype chip (first generation MSS chip; 1G-MSS) which was just a proof of-concept, however, several components had not been optimized. Then, as a second generation MSS chip (2G-MSS), we made several modifications on each part of the MSS design according to the analytical considerations on various geometries of nanomechanical sensors[30] and the know-how on microfabrication developed so far.[9–11,14,16,18] The implemented modifications can be summarized as follows[33]:

(1) *Narrower sensing beams with thinner geometry*
(2) *Thinner passivation layers on piezoresistors*
(3) *Shallower piezoresistors with lower doping level*
(4) *Elimination of negative regions in piezoresistors*

The modifications (1–3) and resulting improvements in sensitivity compared to 1G-MSS are summarized in Table 6.1. Figure 6.3 shows the arrangement of piezoresistors on 2G-MSS, representing the modification 4. The detailed explanations of each modification can be found in the recent publication.[33] The photo images of each MSS chip are shown in Fig. 6.4 with the SEM image of one of the MSS channels on a 2G-MSS chip.

Table 6.1 Summary of the modifications implemented in 2G-MSS with the results of finite element analyses

Modification	1G-MSS	2G-MSS	Δ
• Narrower beams • Thinner geometry	Width of beams for R_2 and R_4		~40%
	36 μm	26 μm	
	Thickness of bridges and membrane		
	3.2 μm	2.5 μm	
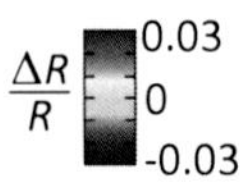	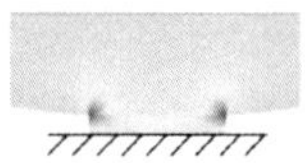	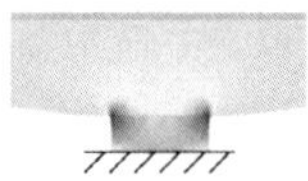	
• Thinner passivation layer on piezoresistors	Thickness of passivation layer on piezoresistors		~40%
	SiO_2 650 nm + Si_3N_4 100 nm	SiO_2 80 nm + Si_3N_4 100 nm	
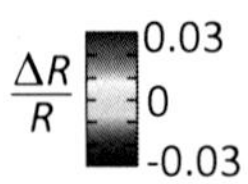	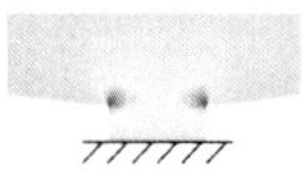	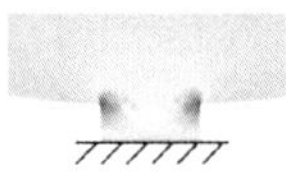	
• Shallower doping with lower doping level	Depth of piezoresistors		~10%
	500 nm	300 nm	
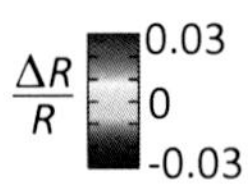	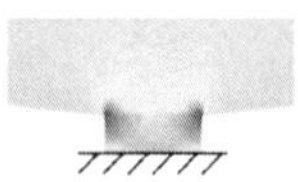	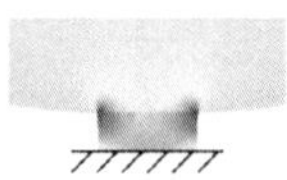	

Note: The improvements in sensitivity compared to 1G-MSS are given in the column on the right side (Δ).

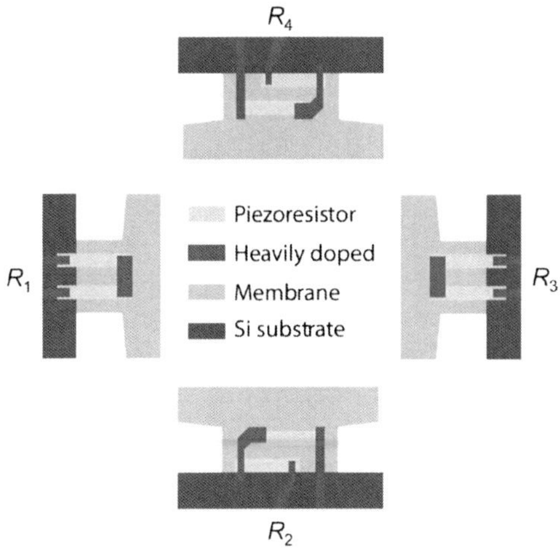

Figure 6.3 The arrangement of piezoresistors on 2G-MSS. The negative regions of piezoresistors are replaced by the heavily doped parts.

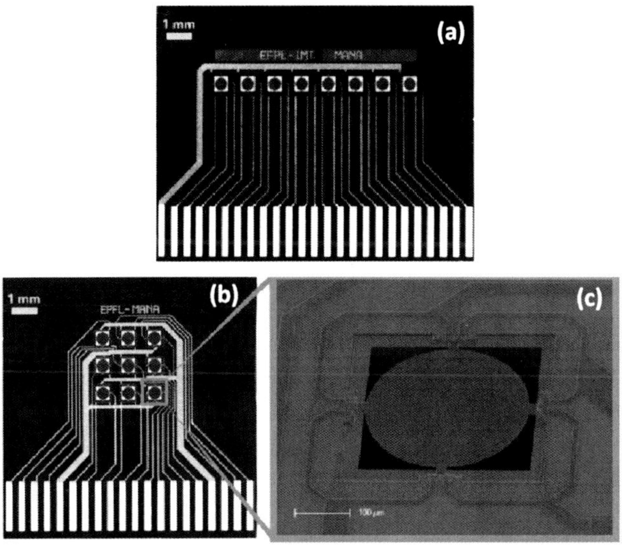

Figure 6.4 The photo images of (a) 1G-MSS with MSS channels in a one dimensional array (1 × 8; in total 8 channels) and (b) 2G-MSS with a two dimensional array (3 × 3; in total 9 channels). (c) Magnified SEM image of an MSS channel on the 2G-MSS chip.

6.2.4 Experimental Verification

To verify the improved sensitivity by the MSS structure, we performed gas sensing experiments with 1G-MSS, 2G-MSS, and a model piezoresistive cantilever designed for atomic force microscopy.[9,20] The sensing parts (i.e., circular membranes of 1G-MSS and 2G-MSS, and the surface of a piezoresistive cantilever) were coated with poly(sodium 4-styrenesulfonate) (PSS) layers by a customized ink-jet spotting system (Microjet Co. Ltd., Model "LaboJet-500SP"). PSS solution with a concentration of 5 g/L was deposited to form a 1 μm-thick PSS layer on the surface. The PSS-coated chips were mounted in a closed chamber and were exposed to 20% water vapor in pure nitrogen carrier gas with a flow rate of 100 mL/min for 3 min followed by nitrogen purging for 3 min. Figure 6.5 shows the obtained output signal (V_{out}) of the 2G-MSS chip compared to a 1G-MSS chip and a model piezoresistive cantilever. The signals were measured with the same electrical setup with a bias voltage of −1.5 V in all cases. The 2G-MSS chip exhibited ~4 and ~100 times larger signals than those of the 1G-MSS chip and the model piezoresistive cantilever, respectively. These experimental results confirm a significantly enhanced sensitivity by means of the MSS structure with the fine-tuning of each component.

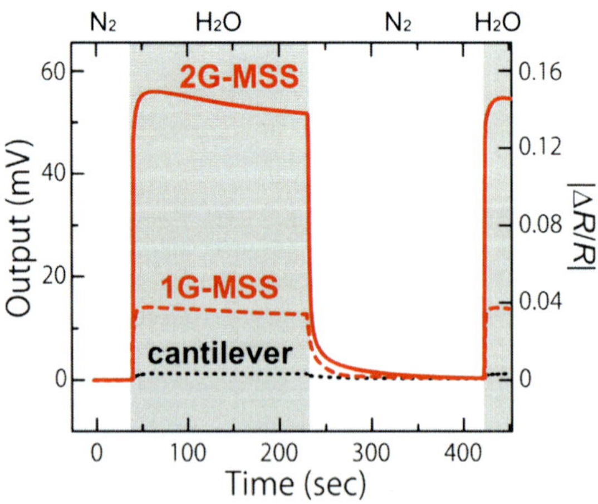

Figure 6.5 Obtained output signals from the 2G-MSS chip (red solid line),[33] 1G-MSS chip (red dashed line),[32] and model piezoresistive cantilever (black dotted line).[9]

6.3 Improvement in User Experience

The better user experience will enlarge the opportunities for MSS to be used for various applications. In this section, the two major issues are addressed: coating of receptor layers and wiring of piezoresistors to an electrical device.

6.3.1 Coating

One of the major issues of the nanomechanical sensors is the difficulty in coating receptor layers on their surfaces. In the case of static mode operation, single-side coating is commonly applied to cantilever-type geometry to have deflection, which is measured by either optical or electrical methods as a signal of the sensors. In this case, however, the coating methods and protocols are restricted to a few specific ones, such as the ink-jet spotting[34] or thiol chemistry with a single-side-coated gold layer.[35] With such specific methods or protocols, most of the standard coating procedures cannot be applied because those procedures usually include the incubation of the substrate in a solution for several hours. In addition, the single-side coating also requires a passivation of the opposite surface to avoid non-specific binding, which results in difficulties in analyzing the observed deflection signal.

Double-side coating is a simple approach to circumvent these problems. Although the double-side coating can be effectively utilized in the case of dynamic mode operation,[36] it is difficult to read out in the case of static mode by the common optical (laser) method because double-side coating results in an in-plane elongation as illustrated in Table 6.2. Alternatively, piezoresistive read-out is an effective method for the measurement of in-plane elongation, more specifically, in-plane stress induced by the surface stresses on both top and bottom sides of a sensing structure. A few studies reported that piezoresistive cantilevers with specific structures and piezoresistors can achieve a better sensitivity by the double-side-coating.[37,38] In the case of cantilever geometry, however, a piezoresistor can detect only a local stress that is induced on the place where the piezoresistor is located as shown in Table 6.2. Another problem arises for measuring isotropic surface stress with a piezoresistive cantilever made of commonly used p-type single crystal Si(100) because of the zero signal problem according to Eq. 6.1 as discussed in Section 6.2.2. Although one can design a larger piezoresistor by means of n-type doping along the entire cantilever to integrate the whole induced stress,[29,39] it requires the optimization of thermal issues and larger passivation or stress compensation layers which reduce the sensitivity.[39] In contrast, the MSS structure can efficiently detect the whole surface stress induced on the center membrane even in the case of double-side coating as shown in Table 6.2.[40]

Table 6.2 Comparison between cantilevers with optical and piezoresistive read-outs and MSS for single- and double-side coatings.

Platform	Single-side-coating	Double-side-coating
Optical cantilever	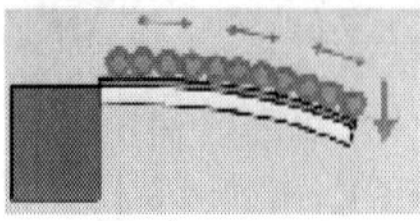	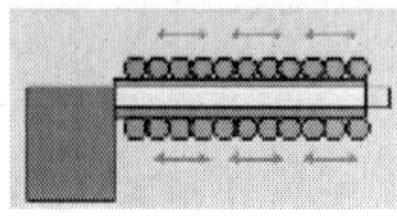
	All of the analytes can be measured as a deflection at the free-end	In-plane elongation → difficult to measure
Piezoresistives cantilever	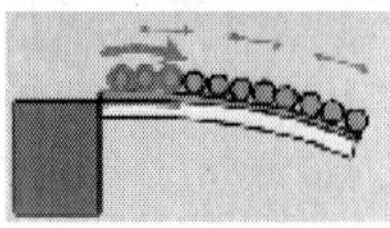	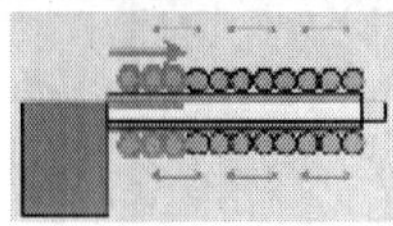
	Only the analytes adsorbed on the piezoresistor can be measured (Difficult to measure with p-type Si(100))	Only the analytes adsorbed on the piezoresistor can be measured (Difficult to measure with p-type Si(100))
MSS	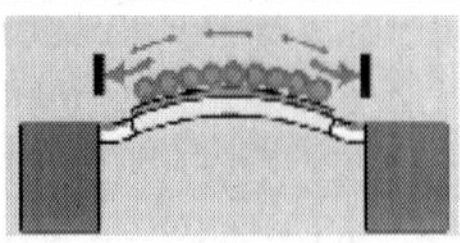	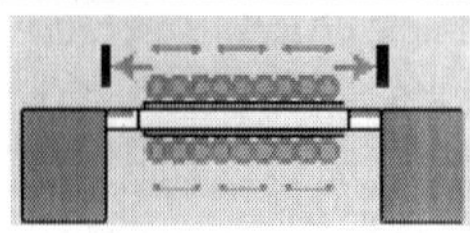
	All of the analytes can be measured as bending and in-plane stresses (Efficiently measured by a full Wheatstone bridge with p-type Si(100))	All of the analytes can be measured as in-plane stress (Efficiently measured by a full Wheatstone bridge with p-type Si(100))

The compatibility with the double-side coating makes it possible for MSS to be coated by virtually any kind of method including dip coating. To demonstrate this feature of MSS, one of the simplest coating methods was performed, that is, a hand-operated dip coating as shown in Fig. 6.6, which also presents the consistent signals obtained by this simple coating method. The dip coating becomes critically important in the case of the coating with self-assembled monolayers, which usually require the incubation in solution for several hours. Such process is rather difficult to perform with single-side coating methods.

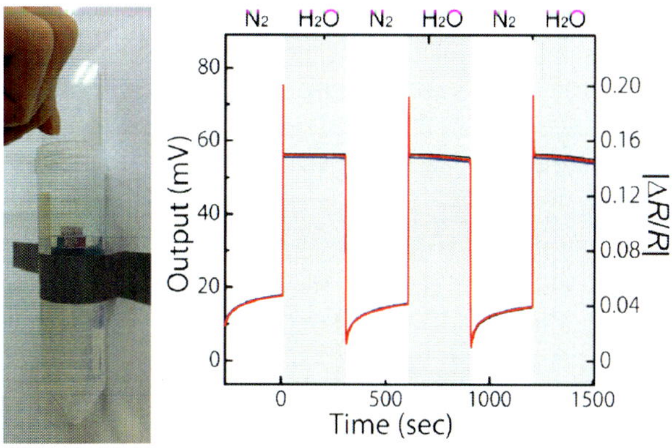

Figure 6.6 (Left) Photograph of the hand-operated dip coating for preparing a double-side-coated MSS chip. Observe that the setup is very simple and practical, and can be used for various coating solutions. (Right) Obtained output signals for three MSS membranes on the same double-side-coated chip prepared by hand-dipping into PVP solution. Note that the three output signals (solid red, black, and blue lines) are almost completely overlapped and appear as almost a single line, demonstrating a high degree of consistency in coating.

In the case of double-side coating methods, all channels on a chip will be coated with the same receptor layer, thereby working as one channel. We call this approach as *one-chip-one-channel*.[40] A chip with multiple membranes can record iterative data that are important in assessing the reproducibility of the measurement. To measure multiple samples, one can prepare multiplc chips coated with different receptors instead of having one chip coated with different types of receptors on the same chip (*one-chip–multiple-channel*), which is rather difficult to prepare in most cases because of the technical problems such as cross-contamination, drying out while coating different channels, and so forth. This *one-chip–one-channel* approach will be also helpful in realizing standard chips by batch fabrication for mass production with acceptable lot-to-lot variation.

It should be noted that the double-side coating sacrifices some sensitivity compared to a single-side coating due to lack of bending stress as shown in Table 6.2. With the experiments and finite element analyses, the signal of the double-side-coated MSS is found to be about 30–40% of that of the single-side-coated MSS.[40] Nevertheless, MSS still yields a considerable amount of signal owing to the substantially enhanced sensitivity. We have also investigated the difference between single- and double-side coatings and have provided some guidelines for the optimization of the geometry for double-side coatings. The MSS with piezoresistors embedded at the neutral axis plane is an example of the optimized structures for the double-side coating because it is almost insensitive to the anisotropic coatings on both surfaces without suffering from the zero output issue. The detailed information is available in the recent publication.[40]

6.3.2 Wiring

In contrast to the optical read-out method, a piezoresistive sensor chip has to have an electric connection between the piezoresistors and the read-out circuit. Most piezoresistive sensors utilize the wire-bonding method to establish the electric connections. Although this approach will be effective for the fully automated production line, it is inconvenient for some experiments, especially for biological species that should be prevented from being dried out. Thus, we designed the MSS chips to be easily connected by a simple insertion into a standard 0.5 mm-pitch connector; this eliminates wire-bonding procedure, offering easy chip exchange. The 0.5 mm-pitch pads for the electrical connection of all channels as well as bias voltage and ground can be observed at the bottom of each MSS chip in Fig. 6.4. The actual procedure of this simple insertion is shown in Fig. 6.7. It takes only a few seconds to insert the chip into a connector, providing all required electrical input/output connections.

Figure 6.7 Photographs of the simple insertion procedure of a 2G-MSS chip into a standard 0.5 mm-pitch connector. It takes only a few seconds, providing all required input/output electrical connections. After insertion, the chip is ready for measurements.

Together with a hand-operated dip coating presented in Section 6.3.1, the preparation of MSS chips becomes very easy and simple. For example, the preparation of the MSS chip used for the measurements shown in Fig. 6.6 took only a few minutes starting from a bare chip taken out from a package.

6.3.3 Comparison with Cantilever-Type Sensors

With the structural optimization and the improvements in the related procedures, the MSS has become one of the most practical platforms among various nanomechanical sensors. Here we compare the MSS with a standard optical read-out cantilever sensor and a model piezoresistive cantilever. The comprehensive comparisons in various aspects are summarized in Table 6.3.

Table 6.3 Comparison between a standard optical read-out cantilever sensor, a model piezoresistive cantilever, and MSS

	Optical read-out cantilever	**Piezoresistive cantilever**	**MSS**
	(Typical cantilever with 500 × 100 × 1 μm^3)	**(Model cantilever reported in [9])**	**(2G-MSS)**
Sensitivity (limit of detection)	0.15–0.90 mN/m	5.0–7.5 mN/m	0.077–0.11 mN/m
Mechanical stability	Low	Low	High
Electrical stability	—	Moderate	High
Thermal stability	Moderate	Low	Moderate
Wiring	Not required	Wire bonding required	Simple insertion
Coating feasibility	Only single-side coating	Partial detection with both Single- and Double-side coatings	Efficient detection with both Single- and Double-side coatings
Coating reproducibility	Low	Low	High
Measurement conditions	Only in optically transparent media	Feasible in opaque liquid	Feasible in opaque liquid
Initial adjustment	Laser alignment required	Bridge offset adjustment (optional)	Bridge offset adjustment (optional)
Measurement system	Complex, large, and expensive	Simple, small, and low-cost	Simple, small, and low-cost
Other requirements	• Reflection layer (ex. Au thin film) • Refraction index correction	• Passivation of electrical components • Electrical noise consideration	• Passivation of electrical components • Electrical noise consideration

6.4 Conclusion

The recent developments of the MSS have been reviewed. With the comprehensive structural optimization for high sensitivity and stability as well as the improvements in related procedures including the coating and wiring, the MSS has become one of the most practical platforms among various nanomechanical sensors. In addition, the MSS is ready for the miniaturization and mass production owing to the compatibility with the CMOS fabrication process. Moreover, the MSS structure has a potential for further enhancement in sensitivity up to several orders of magnitude by changing the dimensions of the adsorbate membrane and sensing beams.[32] However, it is important to note that an effective amplification would be somewhat reduced because of practical requirements, such as properties of piezoresistors or passivation layers on the sensing beams. It is also important to notice that the non-linearity of the bridge circuit or fracture stress of silicon will take non-negligible effect in some extreme cases.

With the MSS structure, the practical sensing element for the static mode operation has been established. Although it is always important to explore a better platform, the MSS can satisfy almost all demands for nanomechanical sensors in static operation. Therefore, it is time to concentrate on the development of effective coating layers with high sensitivity, stability, and reproducibility. The analytical model which we recently formulated shows the dependence of material properties of a receptor layer on the nanomechanical signals, providing a guideline for designing effective coating layers in terms of pure mechanics.[41] It will be also important to find better coating methods and establish a standard protocol for a standard chip with a guaranteed performance. The optimized receptor layers together with the MSS platform will open the door to practical applications of the nanomechanical sensing.

Acknowledgements

The author expresses appreciation to Dr. Heinrich Rohrer for his initial investigations and discussions on piezoresistive surface stress sensing, which led to the invention and development of the MSS concept and technology. The author thanks for the indispensable contributions of Dr. Terunobu Akiyama, Dr. Frederic Loizeau, Dr. Sebastian Gautsch, and Dr. Peter Vettiger, to realize MSS chips and related technologies. The author thanks Dr. Kota Shiba and Dr. Gaku Imamura for the discussion and the development of the receptor layers. The author acknowledges Professor Masakazu Aono and Professor Tomonobu Nakayama from WPI Center for Materials Nanoarchitectonics (MANA), National Institute for Materials Science (NIMS), Tsukuba, Japan and Professor Nico de Rooij from Institute of Microengineering (IMT) of Ecole Polytechnique Federale de Lausanne (EPFL), Neuchatel, Switzerland, for their help and support. The author thanks the members of the Cantilever Array Sensor Group, Department of Physics, University of Basel, Switzerland, for their kind introduction to nanomechanical sensors, valuable discussions, and help and support. The author thanks Ms. Takako Sugiyama, Ms. Keiko Koda, Ms. Yuko Kameyama, Ms. Eri Sakon, and Ms. Ikumi Nakakubo for their technical assistance. The author thanks Mr. Max Palumbo, Mr. Cory J. Y. Lee, Mr. Mayuran Saravanapavanantham, Mr. Ricardo Jose S. Guerrero, Mr. Francis Nguyen, Mr. Kojo Gyamfi Adjei-Afriyie, and Mr. Aravind Ramaraju from University of Waterloo for the discussion and the development of peripheral electronics. The author thanks for the technical support by the staff of Comlab, the CSEM micro- and nanofabrication facility. The author appreciates the technical supports of the MANA foundry, especially Dr. Akihiko Ohi and Ms. Tomoko Ohki. The financial supports for the development of MSS were provided by World Premier International Research Center (WPI) Initiative on Materials Nanoarchitronics (MANA) and by the Grant-in-Aid for Young Scientist (B) 21750083 (2009) and a Grant-in-Aid for Young Scientists (A) 23685017 (2011), Grant-in-Aid for Scientific Research (B) 15H03588, MEXT, Japan; a Nakatani Foundation research grant; Research Foundation for Opto-Sceince and Technology (REFOST); TEPCO Memorial Foundation; and the Japan Science and Technology Agency (JST).

References

1. Lang, H. P., Baller, M. K., Berger, R., Gerber, C., Gimzewski, J. K., Battiston, F. M., Fornaro, P., Ramseyer, J. P., Meyer, E., and Güntherodt, H. J. (1999).

An artificial nose based on a micromechanical cantilever array, *Anal. Chim. Acta*, **393**, 59.

2. Fritz, J., Baller, M. K., Lang, H. P., Rothuizen, H., Vettiger, P., Meyer, E., Guntherodt, H. J., Gerber, C., and Gimzewski, J. K. (2000). Translating biomolecular recognition into nanomechanics, *Science*, **288**, 316.
3. Backmann, N., Zahnd, C., Huber, F., Bietsch, A., Pluckthun, A., Lang, H. P., Güntherodt, H. J., Hegner, M., and Gerber, C. (2005). A label-free immunosensor array using single-chain antibody fragments, *Proc. Natl. Acad. Sci. U. S. A.*, **102**, 14587.
4. Zhang, J., Lang, H. P., Huber, F., Bietsch, A., Grange, W., Certa, U., McKendry, R., Güntherodt, H.-J., Hegner, M., and Gerber, C. (2006). Rapid and label-free nanomechanical detection of biomarker transcripts in human RNA, *Nat. Nanotech.*, **1**, 214.
5. Braun, T., Ghatkesar, M. K., Backmann, N., Grange, W., Boulanger, P., Letellier, L., Lang, H. P., Bietsch, A., Gerber, C., and Hegner, M. (2009). Quantitative time-resolved measurement of membrane protein-ligand interactions using microcantilever array sensors, *Nat. Nanotech.*, **4**, 179.
6. Huber, F., Lang, H. P., Backmann, N., Rimoldi, D., and Gerber, C. (2013). Direct detection of a BRAF mutation in total RNA from melanoma cells using cantilever arrays, *Nat. Nanotech.*, **8**, 125.
7. Gimzewski, J. K., Gerber, C., Meyer, E., and Schlittler, R. R. (1994). Observation of a chemical-reaction using a micromechanical sensor, *Chem. Phys. Lett.*, **217**, 589.
8. Thundat, T., Warmack, R. J., Chen, G. Y., and Allison, D. P. (1994). Thermal and ambient-induced deflections of scanning force microscope cantilevers, *Appl. Phys. Lett.*, **64**, 2894.
9. Aeschimann, L., Meister, A., Akiyama, T., Chui, B. W., Niedermann, P., Heinzelmann, H., De Rooij, N. F., Staufer, U., and Vettiger, P. (2006). Scanning probe arrays for life sciences and nanobiology applications, *Microelectron. Eng.*, **83**, 1698.
10. Akiyama, T., Gautsch, S., de Rooij, N. F., Staufer, U., Niedermann, P., Howald, L., Muller, D., Tonin, A., Hidber, H. R., Pike, W. T., and Hecht, M. H. (2001). Atomic force microscope for planetary applications, *Sens. Actuators A*, **91**, 321.
11. Gautsch, S., Akiyama, T., Imer, R., de Rooij, N. F., Staufer, U., Niedermann, P., Howald, L., Brandlin, D., Tonin, A., Hidber, H. R., and Pike, W. T. (2002). Measurement of quartz particles by means of an atomic force microscope for planetary exploration, *Surf. Interface Anal.*, **33**, 163.

12. Mukhopadhyay, R., Sumbayev, V. V., Lorentzen, M., Kjems, J., Andreasen, P. A., and Besenbacher, F. (2005). Cantilever sensor for nanomechanical detection of specific protein conformations, *Nano Lett.*, **5**, 2385.

13. Tortonese, M., Barrett, R. C., and Quate, C. F. (1993). Atomic resolution with an atomic force microscope using piezoresistive detection, *Appl. Phys. Lett.*, **62**, 834.

14. Yoshikawa, G., Lang, H. P., Akiyama, T., Aeschimann, L., Staufer, U., Vettiger, P., Aono, M., Sakurai, T., and Gerber, C. (2009). Sub-ppm detection of vapors using piezoresistive microcantilever array sensors, *Nanotechnology*, **20**, 015501.

15. Boisen, A., and Thundat, T. (2009). Design & fabrication of cantilever array biosensors, *Mater. Today*, **12**, 32.

16. Hecht, M. H., Marshall, J., Pike, W. T., Staufer, U., Blaney, D., Braendlin, D., Gautsch, S., Goetz, W., Hidber, H. R., Keller, H. U., Markiewicz, W. J., Mazer, A., Meloy, T. P., Morookian, J. M., Mogensen, C., Parrat, D., Smith, P., Sykulska, H., Tanner, R. J., Reynolds, R. O., Tonin, A., Vijendran, S., Weilert, M., and Woida, P. M. (2008). Microscopy capabilities of the Microscopy, Electrochemistry, and Conductivity Analyzer, *J. Geophys. Res.-Planet.*, **113**, DOI: 10.1029/2008JE003077.

17. Hierlemann, A., Lange, D., Hagleitner, C., Kerness, N., Koll, A., Brand, O., and Baltes, H. (2000). Application-specific sensor systems based on CMOS chemical microsensors, *Sens. Actuators B*, **70**, 2.

18. Lutwyche, M., Andreoli, C., Binnig, G., Brugger, J., Drechsler, U., Haberle, W., Rohrer, H., Rothuizen, H., Vettiger, P., Yaralioglu, G., and Quate, C. (1999). 5 × 5 2D AFM cantilever arrays a first step towards a Terabit storage device, *Sens. Actuators A*, **73**, 89.

19. Thaysen, J., Boisen, A., Hansen, O., and Bouwstra, S. (2000). Atomic force microscopy probe with piezoresistive read-out and a highly symmetrical Wheatstone bridge arrangement, *Sens. Actuators A*, **83**, 47.

20. Aeschimann, L., Goericke, F., Polesel-Maris, J., Meister, A., Akiyama, T., Chui, B., Staufer, U., Pugin, R., Heinzelmann, H., Rooij, N. F. D., King, W. P., and Vettiger, P. (2007). Piezoresistive scanning probe arrays for operation in liquids, *J. Phys.*, **61**, 6.

21. Kanda, Y. (1982). A graphical representation of the piezoresistance coefficients in silicon, *IEEE Trans. Electron Devices*, **29**, 64.

22. Kanda, Y. (1991). Piezoresistance effect of silicon, *Sens. Actuators A*, **28**, 83.

23. Pfann, W. G., and Thurston, R. N. (1961). Semiconducting stress transducers utilizing transverse and shear piezoresistance effects, *J. Appl. Phys.*, **32**, 2008.

24. Goericke, F. T., and King, W. P. (2008). Modeling piezoresistive microcantilever sensor response to surface stress for biochemical sensors, *IEEE Sens. J.*, **8**, 1404.

25. He, J. H., and Li, Y. F. (2006). High sensitivity piezoresistive cantilever sensor for biomolecular detection, *J. Phys.*, **34**, 429.

26. Loui, A., Goericke, F. T., Ratto, T. V., Lee, J., Hart, B. R., and King, W. P. (2008). The effect of piezoresistive microcantilever geometry on cantilever sensitivity during surface stress chemical sensing, *Sens. Actuators A*, **147**, 516.

27. Privorotskaya, N. L., and King, W. P. (2008). The mechanics of polymer swelling on microcantilever sensors, *Microsyst. Technol.*, **15**, 333.

28. Yu, H., Braun, P., Yildirim, M. A., Lemmens, I., Venkatesan, K., Sahalie, J., Hirozane-Kishikawa, T., Gebreab, F., Li, N., Simonis, N., Hao, T., Rual, J. F., Dricot, A., Vazquez, A., Murray, R. R., Simon, C., Tardivo, L.; Tam, S., Svrzikapa, N., Fan, C., de Smet, A. S., Motyl, A., Hudson, M. E., Park, J., Xin, X., Cusick, M. E., Moore, T., Boone, C., Snyder, M., Roth, F. P., Barabasi, A. L., Tavernier, J., Hill, D. E., and Vidal, M. (2008). High-quality binary protein interaction map of the yeast interactome network, *Science*, **322**, 104.

29. Rasmussen, P. A., Hansen, O., and Boisen, A. (2005). Cantilever surface stress sensors with single-crystalline silicon piezoresistors, *Appl. Phys. Lett.*, **86**, 203502.

30. Yoshikawa, G., and Rohrer, H. (2010). *Strain Amplification Schemes for Piezoresistive Cantilevers*, paper presented at the 7th International Workshop on Nanomechanical Cantilever Sensors, Banff.

31. Yang, S. M., Yin, T. I., and Chang, C. (2007). Development of a double-microcantilever for surface stress measurement in microsensors, *Sens. Actuators B*, **121**, 545.

32. Yoshikawa, G., Akiyama, T., Gautsch, S., Vettiger, P., and Rohrer, H. (2011). Nanomechanical membrane-type surface stress sensor, *Nano Lett.*, **11**, 1044.

33. Yoshikawa, G., Akiyama, T., Loizeau, F., Shiba, K., Gautsch, S., Nakayama, T., Vettiger, P., Rooij, N. F. D., and Aono, M. (2012). Two dimensional array of piezoresistive nanomechanical membrane-type surface stress sensor (MSS) with improved sensitivity, *Sensors*, **12**, 15873.

34. Bietsch, A., Zhang, J. Y., Hegner, M., Lang, H. P., and Gerber, C. (2004). Rapid functionalization of cantilever array sensors by inkjet printing, *Nanotechnology*, **15**, 873.
35. McKendry, R., Zhang, J. Y., Arntz, Y., Strunz, T., Hegner, M., Lang, H. P., Baller, M. K., Certa, U., Meyer, E., Güntherodt, H. J., and Gerber, C. (2002). Multiple label-free biodetection and quantitative DNA-binding assays on a nanomechanical cantilever array, *Proc. Natl. Acad. Sci. U. S. A.*, **99**, 9783.
36. Henriksson, J., Villanueva, L. G., and Brugger, J. (2012). Ultra-low power hydrogen sensing based on a palladium-coated nanomechanical beam resonator, *Nanoscale*, **4**, 5059.
37. Rasmussen, P. A., Grigorov, A. V., and Boisen, A. (2005). Double sided surface stress cantilever sensor, *J. Micromech. Microeng.*, **15**, 1088.
38. Choudhury, A., Hesketh, P. J., Thundat, T. G., Hu, Z., Vujanic, R., and IEEE, Design and testing of single and double sided cantilevers for chemical sensing. In *2007 IEEE Sensors, Vols 1–3*, IEEE: New York, 2007; pp. 1432.
39. Choudhury, A., Hesketh, P. J., Thundat, T., and Hu, Z. Y. (2007). A piezoresistive microcantilever array for surface stress measurement: Curvature model and fabrication, *J. Micromech. Microeng.*, **17**, 2065.
40. Yoshikawa, G., Loizeau, F., Lee, C. J., Akiyama, T., Shiba, K., Gautsch, S., Nakayama, T., Vettiger, P., de Rooij, N. F., and Aono, M. (2013). Double-side-coated nanomechanical membrane-type surface stress sensor (MSS) for one-chip-one-channel setup, *Langmuir*, **29**, 7551.
41. Yoshikawa, G. (2011). Mechanical analysis and optimization of a microcantilever sensor coated with a solid receptor film, *Appl. Phys. Lett.*, **98**, 173502.

Chapter 7

Mechanical Properties Characterization of PZT Nanofibers

Xi Chen,[a] Nan Yao,[b] and Yong Shi[a]

[a]*Department of Mechanical Engineering, Stevens Institute of Technology, Castle Point on Hudson, Hoboken, NJ 07030, USA*
[b]*Princeton Institute for the Science and Technology of Materials, Princeton University, 70 Prospect Avenue, Princeton, NJ 08540, USA*

yong.shi@stevens.edu

7.1 Introduction

Recently, the development of several piezoelectric nanowires, nanofibers, and nanorods has been reported.[1] For example, zinc oxide nanowire,[2] PZT nanofiber,[3] cadmium sulfide nanowire,[4] barium titanate,[5] and gallium nitride[6] have been demonstrated to convert mechanical energy to electric energy due to the piezoelectric effect of these materials. Among these materials, PZT is a typical piezoelectric material with a higher electromechanical coupling coefficient, higher dielectric constant, and large remnant polarization.[3,7] Electrospun PZT nanofiber developed by Prof. Y. Shi[3] is a potential candidate for building nanofiber devices.

Nanocantilever Beams: Modeling, Fabrication, and Applications
Edited by Ioana Voiculescu and Mona Zaghloul

ISBN 978-981-4613-23-1 (Hardcover), 978-981-4613-24-8 (eBook)
www.panstanford.com

The mechanical properties of as-spun PZT nanofibers are desired to be understood for future sensors and actuators applications. Electric field–induced resonant excitation based on in situ TEM[8,9] or SEM[10] was developed for directly measuring the mechanical properties of individual nanowire structures. Here, we achieve Young's modulus measurement of single PZT nanofibers based on this technology. Young's modulus of PZT nanofiber can be calculated from the resonant frequency excited by an oscillating electric field.

In addition, the ability to precisely control the mechanical properties of piezoelectric nanostructures is highly desirable for certain applications, such as nanocantilevers,[11] nanoresonators,[12] nanogenerators,[13,14] and nanosensors.[15] However, the mechanical properties are especially difficult to control at the nanoscale. For example, Young's modulus of nanowires can vary widely as a result of the structural defects and residual stresses introduced during synthesis. The applications of these nanosensors that are based on the resonant method[16] will be restricted by this structural effect. Young's modulus of individual ZnO nanowires has been shown to be adjustable by using ion implantation to achieve a much lower resonant frequency.[10] However, the implanted Ga^+ ions also change other material properties that can cause partial loss of the functionality of the nanostructures. Here, we demonstrate an alternative method to adjust the stiffness of individual PZT nanofibers by low energy electron beam–induced polarization.

7.2 PZT Nanofiber Preparation[17]

PZT nanofibers were prepared by the electrospinning process.[18] The starting material, PZT (52/48) sol-gel and poly vinyl pyrrolidone (PVP, Aldrich) were obtained from commercial sources. Alcohol was used as a solvent for PVP, while acetic acid was added to stabilize the solution and to control the hydrolysis reaction of the sol-gel precursor.[19] After being stirred vigorously for 2 h at room temperature, the mixture was fed into a microscale metallic tube through a syringe pump. A droplet of the modified sol-gel solution was held at the orifice of the metallic tube due to the high surface tension. A high DC voltage (10 kV) between the tip of the micro metallic tube and the collecting substrate was applied to overcome the surface tension and generate an electrically charged jet of the modified solution. After electrospinning, the as-spun nanofibers were annealed at 650°C to obtain PZT nanofibers with the perovskite phase.

Figure 7.1 shows the morphology and X-ray diffraction pattern of the annealed PZT nanofibers. Figure 7.1a shows the SEM image of a selected nanofiber with a diameter of 70 nm. By varying the concentration of the polymer in the precursor, the average diameters of the nanofibers could be tuned from 52 ± 8 nm to 150 ± 12 nm. The X-ray diffraction (XRD) pattern of the annealed PZT nanofibers shown in Fig. 7.1b indicated that the PZT nanofibers annealed at 650°C had pure perovskite phase. The annealing temperature was lower than that used by most other methods. Transmission electron microscopy (TEM) images of the cross section and surface of PZT nanofibers are shown in Figs. 7.1c,d. The grain size of the PZT nanofibers was about 10 nm. The structure of the nanofiber seemed porous probably because of the polymer precursor solution of the so-gel process.[20] By modifying the precursor and controlling the electrospinning parameters, a denser nanostructure could be obtained.

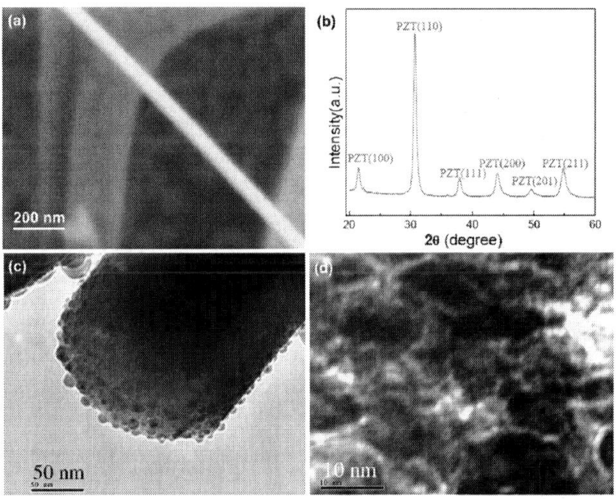

Figure 7.1 (a) SEM image of single PZT nanofiber. (b) X-ray diffraction pattern for the PZT nanofibers. (c) and (d) TEM images of a PZT nanofiber. Reproduced from [3] by permission of the American Institute of Physics (copyright 2009).

A 1 cm × 2 cm silicon wafer with micron size islands, which was fabricated by using deep reactive ion etching (DIRE), was used to collect the nanofibers for mechanical properties characterization (see Fig. 7.2). The diameter of the fabricated PZT nanofiber was controlled between 75 and 85 nm (inset in Fig. 7.2).

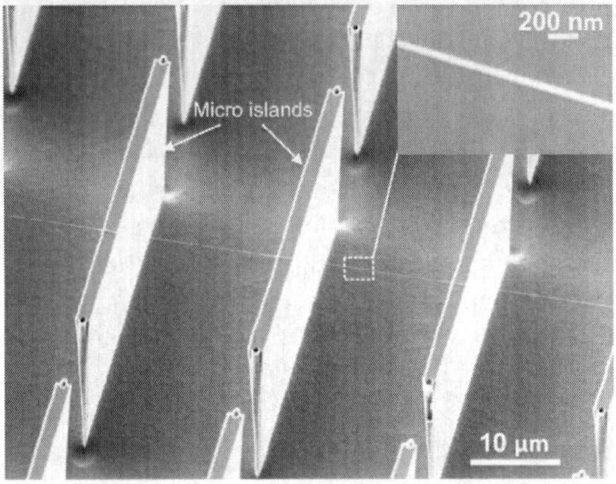

Figure 7.2 SEM image of an individual PZT nanofiber deposited on the micro scale islands. Reproduced from [1] by permission of the American Institute of Physics (copyright 2011).

7.3 Young's Modulus Determination

Traditionally, Young's modulus of a material is measured from the displacement when a force is applied. However, stretching of an individual PZT nanofiber is extremely difficult due to the small size and fracture force. Here, another measurement technique based on resonant frequency was utilized to measure the Yong's modulus of individual electrospun PZT nanofibers.

7.3.1 Electric Field–Induced Resonant Excitation Method

The elastic modulus of individual PZT nanofibers were directly measured using the electric field-induced resonant excitation method. The measurement was performed in the chamber of an SEM equipped with a nanomanipulator. The nanofiber could be activated by the force generated between the surface charge and the applied oscillating electric field. The mechanical resonance was induced when the frequency of the input signal reached the natural frequency of the nanofiber. The largest amplitude of the vibration was observed through the SEM image. By using Bernoulli–Euler analysis, the equation of motion for the forced vibration of a uniform nanofiber is given as[21]

$$YI\frac{\partial^4 w}{\partial x^4}(x,t)+\rho A\frac{\partial^2 w}{\partial t^2}(x,t)=F(x,t), \tag{7.1}$$

where Y is Young's modulus, I is the moment of inertia, ρ is the mass density, A is the cross-sectional area, $w(x, t)$ is the lateral displacement from the equilibrium axis, and $F(x, t)$ is the external force per unit length of nanofiber at position 7. By applying the fixed free boundary condition, the resonant frequency of a uniform cantilever is given by

$$f_{\mathrm{j}}=\left[\beta_{\mathrm{j}}^2 d/8\pi l^2\right]\sqrt{Y/\rho}, \tag{7.2}$$

where f_{j} is the j-th resonant frequency, d is the cross-sectional diameter, l is length and β_{j} ($\beta_1 = 1.875$, $\beta_2 = 4.694$, $\beta_3 = 7.855$...) is a constant for the j-th resonant frequency. Thus, Young's modulus of PZT nanofiber can be simply calculated from Eq. 7.2 as follows:

$$Y=64\pi^2 f_{\mathrm{j}}^2 l^4 \rho/\beta_{\mathrm{j}}^4 d^2 \tag{7.3}$$

Since the external force can shift the resonant frequency of PZT nanofiber, a low amplitude oscillating signal was used to reduce the error of Young's modulus calculated from this simplified model.

7.3.2 Young's Modulus Measurement of PZT Nanofiber

An unpoled PZT nanofiber was ruptured from the collecting substrate and picked up using the tungsten tip of a nanomanipulator under an SEM as shown in Fig. 7.3a. After being fixed on the tungsten tip, the nanofiber was lifted up a certain distance above the silicon substrate. Another tip was used to apply the oscillating electric field by approaching the free end of the PZT nanofiber (inset in Fig. 7.3a). By applying an AC signal of 0.1 V, the fiber reached the first resonant mode at certain frequency and observed through the live SEM image (Fig. 7.3b).

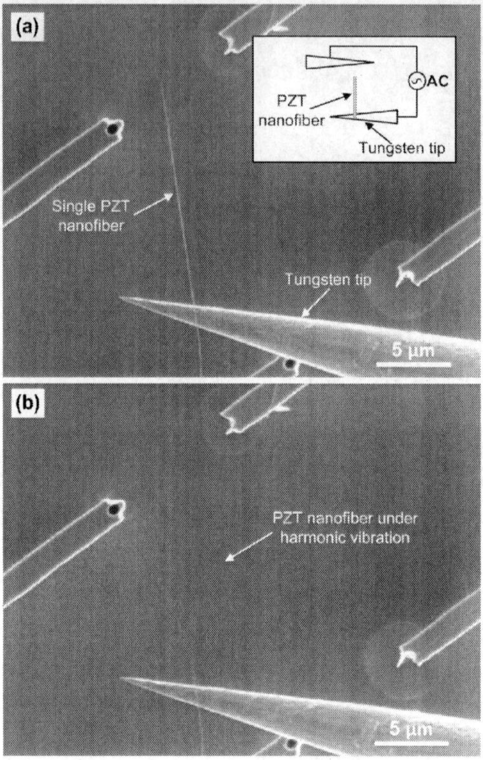

Figure 7.3 (a) The SEM image of a single PZT nanofiber fixed and lifted by a tungsten tip, the inset in (a) shows the experiment set up to induce the vibration of PZT nanofibers, (b) The SEM image of a PZT nanofiber oscillating under an applied electric field. Reproduced from [1] by permission of the American Institute of Physics (copyright 2011).

During the experiment, a network analyzer (4096B) was utilized to precisely find the resonant frequency of the nanofiber from the transfer frequency spectra of the electric field. The setup was illustrated in Fig. 7.4a. The network analyzer was utilized to apply a swept frequency in a certain range through the tungsten tip of the nanomanipulator. The free tip was used to sense the changing of the local electric field by sending the signal back to the network analyzer. The largest amplitude of the vibration was observed through the SEM image and the transfer frequency spectrum was recorded at the same time. The resonant frequency of a single nanofiber was confirmed by both the SEM image and the amplitude peak of the frequency spectrum. Figure 7.4b shows an example of the magnitude peak of frequency spectrum was detected at the resonant frequency. From the frequency spectrum, the exact value of the resonant peak of 1.04 MHz was measured to be used for Young's modulus calculation. The left peak in Fig. 7.4b might have been caused by the first out-of-plane resonant mode, which was difficult to observe in this certain direction from the live SEM image.

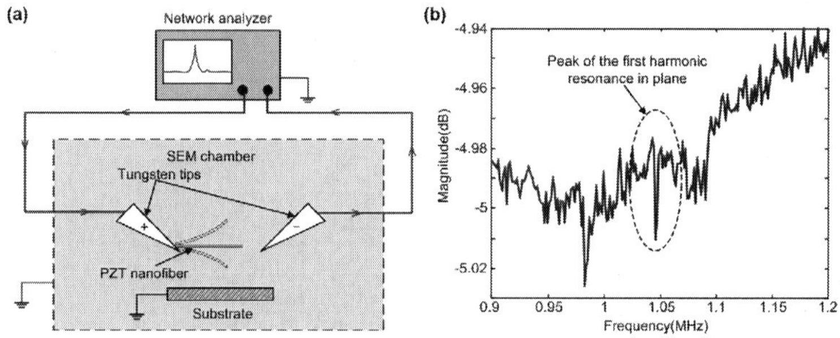

Figure 7.4 (a) Schematic of the network analyzer setup, (b) The transfer frequency spectrum during the resonance excitation. Reproduced from [22] by permission of the American Scientific Publishers (copyright 2011).

Measurements were performed on PZT nanofibers with different lengths. The first-harmonic resonant frequencies and calculated elastic modulus are shown in Fig. 7.5. The upper plot is the first harmonic resonant frequencies for fibers of various lengths. The lower plot is the calculated elastic modulus of PZT nanofibers with various lengths. The elastic modulus of PZT nanofiber varied from 46.4 to 99.3 GPa due to possible reasons such as different residual stresses and surface structures formed during the fabrication process. The average elastic modulus is similar with the value measured for bulk PZT, which ranges from 63 to 94 GPa.[23] However, Young's modulus measured for PZT nanofibers has shown a large variation.

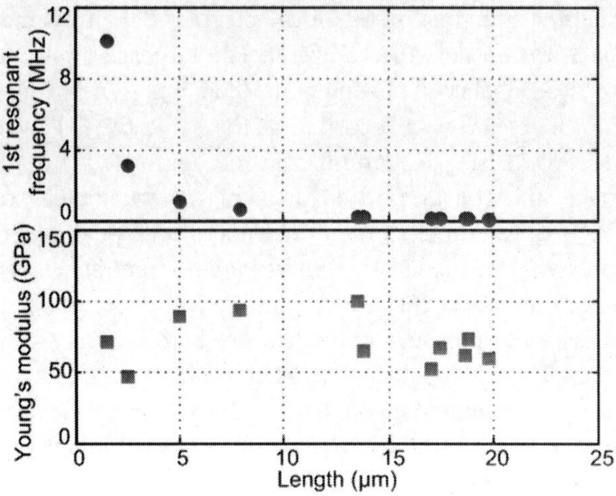

Figure 7.5 The first harmonic resonant frequencies and calculated elastic modulus for fibers of various lengths. Reproduced from [1] by permission of the American Institute of Physics (copyright 2011).

7.4 Stiffness Adjustment of PZT Nanofiber

7.4.1 Electron Beam–Induced Polarization

The ultralow stiffness at the domain boundaries was discovered in bulk PZT material by using piezoresponse force microscopy (PFM).[24] These boundaries exhibit much lower stiffness than the domain structures. In the entire mechanical structure, the stiffness of piezoelectric material could be considered by the average modulus, which could be expressed by[24]

$$\overline{Y} = (1 - R_{DB})Y_D + R_{DB}Y_{DB}, \tag{7.4}$$

where Y_D and Y_{DB} are Young's modulus of the domain and domain boundary, respectively, and R_{DB} is the area ratio of the domain boundary. Thus, the smaller the area ratio of the domain boundary, the higher the stiffness of the piezoelectric nanostructure. The average Young's modulus could be easily adjusted if the area ratio of the domain boundaries were controllable throughout the whole structure. The polarization $\vec{P}$ will affect the ratio of domain boundaries and can be used to adjust the stiffness of the nanostructures.

Electron beam has been used to manipulate the polarization orientation and achieve the nanoscale domain patterning of a PZT thin film.[25] The domain and domain boundary structures of a single PZT nanofiber are illustrated on upper right in Fig. 7.6. The PZT nanofiber has random domain configuration without poling. To induce the polarization, the electron beam of a typical SEM was used to apply an external electric field ($\vec{E}$). Electrons stay on the top surface of nanofiber for a long time due to the high dielectric coefficient of PZT. The depolarization energy of half of a prolate spheroid is given by[26]

$$U_{\mathrm{d}} = \frac{16\pi^2 P_{\mathrm{s}}^2}{3\varepsilon_{\mathrm{a}}}\left[\ln\left(2\frac{l}{r}\sqrt{\frac{\varepsilon_{\mathrm{a}}}{\varepsilon_{\mathrm{c}}}}\right) - 1\right]\frac{r^4}{l}, \tag{7.5}$$

where P_{s} is the magnitude of the spontaneous polarization, r is the domain radius, l is the domain length, and ε_{a} and ε_{c} are the permittivity in the polar direction and in a perpendicular direction, respectively.

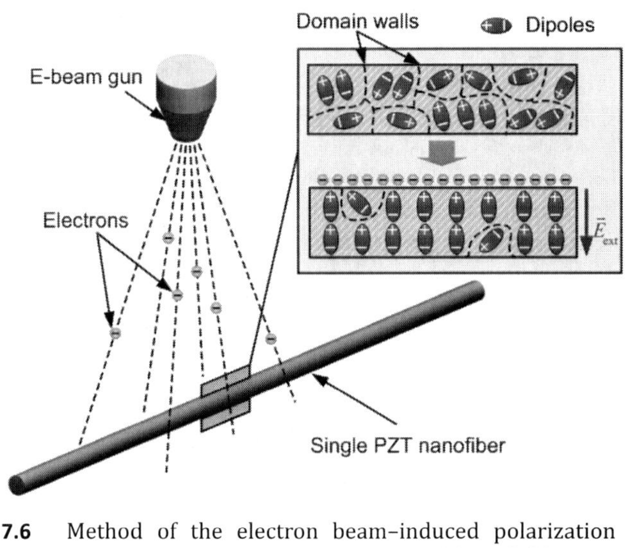

Figure 7.6 Method of the electron beam–induced polarization of an individual PZT nanofiber. Reproduced from [1] by permission of the American Institute of Physics (copyright 2011).

Typically, an external electric field of 4 V/μm is desired for the polarization process for thin film structures. The required electrons are roughly simulated from the electric field across a single nanofiber, which can be simply written as

$$\vec{E}_{\text{ext}} = Q/(2\varepsilon\varepsilon_0 A^*), \tag{7.6}$$

where Q is the charge accumulated on top surface of nanofiber, ε is the relative static permittivity, ε_0 is the absolute permittivity and A^* is considered as the top surface of a single nanofiber exposed under electron beam. The required number of electrons can be approximately calculated as

$$n = Q/e = 2\vec{E}_{\text{ext}}\varepsilon\varepsilon_0 A^*/e, \tag{7.7}$$

where e is the electric charge of a single electron. By using Eq. 7.7, the required number of electrons is calculated as 76 electrons per 100 nm^2 area. The external electric field formed by this number of electron can align the dipoles. Domain boundaries, which determine average Young's modulus of PZT nanofiber, can thus be modified by controlling the exposure time and strength of the electron beam.

7.4.2 Experiment

A single PZT nanofiber, with a diameter of 82 nm and a length of 17 μm, was selected for stiffness modification. Young's modulus of this PZT nanofiber was determined to be 51.2 GPa using the resonant excitation method. The electron beam was first set up with an acceleration voltage of 5 kV and the current on the sample was at picoampere level. The electron beam was focused on the nanofiber and the resonant frequency was measured every minute. The resonant frequency of the PZT nanofiber continually increased with exposure to the electron beam.

As shown in Figs. 7.7a,b, over the course of 30 min, the resonant frequency shifted from 102 to 111.6 kHz, and the corresponding shift of Young's modulus was calculated to be from 51.2 to 64.5 GPa. The induced polarization reduced the domain boundaries in PZT nanofibers and resulted in increasing of the stiffness. The data was fitted to a power function

$$Y(t) = at^b + c, \tag{7.8}$$

where $a = 2.438$, $b = 0.4296$, and $c = 5.092 \times 10^{10}$.

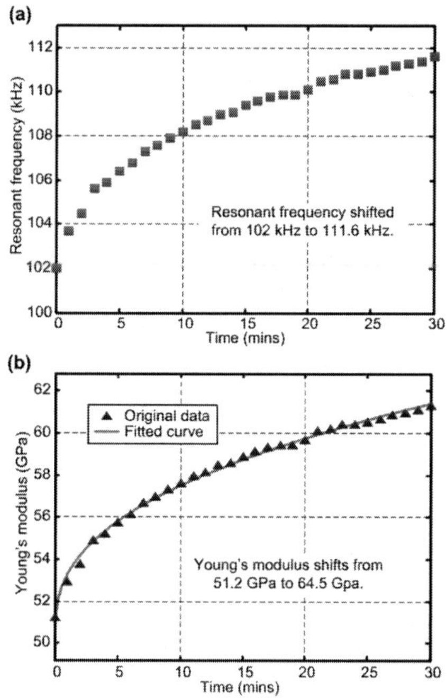

Figure 7.7 (a) The resonant frequencies shift during 30 min of electron beam exposure. (b) The corresponding elastic modulus and the fitted curve to a power function. Reproduced from [1] by permission of the American Institute of Physics (copyright 2011).

Additionally, a second resonant mode at 109.9 kHz (59.4 GPa) was observed after 30 min exposure as shown in Fig. 7.8a. As shown in Fig. 7.8b, the left figure shows the vibrations in the vertical resonant mode. The narrower and sharper arc and asymmetric oscillation pattern indicate that the wire is moving up and down relative to the substrate, with the plane of vibrations more closely parallel to the direction of the electron beam. The right figure shows the same sample vibrating in the horizontal resonant mode. The broad arc and symmetry of the oscillation pattern indicate that the vibrations occur in a plane perpendicular to the direction of the electron beam. Based on images of the resonant modes, it was deduced that the vertical resonant mode corresponded to the lower frequency and the horizontal resonant mode corresponded to the higher frequency (Fig. 7.8b). This two-mode behavior exhibits the anisotropic stiffness in polarized PZT nanofiber according to the theoretical model of bulk PZT materials.[27,28] This phenomenon indicates that the electron deposited on the surface of nanofiber can induce the polarization and reduce the ratio of the domain boundary in PZT nanofiber structures.

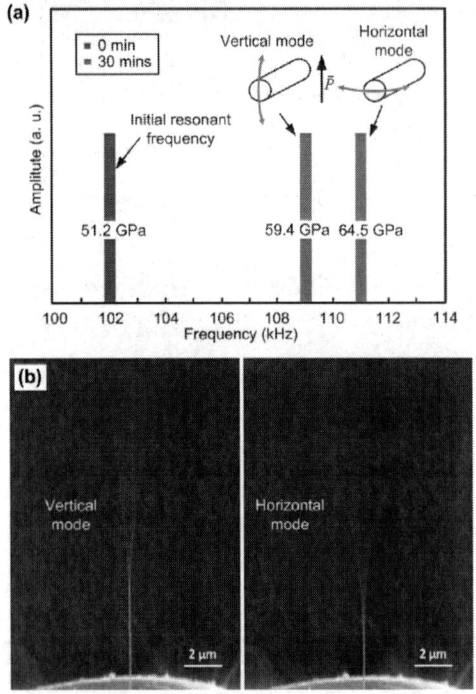

Figure 7.8 (a) The plot shows the two resonant modes after the 30 min exposure, the inset in (a) shows the vibration direction according to the polarized direction. (b) Two different resonant modes due to polarization, scale bar = 2 μm. Reproduced from [1] by permission of the American Institute of Physics (copyright 2011).

The stabilization of the polarized structure might not stable due to the nanoscale diameter of PZT nanofiber. The stabilization of the nanoscale polarized domain structure is related to the critical thickness, which was calculated to be less than 20 nm at room temperature.[29,30] Even for a 4 nm-thick film, it was experimentally shown that polarized domain structures are stable for over 140 h.[31] In order to further study the stability of the poled domain structure and polarizing effects of the electron beam on PZT nanowires, we performed a lengthened experiment. In this experiment (Fig. 7.9), a single sample was exposed to a 5 kV electron beam with same beam intensity as before in three separate sessions (80, 250, and 280 min) spread over three days. The dotted lines in Fig. 7.9 indicate the beginning of a new session after allowing the system to rest overnight with no exposure. At the end of each session, the beam was turned off and the system was left undisturbed until the next session.

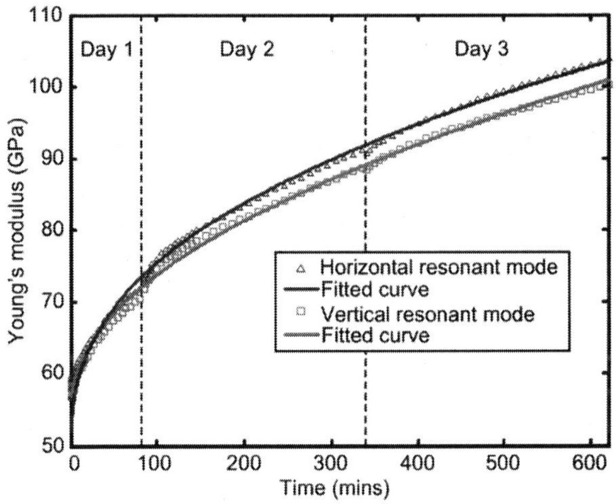

Figure 7.9 Plot of calculated Young's modulus as a function of electron beam exposure time. Reproduced from [1] by permission of the American Institute of Physics (copyright 2011).

From the start of the experiment to the end of the third session (total of 10 h of electron beam exposure), a very large continuous shift from 123.8 kHz (56.9 GPa) to 164.2 kHz (100.1 GPa) was observed, representing a 33% increase in resonance frequency or a 76% increase in Young's modulus. As shown in Fig. 7.9, the continuous Young's modulus between each session demonstrates that the polarization was stable for the time scale of a day in the

PZT nanofiber. Splitting of resonant frequencies due to anisotropic polarization occurred within minutes of exposure, beginning with a 1.7 kHz split that widened to 3.0 kHz by the end of the experiment. Following the power function model (Eq. 7.8), $a_1 = 1.882$, $b_1 = 0.496$ and $c_1 = 5.53 \times 1010$ were used to fit the data of vertical resonant mode and $a_2 = 4.148$, $b_2 = 0.3995$ and $c_2 = 4.93 \times 10^{10}$ for horizontal resonant mode. This result raises the question of the practical limit of stiffening these nanowires using electron beam–induced polarization.

The electrostatic charging and electron beam–induced deposition (EBID) on surface of PZT nanofibers might also have effect on the measured resonant frequencies. To eliminate the influence, the PZT nanofiber was discharged by conductive tungsten tips several times during the experiment. No significant difference was found in the resonant frequencies after discharging. EBID also results in the deposition of tiny fragments on the nanofiber surface due to particulate matter present in the chamber. To reduce this effect, the SEM chamber was pumped overnight to achieve a high vacuum level. On the other hand, this discontinuous deposition is expected to cause a decrease in the resonant frequency due to the increasing of the mass on nanofiber structure.[32] Moreover, the carbon deposited on the top surface will induce the vertical resonant mode appearing at a higher frequency, as the thickness would increase primarily along the vertical direction. The polarization effect is supported by the experimental splitting effect of the resonant frequencies where the vertical mode appeared at a lower frequency. By using this method, the stiffness of single piezoelectric nanostructure can be precisely adjusted by controlling the exposure time under the electron beam.

7.5 Conclusion

Young's modulus of single electrospun PZT nanofibers was measured under harmonic vibration using in situ SEM equipped with a nanomanipulator. SEM live image and transfer frequency spectrum were simultaneously used to find the resonant frequency of individual nanofibers. Young's modulus was determined as ~70 GPa from the first resonant frequency using the Euler–

Bernoulli equation. The surface charge of PZT nanofiber changed the measurement value slightly. A method to adjust the stiffness of individual piezoelectric nanostructures by electron beam–induced polarization under an in situ SEM was presented. PZT nanofibers with a diameter of ~80 nm were fabricated by an electrospinning process. Young's modulus of an individual PZT nanofiber shifted from 56.9 to 100.1 GPa after 10 h of electron beam exposure. Two resonant modes were observed corresponding to the strengthened anisotropic stiffness after the electron beam–induced polarization. The stiffness of single PZT nanofibers could be precisely adjusted up to 76% of the original Young's modulus by the polarization effect. The polarized nanofibers demonstrated a stable Young's modulus after the external electric field was removed. This method of electron beam–induced polarization provides a way to tune the mechanical properties of piezoelectric nanostructures before employing them in devices and other applications.

Acknowledgements

This work was supported in part by the National Science Foundation through Steven Institute of Technology (ECCS-0802168; 7.C., Y.S.,) and through the Princeton Center for Complex Materials (DMR-0819860; A.L., N.Y.). The authors would also like to thank Dr. Y. T. Tsai (Anderson), W. Xu and R. Galos for the help of silicon substrate and useful discussions.

References

1. Chen, X., Li, A., Yao, N., and Shi, Y. (2011). Adjustable stiffness of individual piezoelectric nanofibers by electron beam polarization, *Appl. Phys. Lett.*, **99**, 193102.
2. Wang, Z. L., and Song, J. H. (2006). Piezoelectric nanogenerators based on zinc oxide nanowire arrays, *Science*, **312**, 242–246.
3. Chen, X., Xu, S. Y., Yao, N., Xu, W., and Shi, Y. (2009). Potential measurement from a single lead zirconate titanate nanofiber using a nanomanipulator, *Appl. Phys. Lett.*, **94**, 253113.
4. Lin, Y. F., Song, J. H., Ding, Y., Lu, S. Y., and Wang, Z. L. (2008). Piezoelectric nanogenerator using CdS nanowires. *Appl. Phys. Lett.*, **92**, 022105.

5. Wang, Z. Y., Hu, J., Suryavanshi, A. P., Yum, K., and Yu, M. F. (2007). Voltage generation from individual $BaTiO_3$ nanowires under periodic tensile mechanical load, *Nano Lett.*, **7**, 2966–2969.
6. Su, W. S., Chen, Y. F., Hsiao, C. L., and Tu, L. W. (2007). Generation of electricity in GaN nanorods induced by piezoelectric effect, *Appl. Phys. Lett.*, **90**, 063110.
7. Pandey, S. K., James, A. R., Raman, R., and Chatterjee, S. N. (2005). Structural, ferroelectric and optical properties of PZT thin films, *Physica B*, **369**, 135.
8. Poncharal, P., Wang, Z. L., Ugarte, D., and de Heer, W. A. (1999). Electrostatic deflections and electromechanical resonances of carbon nanotubes, *Science*, **283**, 1513–1516.
9. Bai, 7. D., Gao, P. 7., and Wang, Z. L. (2003). Dual-mode mechanical resonance of individual ZnO nanobelts, *Appl. Phys. Lett.*, **82**, 4806–4808.
10. Cohen-Tanugi, D., Akey, A., Yao, N. (2010). Ultralow superharmonic resonance for functional nanowires, *Nano Lett.*, **10**, 852–859.
11. Ke, C., and Espinosa, H. D. (2004). Feedback controlled nanocantilever device, *Appl. Phys. Lett.*, **85**, 681.
12. Verbridge, S. S., Shapiro, D. F., Craighead, H. G., and Parpia, J. M. (2007). Macroscopic tuning of nanomechanics: substrate bending for reversible control of frequency and quality factor of nanostring resonators, *Nano Lett.*, **7**, 1728.
13. Chen, 7., Xu, S. Y., Yao, N., and Shi, Y. (2010). 1.6 volt nanogenerator for mechanical energy harvesting using PZT nanofibers, *Nano Lett.*, **10**, 2133–2137.
14. Hu, Y., Zhang, Y., Xu, C., Zhu, G., and Wang, Z. L. (2010). High-output nanogenerator by rational unipolar assembly of conical nanowires and its application for driving a small liquid crystal display, *Nano Lett.*, **10**, 5025–5031.
15. Chen, 7., Li, J., Zhang, G., and Shi, Y. (2011). PZT nano active fiber composites for acoustic emission detection, *Adv. Mater.*, **23**, 3965–3969.
16. Gupta, A. K., Nair, P. R., Akin, D., Ladisch, M. R., Broyles, S., Alam, M. A., and Bashir, R. (2006). Anomalous resonance in a nanomechanical biosensor, *Proc. Natl. Acad. Sci.*, **103**, 13362.
17. Chen, 7., Yao, N., and Yong, S. (2011). Energy harvesting based on PZT nanofibers. In: *Energy Efficiency and Renewable Energy through Nanotechnology* (ed. Zhang L.), Chapter 12, Springer Series in Nanoscience and Technology.

18. Xu, S. Y., Shi, Y., and Kim, S. G. (2006). Fabrication and mechanical property of nano piezoelectric fibers, *Nanotechnology*, **17**, 4497–4501.

19. Li, D., and Xia, Y. (2003). Fabrication of titania nanofibers by electrospinning, *Nano Lett.*, **3**, 555–560.

20. Li, D., Wang, Y., and Xia, Y. (2004). Electrospinning nanofibers as uniaxially aligned arrays and layer-by-layer stacked films, *Adv. Mater.*, **16**, 361–366.

21. Rao, S. S. (2004). *Mechanical Vibrations*, Pearson Education, Singapore.

22. Chen, 7., Xu, S. Y., Yao, N., and Shi, Y. (2011). Young's modulus determination of unpolled electrospun PZT nanofibers, *Sci. Adv. Mater.*, in press.

23. Walter, V., Delobelle, P., Le Moal, P., Joseph, E., and Collet, M. (2002). A piezo-mechanical characterization of PZT thick films screen-printed on alumina substrate, *Sensors Actuators A*, **96**, 157.

24. Tsuji, T., Saito, S., Fukuda, K., Yamanaka, K., Ogiso, H., Akedo, J., and Kawakami, Y. (2005). Significant stiffness reduction at ferroelectric domain boundary evaluated by ultrasonic atomic force microscopy, *Appl. Phys. Lett.*, **87**, 071909.

25. Ferris, J. H., Li, D. B., Kalinin, S. V., and Bonnell, D. A. (2004). Nanoscale domain patterning of lead zirconate titanate materials using electron beams, *Appl. Phys. Lett.*, **84**, 774.

26. Landauer, R. (1957). Electrostatic considerations in $BaTiO_3$ domain formation during polarization reversal, *J. Appl. Phys.*, **28**, 227.

27. Brainerd, J. G., et al. (1949). Standard on piezoelectric crystals, 1949, The Institute of Radio Engineers Inc., *Proc. I.R.E.*, **37**, 1378–1395.

28. Jordan, T. L., and Ounaies, Z. (2001). Piezoelectric ceramics characterization, *NASA/CR-2001-211225*, ICASE Report No. 2001-28.

29. Li, S., Eastman, J. A., Vetrone, J. M., Foster, C. M., Newnham, R. E., and Cross, L. E. (1997). Dimension and size effects in ferroelectrics, *Jpn. J. Appl. Phys.*, **36**, 5169.

30. Ahn, C. H., Rabe, K. M., and Triscone, J.-M. (2004). Ferroelectricity at the nanoscale: local polarization in oxide thin films and heterostructures, *Science*, **303**, 488.

31. Tybell, T., Ahn, C. H., and Triscone, J.-M. (1999). Ferroelectricity in thin perovskite films, *Appl. Phys. Lett.*, **75**, 856.

32. Dikin, D. A., Chen, X., Ding, W., Wagner, G., and Ruoff, R. S. (2002). Resonance vibration of amorphous SiO_2 nanowires driven by mechanical or electrical field excitation, *J. Appl. Phys.*, **93**, 226.

Chapter 8

Micro- and Nanomechanical String Resonators

Tom Larsen[a] **and Silvan Schmid**[b]

[a]*Department of Mechanical Engineering, Stanford University, Stanford, CA 94305, USA*
[b]*Department of Micro- and Nanotechnology, Technical University of Denmark, DTU Nanotech, DK-2800 Kgs. Lyngby, Denmark*

tola@stanford.edu

The dynamic properties of doubly clamped micro- or nanomechanical beam resonators become string-like when tensile stress dominates their mechanical behavior. This tensile stress leads to an enhancement of the quality factor (Q). With Q values of up to several millions, string resonators are interesting for fundamental studies as well as for a number of applications. The string-like behavior of tensile stressed beams allows for simple analytical modeling of their dynamic properties: resonance frequency and quality factor. The simple modeling is utilized in string-based mass sensing where both the position and mass of an added point mass can be determined via dual resonant mode detection. The detection limit in air is in the sub-picogram range.

Nanocantilever Beams: Modeling, Fabrication, and Applications
Edited by Ioana Voiculescu and Mona Zaghloul

ISBN 978-981-4613-23-1 (Hardcover), 978-981-4613-24-8 (eBook)
www.panstanford.com

Thermal expansion of a string couples its resonance frequency to its temperatures, which can be utilized for temperature sensing. Low-frequency instability, $\sigma(1s) \approx 5.9 \times 10^{-8}$, leads to an estimated temperature resolution of 2.5×10^{-4}°C. String-based photothermal spectroscopy, utilizing the high temperature sensitivity, allows for spectroscopy analysis of picogram samples.

8.1 Introduction

Room temperature quality factors of more than one million[1,2] makes micro- and nanomechanical string resonators interesting for fundamental studies as well as for a variety of applications. String resonators have been utilized for fundamental studies of, e.g., damping mechanisms[1–7] and nanomechanical classical two-level systems.[8] The use of string resonators for material characterization[9,10] and gas,[11] mass[12,13] and thermal sensing[14–18] has also been demonstrated.

The string-like behavior of a beam is caused by a high built-in level of tensile stress. In this case, the bending rigidity can be neglected and the resonance frequency becomes a function of the tensile stress. Therefore, the resonance frequency can be tuned without changing beam material or dimensions. The tensile stress increases the stored energy, and thus the resonance frequency, while its influence on the intrinsic damping (energy loss) is typically small. The quality factor is defined by the ratio of stored to lost energy during one cycle of vibration. The tensile stress thus leads to high Q values in string resonators. The influence of the tensile stress seen for string resonators has also been observed for membrane resonators.[19–21]

Despite the similarities between string and membrane resonators, we will only treat string resonators in this chapter. The content of this chapter is organized in the following way: The beam equation for a tensile stressed beam and an approximated solution to it is described in Section 8.2. In Section 8.3, the damped vibration of string resonators is described. The use of string resonators for mass sensing, temperature-sensing and photothermal spectroscopy is described in Section 8.4. Section 8.5 is used for conclusions.

8.2 The String Approximation

We will here describe the free undamped vibration of a thin tensile stressed beam. Fig. 8.1 shows a schematic representation of such a beam structure. For linear elastic beams with a large length to thickness ratio the equation of motion can written as[22]:

$$EI_z \frac{\partial^4 u(x,t)}{\partial x^4} - N\frac{\partial^2 u(x,t)}{\partial x^2} + \rho A\frac{\partial^2 u(x,t)}{\partial t^2} = 0, \tag{8.1}$$

where $u(x, t)$ is the displacement function of the beam as a function of position and time, E Young's modulus, I_z the area moment of inertia, ρ the mass density, $A = wh$ the beam cross-sectional area and N the axial force. The axial force can also be written as the axial stress, σ, times the cross sectional area, A, which we will do from here on. The three force terms in Eq. 8.1 represent the flexure, tensile stress, and inertia.

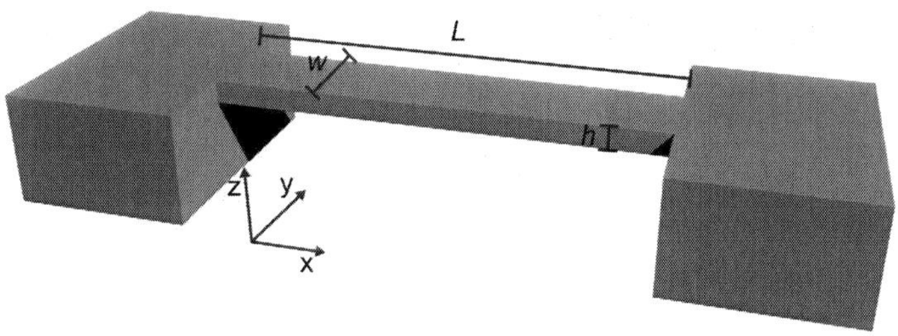

Figure 8.1 Schematic of a doubly clamped beam structure with length ***L***, width ***w***, and height ***h***.

A closed-form analytical solution to Eq. 8.1 cannot be obtained using the boundary conditions for a clamped-clamped beam[23]: $u(0, t) = u(L, t) = 0$ and $u'(0, t) = u'(L, t) = 0$. However, an approximated solution can be obtained by using the boundary conditions of a pinned-pinned beam: $u(0, t) = u(L, t) = 0$ and $u''(0, t) = u''(L, t) = 0$. The drawback is that the shape function becomes less accurate, especially near the clamping regions. Using this approach, the eigenfrequency is given by[22]

$$\omega_n = \frac{\lambda_n^2}{L^2}\sqrt{\frac{EI_z}{\rho A}}\sqrt{1+\frac{\sigma A L^2}{\lambda_n^2 E I_z}}, \tag{8.2}$$

Where $\lambda_n^2 = n\pi (n = 1, 2, 3, ...)$ are solutions to the characteristic equation. It can be seen that the eigenfrequency is increasing for increasing tensile stress. The flexural rigidity of the beam can be neglected when the tensile stress becomes sufficiently high. For a beam with a rectangular cross section, this condition is fulfilled when $12\sigma L^2 >> \lambda_n^2 Eh^2$. Ignoring the flexural rigidity, Eq. 8.2 is reduced to

$$\omega_n = \frac{n\pi}{L}\sqrt{\frac{\sigma}{\rho}} \qquad n = 1, 2, 3, \ldots \tag{8.3}$$

which is the eigenfrequency of a string with the n-th mode shape

$$u_n(x) = a_n \sin\left(\frac{n\pi}{L}x\right), \tag{8.4}$$

where a_n is the vibrational amplitude for the n-th mode.

Figure 8.2 shows the resonance frequency of a 500 μm-long, 5 μm-wide and 500 nm-thick beam as a function of stress, calculated using Eq. 8.2, Eq. 8.3 and FEM simulations. Following values were used for Young's modulus, Poisson's ratio and mass density: 250 GPa, 0.23 and 3000 kg/m^3. The three methods are converting towards each other for high level of stress. The difference between the three is less than 3% when the tensile stress excides 100 MPa. The most frequently used materials for micro- and nano-string resonators are silicon rich silicon nitride and stoichiometric silicon nitride, which are typically tensile stressed to a level well above 100 MPa. Hence, treating thin and long beams of these materials as strings is reasonable in most cases. From Fig. 8.2 it can be seen that the FEM simulations and the analytical model (Eq. 8.2) do not converge towards the same resonance frequency when the tensile stress is approaching zero. The difference between the two can be explained by the use of incorrect boundary conditions when solving Eq. 8.1.

Due to the distinct string-like behavior of tensile stressed beams, higher order resonance modes are harmonics of the fundamental resonance frequency (see Eq. 8.3). Geometrical scaling of the frequency can only be done via the length of the string. Hence, the thickness can be reduced without lowering the frequency. This is useful to achieve, e.g., a low resonator mass for mass sensing. For non-stressed beams, a reduction of beam thickness will lead to a decrease of the resonance frequency. Scaling of the frequency by tuning the tensile stress is convenient. The tensile stress in thin films can often be controlled to a large extent in the deposition/growth process or through post treatment. Combined, this makes it possible to make thin high-frequency resonators without using sub-micrometer-long beams and still have, e.g., a low mass.

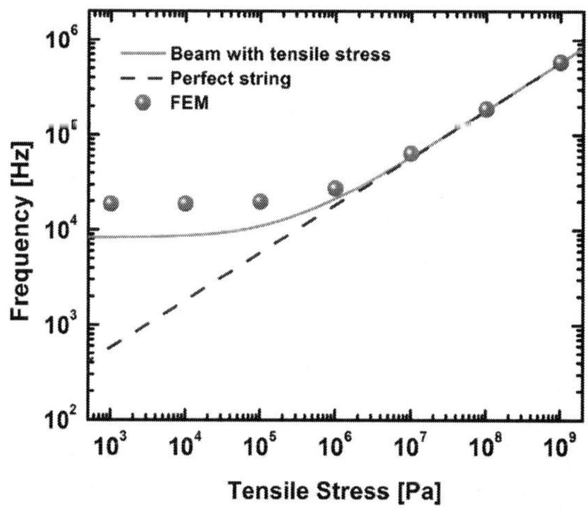

Figure 8.2 Resonance frequency of a 500 μm-long, 5 μm-wide, and 500 nm-thick beam as a function of tensile stress. Young's modulus, Poisson's ratio and mass density of the beam are 250 GPa, 0.23 and 3000 kg/m^3, respectively.

8.3 Quality Factor of String Resonators

The quality factor (Q) is a key figure of merit of micro- and nanomechanical resonators. Q is a measure for the sharpness of the resonance peak and consequently for the precision with which the resonance frequency can be measured. In sensor applications, the smallest detectable unit is directly proportional to the frequency resolution. Resonant sensors are typically driven at the resonance frequency by means of an oscillator circuit, such as a phase-locked loop or positive feedback. In this case, the frequency resolution Δf is proportional to $Q^{-1/2}$.[24,25] Q is also a measure for the thermal decoupling of a mechanical resonator from the environment. Q is inversely proportional to the thermal decoherence rate, which is the inverse time it takes until a single quantum is added from the environment to a mechanical resonator in the ground state. Thus for quantum optomechanics, the decoherence time has to be longer than the time of an oscillation.[26] In both cases, the classical resonant sensor technology and the quantum optomechanics, high-Q micro- and nanomechanical resonators are highly desired.

The quality factor can be defined by the ratio of the resonance frequency to the half-power-bandwidth $\Delta\omega_{-3\,\mathrm{dB}}$.

$$Q = \frac{\omega_\mathrm{r}}{\Delta\omega_{-3\,\mathrm{dB}}} \tag{8.5}$$

For low damping, $\Delta\omega_{-3\,\mathrm{dB}}$ is directly proportional to the damping coefficient. In the case of negligible viscous damping and clamping losses, Q in a relaxed mechanical beam resonator, such as singly clamped cantilevers, is given by intrinsic bulk and surface damping mechanisms.[7] For example, intrinsic damping mechanisms in silicon nitride micro- and nano-cantilevers limit Q to maximal a few 10,000 at room temperature.[27] In a pre-stressed beam on the other hand, the resonance frequency is increased by the tensile stress σ, as described by Eq. 8.2. Assuming that σ does not significantly influence the intrinsic damping, Q will increase with additional tensile stress according to Eq. 8.5. Indeed, Q values of string-like micro- and nanomechanical resonators are up to three orders of magnitude higher than those of similar unstressed cantilever resonators[29,30] and can reach values of up to several millions.[1,2] Similarly, high Q values have been measured with membrane resonators that are the bi-axially pre-stressed equivalent to the uni-axially stressed strings.[20,21] Owing to their exceptionally high quality factors, string and membrane nanomechanical resonators became the centerpiece of recent research in, for example, cavity optomechanics,[30–33] as temperature and mass sensor, or as photothermal IR spectrometer, as discussed in Section 8.5.

In the following, the quality factor of pre-stressed string resonators and its relationship with the intrinsic damping are discussed in more details. The physical definition of Q is given by the ratio of the energy stored versus the energy lost during a cycle of vibration. In a pre-stressed string, energy can be stored and lost in the string bending and elongation. Additionally, energy is stored in the work against the tensile force when the string deflects. Q of a string, called Q_{string}, is then defined by[30]

$$Q_{\text{string}} = 2\pi \frac{W_{\text{tension}} + W_{\text{bending}} + W_{\text{elongation}}}{\Delta W_{\text{bending}} + \Delta W_{\text{elongation}}}. \tag{8.6}$$

It is obvious from Eq. 8.6 that the additional energy term W_{tension} enhances Q_{string}. The common case that the vibrational amplitude is a lot smaller than the string thickness ($a_{\text{n}} << h$), the contribution from the elongation can be neglected and Q_{string} becomes[2]

$$Q_{\text{string}} \approx \frac{W_{\text{tension}}}{W_{\text{bending}}} Q_{\text{bending}} = \frac{\frac{1}{2}\sigma A \int_0^L \left(\frac{\partial}{\partial x}u(x)\right)^2 dx}{\frac{1}{2}EI_{\text{z}} \int_0^L \left(\frac{\partial^2}{\partial x^2}u(x)\right)^2 dx} Q_{\text{bending}} \tag{8.7}$$

From Eq. 8.7, it can be seen that the bending energy is stored in the curvature of the string, which is in the main bending of the antinodes and in the local bending at the clamping. It was found that the local curvature at the clamping sites contributes significantly to the total stored bending energy and therefore is crucial for the calculation of Q_{string}.[34]

The curvature of half a string ($x < L/2$) can then be described by a sinusoidal curve, according to Eq. 8.4, with an exponential correction accounting for the string clamping[19]

$$u_{\text{n}}(x) = a_{\text{n}} \left[\sin\left(\frac{n\pi}{L}x\right) + n\pi\lambda \left(\exp\left(-\frac{x}{\lambda L}\right) - \cos\left(\frac{n\pi}{L}x\right)\right)\right] \tag{8.8}$$

with the relative exponential clamping length

$$\lambda = \frac{1}{L}\sqrt{\frac{EI_{\text{z}}}{\sigma A}}. \tag{8.9}$$

Solving (Eq. 8.7) for the curvature (Eq. 8.8) results in the string quality factor

$$Q_{\text{string}} \approx Q_{\text{bending}} [2\lambda + (n\pi\lambda)^2]^{-1}. \tag{8.10}$$

The two terms in the brackets of Eq. 8.10 represent the effect of the bending at the clamping (left term) and the bending of the anti-nodes (right term). For long and thin strings, the clamping related term is dominating over the contribution from the anti-node bending. The right term can start to play a significant role at high mode numbers. Figure 8.3 shows the measured Q values of low-stress silicon nitride micro-string resonators as a function of the string length. It can be seen that the measured Q values are a linear function of the string length, as predicted by Eq. 8.10. It can also be seen, that Q is lower for wider strings. This can be explained by an increased clamping loss of wide strings. The dashed line represents the theoretical model (Eq. 8.10) for a bending related $Q_{\text{bending}} = 9000$. This value corresponds well with Q values measured with relaxed singly clamped silicon nitride cantilevers of similar thickness.[27]

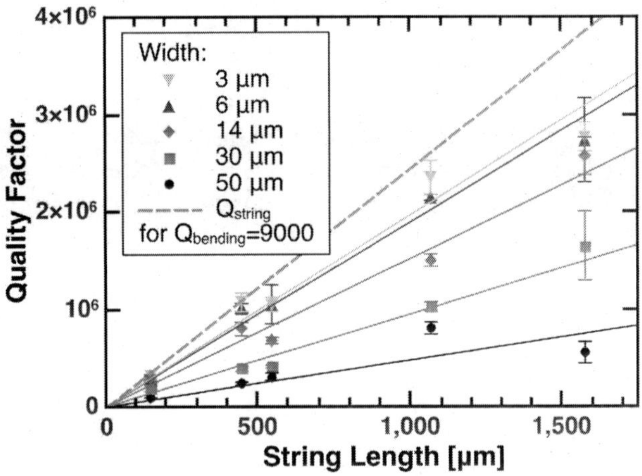

Figure 8.3 Quality factors of low-stress ($\sigma \approx 200$ MPa) silicon nitride micro string resonators ($h = 177$ nm) with varying width as a function of string length. Reprinted and adapted with permission from Ref. 2. © 2011 by the American Physical Society.

8.4 String Resonator Applications

The excellent dynamic properties of string resonators and their simple modeling make them interesting for a number of applications as described in Section 8.1. For now, we will focus on the use of string resonators as mass and thermal sensors.

8.4.1 Mass Sensing

Mass sensing of microscopic substances is a classic application of micro- and nanomechanical resonators. As the technology has evolved, mass sensing of single cells,[35] biomolecules,[36,37] and single hydrocarbon molecules[38] ($C_{10}H_8$) has become possible. An impressive mass resolution of 1 yg has been achieved by using high frequency, 2 GHz, carbon nanotube resonators with an intrinsic mass of only ~0.3 ag.[38] The ultrahigh sensitivity of these sensors has been used to create microscale mass-spectrometers,[11,35,36] chemisorption sensors,[39,40] and aerosol detectors.[13]

The sensing principle of resonating mass sensors relies on the shift in resonance frequency induced by an added mass on the resonator. For single particle or molecule detection, the location of the added mass has a great impact on the induced frequency shift. The induced shift is maximal when the added mass is located where the vibrational amplitude is largest and vice versa. For detection of substances of unknown mass, this problem can be addressed by using a statistical approach[36] or multi-mode detection.[12,36,41] In multi-mode detection, the shift of multiple modes, normally only the first and second resonant modes, are used to determine both the mass and the position of the added mass. For single and double clamped beam structures, numerical analysis is needed to determine both. However, the simple sinusoidal displacement function of string resonators allows for deriving an analytical expression relating the added mass and its position to the shift of the two first resonance modes.[12] Here, we will derive the analytical model, from Ref. 12, utilized in multi-mode string-based mass sensing.

We here consider a string with a mass m_0 loaded by a point mass Δm positioned at $x_{\Delta\mathrm{m}}$. For small vibrational amplitudes and high tensile stress, the string approximation allows us to neglect the energy stored in the beam bending. Hence, at resonance, the average kinetic energy of the string and the loaded mass equals the time average strain energy. The time average kinetic energy of a string can be written as

$$E_{\mathrm{kin}} = \iiint \frac{1}{2}\rho\omega_{\mathrm{n},\,\Delta\mathrm{m}}^{2} u_{\mathrm{n}}^{2}(x)dV, \tag{8.11}$$

where ρ is the mass density, $\omega_{n,\Delta m}$ the resonance frequency of the loaded string, and $u_n(x)$ the displacement function of the n-th resonance mode. Inserting the displacement function, Eq. 8.4, and solving the integral Eq. 8.11 gives

$$E_{\text{kin}} = \frac{1}{4} m_0 \omega_{n,\Delta m}^2 a_n^2, \tag{8.12}$$

where m_0 is the intrinsic mass of the string and a_n the vibrational amplitude. The kinetic energy of the point mass is given by

$$E_{\text{kin},\Delta m} = \frac{1}{2} \Delta m \omega_{n,\Delta m}^2 u_n^2 (x_{\Delta m}). \tag{8.13}$$

Assuming that the shape form of the string is not altered when adding the mass, $m_0 >> m_{\Delta m}$, the strain energy can be found by solving Eq. 8.11, for the unloaded string:

$$E_{\text{strain}} \approx \frac{1}{4} m_0 \omega_n^2 a_n^2. \tag{8.14}$$

The resonance frequency of the loaded string can now be found by setting the strain energy equal to the sum of the kinetic energy of the string and the point mass:

$$\omega_{n,\Delta m}^2 = \omega_n^2 \left(1 + 2\frac{\Delta m}{m_0} \varphi_n^2 (x_{\Delta m})\right)^{-1}, \tag{8.15}$$

where $\varphi_n(x_{\Delta m}) = u_n(x_{\Delta m})/a_n$. The unknowns in this equation system are the load mass and its position. For small masses, Eq. 8.15 can be simplified with a Taylor series approximation:

$$\omega_{n,\Delta m} = \omega_n \left(1 - \frac{\Delta m}{m_0} \varphi_n^2 (x_{\Delta m})\right) \tag{8.16}$$

An expression for the position can be obtained by solving Eq. 8.16 for the first resonance mode:

$$x_{\Delta m} = \frac{L}{\pi} \sin^{-1} \sqrt{\frac{m}{\Delta m_0} \left(1 - \frac{\omega_{1,\Delta m}}{\omega_1}\right)} \tag{8.17}$$

The position of the load mass can only take values between $0 < x_{\Delta m} < L/2$ as the first mode displacement function of a string is symmetrical. The induced frequency shift is therefore independent of which half side the load mass is positioned. By inserting Eq. 8.17 into Eq. 8.16 and solving for the second mode, $n = 2$, a closed form solution for the added mass can be found:

$$\Delta m = m_0 \frac{\left(1 - \frac{\omega_{1,\Delta m}}{\omega_1}\right)^2}{\left(1 - \frac{\omega_{1,\Delta m}}{\omega_1}\right) - \frac{1}{4}\left(1 - \frac{\omega_{2,\Delta m}}{\omega_2}\right)} \tag{8.18}$$

It is here utilized that $(\sin(2\ \sin^{-1}\ y)) = 2y\sqrt{1-y^2}$. This theorem is only true if $|y| < 1$, which is the case as $x_{\Delta m} < L/2$. With Eq. 8.16 and Eq. 8.17 in hand, dual-mode detection allows us to calculate both the position and the mass of a point mass placed on a string with a total mass m_0.

String-based mass sensing is experimentally demonstrated in Ref. 12. The mass and position of microparticles placed on micro silicon nitride strings are determined from the measured shift of the first two resonant modes. Figure 8.4a presents the measured relative shifts of the first and second resonant mode due to an added mass and Fig. 8.4b the calculated positions and mass ratios. Three different particles were used in the experiments: 2 μm (4.8 ± 0.18 pg) and 6 μm (111.8 ± 14.4 pg) polystyrene beads and 2.8 μm (14.9 ± 5.4 pg) magnetic beads. The used silicon nitride strings were 216.4 μm long, 2.7 μm wide, and 340 nm thick and had a build-in tensile stress of 220 MPa. The measured mass ratios agree well with the expected values.

One of the assumptions made when describing a tensile stressed beam as a string is that the mode shape is sinusoidal. As pointed out earlier, this approximation introduces an error that is most pronounced near the clamping regions. For mass sensing, the error of the calculated mass and position is therefore also largest for particles placed close the one of the two anchor regions. This type of error can be minimized by using high-aspect-ratio strings and focusing the particle deposition around the center of the string. Another way to minimize this error is to dismiss all particles that are located close to the clamping. The particle location can be calculated with Eq. 8.17. In Fig. 8.4, all particles with $x_{\Delta m} < 0.2L$ were dismissed and colored green. It can be seen in Fig. 8.4b that the mass calculations of the dismissed measurements indeed show large deviations from the expected values.

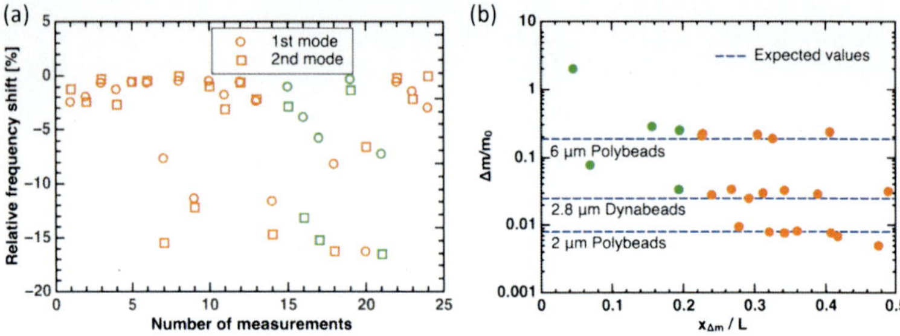

Figure 8.4 (a) Measured relative frequency shifts of the first and second bending mode of string resonators induced by an added mass. (b) Calculated mass ratio (Eq. 8.18) for the three different types of microparticles. The measurements close to the anchor ($x_{\Delta m} < 0.2L$) that were dismissed are colored green. Data reprinted and adapted from Ref. 12.

True nanoparticle detection in air using string resonators is demonstrated in Ref. 13. Here, the string itself is used as a single filter fiber to filter out nanoparticles from an aerosol passing it. By letting the aerosol passing the string at high velocity, non-diffusion limited sampling can be achieved. Hence, mass sensing of nanoparticles at a short time scale is possible.

The estimated mass resolution, based on experimental values, of string-based mass sensors in air is 740 ag.[13] In comparison, Li et al. have demonstrated a mass resolution of ~1 ag in air using high frequency, 127 MHz, nanomechanical cantilevers.[39] By downscaling the dimensions of the string, to achieve a lower intrinsic mass, a similar mass resolution should be achievable.

The simplicity of string-based mass sensing and the ability for non-diffusion–based sampling make string-based mass sensing a potential tool for nanoparticle counting in research, industry and urban environments.

8.4.2 Temperature Sensing

Temperature sensing at the micro- and nanoscale can provided an insight into thermal processes at the single-cell level,[42] enable uncooled IR detection[43–45] and photothermal spectroscopy of picogram samples.[18] Common for microscale temperature sensors is that they have a low thermal mass and a short thermal time constant. The use of suspended structures can provide excellent thermal isolation from the surroundings leading to sub-picowatt power resolution.[46] The temperature detection scheme for micro- and nanomechanical temperature sensors is often based on measuring changes in static (deflection) or dynamic (frequency and/or damping) properties.

The temperature-dependent resonance frequency of string resonators has been studies in a number of publications.[14–18] Here, we will describe an analytical model for the temperature-dependent resonance frequency of a string, give a brief review of the published studies and compare the strings to other types of temperature-sensing resonators.

Differential thermal expansion or contraction of a string and the frame spanning it leads to an effective change in the tension of the string, which is reflected in its resonance frequency. A measure of the temperature or changes of it can therefore be obtained by monitoring the resonance frequency. The temperature-dependent strain of a string supported by a material with a different coefficient of thermal expansion is given by

$$\varepsilon(T) = \varepsilon_0 - (\alpha_{\mathrm{Str}} - \alpha_{\mathrm{Sup}})(T - T_0), \tag{8.19}$$

where ε_0 is the strain at temperature T_0, T the actual temperature, and α_{Str} and α_{Sup} the coefficient of thermal expansion of the string and its support, respectively. Assuming linear elastic materials, the temperature-dependent stress is given by Hooke's law:

$$\sigma(T) = E(\varepsilon_0 - (\alpha_{\mathrm{Str}} - \alpha_{\mathrm{Sup}})(T - T_0))$$

$$\sigma(T) = \sigma_0 - E(\alpha_{\mathrm{Str}} - \alpha_{\mathrm{Sup}})(T - T_0). \tag{8.20}$$

Inserting Eq. 8.20 into Eq. 8.3 and dividing by 2π, the temperature-dependent resonance frequency, in hertz, can be written as

$$f_{\mathrm{n}}(T) = \frac{n}{2L}\sqrt{\frac{\sigma_0 - E(\alpha_{\mathrm{Str}} - \alpha_{\mathrm{Sup}})(T - T_0)}{\rho}}. \tag{8.21}$$

Temperature induced changes in length, Young's modulus, coefficients of thermal expansion and mass density are neglected in this model. The temperature sensitivity can be written as

$$S_{\mathrm{T}} = \frac{df_{\mathrm{n}}(T)}{dT} = \frac{-nE(\alpha_{\mathrm{Str}} - \alpha_{\mathrm{Sup}})}{4L\sqrt{\rho(\sigma_0 - E(\alpha_{\mathrm{Str}} - \alpha_{\mathrm{Sup}})(T - T_0))}}. \tag{8.22}$$

The sensitivity is increasing for decreasing length, mass density, and initial stress and increasing resonant mode, Young's modulus, and difference in thermal expansion. For low initial stress, the sensitivity becomes strongly dependent on the temperature.

Figure 8.5a shows the measured sensitivity of a number of different string resonators and Fig. 8.5b the frequency of the first three resonant modes of an aluminum string as a function of temperature. It can be seen that the trends from the measured values are in accordance with Eq. 8.22.

When comparing the sensitivities plotted in Fig. 8.5, the silicon nitride strings show the lowest values. However, their high-frequency stability leads to the best temperature resolution of the three tested string types. An Allan deviation of $\sigma(1s) \approx 5.9 \times 10^{-8}$ has been measured for 114 μm-long, 15 μm-wide, and 340 nm-thick silicon nitride strings, which leads to an estimated temperature resolution of 2.5×10^{-4}°C (signal-to-noise ratio equals 3).[17] Temperature resolutions reported for other resonating temperature sensors are 4.6×10^{-4}°C for "beat mode based" silicon resonators,[47] 5.1×10^{-3}°C for "*Q*-factor based" silicon resonator,[48] 1.2×10^{-2}°C for micromachined quartz resonators,[49] 1.6×10^{-3}°C silicon cantilever resonators,[42] and 2.2×10^{-3}°C and 1.3×10^{-2}°C for the flexural and torsional modes of a paddle resonator.[43] A signal-to-noise ratio of 3 has been used in all cases. Comparing the values, the silicon nitride strings show the best temperature resolution.

The string-based temperature sensor described so far relies on the thermal expansion of both the string and frame. The drawback of this sensing principle is the large thermal mass of the frame. However, for certain applications, e.g., photothermal spectroscopy, the frame can be used as a heat sink. Hence, only the string acts as the temperature-sensing element making the sensing principle more interesting for monitoring thermal processes at the micro- and nano-scale.

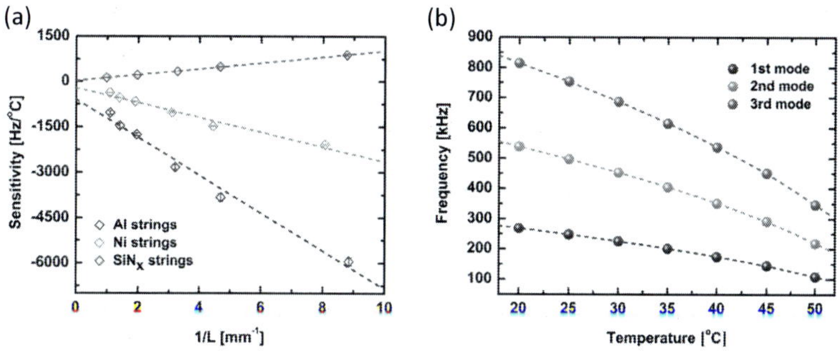

Figure 8.5 (a) The measured temperature sensitivity of silicon rich silicon nitride (SiN_x), nickel (Ni) and aluminum (Al) strings at 20°C as a function of the inverse length. The silicon nitride strings are 340 nm thick and 15 μm wide, the nickel strings 100 nm thick and 16.5 μm wide, and the aluminum strings 150 nm thick and 7.5 μm long. The pre-stress at 20°C for the silicon nitride, nickel and aluminum strings are 210 MPa, 285 MPa and 31 MPa, respectively. The error bars indicate the standard deviation of the measured values. (b) The frequency of the first three resonant modes of an aluminum string (L = 213 μm, w = 7.5 μm and t = 150 nm) as a function of temperature. Reprinted and adapted with permission from Ref. 15 and Ref. 17. © 2011 and © 2013, AIP Publishing LLC.

8.4.3 String-Based Photothermal Spectroscopy

A large number of highly sensitive micro- and nanomechanical sensors have been described in the literature. However, achieving high resolution and high selectivity at the same time has always been challenging. For instance, the limit for mass sensing has been pushed to the extreme, yoctogram resolution,[38] but the selectivity is non-existing. To increase the selectivity, selective coatings have been used, but in most cases at the expense of versatility.[50] Particular applications require specially designed coatings.

Spectroscopic techniques of various kinds are frequently used in science and industry due to the high selectivity. Combining the best from spectroscopic techniques, selectivity, with the best from micromechanical sensors, sensitivity, is an attractive approach. Microscale photothermal spectroscopy is one example where this approach is used. Bimorph cantilevers have previously been used as the thermal sensor in photothermal spectroscopy for detecting and/or characterizing chemicals and thin films.[51–57] The advantage of using strings instead of cantilevers is that the detection scheme becomes frequency based instead of deflection based. Frequency based detection is not directly subjected to thermomechanical noise and external vibrations.

The used of strings as the thermal sensing element in a micromechanical photothermal spectroscopy has been demonstrated in Refs. 16 and 18. A substance placed on a string is irradiated by light of varying wavelengths. Non-radiative relaxation leads to heating of the string, which lowers the tensile stress and hence the resonance frequency. In this configuration, the strings serve as a high-resolution temperature sensor with a short characteristic thermal time constant, ~100 ms, and a microscopic thermal mass, ~1 nJ/°C. Yamada et al. have presented an analytical model for the resonance frequency of a string with local heating at the string center.[18] Here, we will briefly describe how the model is derived and then present the final analytical expression. A detailed description can be found in the supplementary information of Ref. 18.

When demonstrating the string-based photothermal spectrometer the strings were operated in high-vacuum condition. This means that the heat transfer to the surroundings is primarily via conduction through the two anchor regions. The average temperature increase of a string due to local heating can be found by treating the system as a quasi one-dimensional problem where a power, P, is applied at the center of the string while keeping its ends at a fix temperature. For the steady-state condition, the average temperature difference with respect to the ambient temperature is given by[58]

$$\Delta T = \frac{PL}{4\kappa A}, \tag{8.23}$$

where κ is the thermal conductivity of the string. By setting Eq. 8.23 into Eq. 8.21 and neglecting expansion of the frame, the resonance frequency as a function of power influx is given by

$$f_{\mathrm{n}}(P) = \frac{n}{2L}\sqrt{\frac{\sigma_0}{\rho} - \frac{E\alpha_{\mathrm{Str}}PL}{8\rho\kappa A}}. \tag{8.24}$$

The power sensitivity, in the case of heating at the center, equals

$$S_{\mathrm{P}} = \frac{df_{\mathrm{n}}(P)}{dP} = \frac{-nE\alpha_{\mathrm{Str}}}{32\rho\kappa A\sqrt{\frac{\sigma_0}{\rho} - \frac{E\alpha_{\mathrm{Str}}PL}{8\rho\kappa A}}}. \tag{8.25}$$

The power sensitivity is increasing for increasing resonant mode number, Young's modulus, and coefficient of thermal expansion and decreasing mass density, thermal conductivity, cross-sectional area, and initial stress. Owing to the unique dynamic properties of string resonators, it is possible to geometrically minimize the heat transfer to the support while keeping a high frequency and quality factor. A reduction in cross-sectional area of a non-stress beam resonator will also lead to a lowering of the resonance frequency.

String-based photothermal spectroscopy[58] has been used for detection and characterization of single[16] and multiple nanoparticles.[17] For single-particle detection and characterization, micrometer- and nanometer-sized particles were placed on 180 nm-thick, 895 μm-long, and 3–50 μm-wide silicon nitride strings and irradiated with visible light. It was demonstrated that the induced frequency shift due to the presence of a particle increases for decreasing string width and increasing irradiation power, in accordance to Eq. 8.25. Detection of individual blue 170 nm in diameter polystyrene particles was demonstrated using red light (635 nm) for the irradiation. It was estimated that a ~40 nm (~40 ag) blue polystyrene particle would induce a frequency shift that equals three times the noise floor of the measurement.

Figure 8.6 shows the infrared absorption spectrum of PVP (polyvinylpyrrolidone) obtained by conventional FTIR (Fourier transform infrared spectroscopy) and string-based photothermal spectroscopy. For the photothermal spectrum, PVP nanoparticles were deposited using the string filter-fiber technique described earlier. The estimated mass of the PVP nanoparticles is 580 pg. The four major peaks are consisted between the two spectra, which demonstrates that true spectroscopic data can be obtained using this technique. The broad absorption peak at 3428 cm^{-1} seen in the FTIR spectrum is due to the presence of water vapor.

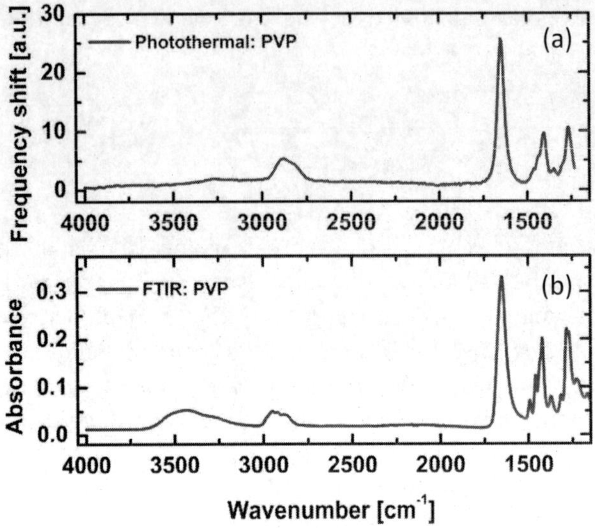

Figure 8.6 (a) Photothermal IR spectrum of 580 pg PVP loaded onto a 506 μm-long 3 μm-wide and 157 nm-thick silicon nitride string. (b) IR spectrum of PVP measured by a conventional FTIR spectrometer. Reprinted and adapted with permission from Ref. 18. © 2013 American Chemical Society.

The estimated power resolution of the string-based photothermal spectrometer is ~20 pW. Sub-picowatt power resolution has been reported for bimorph cantilever structures.[46] However, this power resolution is obtained using a 100 sec integration time, which is two orders of magnitude higher than the one used in the string experiments. Improvement in power resolution leading to a reduction in the required sample mass can be achieved by optimizing the string structure. A further reduction in sample mass can be achieved by using a more powerful irradiation source, e.g., a tunable quantum cascade laser. Doing this could enable photothermal spectroscopy of zeptogram samples—single protein range—in the future.

8.5 Conclusions

In this chapter, we have described the concept of micro- and nanomechanical string resonators and some of their applications. First, the beam equation of an axial stressed beam was described and an approximated solution to it. For sufficiently high tensile stress, it is reasonable to describe these structures as simple string-like resonators. Besides increasing the resonance frequency, the tensile stress does also have an enhancing effect on the quality factor. Quality factors of up to 7 million have been reported. The string description allows for a simple modeling of the frequency shift induced when a point mass added on the string. The mass and the position of it can be determined by measuring the shift of the first two resonant modes. Utilizing this and using the strings as single filter fibers allows for single nanoparticles mass sensing at atmospheric pressure. The use of string resonators for thermal sensing has also been described. The detection scheme relies on temperature-induced changes in the tensile stress. Again the string description allows for simple modeling. The high performance as a temperature sensor has been utilized in a string-based photothermal spectrometer, which shows picowatt power resolution. For the interested reader, a good starting point for a more detailed description of string resonators and their applications is given in references 1–18.

Acknowledgment

The authors would like to thank Shoko Yamada for making experimental data available for this chapter. T. L. acknowledges financial support from the Stanford BioX. S. S. acknowledges financial support from the Villum Foundation under grant agreement VKR023125.

References

1. Verbridge, S. S., Craighead, H. G., and Parpia, J. M. (2008). A megahertz nanomechanical resonator with room temperature quality factor over a million, *Applied Physics Letters*, **92**(1), 013112.
2. Schmid, S., Jensen, K. D., Nielsen, K. H., and Boisen, A. (2011). Damping mechanisms in high-Q micro and nanomechanical string resonators, *Physical Review B*, **84**(16), 165307.
3. Verbridge, S. S., Parpia, J. M., Reichenbach, R. B., Bellan, L. M., and Craighead, H. G. (2006). High quality factor resonance at room temperature with nanostrings under high tensile stress, *Journal of Applied Physics*, **99**(12), 124304.

4. Verbridge, S. S., Shapiro, D. F., Craighead, H. G., and Parpia, J. M. (2007). Macroscopic tuning of nanomechanics: Substrate bending for reversible control of frequency and quality factor of nanostring resonators, *Nano Letters*, **7**(6), 1728–1735.
5. Southworth, D. R., Barton, R. A., Verbridge, S. S., Ilic, B., Fefferman, A. D., Craighead, H. G., and Parpia, J. M. (2009). Stress and silicon nitride: A crack in the universal dissipation of glasses, *Physical Review Letters*, **102**(22), 225503.
6. Schmid, S., and Hierold, C. (2008). Damping mechanisms of single-clamped and prestressed double-clamped resonant polymer microbeams, *Journal of Applied Physics*, **104**(9), 093516.
7. Villanueva, L. G., and Schmid, S. (2014). Evidence of surface loss as ubiquitous limiting damping mechanism in SiN micro- and nanomechanical resonators, *Physical Review Letters*, **113**, 227201, 1–6.
8. Faust, T., Rieger, J., Seitner, M. J., Kotthaus, J. P., and Weig, E. M. (2013). Coherent control of a classical nanomechanical two-level system, *Nature Physics*, **9**(8), 485–488.
9. Kim, Y. J., and Allen, M. G. (1999). In situ measurement of mechanical properties of polyimide films using micromachined resonant string structures, *Components and Packaging Technologies, IEEE Transactions on*, **22**(2), 282–290.
10. Schmid, S. (2009) *Electrostatically Actuated All-Polymer Microbeam Resonators, Scientific Reports on Micro and Nanosystems,* vol. 6 (Der AndereVerlag, Tönning).
11. Henriksson, J., Villanueva, L. G., and Brugger, J. (2012). Ultra-low power hydrogen sensing based on a palladium-coated nanomechanical beam resonator, *Nanoscale*, **4**(16), 5059–5064.
12. Schmid, S., Dohn, S., and Boisen, A. (2010). Real-time particle mass spectrometry based on resonant micro strings, *Sensors (Basel, Switzerland)*, **10**(9), 8092.
13. Schmid, S., Kurek, M., Adolphsen, J. Q., and Boisen, A. (2013). Real-time single airborne nanoparticle detection with nanomechanical resonant filter-fiber, *ScientificReports*, 3, doi:10.1038/srep01288.
14. Pandey, A. K., Gottlieb, O., Shtempluck, O., and Buks, E. (2010). Performance of an AuPd micromechanical resonator as a temperature sensor, *Applied Physics Letters*, **96**(20), 203105.
15. Larsen, T., Schmid, S., Grönberg, L., Niskanen, A. O., Hassel, J., Dohn, S., and Boisen, A. (2011). Ultrasensitive string-based temperature sensors, *Applied Physics Letters*, **98**(12), 121901.

16. Larsen, T., Schmid, S., Villanueva, L. G., and Boisen, A. (2013). Photothermal analysis of individual nanoparticulate samples using micromechanical resonators, *ACS Nano*, **7**(7), 6188–6193.

17. Larsen, T., Schmid, S., and Boisen, A. (2013). Micro string resonators as temperature sensors, *AIP Conference Proceedings Series,* **15**(52), 931–936.

18. Yamada, S., Schmid, S., Larsen, T., Hansen, O., and Boisen, A. (2013). Photothermal infrared spectroscopy of airborne samples with mechanical string resonators, *Analytical Chemistry,* **85**(21), 10531–10535.

19. Yu, P. L., Purdy, T. P., and Regal, C. A. (2012). Control of material damping in high-*Q* membrane microresonators, *Physical Review Letters*, **108**(8), 083603.

20. Zwickl, B. M., Shanks, W. E., Jayich, A. M., Yang, C., BleszynskiJayich, A. C., Thompson, J. D., and Harris, J. G. E. (2008). High quality mechanical and optical properties of commercial silicon nitride membranes, *Applied Physics Letters*, **92**(10), 103125.

21. Wilson, D. J., Regal, C. A., Papp, S. B., and Kimble, H. J. (2009). Cavity optomechanics with stoichiometric SiN films, *Physical Review Letters*, **103**(20), 207204.

22. Timoshenko, S. P., Young, D. H., and Weaver, W. (1974). *Vibration Problems in Engineering*, John Wiley and Sons, USA.

23. Tilmans, H. A., Elwenspoek, M., and Fluitman, J. H. (1992). Micro resonant force gauges, *Sensors and Actuators A: Physical*, **30**(1), 35–53.

24. Ekinci, K. L., Yang, Y. T., and Roukes, M. L. (2004). Ultimate limits to inertial mass sensing based upon nanoelectromechanical systems, *Journal of Applied Physics*, **95**(5), 2682–2689.

25. Albrecht, T. R., Grütter, P., Horne, D., and Rugar, D. (1991). Frequency modulation detection using high-*Q* cantilevers for enhanced force microscope sensitivity, *Journal of Applied Physics*, **69**, 668.

26. Aspelmeyer, M., Kippenberg, T. J., and Marquardt, F. (2013). Cavity Optomechanics, arXiv:1303.0733.

27. Yasumura, K. Y., Stowe, T. D., Chow, E. M., Pfafman, T., Kenny, T. W., Stipe, B. C., and Rugar, D. (2000). Quality factors in micron-and submicron-thick cantilevers, *Microelectromechanical Systems, Journal of*, **9**(1), 117–125.

28. Verbridge, S. S., Parpia, J. M., Reichenbach, R. B., Bellan, L. M., and Craighead, H. G. (2006). High quality factor resonance at room temperature with nanostrings under high tensile stress, *Journal of Applied Physics*, **99**(12), 124304.

29. Schmid, S., and Hierold, C. (2008). Damping mechanisms of single-clamped and prestressed double-clamped resonant polymer microbeams, *Journal of Applied Physics*, **104**(9), 093516.
30. Thompson, J. D., Zwickl, B. M., Jayich, A. M., Marquardt, F., Girvin, S. M., and Harris, J. G. E. (2008). Strong dispersive coupling of a high-finesse cavity to a micromechanical membrane, *Nature*, **452**(7183), 72–75.
31. Purdy, T. P., Peterson, R. W., and Regal, C. A. (2013). Observation of radiation pressure shot noise on a macroscopic object, *Science*, **339**(6121), 801–804.
32. Gavartin, E., Verlot, P., and Kippenberg, T. J. (2012). A hybrid on-chip optomechanical transducer for ultrasensitive force measurements, *Nature Nanotechnology*, **7**(8), 509–514.
33. Faust, T., Krenn, P., Manus, S., Kotthaus, J. P., and Weig, E. M. (2012). Microwave cavity-enhanced transduction for plug and play nanomechanics at room temperature, *Nature Communications*, **3**, 728.
34. Unterreithmeier, Q. P., Faust, T., and Kotthaus, J. P. (2010). Damping of nanomechanical resonators, *Physical Review Letters*, **105**(2), 027205.
35. Ilic, B., Czaplewski, D., Zalalutdinov, M., Craighead, H. G., Neuzil, P., Campagnolo, C., and Batt, C. (2001). Single cell detection with micromechanical oscillators, *Journal of Vacuum Science & Technology B: Microelectronics and Nanometer Structures*, **19**(6), 2825–2828.
36. Naik, A. K., Hanay, M. S., Hiebert, W. K., Feng, X. L., and Roukes, M. L. (2009). Towards single-molecule nanomechanical mass spectrometry, *Nature Nanotechnology*, **4**(7), 445–450.
37. Hanay, M. S., Kelber, S., Naik, A. K., Chi, D., Hentz, S., Bullard, E. C., Colinet, E., Duraffourg, L., and Roukes, M. L. (2012). Single-protein nanomechanical mass spectrometry in real time, *Nature Nanotechnology*, **7**(9), 602–608.
38. Chaste, J., Eichler, A., Moser, J., Ceballos, G., Rurali, R., and Bachtold, A. (2012). A nanomechanical mass sensor with yoctogram resolution, *Nature Nanotechnology*, **7**(5), 301–304.
39. Li, M., Tang, H. X., and Roukes, M. L. (2007). Ultra-sensitive NEMS-based cantilevers for sensing, scanned probe and very high-frequency applications, *Nature Nanotechnology*, **2**(2), 114–120.
40. Bargatin, I., Myers, E. B., Aldridge, J. S., Marcoux, C., Brianceau, P., Duraffourg, L., Colinet, E., Hentz, S., Andreucci, P., and Roukes, M. L. (2012). Large-scale integration of nanoelectromechanical systems for gas sensing applications, *Nano Letters*, **12**(3), 1269–1274.

41. Dohn, S., Sandberg, R., Svendsen, W., and Boisen, A. (2005). Enhanced functionality of cantilever based mass sensors using higher modes, *Applied Physics Letters*, **86**(23), 233501.

42. Inomata, N., Toda, M., Sato, M., Ishijima, A., and Ono, T. (2012). Pico calorimeter for detection of heat produced in an individual brown fat cell, *Applied Physics Letters*, **100**(15), 154104.

43. Zhang, X. C., Myers, E. B., Sader, J. E., and Roukes, M. L. (2013). Nanomechanical torsional resonators for frequency-shift infrared thermal sensing, *Nano Letters*, **13**(4), 1528–1534.

44. Oden, P. I., Datskos, P. G., Thundat, T., and Warmack, R. J. (1996). Uncooled thermal imaging using a piezoresistive microcantilever, *Applied Physics Letters*, **69**(21), 3277–3279.

45. Steffanson, M., Gorovoy, K., Ramkiattisak, V., Ivanov, T., Król, J., Hartmann, H., and Rangelow, I. W. (2012). ARCH-type micro-cantilever FPA for uncooled IR detection, *Microelectronic Engineering*, **98**, 614–618.

46. Canetta, C., and Narayanaswamy, A. (2013). Sub-picowatt resolution calorimetry with a bi-material microcantilever sensor, *Applied Physics Letters*, **102**(10), 103112.

47. Jha, C. M., Bahl, G., Melamud, R., Chandorkar, S. A., Hopcroft, M. A., Kim, B., Agarwal, M., Salvia, J., Mehta and Kenny, T. W. (2007). CMOS-compatible dual-resonator MEMS temperature sensor with milli-degree accuracy, *Solid-State Sensors, Actuators and Microsystems Conference, 2007. TRANSDUCERS 2007. International* (pp. 229–232). IEEE.

48. Hopcroft, M. A., Kim, B., Chandorkar, S., Melamud, R., Agarwal, M., Jha, C. M., Bahl, G., Salvia, J., Mehta, H., Lee, H. K., Chandler, R. N. Kenny, T. W. (2007). Using the temperature dependence of resonator quality factor as a thermometer, *Applied Physics Letters*, **91**(1), 013505.

49. Ren, K., Kao, P., Pisani, M. B., and Tadigadapa, S. (2011). Monitoring biochemical reactions using Y-cut quartz thermal sensors, *Analyst*, **136**(14), 2904–2911.

50. Boisen, A., Dohn, S., Keller, S. S., Schmid, S., and Tenje, M. (2011). Cantilever-like micromechanical sensors, *Reports on Progress in Physics*, **74**(3), 036101.

51. Barnes, J. R., Stephenson, R. J., Woodburn, C. N., Oshea, S. J., Welland, M. E., Rayment, T., Gimzewski, J. K., and Gerber, C. (1994). A femtojoule calorimeter using micromechanical sensors, *Review of Scientific Instruments*, **65**(12), 3793–3798.

52. Barnes, J. R., Stephenson, R. J., Welland, M. E., Gerber, C., and Gimzewski, J. K. (1994). Photothermal spectroscopy with femtojoule sensitivity using a micromechanical device, *Nature*, **372**(6501), 79–81.

53. Datskos, P. G., Rajic, S., Sepaniak, M. J., Lavrik, N., Tipple, C. A., Senesac, L. R., and Datskou, I. (2001). Chemical detection based on adsorption-induced and photoinduced stresses in microelectromechanical systems devices, *Journal of Vacuum Science & Technology B: Microelectronics and Nanometer Structures*, **19**(4), 1173–1179.

54. Wig, A., Arakawa, E. T., Passian, A., Ferrell, T. L., and Thundat, T. (2006). Photothermal spectroscopy of Bacillus anthracis and Bacillus cereus with microcantilevers, *Sensors and Actuators B: Chemical*, **114**(1), 206–211.

55. Krause, A. R., Van Neste, C., Senesac, L., Thundat, T., and Finot, E. (2008). Trace explosive detection using photothermal deflection spectroscopy, *Journal of Applied Physics*, **103**(9), 094906.

56. Kim, S., Lee, D., Liu, X., Van Neste, C., Jeon, S., and Thundat, T. (2013). Molecular recognition using receptor-free nanomechanical infrared spectroscopy based on a quantum cascade laser, *Scientific Reports,*, **3**, doi:10.1038/srep01111.

57. Yun, M., Kim, S., Lee, D., Jung, N., Chae, I., Jeon, S., and Thundat, T. (2012). Photothermal cantilever deflection spectroscopy of a photosensitive polymer, *Applied Physics Letters*, **100**(20), 204103.

58. Schmid, S., Wu, K., Larsen, P. E., Rindzevicius, T., and Boisen, A. (2014). Low-power photothermal probing of single plasmonic nanostructures with nanomechanical string resonators, *Nano Letters*, **14**, 2318–2321.

Chapter 9

Optical Transduction and Actuation of Subwavelength Nanomechanical Resonators

Eduardo Gil-Santos, Valerio Pini, Álvaro San Paulo, Montserrat Calleja, Javier Tamayo, and Daniel Ramos

Instituto de Microlectrónica de Madrid, CSIC, Isaac Newton 8 (PTM), Tres Cantos, Madrid 28760, Spain

9.1 Introduction

Nanomechanical sensors require transduction techniques that allow the monitoring of their displacement in real time and provide good spatial resolution at the nanometer scale or even below. Therefore, the majority of the effort applied to the development of a sensor is intended to improve the signal reading process. Such position-sensing devices must be able to measure different parameters related to the deflection of the lever, including, among others, the position of the free end, the spatial orientation, the radius of curvature, and the intrinsic stress of the cantilever. Furthermore, there are special requirements that depend on both

Nanocantilever Beams: Modeling, Fabrication, and Applications
Edited by Ioana Voiculescu and Mona Zaghloul

ISBN 978-981-4613-23-1 (Hardcover), 978-981-4613-24-8 (eBook)
www.panstanford.com

the device itself, e.g., the cantilever material or the geometrical dimensions, and the use to which it will take, that is, the time-domain measurements (static deflection or resonance frequency), the magnitude of the expected response, and the environment in which it will be used (vacuum, air, or liquids).

Most of the experimental systems used in the literature can be grouped into two different sets: optical and electrical methods. Each of them has several advantages and disadvantages. For example, the optical methods, which are the scope of this chapter, have excellent spatial resolution; however, they require the use of cantilevers as reflective mirrors whose surface should be larger than the wavelength of the shining light to avoid diffraction. Therefore, these techniques are not, in principle, suitable for measuring nanometric size levers, although we will describe an evanescent field based method to overcome this limitation. On the other hand, the electrical methods rely on the electrical nature of the cantilever material and the surrounding medium. Therefore, the implementation of these techniques in aqueous solutions is still a big technical challenge because the electrolyte ions present in the liquid tend to cause eddy currents.

9.2 Optical Methods

The nanomechanical sensor devices based on micro-electro-mechanical systems have not only inherited from the atomic force microscope (AFM) the benefits of using microcantilevers, they also borrow the method of "optical beam deflection" to detect the induced deflections in cantilevers and their resonance frequency. Optical transduction methods include optical beam deflection method[1] and those based on optical interferometry.[2,3]

9.2.1 Measuring Static Deflections: Optical Beam Deflection

The optical beam deflection method is the most used method in both AFM and nanomechanical sensors for measuring the deflections of the lever, both dynamic and static. This method is based on the reflection of a light beam on the surface of the cantilever that is collected by a segmented photodiode or position sensitive detector

(PSD). The light sources commonly used for this type of action are light emitting diodes (LED) and diode lasers.

Photodiodes divided into two or four quadrants are responsible for the conversion of light energy of the laser spot that reaches each segment into a measurable electric signal, which can be filtered, amplified, compared with the signal coming out from the other segments, and finally monitored. The deflection of the cantilever changes the position of the reflected laser spot on the photodiode surface (due to changes in the angle of reflection), and consequently, the amount of light reaching each segment, correspondingly varying the differential electric signal. By the appropriate selection of the sum scheme along the four quadrant segments of the photodiodes it is possible to discern, in principle, displacements in two orthogonal axis, which can be useful to independently measure two kinds of modes of vibration of the cantilever: flexural and torsional vibrations.

In typical experimental conditions for static measurements, the reflected light spot on the surface of the cantilever is centered on the photodiode, so that all segments receive the same amount of light at the beginning of each measurement. Thus, when the cantilever is bent due to the differential surface stress caused by physical or chemical changes on the cantilever surface, the laser spot displaces over the surface of the segmented photodiode. The position of the laser spot and therefore the deflection of the cantilever can be determined using the outputs of each of the quadrants.

A different approach of the optical beam deflection method changes the segmented photodetector by PSDs that are monolithic PIN diodes ("Positive Intrinsic Negative") with uniform resistance in one or two dimensions. The light incident on the photosensitive region of the PSD generates two photocurrents; each of them is inversely proportional to the distance of the spot with respect to the end of said region. The difference between the photocurrents is converted into a voltage, which is amplified. These devices have a good position resolution and the response speed is determined by their control electronics. The principal advantage of these devices is that they present a linear response over the entire active surface.

The optical beam deflection technique is currently one of the most sensitive, inexpensive and extended methods to measure both the static deflection and the resonance frequency of the cantilever. The resolution of this technique has been settled in the order of angstroms. However, the geometrical constrains complicates the on-chip integration of this measurement technique because the laser and the detector must be outside the cell that contains the cantilever, which can be submerged into a liquid or gas to be analyzed. That means that this detection technique cannot be used when the microcantilever is placed inside a medium that strongly disperses or absorbs light, such as a solution or suspension of microparticles, or undiluted colored liquids such as blood.[4] Besides these difficulties, there is a physical limitation in the size of the microcantilever, the diffraction limit. Since the optical beam deflection technique deals with the specular reflection of a laser spot onto the cantilever surface, if the reflecting surface is smaller than the wavelength the reflected light dramatically drops. Furthermore, the dimensions of the laser and detector are much higher than those of the microcantilever to analyze, compromising the portability of the device.

It has been demonstrated that this technique allows for the measurement of arrays of cantilevers. However, despite being proposed a lot of different designs, only two proposals in the literature are commonly adopted. The first approach is to have different light sources and detectors, as many as the number of cantilevers in the array to be measured. This approach allows quasi-simultaneous measurement of all cantilevers, however, requires the integration of a large number of independent light sources, in addition to all the detectors, which makes the signal processing complex and poorly versatile. The cost can be significantly reduced using a single laser source sequentially focused onto each cantilever and collecting all the reflected beams into a single detector. The way to achieve this is to use a laser beam, which moves in the direction perpendicular to the long axis of the cantilever.[5,6] Figure 9.1 shows a schematic depiction of the principle of operation of the device, where it is possible to observe the direction of the laser sweep and, for a matrix of five cantilevers, the experimental intensity profile of light collected by the PSD after a laser cross sweep.

By changing the sweep direction, along the cantilever long axis, it is possible to extract the information about the displacement profile of the cantilever. In order to do this we have to take into account the angle subtended by the cantilever plane and the PSD surface as it is shown in Fig. 9.2. Once we have determined the angle, the deflection of the cantilever is recovered by simply integrating the PSD signal.

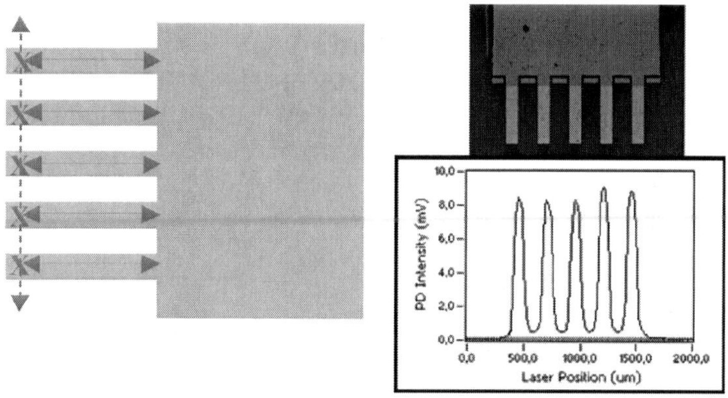

Figure 9.1 Schematic depiction showing the laser sequential scanning over five different cantilevers. The optical micrograph shows a typical chip used in the experiments and the intensity profile measured by the photodetector.

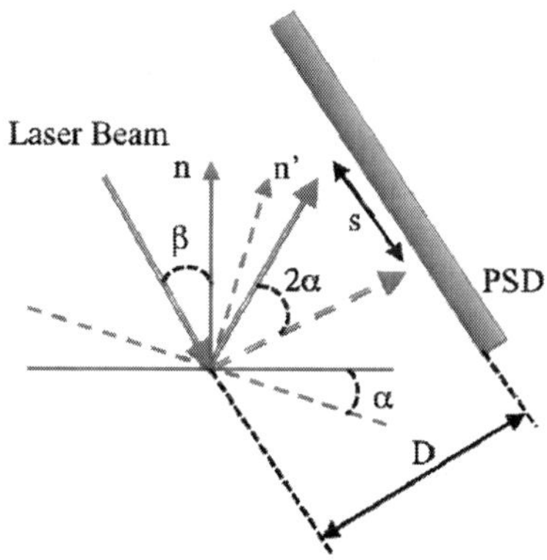

Figure 9.2 Schematic depiction of the change in the laser spot position on the PSD surface induced by a change in the cantilever angle.

As it is shown in the previous figure, a change in the cantilever slope is translated into a change in the position of the laser spot on the surface of the PSD, $s(x)$. Therefore, the displacement of the spot during the scan along the cantilever long axis can be described as

$$s(x) \cong 2D\frac{dz}{dx}(x) + x\cos\beta, \tag{9.1}$$

where $z(x)$ is the cantilever profile along its longitudinal axis, D is the distance between the cantilever and PSD, and β is the angle between the incident laser beam and the cantilever normal at its rest position (Fig. 9.2). Second summand accounts for the effect of the laser beam displacement, which is generally small compared to the first summand associated with the slope variations. The cantilever profile is obtained by integrating Eq. 9.1, as follows:

$$z(x) = \frac{1}{2D}\int_0^x s(x')dx' - \frac{1}{4}\frac{x^2}{D}\cos\beta \tag{9.2}$$

Figure 9.3 shows an optical micrograph of a gold-coated cantilever, with a laser spot on it. Note that the resolution of the measurement depends on the spot initial size.

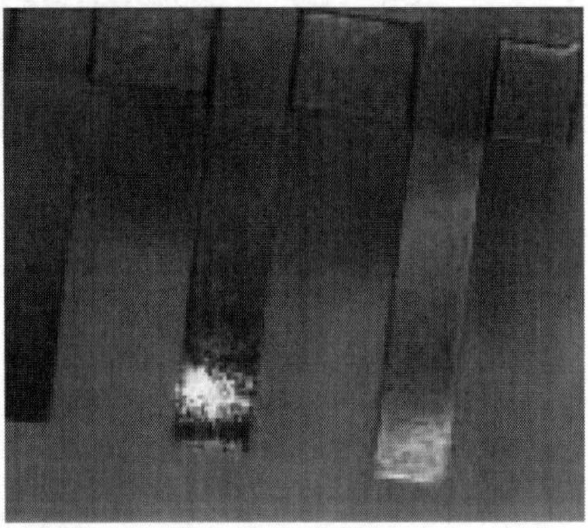

Figure 9.3 Optical micrograph showing a gold-coated cantilever being scanned by a laser spot.

The resulting cantilever profiles measured[6] in water before and after exposures of the cantilevers to mercaptohexanol (MCH) are shown in Fig. 9.4a. The scan speed was set to 500 μm/s. The cantilever free end was established at the positions where the total light intensity collected by the PSD falls to the half. The cantilevers displace upward (towards the silicon side). The displacement arises from the compressive surface stress on the gold due to the strong attachment of the MCH molecules, which expands the gold surface with respect to the silicon side. From the profile data, the displacement of the cantilever free end is obtained and monitored in real time to follow the adsorption process (Fig. 9.4b). The cantilevers significantly displace upward during the first seconds of exposure to MCH indicating that a highly packed monolayer is quickly formed. Then the cantilevers slowly displace downward during several minutes. This is attributed to the detachment of loosely bound molecules as well as long-term reorganization process in the monolayer.

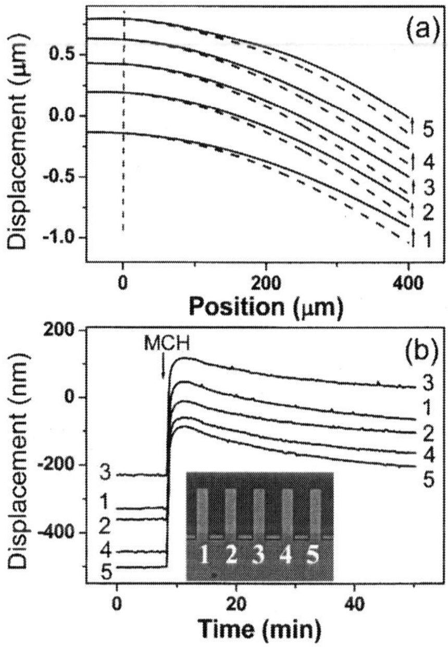

Figure 9.4 (a) Displacement profile of five gold-coated silicon cantilevers of an array before (dotted lines) and after adsorption (solid lines) of MCH. The vertical dashed line indicates the clamping position. (b) Real-time measurement of the displacement of the free end of the five microcantilevers. The arrow indicates the moment when the MCH begins to adsorb on the cantilevers. The measurements were performed in water and the cantilevers length, width, and thickness were 400, 100, and 1 μm, respectively. The array pitch distance was 250 μm. Reproduced from Ref. [6] Mertens, J., Álvarez, M., and Tamayo, J. (2005) Real-time profile of microcantilevers for sensing applications, *Appl. Phys. Lett.*, **87**, 234102. Copyright 2005 AIP.

9.2.2 Dynamic Measurement: Interferometric Method

The interferometric technique is based on the constructive or destructive interferences of a collimated laser beam reflected by the cantilever. Depending on the reference beam, we can distinguish between two different schemes: Michelson when the reference light is coming from an orthogonal path, or Fabry–Perot when the reference light is coming from the same optical path (Fig. 9.5).

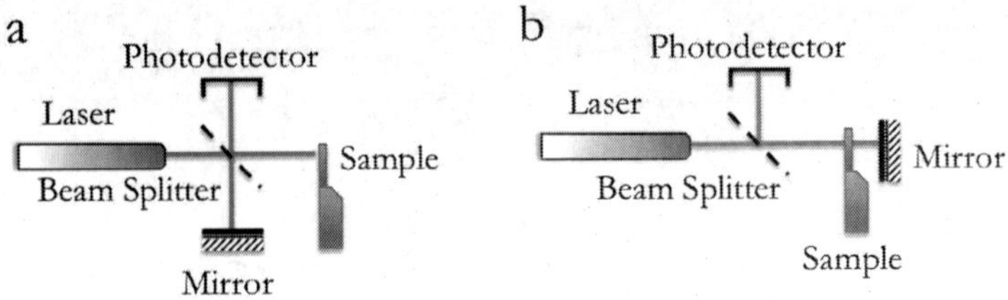

Figure 9.5 Schematic depiction of different configurations for an interferometric reading system. (a) Michelson configuration with a reference arm provided by a mirror. (b) Fabry–Perot configuration where the reference is provided by a mirror placed in the same optical path.

The first implementations of the optical interferometry[7,10] as a method for the measurement of a nanomechanical resonator were as a Fabry–Perot configuration. This approach uses the interference of the light bouncing off from a silicon substrate and the light from the cantilever surface. Since this method is based on the specular reflection on the cantilever, it is not suitable for subwavelength structures. In the context of future applications of nanomechanical sensors, which will ultimately require highly sensitive displacement detection, it is both interesting and important to elucidate the optical mechanism responsible for characterization of nanobeam mechanical motion. As the lateral dimensions of the nanobeams are shrunk down to be on-the-order-of or smaller than the wavelength of the probe laser, and the fact that in plane motion of cantilevers can be observed, more than simple scattering of light is required to detect the levels of movement measured. Intuitively, it is possible to implement a shorter wavelength probe to enhance interferometric displacement detection, and indeed, this is the case in the vast majority of prior work on nanomechanical characterization by optical interferometry, where a visible band (typically red) probe laser is used. However, this would likely complicate studies focused on performing visible spectroscopy or fluorescence dyes, while also optically probing the mechanical motion. Therefore, a detailed study on how the interferometric displacement detection may be optimized as a function of cantilever cross-sectional geometry should be performed.

9.2.2.1 Interferometry with leaky modes

One-dimensional nanomechanical resonators based on nanowires or nanotubes have emerged as promising candidates for mass sensors. However, their poor back-reflected light due to the strong scattering at subwavelength typically available in the laboratory handicaps the use of these sensitive resonators. We will explain in detail the use of confined electromagnetic modes inside a dielectric structure to efficiently extend optomechanics to sub-wavelength structures such as silicon nanowires.

Silicon nanowires (SiNWs) were grown via the vapor-liquid-solid mechanism in an atmospheric-pressure chemical vapor deposition (CVD) reactor.[11–14] The nanowires were horizontally grown on the sidewalls of microtrenches fabricated in silicon on insulator (SOI) substrates by photolithography and reactive ion etching. The vertical separation between the nanowire and the substrate underneath ranges from 1.0 to 1.3 μm.

9.2.2.1.1 *Experimental setup*

The experimental measurements of the tapered NWs resonance frequencies were performed using a homemade[15–17] hybrid-interferometric optical system with a He-Ne laser (5 mW, 633 nm, Thorlabs, Inc). The laser beam is focused on the free end of the nanowire by means of a long working distance Mitutoyo objective (50×, N.A. = 0.55) that produces a spot size of about 2 μm. The sample is maintaining in a high-vacuum environment (10^{-7} mbar). In order to reach those environmental conditions, a high-vacuum chamber is pumped down by means of a rotatory and a turbomolecular pump (Varian Inc.). A Faraday isolator is used to avoid backscattering reflections that could damage the laser cavity. By using a polarizing plate after a quarter-wave plate, the laser beam is linearly polarized in the desired angle. A collimated LED white source is used to image the laser spot on the chip by means of a CCD camera, whereas a non-polarized beam splitter split the light focused on the chip and the reflected one. The position of the chip is controlled by a three-dimensional closed loop nanopositioning stage (Attocube Systems, AG). The signal coming out from a silicon-amplified photodetector (Thorlabs, Inc.) is acquired by a DAQ board (National Instruments Corp.) and finally analyzed.

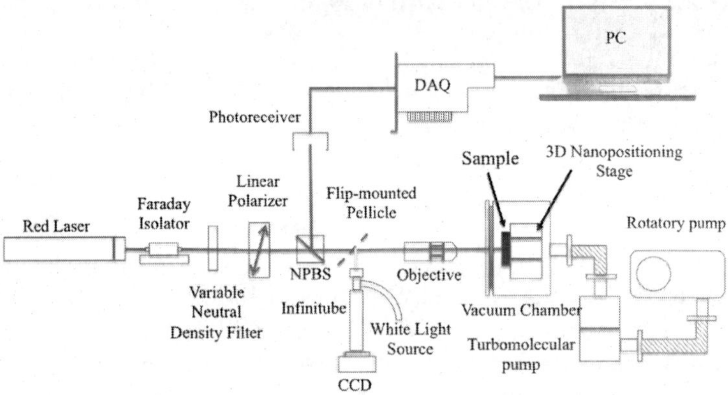

Figure 9.6 Block diagram of the experimental setup used in the experiments. Adapted from Ref. [17], Gil-Santos, E., Ramos, D., Pini, V., Llorens, J., Fernandez-Regulez, M., Calleja, M., Tamayo, J., and San Paulo, A. (2013) Optical back-action in silicon nanowire resonators: bolometric versus radiation pressure effects, *New J. Phys.*, **15**, 035001. © IOP Publishing & Deutsche PhysikalischeGesellschaft. CC BY-NC-SAdoi:10.1088/1367-2630/15/3/035001.

9.2.2.1.2 *Far-field FEM calculations*

Facing the problem of the origin of the optomechanical mechanism responsible for the transduction of the nanowire mechanical displacement requires the determination of the system geometrical symmetries. The laser beam coming from the normal direction to the substrate shines the horizontally grown nanowires, which means that the laser only probes the nanowire cross section. By assuming that the length of the wires greatly exceeds their diameter (usually 100:1 length to diameter ratio), we can reduce the dimensionality of the problem to a cross-sectional 2D study.

By taking into account that the probe laser has a wavelength of 632.8 nm, the fabricated silicon nanowires are sub-wavelength in their dominant length scale, which is their nominal diameter, ranging from 40 to 150 nm; therefore, we can assume that the detected interference pattern likely comes from laser light scattered by the moving nanostructures.

Finite element simulations, FEM (COMSOL Multiphysics), were performed in order to solve Maxwell's equations in the near environment of the nanowire, showing a spatial confinement of the electric field intensity within the dielectric cross section. From these simulations, it is clear that the nanowire size actively selects the amount of light confined within the nanostructure, depending on the incident laser wavelength. The use of similar confined electromagnetic modes in sub-wavelength structures such as diamond triangular nanobeams[18] was recently reported in the literature. Evanescent fields emanating from these confined modes overlap with the standing wave generated by reflections from the silicon surface. Therefore, the phase contrast caused by nanowire vibrations produces a detectable signal. To confirm the effect of the electromagnetic modes confined by the dielectric structure in the detection of the mechanical displacement field, we have to calculate the far field profile of the laser light collected by the objective.

The far electromagnetic field from the scattering nanowire is calculated by means of the near field propagation in free space. Since the system was reduced to a 2D scattering problem the far-field electric field, $\overrightarrow{E_\mathrm{p}}$, calculation is simplified to the Stratton–Chu formula[19]

$$\overrightarrow{E_\mathrm{p}} = \sqrt{\lambda}\frac{ik}{4\pi}\overrightarrow{r_0} \times \int [\vec{n} \times \vec{E} - \eta\overrightarrow{r_0} \times (\vec{n} \times \vec{H})] e^{(ik\vec{r}\cdot\overrightarrow{r_0})}\, dS, \tag{9.3}$$

where the calculated far-field electric field in the direction from the origin towards point p is taken at infinity but with a well-defined angular position (θ, ϕ). In this formula, the scattering object (nanowire cross section) is supposed to be located at the origin of coordinates. $\vec{E}$ and $\vec{H}$ are the electric and magnetic fields on the surface enclosing the scatters S, $\overrightarrow{r_0}$ is the unit vector pointing from the origin to the field point p, $\vec{n}$ is the unit vector normal to the surface S, $\eta = \sqrt{\mu/\varepsilon}$ is the impedance referred as a function of the permeability, μ, and the permittivity, ε, k, and λ are, respectively, the wave number and the wavelength, and $\vec{r}$ is the position vector of the surface S. In this case, the surface S should enclose, not only the nanowire cross section but also the reflecting surface (Fig. 9.7). A phase match layer (PML) encloses the calculation domain in order to avoid undesired laser reflection, which could affect the final result.

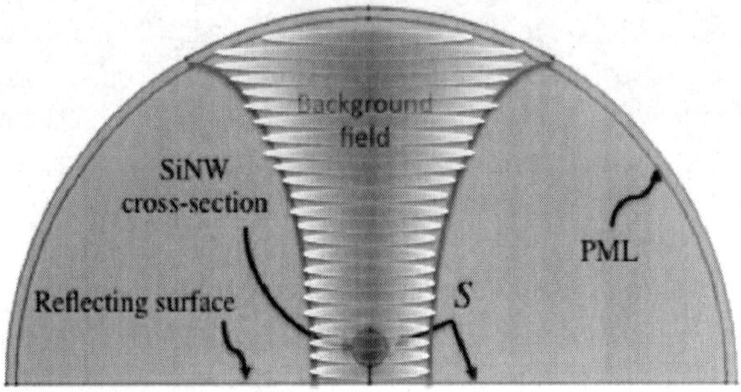

Figure 9.7 Calculation scheme.

To confirm the effect of the electromagnetic modes confined by the dielectric structure do in fact yield an optomechanical coupling, the far-field profile of reflected laser light is first calculated for nanowires with different radii suspended 1.0 μm above the substrate, as shown in Fig. 9.8. In this case, it is clear the electric far-field maximum is located at direction normal to the substrate surface nanowire. Thus, the signal resulting from the thinnest nanowires is largely due to laser light reflected by the substrate. However, with increasing nanowire diameter, the far field profile is significantly modified by emerging side lobes, which may be attributed to scattering of the incident laser beam by electromagnetic modes confined in the dielectric cross section. This finding can be explained on the basis of the Mie scattering theory.[20,21] Thus, it is evident that the far field profile is a function of the combination of the nanowire diameter and the probe laser wavelength for a fixed substrate distance.

To further illustrate this dependence, the squared electric field intensity in the far field was integrated over the objective lens numerical aperture[22] used in the actual characterization setup (Fig. 9.6) (50× long working distance objective, N.A. 0.55, i.e., the angular integration is of ±30°), as a function of the nanowire radius and the distance to the substrate (Fig. 9.9). Thus, the resulting matrix of values is proportional to the photoreceiver signal. As it was stated before, the less interaction between the laser light and the nanowire, the higher the reflected light; therefore, the higher values (reddish areas) correspond with the backscattered laser. This implies that the nanowire interaction is viewed as a deep in the DC component of the photodetector signal (bluish areas).

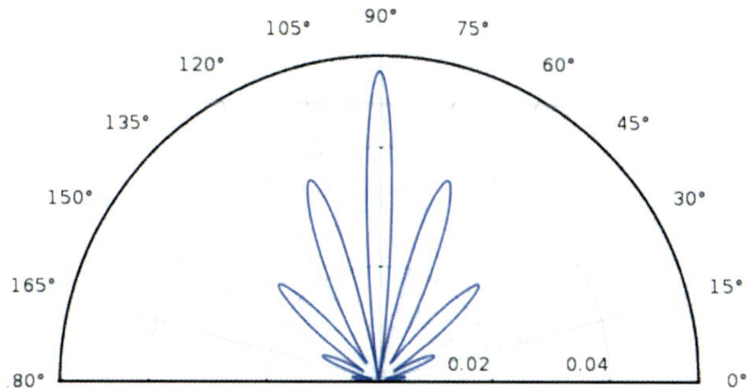

Figure 9.8 Polar plot of the angular dependence of the squared electric field intensity in the far field for a nanowire suspended above a reflecting substrate at 1 μm and a diameter of 120 nm.

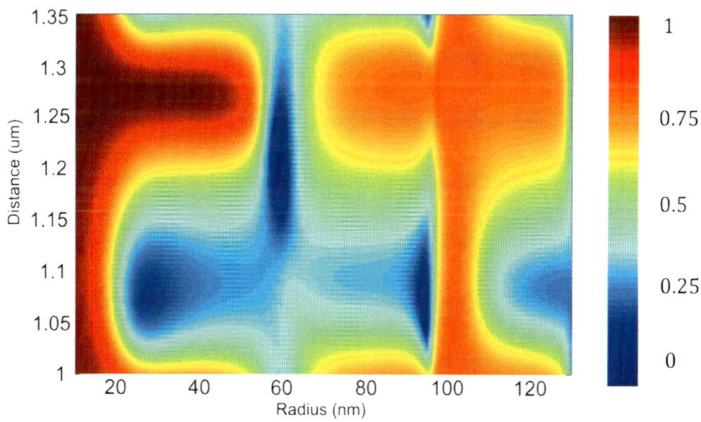

Figure 9.9 Signal strength in color (linearly normalized from blue to red) as a function of the nanowire radius and the substrate distance for TM polarization, i.e., the one in which the electric field propagates in the plane containing the nanowire long axis or "parallel" polarization.

According to previous studies in literature,[23,24] in order to maximize the output signal, the polarization angle is the one in which the electric field propagates in the plane containing the nanowire long axis, or transversal magnetic, "TM." Taking the one-dimensional gradient by respect to the substrate distance results in the measurement sensitivity (units of power per meter, or W/m), as represented by the color map shown in Fig. 9.10.

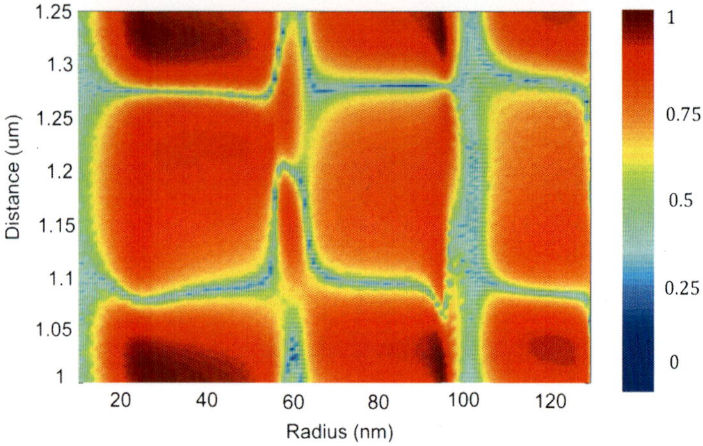

Figure 9.10 One-dimensional gradient of the signal strength (in color linear scale from blue to red) as a function of the nanowire radius and the substrate distance for TM polarization.

As expected, these calculations suggest that the measurement sensitivity depends on both the radius and the substrate distance, thus confirming that laser light confined by the nanowire cross section yields an optomechanical coupling, which transduces the mechanical motion. All results computed by FEM have been contrasted with analytical calculations obtained by expanding the electric and magnetic fields in cylindrical harmonics following the derivation of Videen and Ngo.[25] The far-field distribution is obtained as scattering amplitudes, which are derived by recalling the asymptotic behavior of the cylindrical harmonics.[20]

More intriguing is the dependence of the absorption on the nanowire diameter at a fixed substrate distance *D*. The absorption is significantly enhanced at certain diameters. As it can be inferred from the stated above about the scattering signal, the series of peaks found are in good agreement with the morphology-dependent optical resonances,[26] or Mie resonances, of a bare Si infinite cylinder illuminated by a plane wave. These resonances can be easily computed by finding the poles of the scattering coefficients resulting of the expansion of the scattering field on cylindrical harmonics.[20] For TM polarization and normal incidence, the optical resonances of a cylinder of diameter d and refractive index n is determined by the complex roots $k_0\frac{d}{2}$

$$nJ'_m\left(k\frac{d}{2}\right)H_m^{(1)}\left(k_0\frac{d}{2}\right)-J_m\left(k\frac{d}{2}\right)H_{m-1}^{(1)'}\left(k_0\frac{d}{2}\right)=0, \tag{9.4}$$

where $k_0 = 2\pi/\lambda$, $k = nk_0$ and the functions $J_m(z)$ and $H_m^{(1)}(z)$ are the m-th order Bessel and Hankel functions, respectively. For each m, there exists a family of roots that can be labeled according to the whole number l. Thus the resonances value can be identified by the tuple[26] (m, l). The real part of the root provides the position of the resonances and the imaginary part their quality factor: $Q=\lambda/\Delta\lambda=-\mathrm{Re}\left(k_0\frac{d}{2}\right)/2\mathrm{Im}\left(k_0\frac{d}{2}\right)$.

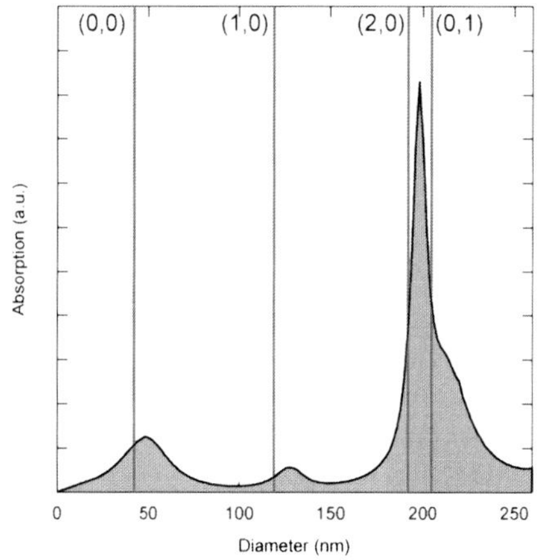

Figure 9.11 FEM calculation (gray area) of the absorption as a function of the nanowire diameter together with the position of the Mie resonances obtained from the roots of Eq. 9.4, red solid line. The labels are explained in the text. Reproduced from supplementary information of Ref. [16] Ramos, D., Gil-Santos, E., Pini, V., Llorens, J. M., Fernandez-Regulez, M., San Paulo, A., Calleja, M., and Tamayo, J. (2012) Optomechanics with silicon nanowires by harnessing confined electromagnetic modes, *Nano Lett.*, **12**, 932–937.

Figure 9.11 shows the comparison between the numerical absorption spectrum and the analytical Mie scattering resonances. The analytical resonances calculated from the Eq. 9.4 are in a very good agreement with the numerical calculations.

9.2.2.1.3 *Near-field FEM calculations*

In order to obtain insight on how the optical resonances of the nanowires interact with the interference pattern field, we have used the FEM software to calculate the electromagnetic field distribution for the first Mie mode for four different separations between the nanowire and the substrate underneath (800 nm, 875 nm, 950 nm, 1025 nm, and 1100 nm) at the near-field. Prior to that, it is interesting to observe the fundamental Mie mode excited by the normal laser shining when there is not substrate underneath. The TM (0,0) mode is excited for a nanowire diameter of 40 nm, as shown in Fig. 9.12.

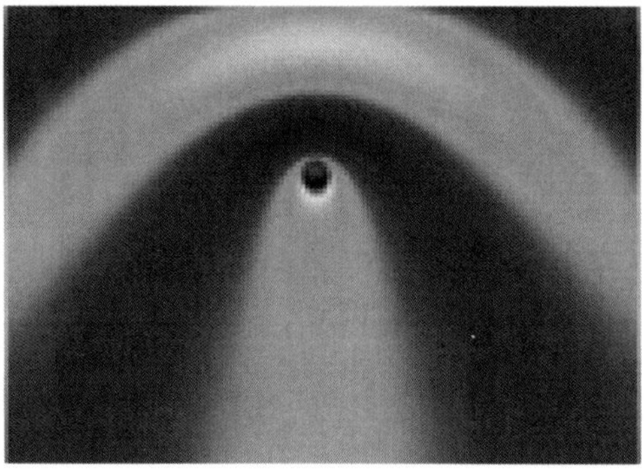

Figure 9.12 Near-field calculations showing the TM (0,0) mode excited for a nanowire of 40 nm of diameter when there is not substrate underneath.

The interaction with the standing wave field shifts the resonant diameter (40 nm when there is no substrate) between 10 and 24 nm depending on the separation between the substrate and the nanowire. This outstanding shift is a direct confirmation of the extraordinary optomechanical coupling coefficient,[27,28] g_{OM}, presented by the system. The g_{OM} is defined as the variation of the optical mode as a function of the mechanical displacement. This parameter can be viewed as the manifestation of the interaction of the leaky-mode evanescent field with the surrounding standing wave at the near field; consequently, the radial symmetry of the each optical mode plays an important role. Therefore, depending on the specific coupling of each optical mode, the sensitivity to the mechanical displacement may vary. As it is shown in the simulations of the Fig. 9.13, the spatial distribution of the electric field in the TM (0,0) mode is completely distorted by the surrounding electromagnetic field distribution. The simulations show a nanowire of 40 nm of diameter placed at four different distances from the substrate (800 nm, 875 nm, 950 nm, 1025 nm, and 1100 nm). All the color maps are plotted at the same scale, showing how the maximum of the light intensity confined inside the nanowire is placed at the points of maximum slope of the standing wave.

Figure 9.13 Near-field calculations showing the coupling of the leaky confined mode TM (0,0) with the surrounding standing wave for four different substrate separations (800 nm, 875 nm, 950 nm, 1025 nm, and 1100 nm).

9.2.2.1.4 *Optical back-action*

As the size of the resonator approaches the optical regime, the intimate connection between the probe and target, or observer and observable, also known as back-action, raises. Therefore, last years have witnessed a crescent interest in the community in the non-linear potential imposed by an optical cavity due to its capabilities in the noise control of nanostructures or the tuning of the parameters of nanomechanical resonators. Here we can distinguish two different effects: optical spring or the variation of the spring constant of nanomechanical resonators, and the optomechanical amplification or cooling of mechanical modes.

9.2.2.1.4.1 *Optical spring: bolometric versus radiation pressure effects*

By measuring the mechanical resonance of more than 20 NWs as a function of the incident laser power it is possible to distinguish two different behaviors: the mechanical resonance can be shifted to higher or lower frequencies. The mechanical quality factors of the peak resonances are of about 3000, being the same for both peaks conforming the fundamental resonance mode, hereinafter referred as s-mode and the f-mode. Figures 9.14a,b show the spectra of two representative examples of NWs for increasing laser power ranging from 10 to 500 μW. These results have been reproducible in all the observed NWs. The resonance

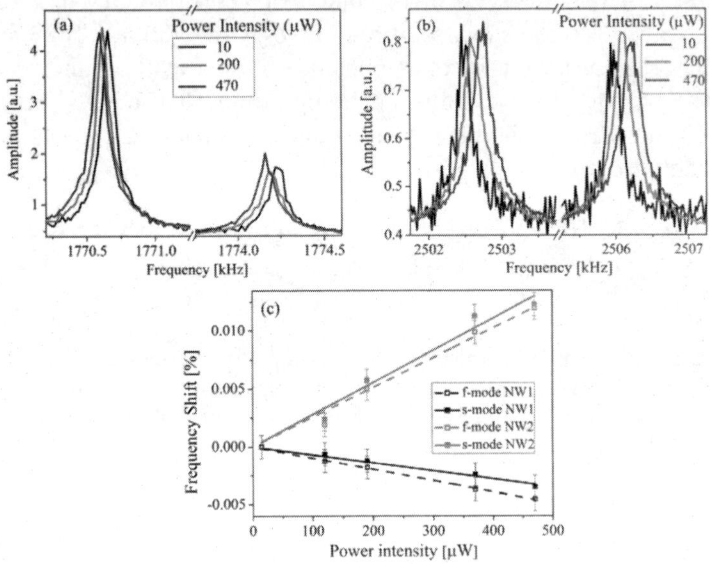

Figure 9.14 Back-action measurements. (a) A representative example of a thermomechanical noise spectrum measurement for different laser beam power intensities for a NW with resonance frequencies of the s- and f-modes that decrease for increasing power (L = 12.7 μm; d = 165 nm). (b) A representative example of a thermomechanical noise spectrum measurement for different laser beam power intensities for a NW with resonance frequencies of the s- and f-modes that increase for increasing power (L = 9.3 μm; d = 90 nm). (c) Resonant frequencies of each mode for both NWs as a function of the power intensity; solid lines are the linear fittings providing the values for the sensitivity. Reproduced from Ref. [17] Gil-Santos, E., Ramos, D., Pini, V., Llorens, J., Fernandez-Regulez, M., Calleja, M., Tamayo, J., and San Paulo, A. (2013) Optical back-action in silicon nanowire resonators: bolometric versus radiation pressure effects, *New J. Phys.*, **15**, 035001.©IOPPublishing&DeutschePhysikalischeGesellschaft. CC BY-NC-SAdoi:10.1088/1367-2630/15/3/035001.

frequencies of both modes of the NW of Fig. 9.14a are shifted to lower values as the power increases, while the opposite behavior is observed for the case shown in Fig. 9.14b. Figure 9.14c shows the changes in resonance frequencies for these two NWs as a function of the incident power. Each point in this graph is the result of the average of 100 independent measurements. The lack of symmetry in the obtained results for both NWs strongly suggests that different back-action mechanisms may be dominating in each case.

9.2.2.1.4.1.1 *Theoretical analysis*

The nanowire can be modeled as a lumped harmonic oscillator

$$m_{\text{eff}}\ddot{z} + \gamma_{\text{eff}}\dot{z} + k_{\text{eff}}z = F_{\text{th}} + F', \tag{9.5}$$

where m_{eff}, γ_{eff}, and k_{eff}, respectively, are the effectives mass, damping, and stiffness of the first flexural mode of the nanowire, F_{th} the thermomechanical forces that sustain its movement, and F' is an external force. On the one hand, the absorbed optical power heats up the nanowire. The inversely proportional dependency of the material Young's modulus on the temperature leads to the variation of the effective resonator stiffness, which shifts the resonance to lower frequencies. On the other hand, it is well known that the radiation pressure is a position dependent force and, therefore, it can be viewed as a change in the effective spring constant and written as a Taylor series, which at first order is expressed as $F' = F_0 + \frac{\partial F}{\partial z}z$, where F_0 is a static contribution that can be neglected because it does not produce any significant effect on the resonance frequency. Taking into account that the low finesse of our cavity does not introduce any delay between the acting force and the resonator displacement, which means the force is real; we finally obtain the modified harmonic oscillator equation;

$$m_{\text{eff}}\ddot{z} + \gamma_{\text{eff}}\dot{z} + \left(k_{\text{eff}}(1 - \propto T_{\text{eff}}) - \frac{\partial F}{\partial z}\right)z = F_{\text{th}}, \tag{9.6}$$

where T_{eff} is the first mode effective temperature as defined below. As we can see in this expression, the contribution of the temperature will shift the resonances of the oscillators to lower frequencies because the temperature gradient is always positive, while the radiation pressure effect can induce a shift both negative or positive depending on the sign of the optical force gradient.

The absorbed power by a physical system is calculated by means of the scattered electric field and the imaginary part of its permittivity[29]

$$P_{\text{abs}} = \frac{1}{2}\omega |E|^2 \,\text{Im}(\varepsilon), \tag{9.7}$$

where ω is the optical angular frequency, E is the electric field and ε is the system's permittivity. The power absorbed by the nanowire depends on both the diameter and the distance between the nanowire and the substrate. As it was shown above, the diameter dependence is related with the distribution of the confined electromagnetic modes inside the semiconductor structures, while the distance dependence is related with the coupling between these modes and the standing wave that arises from the interference between the incident and the reflected beam from the substrate.

Once we have the optical power absorbed we calculate the temperature distribution in the nanowire applying the heat diffusion equation by assuming the following: the penetration depth of the absorbed power is several orders of magnitude larger than the nanowire typical thickness and there are not any eventual heat convective losses as well as irradiative power losses. The temperature distribution $T(x)$, determined with respect to a given background temperature, is finally derived by applying the one-dimensional heat diffusion equation:

$$A\kappa_{\text{e}} \frac{\partial T(x)}{\partial x^2} + p(x) = 0, \tag{9.8}$$

where A is the cross-sectional area, κ_{e} is the effective thermal conductivity, and $p(x)$ is the absorbed power density produced by the Gaussian laser beam that is described by

$$p(x) = \frac{P_{\text{abs}}}{\sqrt{\frac{\pi}{2}} w_0} \exp\left(-\frac{2(x - x_0)^2}{w_0^2} \right), \tag{9.9}$$

where P_{abs} is the total power absorbed, w_0 is the laser beam waist and x_0 is the laser spot coordinate along the nanowire axis. The boundary conditions imposed by the geometry imply a zero heat flux at the nanowire free end and a null increment of temperature at the clamping region.

By using the calculated temperature distribution along the nanowire, we can obtain the equivalent effective temperature for the first flexural mode by using the Rayleigh method[30]:

$$T_{\text{eff}} = \frac{\int_0^L T(x)\left(\frac{\partial^2 \psi(x)}{\partial x^2}\right)^2 dx}{\int_0^L \left(\frac{\partial^2 \psi(x)}{\partial x^2}\right)^2 dx} \tag{9.10}$$

This effective temperature allow us to directly obtain the resonance frequency by applying

$$\omega_0' = \omega_0 \sqrt{1 - T_{\text{eff}}\alpha}, \tag{9.11}$$

where ω_0 is the initial angular frequency of the first resonant mode and α is the thermal expansion coefficient.

On the other hand, the radiation pressure on the silicon nanowires can be calculated by means of the asymmetries of the Maxwell stress tensor. This requires the knowledge of the scattered electric $E(x, y)$ and magnetic $H(x, y)$ fields, which are performed also by using the FEM software. The net force acting on a closed domain can be computed by integrating Maxwell's stress tensor over a close path[29]:

$$F_\alpha = \sum_\beta \oint T_{\alpha\beta} n_\beta \, dl, \tag{9.12}$$

where Maxwell's stress tensor is given by

$$T_{\alpha\beta} = \varepsilon E_\alpha E_\beta^* + \mu H_\alpha H_\beta^* - \frac{1}{2}\delta_{\alpha\beta}(\varepsilon |E|^2 + \mu |H|^2) \tag{9.13}$$

$\hat{n}$ is a unit vector normal to the integration path pointing outwards and μ is the permeability. The electric and magnetic fields are those obtained by solving Maxwell's equations. As has been stated before, the distance dependence of the optical force induces a frequency shift proportional to the gradient of the radiation pressure. Therefore, we can calculate the frequency shift as

$$\omega_0' = \omega_0 \sqrt{1 - \frac{1}{k_{\text{eff}}} \frac{\partial F}{\partial x}} \tag{9.14}$$

9.2.2.1.4.2 *Optomechanical amplification and cooling of mechanical modes*

In order to obtain the general solution of the frequency-dependent temperature distribution we have to modify the heat transfer equation, Eq. 9.8, including the temporal dependence of temperature

$$A\rho_e c_{pe} \frac{\partial T(x,t)}{\partial t} = A\kappa_e \frac{\partial T(x,t)}{\partial x^2} + p(x,t) \tag{9.15}$$

where ρ_e is effective material density and c_{pe} is the effective heat capacity. The general solution has complex values, giving us important information. Writing the complex value of temperature in polar form, its norm corresponds to the temperature raise induced by the optical source, while the phase gives us information about the delay between the optical source and the temperature field.

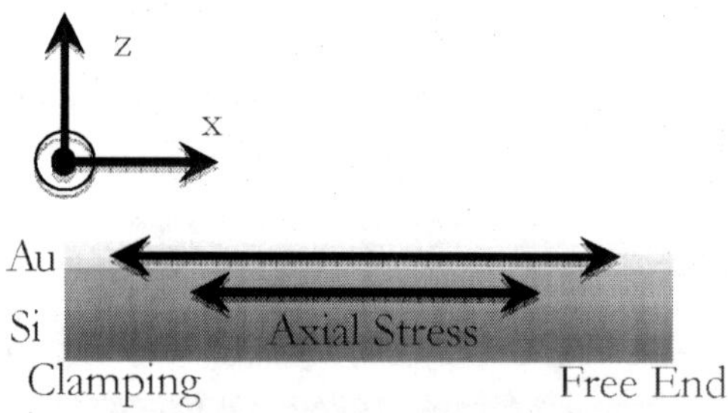

Figure 9.15 Schematic drawing of a nanowire subject to a differential thermal stress generated by a Gaussian optical absorption.

As it is known in literature,[31] the photothermal force is the result of the differential stress generated in the resonator cross sectional area. Thus, it is greatly enhanced by using different material layers conforming the cross section; therefore, we have deposited a thin gold layer (20 nm of gold and 2 nm of chromium as adhesion layer). Figure 9.15 shows a schematic depiction of the cross section of the nanowire after deposition of the gold layer on top of it.

It can be demonstrated[16] that the photothermal force in a gold-coated silicon nanowire is of about three orders of magnitude than the corresponding radiation pressure, calculated from Eq. 9.12. Therefore, hereinafter we will be only focused on the photothermal effect.

As a result of the absorption-induced temperature distribution, a stress field appears in the structure.[31] Since the silicon substrate and the gold coating have different thermal expansion coefficients and they are in contact, its different stress components give rise to a gradient of the y-component of the bending moment; which can be calculated by integrating in the nanowire cross section as

$$\nabla M(x) = \iint_S \nabla T(x)\sigma_{\text{th}} z dS \tag{9.16}$$

where dS is the differential element of the cross section, and σ_{th} is the differential thermal stress that can be evaluated as and $\sigma_{\text{th}} = \frac{\alpha_i E_i}{1 - \nu_i}$, being α the thermal expansion coefficients, E the Young's modulus, ν the Poisson coefficients, where the subscript i refers to each layer (silicon and gold).

The gradient of the mode-dependent effective force that is acting onto the cantilever can be evaluated at each nanowire-substrate distance by integrating the projection of the gradient of the bending moment onto the curvature mode

$$\nabla F_{\text{phot}} = -\int_0^L \nabla M(x) \frac{d^2\varphi_0(x)}{dx^2} dx \tag{9.17}$$

The imaginary part affects the nanowire quality factor, whose sign determines the self-excitation or cooling regime. The feedback gain defined as the amplitude of a force proportional to the oscillation velocity[32] is given by

$$g(z) = \text{Im}[\nabla F_{\text{phot}}] \tag{9.18}$$

In our experimental framework, the sign of the gain depends on the sign of the slope of the optical absorption coefficient. A positive slope coefficient corresponds to the optomechanical amplification regime, while negative slope induces a cooling regime. To further gain insight in physical meaning of the feedback gain let us revisit the harmonic oscillator equation of movement, Eq. 9.5, substituting the external force F' by the feedback force proportional to the velocity.

$$m_{\text{eff}}\ddot{x}+\gamma_{\text{eff}}\dot{x}+k_{\text{eff}}x=F_{\text{th}}+g\frac{1}{\omega_0}\dot{x} \tag{9.19}$$

As can be seen by simply substituting in Eq. 9.19, the optomechanical feedback force modifies the damping coefficient to

$$\gamma_{\text{eff}}=\gamma-\frac{g}{\omega_0} \tag{9.20}$$

Therefore, the feedback force modifies the oscillation amplitude at resonance for a measurement bandwidth $\Delta\omega$,

$$A_0=2\frac{\sqrt{\gamma k_{\text{B}}T\Delta\omega}}{\omega_0\left(\gamma-\dfrac{g}{\omega_0}\right)} \tag{9.21}$$

The modification of the damping coefficient by the optomechanical feedback gain (Eq. 9.20), leads to the existence of an optical threshold power. Above this critical power, the damping can be suppressed or enhanced depending on the sign of g. Figure 9.16 shows the amplitude at resonance as a function of the laser power (wavelength of 633 nm) for two different nanowires placed at the same distance to the substrate with different diameters: 120 and 180 nm. The thinner nanowire exhibits optical cooling, whereas the other shows optical amplification. The experimentally determined optical power threshold is of about 310 μW.

These experimental results are consistent with our hypothesis that the confined optical resonances supported by the nanowire can be used to develop an efficient optomechanical system. To obtain further demonstration of this concept, we performed numerical calculations to obtain the optomechanical feedback gain, Eq. 9.19. Figure 9.17 shows the color intensity maps of the calculated optomechanical feedback gain per unit of light power as a function of the nanowire diameter and the distance to the substrate, both normalized to the wavelength of 633 nm.

A positive value of the gain leads to the amplification of the fundamental vibration mode, whereas a negative value leads to its cooling. In addition, we have plotted the experimental values of diameter and distance coordinates of the nanowires that exhibited optical amplification (red circles) and cooling (cyan circles). We find a very good agreement between the experimental data and the proposed theoretical model.

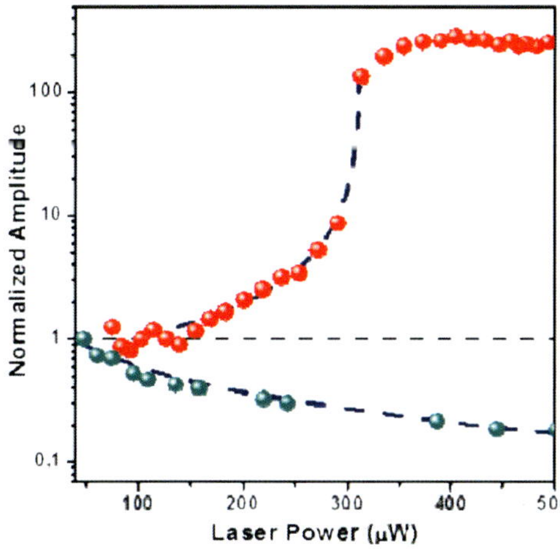

Figure 9.16 Amplitude at resonance of two nanowires as a function of the laser power intensity (symbols). The fundamental vibration mode of one nanowire is optically amplified (red symbols) but is cooled in the other case (cyan symbols). In the case of optical amplification, the amplitude moderately grows with the laser intensity until it undergoes an abrupt increase of about 20 times when the laser power is 310 μW. Above this critical laser power, the nanowire exhibits self-sustained oscillation. The blue dashed lines represent the fitting to the theory. Adapted from Ref. [16] Ramos, D., Gil-Santos, E., Pini, V., Llorens, J. M., Fernandez-Regulez, M., San Paulo, A., Calleja, M., and Tamayo, J. (2012) Optomechanics with silicon nanowires by harnessing confined electromagnetic modes, *Nano Lett.*, **12**, 932–937.

In the self-sustained oscillation regime due to the optomechanical amplification the transient oscillation grows exponentially until the non-linear saturation mechanisms of the mechanical system comes into play and the oscillation reaches a steady state with virtually no noise in the amplitude and small phase fluctuations of diffusive nature.[16,32] The ultimate limit in the frequency noise is imposed by the phase diffusion that follows the Einstein equation for diffusion process.[32] The corresponding frequency noise is given by

$$\frac{\delta\omega_0}{\omega_0} = 0.341\sqrt{\frac{k_B T}{Q k_{\text{eff}} A_0 \omega_0 \tau}} \tag{9.22}$$

where Q is the mechanical quality factor and τ is the averaging time. Therefore, the frequency fluctuations of the mechanical resonators in the self-sustained regime are decreased, being promising candidates for ultrasensitive mass sensors.

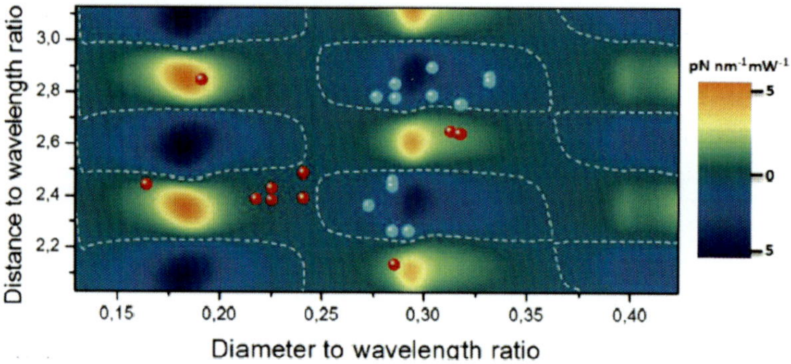

Figure 9.17 Calculated color map of the optomechanical feedback gain, the symbols represent experimental measurements showing amplification (red circles) or cooling (cyan circles). Adapted from Ref. [16] Ramos, D., Gil-Santos, E., Pini, V., Llorens, J. M., Fernandez-Regulez, M., San Paulo, A., Calleja, M., and Tamayo, J. (2012) Optomechanics with silicon nanowires by harnessing confined electromagnetic modes, *Nano Lett.*, **12**, 932–937. Copyright 2012 American Chemical Society.

References

1. Meyer, G., and Amer, N. M. (1988). Novel optical approach to atomic force microscopy, *Appl. Phys. Lett.*, **53**, 1045.
2. Rugar, D., Mamin, H. J., and Guethner, P. (1989). Improved fiber optic interferometer for atomic force microscopy, *Appl. Phys. Lett.*, **55**, 2588.

3. Hoummady, M., Farnault, E., Yahiro, T., and Kawakatsu, H. (1997). Simultaneous optical detection techniques, interferometry, and optical beam deflection for dynamic mode control of scanning force microscopy, *J. Vac. Sci. Technol. B*, **15**, 1539.
4. Bottomley, L. A., Poggi, M. A., and Shen, S. (2004). Impact of nano- and mesoscale particles on the performance of microcantilever-based sensors, *Anal. Chem.*, **76**, 5685.
5. Álvarez, M., and Tamayo, J. (2005). Optical sequential readout of microcantilever arrays for biological detection, *Sens. Actuators B*, **106**, 687.
6. Mertens, J., Álvarez, M., and Tamayo, J. (2005). Real-time profile of microcantilevers for sensing applications, *Appl. Phys. Lett.*, **87**, 234102.
7. Rugar, D., Mamin, H. J., Erlandsson, R., Stern, and J. E., Terris, B. D. (1988). Force microscope using a fiber-optic displacement sensor, *Rev. Sci. Instrum.*, **59**(11), 2337–2340.
8. Schonenberger, C., and Alvarado, S. F. (1989). A differential interferometer for force microscopy, *Rev. Sci. Instrum.*, **60**(10), 3131–3134.
9. Carr, D. W., and Craighead, H. G. (1997). Fabrication of nano-electromechanical systems in single crystal silicon using silicon on insulator substrates and electron beam lithography, *J. Vac. Sci. Technol. B*, **15**, 2760.
10. Carr, D. W., Sekaric, L., and Craighead, H. G. (1998). Measurement of nanomechanical resonant structures in single-crystal silicon, *J. Vac. Sci. Technol. B*, **16**, 3821.
11. San Paulo, A., Arellano, N., Plaza, J. A., He, R. R., Carraro, C., Maboudian, R., Howe, R. T., Bokor, J., and Yang, P. D. (2007). Suspended mechanical structures based on elastic silicon nanowire arrays, *Nano Lett.*, **7**, 4.
12. Malvar, O., Gil-Santos, E., Ruz, J. J., Ramos, D., Pini, V., Fernandez-Regulez, M., Calleja, M., Tamayo, J., and San Paulo, A. (2013). Tapered silicon nanowires for enhanced nanomechanical sensing, *Appl. Phys. Lett.*, **103**, 033101.
13. Hannon, J. B., Kodambaka, S., Ross, F. M., and Tromp, R. M. (2006). The influence of the surface migration of gold on the growth of silicon nanowires, *Nature*, **440**, 69–71.
14. Wang, Y., Schmidt, V., Senz, S., and Gösele, U. (2006). Epitaxial growth of silicon nanowires using an aluminium catalyst, *Nat. Nanotechnol.*, **1**, 214–220.

15. Gil-Santos, E., Ramos, D., Martinez, J., Fernandez-Regulez, M., Garcia, R., San Paulo, A., Calleja, M., and Tamayo, J. (2010). Nanomechanical mass sensing and stiffness spectrometry based on two-dimensional vibrations of resonant nanowires, *Nat. Nanotechnol.*, **5**(9), 641.

16. Ramos, D., Gil-Santos, E., Pini, V., Llorens, J. M., Fernandez-Regulez, M., San Paulo, A., Calleja, M., and Tamayo, J. (2012). Optomechanics with silicon nanowires by harnessing confined electromagnetic modes, *Nano Lett.*, **12**, 932–937.

17. Gil-Santos, E., Ramos, D., Pini, V., Llorens, J., Fernandez-Regulez, M., Calleja, M., Tamayo, J., and San Paulo, A. (2013). Optical back-action in silicon nanowire resonators: Bolometric versus radiation pressure effects, *N. J. Phys.*, **15**, 035001.

18. Burek, M. J., Ramos, D., Patel, P., Frank, I. W., and Lončar, M. (2013). Nanomechanical resonant structures in single-crystal diamond, *Appl. Phys. Lett.*, **103**, 131904.

19. Volakis, J. L., and Sertel, K. (2012). *Integral Equation Methods for Electromagnetics* (Kay, D. R., ed.), SciTech Publishing, Inc. Raleigh, NC 27615.

20. Bohren, C. F., and Huffman, D. R. (1998). *Absorption and Scattering of Light by Small Particles*; Wiley-VCH: Berlin.

21. Brönstrup, G., Jahr, N., Leiterer, C., Csáki, A., Fritzsche, W., and Christiansen, S. (2010). Optical properties of individual silicon nanowires for photonic devices, *ACS nano*, **4**, 7113–7122.

22. http://www.luxpop.com.

23. Nichol, J. M., Hemesath, E. R., Lauhon, L. J., and Budakian, R. (2008). Displacement detection of silicon nanowires by polarization-enhanced fiber-optic interferometry, *Appl. Phys. Lett.*, **93**(19), 193110–193113.

24. Belov, M., Quitoriano, N., Sharma, S., Hiebert, W., Kamins, T., and Evoy, S. (2008). Mechanical resonance of clamped silicon nanowires measured by optical interferometry, *J. Appl. Phys.*, **103**(7), 074304–074307.

25. Videen, G., and Ngo, D. (1997). Light scattering from a cylinder near a plane interface: Theory and comparison with experimental data, *J. Opt. Soc. Am. A*, **14**, 70–78.

26. Cao, L., White, J. S., Park, J.-S., Schuller, J. A., Clemens, B. M., and Brongersma, M. L. (2009). Engineering light absorption in semiconductor nanowire devices, *Nat. Mater.*, **8**(8), 643–647.

27. Eichenfield, M., Camacho, R., Chan, J., Vahala, K. J., and Painter, O. (2009). A picogram-and nanometre-scale photonic-crystal optomechanical cavity, *Nature*, **459**, 550–555.

28. Aspelmeyer, M., Kippenberg, T. J., and Marquardt, F. (2013). Cavity Optomechanics, *arXiv,* 1303.0733v1.

29. Jackson, J. D. (1975). *Classical Electrodynamics* (New York: Wiley).

30. Ramos, D., Tamayo, J., Mertens, J., Calleja, M., and Zaballos, A. (2006). Origin of the response of nanomechanical resonators to bacteria adsorption, *J. Appl. Phys.,* **100**, 106105.

31. Ramos, D., Tamayo, J., Mertens, J., and Calleja, M. (2006). Photothermal excitation of microcantilevers in liquids, *J. Appl. Phys.,* **99**, 124904.

32. Tamayo, J., Calleja, M., Ramos, D., and Mertens, J. (2007). Underlying mechanisms of the self-sustained oscillation of a nanomechanical stochastic resonator in a liquid, *Phys. Rev. B,* **76**, 180201.

Chapter 10

Cantilever Resonance Detection Using Nanophotonic Structures

Vincent T. K. Sauer,[a,b] Zhu Diao,[a,c] and Wayne K. Hiebert[a,c]

[a]*National Institute for Nanotechnology, 11421 Saskatchewan Drive, T6G 2M9, Edmonton, Canada*
[b]*Department of Electrical and Computer Engineering, University of Alberta, Edmonton, T6G 2V4, Canada*
[c]*Department of Physics, University of Alberta, Edmonton, T6G 2E1, Canada*

wayne.hiebert@nrc-cnrc.gc.ca

10.1 Introduction

Nanomechanical beam resonators have two qualities which make them extremely promising for detecting ultra-small masses.[1] First, they have exceptionally small masses themselves. Second, their high mechanical quality factors and high signal-to-noise ratios allow for precise measurements of their resonance frequencies. As a result, they have been able to reach mass sensitivities in the zeptogram range,[2,3] they have detected masses of single atoms,[4–6] and they have even reached single-proton sensitivity levels.[7,8] This gravimetric mass sensing has also demonstrated single-molecule mass spectrometry,[9–11] single-particle measurement in both gas phase[12,13] and fluid,[14–19] as well as operation in

Nanocantilever Beams: Modeling, Fabrication, and Applications
Edited by Ioana Voiculescu and Mona Zaghloul

ISBN 978-981-4613-23-1 (Hardcover), 978-981-4613-24-8 (eBook)
www.panstanford.com

gas chromatography.[20,21] These advances in nanomechanical sensor technology have it well on its way toward competing with both conventional time-of-flight mass spectrometers and gas chromatographers, but in a package that can be integrated onto a single chip.

The push for higher sensitivity has driven these devices to become smaller in size since smaller resonator masses can lead to higher mass sensitivity.[1] However, transducing the nanomechanical beam signal becomes more and more difficult with smaller devices as, generally, their signals also decrease with device size. Nanophotonics has emerged as a viable option for addressing the transduction issues of nanomechanical beams of decreasing size.

Nanophotonics involves the study of light interactions in the optical near field. It is a subset of the integrated photonics field, which studies the combination of waveguide optics with other optical disciplines such as electro-optics, acousto-optics, non-linear optics and opto-electronics. Integrated photonics aims to miniaturize optical systems onto an optical chip and involves integrated waveguides guiding light along the chip to interact with various on-chip optical devices.[22] In the context of this chapter, nanophotonics will refer to the interaction of the evanescent fields of these waveguides with various integrated optical and/or nanomechanical devices.

Pursuing smaller nanomechanical devices lead to several challenges. One, smaller devices will require very high displacement sensitivities in order to detect smaller ranges of motion. Next, conventional free-space optics are diffraction limited, and this creates a minimum size constraint when probing devices.[23,24] Finally, nanomechanical devices tend to operate at higher frequencies as they become smaller,[25] and this causes issues in electronic readout due to RLC roll-off. These challenges can be thoroughly met using integrated nano-optomechanical systems (NOMS) which have previously demonstrated unprecedented displacement sensitivity.[26,27] This follows from the availability of on-chip optical cavities which allow photons trapped inside the cavity to interact multiple times with a mechanical device before the photons dissipate.[28] They operate in the optical near field and so are not constrained by diffraction limits, and they are very well suited to high frequency operation as they are not frequency nor bandwidth limited.[26]

10.2 Nanophotonic Devices

Nanophotonic detection of nanomechanical cantilevers can use many different on-chip optical structures in the system. These include devices to channel light into and out of the chip, waveguides to move the light along the chip, and various optical structures to interact directly with nanomechanical devices. These structures were initially developed for the photonics industry to create optical integrated circuits[29,30] but have since become important components in the field of nanomechanics as well.[31]

Many standardized processes and materials have been established in the integrated photonics industry, and these are also used in nano-optomechanical system devices. Silicon-on-insulator substrates, consisting of a 220 nm Si device layer and 2 μm buried oxide layer, are used in conjunction with laser wavelengths in the infrared optical telecommunication C-band having a vacuum wavelength range between 1530 and 1565 nm. The thickness of the silicon device layer is chosen so the film thickness is equal to half the laser wavelength inside the medium. Silicon strip waveguides, described in Section 10.2.2, are typically fabricated around 500 nm in width. This allows for both low loss propagation and an evanescent tail around the waveguide. The evanescent tail permits the interaction of the waveguide with adjacent photonic or mechanical devices.[29,30]

10.2.1 Fiber-Chip Light Coupling

There are several methods available for coupling light onto a chip in a nanophotonic device. This coupling generally occurs from a single-mode optical fiber to an integrated optical waveguide. Some methods have been developed to couple light onto an entire chip, while others to only address individual nanophotonic structures.

Traditional end-fire coupling involves aligning a single-mode fiber to the butt end of a waveguide at the edge of a photonic chip. A taper on the waveguide allows for better optical mode matching between the external fiber and the on-chip waveguide. This reduces losses and back reflections upon coupling.[32,33] Although simple to design and implement, a disadvantage of this method is that alignment can only take place at the edge of the chip. It therefore limits device locations and densities on the sample.

Coupling light onto a chip can also be accomplished using optical gratings.[34–39] Gratings allow light to be coupled perpendicularly from a fiber onto a chip, and this removes the edge alignment requirement as in end-fire coupling. This leads to the importance of grating couplers in large-scale device integration. The density of addressable coupling locations is no longer dependent on the perimeter length of the chip. The grating consists of alternating regions with high and low indices of refraction. These can be fabricated either by etching the waveguide[34–36] or by depositing a periodic array of metal material.[37] Light is channeled into the waveguide at nearly a 90° angle due to diffraction, and it can do this with both high efficiency and large bandwidth.

Normally, the grating requires an adiabatically tapered waveguide to connect the 10 μm-wide grating to the 500 nm-wide waveguide. The grating must be large due to the size mismatch between the optical modes of a single-mode fiber and an integrated waveguide. Focused gratings can also be designed to remove the need for this taper and decrease the device footprint.[38] This permits the dense packing of nanophotonic devices across an entire wafer, which is highly desirable for integrated photonic circuits. Multi-axis fiber positioners are often used to align the two components, as coupling efficiency is dependent on both the position and input angle incident on the grating.[36] Grating couplers offer an effective way to couple light into high-density structures, and, consequently, it is one of the most commonly used methods.

Looking next at a technique to address single devices, dimpled fiber taper probe coupling[40,41] is a simple non-destructive method to locally probe individual devices on a larger optical chip. This method has been adapted from straight[42,43] and curved[44,45] fiber taper probes to address the issues of parasitic loss and mechanical instability experienced by these methods, respectively. The dimpled fiber is created by simultaneously heating and pulling a telecommunications fiber to first create a straight single-mode tapered fiber. Pulling the fiber adiabatically reduces the fiber diameter until it is small enough to become a single-mode fiber with evanescent tails extending significantly into the area surrounding the waveguide. The dimpled fiber is then created from this straight fiber by placing it into a mould with the desired fiber radius. The fiber retains the shape of the mold as it is subsequently heated and cooled. In this way, the fiber can be mounted with high tension to achieve good mechanical stability, while still having a smaller probing area to easily probe individual, high-density devices with lower parasitic losses. The dimpled fiber tapered probe can be mounted on a three-axis stage to position it in relation to the desired optical device, or the device itself can be aligned to the tapered fiber.[46]

For the highest sensitivities, nanomechanical resonators must operate in vacuum to eliminate air damping. Therefore, fiber positioning/alignment must occur in vacuum. This complicates the experimental setup which then requires a larger vacuum chamber with vacuum safe positioning systems. The alternative is to suffer a loss of positioning flexibility if alignment is completed prior to chamber pump down.[47] To circumvent these issues, a free-space confocal laser scanner system can be used to align both the position and angle of incident light on a photonic chip.[48] This system is not much more complicated than a free-space interferometric setup[24,49,50] since the photonic chip sample is the only component required to be placed under vacuum.

The confocal lens system allows for the adjustment of the position and input angle (k-vector) of a laser onto the entrance pupil of a microscope objective. Since a set of Fourier transform planes is formed by this entrance pupil and the microscope objective's focus plane, a change in the position on the entrance pupil defines the k-vector of the output beam and hence the angle of incident light onto the chip. Conversely, the k-vector of the input beam defines the location of the beam at the output focus plane. As such, the confocal system allows the position and input angle on the photonic chip to be controlled independently. A fiber collimator is used to input the laser light into the free-space confocal lens system. Light from the output grating coupler is captured by this microscope objective to couple back into the system. A diagram of this flexible free-space confocal lens setup is shown in Fig. 10.1.[48] This type of system gives the user both the efficiency of input grating couplers with the flexibility of a free-space optical design for probing nanophotonic devices in vacuum.

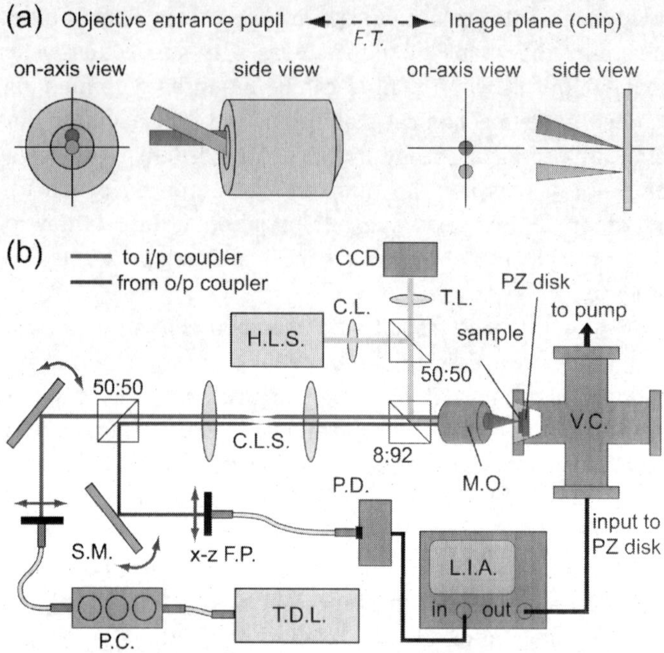

Figure 10.1 (a) Fourier transform planes in a microscope objective. (b) A diagram showing the confocal laser scanner-based NOMS measurement setup. C.L.: collector lens, C.L.S.: confocal lens system, H.L.S.: halogen light source, L.I.A.: lock-in amplifier, M.O.: microscope objective, P.C.: polarization controller, P.D.: photodetector, S.M.: steering mirror, T.D.L.: tunable diode laser, T.L.: tube lens, V.C.: vacuum chamber, and x–z F.P.: x–z fiber positioner. From reference 48.

10.2.2 Light Propagation

Light must be transported to various optical devices once it is coupled into the optical chip. It is transported using an optical waveguide, the most basic integrated optical component, which allows controlled propagation of light along the chip. In this chapter we will be discussing dielectric waveguides. In dielectric waveguides, this controlled propagation is generally accomplished using total internal reflection. There are two basic types of waveguides for propagating light: planar waveguides and channel waveguides. They operate using the same principles but confine light in either one or two dimensions, respectively.[22,51]

A planar waveguide is created when a slab of higher refractive index material (core) is surrounded by lower index media (cladding). Light will only be confined in the direction perpendicular to the layer interfaces. A symmetric waveguide will have top and bottom cladding layers of the same refractive index, whereas an asymmetric waveguide has top and bottom cladding layers of different refractive indices. In the asymmetric waveguide case, both will have a lower index value than that of the core. Light travels along the higher index core and at each interface is reflected back due to total internal reflection. This total internal reflection follows from the boundary conditions of Maxwell's equations, which create the requirement for lower index cladding layers. Solving the boundary equations will depend on the polarization of light travelling in the waveguide. This creates independent solutions for transverse electric (TE) and transverse magnetic (TM) fields. A full analysis of light propagation in a planar waveguide can be found in general photonics textbooks such as references 22, 51–53.

For the purposes of nanophotonic readout of nanomechanical systems, the two-dimensional confinement case of a channel waveguide is much more interesting. With a channel waveguide, light can be guided along the surface of the optical chip to interact with a series of optical devices. As in a slab waveguide, an upper and lower cladding confine light in the direction perpendicular to the chip surface, but in a channel waveguide, there is also an index change parallel to the chip surface. This confines the light laterally, as well, allowing a path to be patterned which the light must follow. There are several different types of channel waveguides as shown in Fig. 10.2.[51]

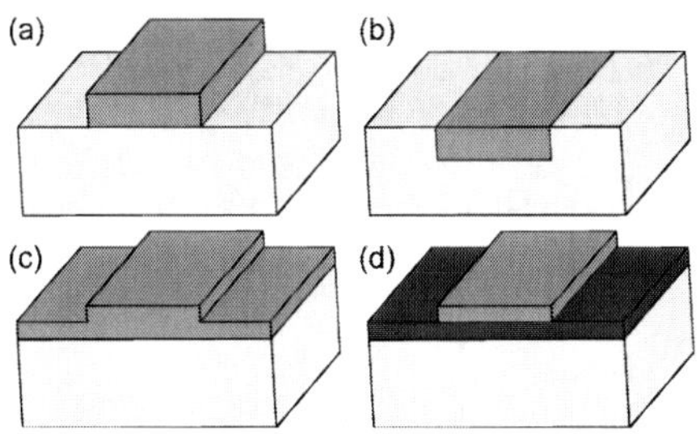

Figure 10.2 Various types of channel waveguides: (a) strip waveguide, (b) embedded strip waveguide, (c) rib waveguide, (d) loaded rib waveguide. The darker shading indicates a higher refractive index.

In silicon photonics, a common channel waveguide implementation is the strip waveguide. These are easy to fabricate using SOI substrates and consist of the top silicon device layer patterned as the waveguide with an upper cladding of air and lower cladding of SiO_2 (the buried oxide layer). TE-like mode propagation is defined as the electric field component parallel to the surface of the substrate, while TM-like mode propagation has the magnetic field component parallel to the wafer surface. The use of SOI substrates allows for easy integration with nanomechanical beam fabrication, as SOI substrates are often used in preparing nanomechanical resonators with SiO_2 as the sacrificial layer.

With integrated optical waveguides, the size of the core is on the same order as the wavelength of propagating light. This causes a discretization in the angle where total internal reflection can occur and, hence, light can be confined in the waveguide. These discretized angles represent the modes of propagation along the waveguide. The number of supported modes is dependent on the index contrast as well as the size of the core. Integrated optics usually operate using single-mode waveguides. With multi-mode waveguides, power is divided among the modes, and these modes will travel along the waveguide at different speeds. This creates mode dispersion in the waveguide, which distorts the signal due to the different modes arriving at the point of interest with different phases.

The point in which a certain mode is no longer supported is called the mode cut-off. It is important to note that in a symmetric waveguide, there is no cut-off for the lowest order mode, but in an asymmetric waveguide there is. This means in an asymmetric waveguide, such as a strip waveguide, the core region can become too small to carry a propagating mode, and this must be considered when designing the device. Standardized SOI wafers are used in the integrated photonics industry, which prevents the cut-off condition from accidentally being met.[22,52]

A special case of light propagation occurs when confinement is based upon Bragg reflection instead of total internal reflection. These devices are known as photonic crystals.[54] They have received much attention in both the photonics and nano-optomechanics communities due to their ability to confine light in a small area with a high field in an optical cavity.[31,55] They can be implemented to guide light as well. Generally, a two-dimensional periodic array of high contrast refractive index materials creates the Bragg grating. On an SOI substrate, this can be accomplished using a 2D array of holes in the silicon device layer. The periodic perturbation of light from the array creates a partial photonic band gap (PBG). This is due to the destructive interference of light reflecting off the various layers of the array. The partial PBG represents a range of wavelengths where light is prohibited within the photonic crystal regardless of the light's k-vector. A complete PBG also prohibits light regardless of its polarization, but this is much more difficult to engineer, so most photonic crystals have a partial PBG.

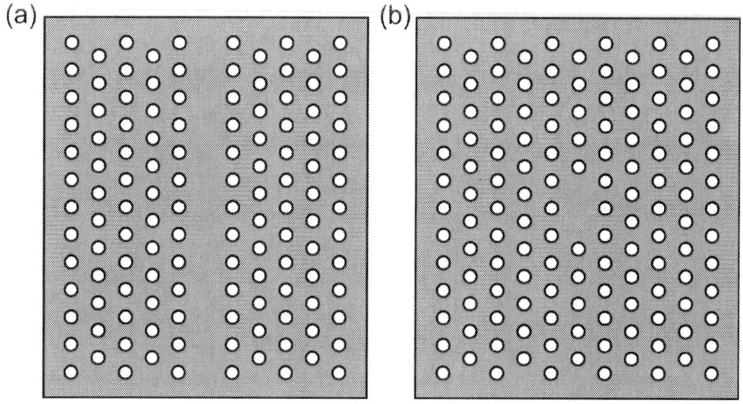

Figure 10.3 (a) Line defect cavity waveguide inside a 2D photonic crystal. (b) Point defect cavity inside a 2D photonic crystal.

To create either a waveguide or optical cavity, defects are introduced into the photonic crystal structure such as those shown in Fig. 10.3. This creates additional states in the photonic band structure, and it is possible to create states where the wavelength of interest can propagate. In this way, line defects can be introduced into the photonic crystal to allow light to propagate through it in a waveguide. Point defects can also be created which can trap light in a very small area. Due to the k-vector independence of a photonic crystal, much smaller cavities and waveguide bends can be achieved compared to waveguides relying on total internal reflection. However, photonic cavity structures are more difficult to fabricate with low losses, which makes them more difficult to implement than typical channel waveguides. A more extensive review on photonic crystals and their uses can be found in reference 54.

10.2.3 Optical Devices Used for Mechanical Beam Interaction

Lastly, we will look at certain nanophotonic devices which have been used to interact directly with nanomechanical beams. These include waveguide-based devices such as directional couplers or Mach–Zehnder interferometers (MZI), or cavity based devices such as optical disks, ring resonators, or photonic crystal cavities. In both cases, the optical device may move freely to act as the mechanical resonator itself, or the mechanical resonator may be placed within an adjacent optical device's evanescent field.

Directional couplers have been used to detect doubly clamped beam motion.[56] A directional coupler consists of two waveguides spaced sufficiently close together so the evanescent tails of the guided modes overlap. This allows photons to tunnel from one waveguide to the other, therefore transferring power from one waveguide to the other. Looking at coupled mode theory, the coupling can be regarded as a scattering effect by the adjacent waveguide. The scattering creates a second source in the adjacent waveguide which changes the overall amplitude in this second waveguide.[51] This effect occurs in each of the two waveguides, and light transfers periodically between them along the coupling region. The amount of light transferred between any given two waveguides will depend on both the coupling length and the waveguide spacing.[57] With a predetermined coupling length, this spacing can be determined by measuring the power transferred between the two waveguides. If an oscillating mechanical beam modulates this waveguide spacing, it will cause the output power on either side of the coupler to modulate as well.[56]

Two adjacent waveguides can also create a slot waveguide if they are placed sufficiently close together. Here, the optical power does not travel within the high index region, but instead in the region between them.[32,58] This occurs when the width of the slot between the high index regions is smaller than the decay length within the slot. The field remains high in this region and proceeds to travel along the slot. If the slot waveguide is suspended it can also be thought of as two adjacent mechanical doubly clamped beams. As the beams oscillate, the gap between them will change which in turn changes the effective index of the slot mode. This index change may be detected in an optical device such as an optical ring resonator.[59]

The MZI is another waveguide-based device used to detect nanomechanical motion. An MZI splits the light from a single waveguide into two separate paths: a reference path and a detection path. The two paths recombine and interfere after a phase shift is produced in the detection path due to a difference in the optical path length. Hence, the amount of phase shift determines the output power after the two paths recombine. Nanomechanical devices can interact with one of these paths (either integrated directly as an oscillating waveguide or as an independent device adjacent to the waveguide) and modulate the output as the mechanical motion modulates the optical path length and resultant phase shift.[3,26,60]

Optical disk and ring resonator cavities are also used to interact with nanomechanical devices. In these optical structures, light tunnels into the cavity from an adjacent waveguide and is trapped inside the cavity as it oscillates in the structure with low losses in a standing wave pattern. The frequency of light that can enter the cavity is determined by the physical characteristics of the cavity such as size and refractive index.[61–63] This optical frequency changes as part of the ring or disk's index is changed due to interaction with a nanomechanical structure. In a NOMS device it is this optical frequency change which is monitored. As with the MZI device, the optical cavity can directly contain an oscillating nanomechanical waveguide or an external nano-mechanical beam may interact with the cavity's evanescent fields. This interaction has been detected with very high accuracy.[27,46,59,64]

More recently, cavity structures using photonic crystals have been used to interact with mechanical devices. The general principle is the same as with the ring cavities where the modulation of the mechanical element perturbs the mode of the cavity.[31,65] Photonic crystals cavities have inherent advantages over total internal reflection–based cavities due to the ability to create strong field gradients confined in small mode volumes. As such they are becoming more popular for investigating nanomechanical systems.[65–67]

10.3 Nano-Optomechanical Systems

Nano-optomechanical systems combine the fields of both nanophotonics and nanomechanics. Nanomechanical structures are actuated and/or their motion is detected using the optical fields associated with a nanophotonic structure.[26,27,41] This can be done by directly modifying the alignment between waveguides using a mechanical beam's motion,[68–71] or otherwise using an interaction with an optical structure's evanescent field.[3,26,41,56,59,60,72–76]

The first type of device we will look at relies simply on the change in power transmission between two waveguides due to their alignment. This method does not necessarily rely on interference between two waves, so an incoherent light source may be used. In these examples, two waveguides are separated by a small gap, and at least one of the waveguides is released to also become a vibrating mechanical beam.[68–71] Light exits one of the waveguides and tunnels across the gap to the receiving waveguide. The transmission is dependent on the alignment between the two waveguides, and therefore modulates with the beam motion. The output waveguide can be connected to a photo-detector, and a frequency spectrum can be taken to find the mechanical resonance of the device.

The second type of device uses the mechanical motion of a resonator to interact with a nanophotonic waveguide's evanescent field. The mechanical beam can function itself as a waveguide,[26,59,72,77] or it can be placed within the range of an adjacent photonic element's evanescent field.[3,27,46,60,64,74–76] As above, they rely on a power modulation caused by the mechanical element's motion. This power change can be caused either by an interferometric phase change induced by changing the optical path length or by simply increasing the scattering of the waveguide.[75,76]

Interferometric phase detection is the most common form of nanophotonic probing. The refractive index of a waveguide is dependent on both the materials and the dimensions of the waveguide and surrounding media.[22] In a NOMS device, the mechanical motion of the beam changes these properties (by changing the spacing between a waveguide and adjacent structure), and this changes the refractive index which can then be detected interferometrically. This may be done using an MZI[3,60,73] or by using a photonic cavity,[27,46,59,64,74,78] as described previously. The interaction between a mechanical device and an evanescent field is very sensitive inside an optical cavity, and this makes it an ideal platform for ultrasensitive measurements.[31,79]

Finally, scattering can be used for resonant beam detection in a NOMS device.[75,76] If placed within its evanescent field, a nanomechanical cantilever causes scattering in an optical fiber taper waveguide. The scattering increases the closer the two objects are, so a power modulation occurs in phase with the nanomechanical motion. Since this is a non-interferometric technique, non-coherent light sources may be used. Although power loss is involved in the process and its ultimate sensitivity is lower compared to phase sensitive techniques, an added advantage of this method is that it is easy to implement into a multiplexed system to detect an array of cantilevers. A single fiber has demonstrated the detection of 63 mechanical resonator devices simultaneously.[76] This multiplexing capability will prove useful in implementing large arrays of nanomechanical beams in a sensing system.

10.4 Nanophotonic Probing of Cantilevers

Many initial NOMS devices consisted of nanophotonic strip waveguides, which were then suspended to create free-standing nanomechanical beams.[26,56,71,80] This allowed for easy integration between the nanophotonic and nanomechanical fields, as a standard strip waveguide needed only a simple etch step to be released, and it achieved some spectacular results. However, an obvious drawback to a free-standing nanomechanical beam/waveguide is the restriction on the size of the mechanical element involved. The ultimate limitation on size is dependent on the waveguide's ability to carry an optical mode. This size limitation follows from a strip waveguide's asymmetric cladding, which can no longer support an optical mode at certain core thickness. As this limit is approached, another issue is the increased loss caused by scattering due to the waveguide's sidewall roughness. A greater percentage of the optical mode energy becomes present in the evanescent tail, and in the tail it is more susceptible to sidewall scattering.[81] Fortunately, it is not very difficult to circumvent this limitation.

To remove this size limitation, one possible option is to separate the mechanical and photonic elements by fabricating independent nanomechanical and nanophotonic structures in close proximity to each other. In this way, the mechanical element can interact with the evanescent tail of a waveguide without needing to carry an optical mode itself. The nanomechanical device may interact with either a simple strip waveguide[3,60] or more complex optical cavity structures such as nanophotonic disk and ring resonators.[27,64,74]

10.4.1 Mach–Zehnder Interferometer Readout

First, we will look a little more closely at the case of nanomechanical cantilever detection using a strip waveguide integrated into an MZI device. An example is shown in Fig. 10.4a.[3] The cantilever will modulate the index of refraction of the waveguide directly adjacent to it, and a simulation of the optical mode shape and the effective index with a varying gap spacing, *z*, is shown in Figs. 10.4b,c. This slight change in the index of refraction can be easily detected using an MZI.[3,26,60]

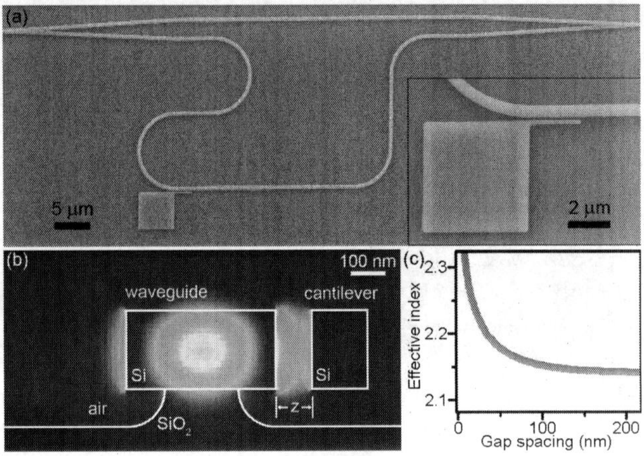

Figure 10.4 (a) SEM image of nanophotonic MZI device with adjacent nanomechanical cantilever. The inset is a zoomed image of the same cantilever. (b) Simulated cross sectional view of the optical mode of a waveguide with an adjacent free-standing nanomechanical beam. The waveguide is undercut when the beam is released.[74] (c) Simulated effective index of the optical mode in (b), with a varying gap spacing, *z*.

To achieve the largest signal from the nanomechanical beam's motion, the MZI must be tuned so the optical path length of each arm is 90° out of phase with the other. The simplest way to accomplish this is to use a tunable laser to scan the probe wavelength across the device, and then set the laser at the appropriate wavelength for the measurements (as indicated by the inflection point of a DC optical output scan). A sample scan is shown in Fig. 10.5a, with the 90° phase difference inflection point for the largest fringe marked by the red circle. The overall envelope of the scan is caused by the bandwidth of the input/output grating couplers. As the index of refraction changes on a single arm of the MZI, the location of these fringes will shift. Therefore, if the probe wavelength is set at the inflection point, the largest change in output power will occur as this is also the location with the steepest slope. The measured signal strength is proportional to the DC optical output scan slope as shown in Fig. 10.5b.[3] Ideally, the mechanical resonator should be as close to the length of the MZI probe arm as possible. This creates the largest optical path length change since the gap is changing along the (entire) length of the probe arm. Due to the sensitivity of an MZI structure, however, even a mechanical oscillator much smaller than the MZI probe arm can still achieve favorable results.

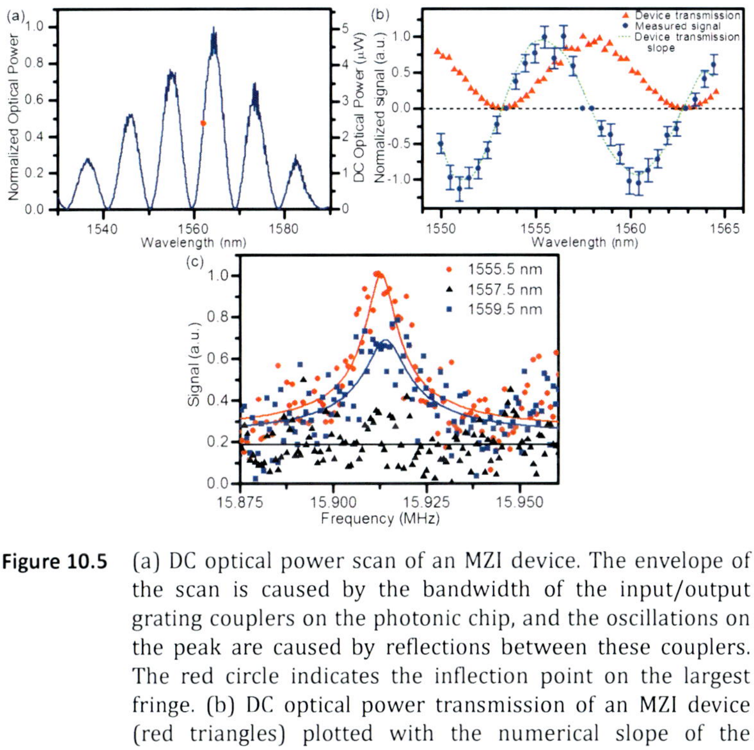

Figure 10.5 (a) DC optical power scan of an MZI device. The envelope of the scan is caused by the bandwidth of the input/output grating couplers on the photonic chip, and the oscillations on the peak are caused by reflections between these couplers. The red circle indicates the inflection point on the largest fringe. (b) DC optical power transmission of an MZI device (red triangles) plotted with the numerical slope of the smoothed data (green dash). The measured amplitude of a nanomechanical cantilever signal (blue dots) overlays well with the slope curve.[3] (c) Mechanical signal peak heights corresponding to different probe wavelengths. Solid lines represent the Lorentzian fit to the data. A low driving signal causes the high noise level in the response.

10.4.2 Optical Cavity Readout

For the highest sensitivity in transducing nanomechanical motion, a high finesse optical cavity should be used as the nanophotonic structure. Finesse is defined by the number of times trapped light travels around an optical cavity. The basic operating principle is the same as the MZI device. The cantilever modulates its position relative to the cavity, which in turn modulates the refractive index of the area of the cavity adjacent to the cantilever. The advantage of the optical cavity arises from the principle that light becomes trapped within it. In this way, the light will interact with the mechanical device many times as it travels around the cavity.[28] Low finesse cavities also allow for nanophotonic detection,[26] but high finesse cavities achieve better results.

We will now focus on the transduction theory of nanomechanical cantilevers using optical ring or racetrack structures like that shown in Fig. 10.6a.[74] We focus on ring structures instead of MZI structures as an optical cavity with a finesse greater than 2.5 will yield superior transduction properties than an MZI.[47]

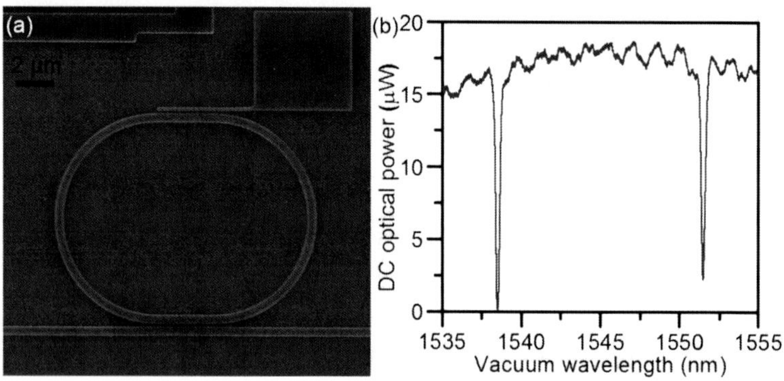

Figure 10.6 (a) Colorized SEM image of an all-pass racetrack resonator used to transduce the motion of an adjacent nanomechanical cantilever. (b) DC optical power wavelength scan of an all-pass racetrack resonator device similar to that shown in (a).

First, we will look at the phase change resultant from a nanomechanical cantilever moving toward a waveguide.[47]

$$\Delta\phi = \Delta n_{\text{eff}} \frac{2\pi}{\lambda} L = \Delta g \frac{\partial n_{\text{eff}}}{\partial g} \frac{2\pi}{\lambda} L, \tag{10.1}$$

where n_{eff} is the effective index of refraction, λ is the free space probe wavelength, L is the length of the cantilever and g is the gap spacing between the cantilever and the waveguide. $\partial n_{\mathrm{eff}}/\partial g$ is the change in effective index of the waveguide due to a changing gap with the cantilever, and can be easily simulated with accurate results knowing the device dimensions. We can extend this to define the transduction coefficient, $\partial T(\phi)/\partial g$, which is equal to the change in transmission power through the optical device due to the phase change caused by the changing waveguide-cantilever gap.

$$\frac{\partial T(\phi)}{\partial g} = \left(\frac{\partial T(\phi)}{\partial \phi}\right) \cdot \left(\frac{\partial \phi}{\partial g}\right) \tag{10.2}$$

Here, $\partial T(\phi)/\partial \phi$ is the phase sensitivity and is defined by the change in output power of the optical device due to a change in the phase. The second term, $\partial \phi/\partial g$, is the phase change caused by the change in gap spacing and can be derived from Eq. 10.1. The former term is dependent on the optical cavity properties and varies with the optical device. Using experimental parameters, this can be expanded to

$$\frac{\partial T(\phi)}{\partial g} = \left[\left(\frac{\partial T}{\partial \lambda}\right) \cdot \left(\frac{FSR}{2\pi}\right)\right] \cdot \left[\left(\frac{2\pi}{\lambda}\right) \cdot \left(\frac{\partial n_{\mathrm{eff}}}{\partial g}\right) \cdot \beta \cdot L\right], \tag{10.3}$$

where $\partial T/\partial \lambda$ is the slope of the DC optical power output at the probe wavelength, λ. It can be measured with a wavelength scan across the device similarly to Fig. 10.6b. The free spectral range (FSR) can also be measured here, and is defined as the wavelength spacing between two adjacent optical cavity modes. β is required to account for the varying gap across the length of the bending cantilever, and can be estimated by the average displacement of the mode shape (0.394 for the first flexural mode of a cantilever).[26]

Maximum signals are achieved when the probe wavelength is set at the greatest $\partial T/\partial \lambda$ value, which yields the greatest phase sensitivity. This phase sensitivity can be expressed in general terms relating to the optical cavity properties. Looking at the normalized transmission (indicated by the hat notation) of an all-pass ring resonator[82]:

$$\hat{T}(\phi)=\left(\frac{\hat{T}_{x,\min}+(2\mathcal{F}/\pi)\sin(\phi/2)^2}{1+(2\mathcal{F}/\pi)\sin(\phi/2)^2}\right), \tag{10.4}$$

where $\hat{T}_{x,\min}$ is the minimum of the normalized transmission at the bottom of the optical dip, and $\mathcal{F}$ is the finesse of the cavity, the maximum slope (the phase sensitivity) can be estimated as[47]

$$\left(\frac{\partial\hat{T}(\phi)}{\partial\phi}\right)_{\max}=\frac{3\sqrt{3}}{8\pi}(1-\hat{T}_{x,\min})\mathcal{F} \tag{10.5}$$

This gives the maximum transduction coefficient for normalized transmission as

$$\left(\frac{\partial\hat{T}(\phi)}{\partial g}\right)_{\max}\approx\left(\frac{3\sqrt{3}}{8\pi}(1-\hat{T}_{x,\min})\,\mathcal{F}\right)\cdot\left(\frac{2\pi}{\lambda}\right)\cdot\left(\frac{\partial n_{\text{eff}}}{\partial g}\right)\cdot\beta\cdot L \tag{10.6}$$

So for a non-normalized transmission with T_0 equal to the power entering the ring, and α equal to the extinction ratio of the dip measured in dB:

$$\left(\frac{\partial T(\phi)}{\partial g}\right)_{\max}\approx\frac{3\sqrt{3}}{8\pi}T_0(1-10^{-\alpha/10})\,\mathcal{F}\cdot\left(\frac{2\pi}{\lambda}\right)\cdot\left(\frac{\partial n_{\text{eff}}}{\partial g}\right)\cdot\beta\cdot L \tag{10.7}$$

With the above equations, it becomes a simple matter to calculate the displacement of the beam. The total power of the signal (P_{total}) can be used to estimate the maximum cantilever tip displacement, z_{peak}, as follows:

$$P_{\text{total}}\approx z_{\text{peak}}\frac{3\sqrt{3}}{8\pi}T_0(1-10^{-\alpha/10})\,\mathcal{F}\cdot\left(\frac{2\pi}{\lambda}\right)\cdot\left(\frac{\partial n_{\text{eff}}}{\partial g}\right)\cdot\beta\cdot L \tag{10.8}$$

Although derived for an all pass ring resonator, the same series of steps can be used to determine general results for any type of optical sensing structure. The variation in transduction properties arise from the different phase sensitivities of each type of optical structure. Practically speaking, it is fairly simple to experimentally measure the phase sensitivity of any device using DC optical power wavelength scan (Figs. 10.5a,b), so even a complex optical structure can be used for nanomechanical sensing with little difficulty.

Looking at the device in Fig. 10.6a (4.86 μm long, 220 nm deep and 160 nm wide), the thermomechanical (TM) noise spectrum of the cantilever can be easily measured. This response is shown in Fig. 10.7. Using experimentally measured parameters, Eq. 10.3 can be used to estimate the displacement of the beam by measuring the total energy of the signal. Regardless, the high signal-to-noise ratio and sensitivity of the system allow for easy analysis of a cantilever's mechanical response, even at the thermomechanical noise limit.

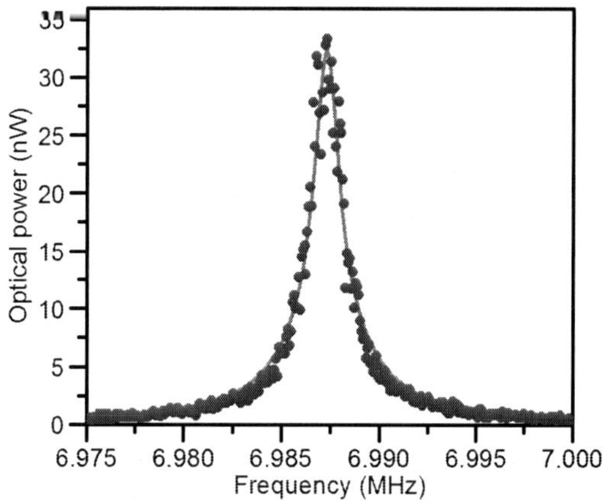

Figure 10.7 Thermomechanical noise spectrum of a 4.86 μm cantilever device detected using a racetrack resonator. The measurement was taken with a 1 kHz bandwidth and 16× averaging.

10.4.3 Device Optimization

One of the greatest advantages of using nanophotonic measurement for nanomechanical systems is the ability to scale down the nanomechanical device without necessarily sacrificing transduction capability. With either free-space optical, or capacitive and magnetomotive transduction mechanisms, smaller devices give smaller signals and it becomes difficult to overcome system noise levels for adequate transduction.[83] On the other hand, scaling down offers some advantages that can be exploited in a nanophotonic system. Device scaling becomes very important in high sensitivity nanomechanical measurements with smaller and smaller devices required to achieve higher sensitivities.[1,4–6]

Unlike free-space optical, which is limited by the diffraction limit of light,[23] magnetomotive,[84] which depends on the area within the beam displacement, piezoresistve,[85,86] which has limits due to impedance matching and background resistance, or capacitive,[87] which is swamped by parasitics at nanosize scales, the transduction coefficient can actually increase with a smaller system of devices. This draws from the fact that any reduction in cantilever size will also decrease the expected displacement of the cantilever. As such, the mechanical device can be placed in closer proximity to the adjacent optical device, which allows the transduction coefficient to increase due to an increased $\partial n_{\mathrm{eff}}/\partial g$. This increase follows from the exponential decay of the evanescent field, which $\partial n_{\mathrm{eff}}/\partial g$ follows in a similar manner.

High finesse optical cavities create a second advantage to offset the difficulties in scaling down nanomechanical devices. As seen in Eq. 10.7, the finesse is directly proportional to the transduction coefficient. Optical devices with very high finesse can be engineered[88] which can offset any losses in displacement sensitivity by decreasing the device size. These reasons make nanophotonic transduction an excellent platform for nanomechanical sensing, especially at the very small scale.

These gains in sensitivity have been demonstrated using a direct comparison between traditional free-space optical measurement and integrated nanophotonic detection using a racetrack resonator.[48] A doubly clamped beam device embedded in the racetrack was measured using both systems, and a 150 × greater signal was measured using the nanophotonic setup. Practically, the noise on the magnitude and phase of the signal was much, much smaller using nanophotonic detection making them much easier to measure. The thermomechanical noise was visible using nanophotonic detection, but it was not using the free-space setup. More generally, the displacement sensitivities around and below ~100 fm $\mathrm{Hz}^{-1/2}$ have been demonstrated, and more impressively these structures can be used to investigate objects near their quantum ground state.[31]

10.5 Conclusion

The field of nanophotonics has emerged as a new paradigm for detecting nanomechanical motion. It retains the measurement bandwidth and displacement sensitivity of traditional optical mechanics, but circumvents the size limits due to diffraction effects by working in the optical near field. As an integrated system, it also retains some advantages of nanoelectromechanical systems with the ability to fabricate large sets of high-density devices cheaply on a standard electronics wafer. With these combined features, nano-optomechanical system resonators have a very bright future in state-of-the-art sensing systems.

References

1. Ekinci, K., Huang, X., and Roukes, M. (2004). Ultrasensitive nanoelectromechanical mass detection, *Applied Physics Letters*, **84**(22), 4469.
2. Yang, Y., Callegari, C., Feng, X., Ekinci, K., and Roukes, M. (2006). Zeptogram-scale nanomechanical mass sensing, *Nano Letters*, **6**(4), 583–586.
3. Sauer, V. T. K., Diao, Z., Freeman, M. R., and Hiebert, W. K. (2012). Nanophotonic detection of side-coupled nanomechanical cantilevers, *Applied Physics Letters*, **100**(26), 261102.
4. Jensen, K., Kim, K., and Zettl, A. (2008). An atomic-resolution nanomechanical mass sensor, *Nature Nanotechnology*, **3**(9), 533–537.
5. Chiu, H. Y., Hung, P., Postma, H. W. C., and Bockrath, M. (2008). Atomic-scale mass sensing using carbon nanotube resonators, *Nano Letters*, **8**(12), 4342–4346.
6. Lassagne, B., Garcia-Sanchez, D., Aguasca, A., and Bachtold, A. (2008). Ultrasensitive mass sensing with a nanotube electromechanical resonator, *Nano Letters*, **8**(11), 3735–3738.
7. Chaste, J., Eichler, A., Moser, J., Ceballos, G., Rurali, R., and Bachtold, A. (2012). A nanomechanical mass sensor with yoctogram resolution, *Nature Nanotechnology*, **7**, 301–304.
8. Hiebert, W. (2012). Mass sensing: Devices reach single-proton limit, *Nature Nanotechnology*, **7**(5), 278–280.
9. Naik, A., Hanay, M., Hiebert, W., Feng, X., and Roukes, M. (2009). Towards single-molecule nanomechanical mass spectrometry, *Nature Nanotechnology*, **4**, 445–450.
10. Hanay, M. S., Kelber, S., Naik, A. K., Chi, D., Hentz, S., Bullard, E. C., Colinet, E., Duraffourg, L., and Roukes, M. L. (2012). Single-protein nanomechanical mass spectrometry in real time, *Nature Nanotechnology*, **7**(9), 602–608.

11. Schmid, S., Dohn, S., and Boisen, A. (2010). Real-time particle mass spectrometry based on resonant micro strings, *Sensors*, **10**(9), 8092–8100.

12. Schmid, S., Kurek, M., and Boisen, A. (2013). Towards airborne nanoparticle mass spectrometry with nanomechanical string resonators. (George, T., Islam, M. S., and Dutta, A. K. eds.), *SPIE Defense, Security, and Sensing*, International Society for Optics and Photonics, 872525–872528.

13. Schmid, S., Kurek, M., Adolphsen, J. Q., and Boisen, A. (2013). Real-time single airborne nanoparticle detection with nanomechanical resonant filter-fiber, *Scientific Reports*, **3**, 1288.

14. Burg, T. P., Godin, M., Knudsen, S. M., Shen, W., Carlson, G., Foster, J. S., Babcock, K., and Manalis, S. R. (2007). Weighing of biomolecules, single cells and single nanoparticles in fluid, *Nature*, **446**(7139), 1066–1069.

15. Lee, J., Shen, W., Payer, K., Burg, T. P., and Manalis, S. R. (2010). Toward attogram mass measurements in solution with suspended nano-channel resonators, *Nano Letters*, **10**(7), 2537–2542.

16. Barton, R. A., Ilic, B., Verbridge, S. S., Cipriany, B. R., Parpia, J. M., and Craighead, H. G. (2010). Fabrication of a nanomechanical mass sensor containing a nanofluidic channel, *Nano Letters*, **10**(6), 2058–2063.

17. Agache, V., Blanco-Gomez, G., Baleras, F., and Caillat, P. (2011). An embedded microchannel in a MEMS plate resonator for ultrasensitive mass sensing in liquid, *Lab on a Chip*, **11**(15), 2598–2603.

18. Toda, M., Otake, T., Miyashita, H., Kawai, Y., and Ono, T. (2012). Suspended bimaterial microchannel resonators for thermal sensing of local heat generation in liquid, *Microsystem Technologies*, **19**(7), 1049–1054.

19. Khan, M. F., Schmid, S., Larsen, P. E., Davis, Z. J., Yan, W., Stenby, E. H., and Boisen, A. (2013). Online measurement of mass density and viscosity of pL fluid samples with suspended microchannel resonator, *Sensors and Actuators B Chemical*, **185**, 456–461.

20. Bargatin, I., Myers, E. B., Aldridge, J. S., Marcoux, C., Brianceau, P., Duraffourg, L., Colinet, E., Hentz, S., Andreucci, P., and Roukes, M. L. (2012). Large-scale integration of nanoelectromechanical systems for gas sensing applications, *Nano Letters*, **12**(3), 1269–1274.

21. Li, M., Myers, E. B., Tang, H. X., Aldridge, S. J., McCaig, H. C., Whiting, J. J., Simonson, R. J., Lewis, N. S., and Roukes, M. L. (2010).

Nanoelectromechanical resonator arrays for ultrafast, gas-phase chromatographic chemical analysis, *Nano Letters*, **10**(10), 3899–3903.

22. Lifante, G. (2003). *Integrated Photonics: Fundamentals (1st ed.)* Chichester, West Sussex: Wiley.
23. Kouh, T., Karabacak, D., Kim, D. H., and Ekinci, K. L. (2005). Diffraction effects in optical interferometric displacement detection in nanoelectromechanical systems, *Applied Physics Letters*, **86**(1), 013106.
24. Karabacak, D., Kouh, T., and Ekinci, K. L. (2005). Analysis of optical interferometric displacement detection in nanoelectromechanical systems, *Journal of Applied Physics*, **98**(12), 124309.
25. Ekinci, K. L., and Roukes, M. L. (2005). Nanoelectromechanical systems, *Review of Scientific Instruments*, **76**(6), 061101.
26. Li, M., Pernice, W. H. P., Xiong, C., Baehr-Jones, T., Hochberg, M., and Tang, H. X. (2008). Harnessing optical forces in integrated photonic circuits, *Nature*, **456**(7221), 480–484.
27. Anetsberger, G., Arcizet, O., Unterreithmeier, Q. P., Rivière, R., Schliesser, A., Weig, E. M., Kotthaus, J. P., and Kippenberg, T. J. (2009). Near-field cavity optomechanics with nanomechanical oscillators, *Nature Physics*, **5**(12), 909–914.
28. Bogaerts, W., De Heyn, P., Van Vaerenbergh, T., De Vos, K., Kumar Selvaraja, S., Claes, T., Dumon, P., Bienstman, P., Van Thourhout, D., and Baets, R. (2012). Silicon microring resonators, *Laser and Photonics Reviews*, **6**(1), 47–73.
29. Bogaerts, W., Taillaert, D., Luyssaert, B., Dumon, P., Van Campenhout, J., Bienstman, P., Van Thourhout, D., Baets, R., Wiaux, V., and Beckx, S. (2004). Basic structures for photonic integrated circuits in Silicon-on-insulator, *Optics Express*, **12**(8), 1583.
30. Bogaerts, W., Baets, R., Dumon, P., Wiaux, V., Beckx, S., Taillaert, D., Luyssaert, B., Van Campenhout, J., Bienstman, P., and Van Thourhout, D. (2005). Nanophotonic waveguides in silicon-on-insulator fabricated with CMOS technology, *Lightwave Technology, Journal of*, **23**(1), 401–412.
31. Chan, J., Alegre, T. P. M., Safavi-Naeini, A. H., Hill, J. T., Krause, A., Gröblacher, S., Aspelmeyer, M., and Painter, O. (2011). Laser cooling of a nanomechanical oscillator into its quantum ground state, *Nature*, **478**(7367), 89–92.
32. Almeida, V. R., Panepucci, R. R., and Lipson, M. (2003). Nanotaper for compact mode conversion, *Optics Letters*, **28**(15), 1302.

33. Sure, A., Dillon, T., Murakowski, J., Lin, C., Pustai, D., and Prather, D. (2003). Fabrication and characterization of three-dimensional silicon tapers, *Optics Express*, **11**(26), 3555.

34. Taillaert, D., Bogaerts, W., Bienstman, P., Krauss, T. F., Van Daele, P., Moerman, I., Verstuyft, S., De Mesel, K., and Baets, R. (2002). An out-of-plane grating coupler for efficient butt-coupling between compact planar waveguides and single-mode fibers, *IEEE Journal of Quantum Electronics*, **38**(7), 949–955.

35. Taillaert, Dirk, Bienstman, P., and Baets, R. (2004). Compact efficient broadband grating coupler for silicon-on-insulator waveguides, *Optics Letters*, **29**(23), 2749.

36. Taillaert, Dirk, Van Laere, F., Ayre, M., Bogaerts, W., Van Thourhout, D., Bienstman, P., and Baets, R. (2006). Grating Couplers for Coupling between Optical Fibers and Nanophotonic Waveguides, *Japanese Journal of Applied Physics*, **45**(8A), 6071–6077.

37. Scheerlinck, S., Schrauwen, J., Van Laere, F., Taillaert, D., Van Thourhout, D., and Baets, R. (2007). Efficient, broadband and compact metal grating couplers for silicon-on-insulator waveguides, *Optics Express*, **15**(15), 9625.

38. Van Laere, F., Roelkens, G., Ayre, M., Schrauwen, J., Taillaert, D., Van Thourhout, D., Krauss, T. F., and Baets, R. (2007). Compact and highly efficient grating couplers between optical fiber and nanophotonic waveguides, *Journal of Lightwave Technology*, **25**(1), 151–156.

39. Roelkens, G., Vermeulen, D., Van Thourhout, D., Baets, R., Brision, S., Lyan, P., Gautier, P., and Feìdeìli, J.-M. (2008). High efficiency diffractive grating couplers for interfacing a single mode optical fiber with a nanophotonic silicon-on-insulator waveguide circuit, *Applied Physics Letters*, **92**(13), 131101.

40. Michael, C. P., Borselli, M., Johnson, T. J., Chrystal, C., and Painter, O. (2007). An optical fiber-taper probe for wafer-scale microphotonic device characterization, *Optics Express*, **15**(8), 4745.

41. Eichenfield, M., Michael, C. P., Perahia, R., and Painter, O. (2007). Actuation of micro-optomechanical systems via cavity-enhanced optical dipole forces, *Nature Photonics*, **1**(7), 416–422.

42. Barclay, P. E., Srinivasan, K., Borselli, M., and Painter, O. (2004). Probing the dispersive and spatial properties of photonic crystal waveguides via highly efficient coupling from fiber tapers, *Applied Physics Letters*, **85**(1), 4.

43. Barclay, P. E., Srinivasan, K., Borselli, M., and Painter, O. (2004). Efficient input and output fiber coupling to a photonic crystal waveguide, *Optics Letters*, **29**(7), 697.

44. Hwang, I. K., Kim, S. K., Yang, J. K., Kim, S. H., Lee, S. H., and Lee, Y. H. (2005). Curved-microfiber photon coupling for photonic crystal light emitter, *Applied Physics Letters*, **87**(13), 131107.

45. Grillet, C., Monat, C., Smith, C. L., Eggleton, B. J., Moss, D. J., Frédérick, S., Dalacu, D., Poole, P. J., Lapointe, J., Aers, G., and Williams, R. L. (2007). Nanowire coupling to photonic crystal nanocavities for single photon sources, *Optics Express*, **15**(3), 1267.

46. Kim, P. H., Doolin, C., Hauer, B. D., MacDonald, A. J. R., Freeman, M. R., Barclay, P. E., and Davis, J. P. (2013). Nanoscale torsional optomechanics, *Applied Physics Letters*, **102**(5), 053102.

47. Roels, J. (2011). *Actuation of Integrated Nanophotonic Devices Through the Optical Gradient Force*, Ghent University.

48. Diao, Z., Losby, J. E., Sauer, V. T. K., Westwood, J. N., Freeman, M. R., and Hiebert, W. K. (2013). Confocal scanner for highly sensitive photonic transduction of nanomechanical resonators, *Applied Physics Express*, **6**(6), 065202.

49. Carr, D. W., and Craighead, H. G. (1997). Fabrication of nanoelectromechanical systems in single crystal silicon using silicon on insulator substrates and electron beam lithography, *Journal of Vacuum Science and Technology B: Microelectronics and Nanometer Structures*, **15**(6), 2760.

50. Hiebert, W. K., Vick, D., Sauer, V., and Freeman, M. R. (2010). Optical interferometric displacement calibration and thermomechanical noise detection in bulk focused ion beam-fabricated nanoelectromechanical systems, *Journal of Micromechanics and Microengineering*, **20**(11), 115038.

51. Saleh, B. E. A., and Teich, M. C. (1991). *Fundamentals of Photonics* (*1st ed.*) Wiley.

52. Iizuka, K. (2002). *Elements of Photonics, Volume II: For Fiber and Integrated Optics*. New York: Wiley-Interscience.

53. Kawano, K., and Kitoh, T. (2001). *Introduction to Optical Waveguide Analysis: Solving Maxwell's Equation and the Schrdinger Equation*. New York: Wiley.

54. Notomi, M. (2010). Manipulating light with strongly modulated photonic crystals, *Reports on Progress in Physics*, **73**(9), 096501.

55. Eichenfield, M., Chan, J., Camacho, R. M., Vahala, K. J., and Painter, O. (2009). Optomechanical crystals, *Nature*, **462**(7269), 78–82.

56. De Vlaminck, I., Roels, J., Taillaert, D., Van Thourhout, D., Baets, R., Lagae, L., and Borghs, G. (2007). Detection of nanomechanical motion by evanescent light wave coupling, *Applied Physics Letters*, **90**(23), 233116.

57. Hunsperger, R. G. (2009). *Integrated Optics: Theory and Technology* (*6th ed.*) Springer.

58. Baehr-Jones, T., Hochberg, M., Walker, C., and Scherer, A. (2005) High-Q optical resonators in silicon-on-insulator-based slot waveguides, *Applied Physics Letters*, **86**(8), 081101.

59. Li, M., Pernice, W. H. P., and Tang, H. X. (2010). Ultrahigh-frequency nano-optomechanical resonators in slot waveguide ring cavities, *Applied Physics Letters*, **97**(18), 183110.

60. Fong, K. Y., Pernice, W. H. P., Li, M., and Tang, H. X. (2010). High Q optomechanical resonators in silicon nitride nanophotonic circuits, *Applied Physics Letters*, **97**(7), 073112.

61. Little, B. E., Foresi, J. S., Steinmeyer, G., Thoen, E. R., Chu, S. T., Haus, H. A., Ippen, E. P., Kimerling, L. C., and Greene, W. (1998). Ultra-compact Si-SiO_2 microring resonator optical channel dropping filters, *IEEE Photonics Technology Letters*, **10**(4), 549–551.

62. Vahala, K. J. (2003). Optical microcavities, *Nature*, **424**(6950), 839–846.

63. Armani, D. K., Kippenberg, T. J., Spillane, S. M., and Vahala, K. J. (2003). Ultra-high-Q toroid microcavity on a chip, *Nature*, **421**(6926), 925–928.

64. Srinivasan, K., Miao, H., Rakher, M, T., Davanço, M., and Aksyuk, V, (2011). Optomechanical transduction of an integrated silicon cantilever probe using a microdisk resonator, *Nano Letters*, **11**(2), 791–797.

65. Safavi-Naeini, A. H., Mayer Alegre, T. P., Chan, J., Eichenfield, M., Winger, M., Lin, Q., Hill, J. T., Chang, D. E., and Painter, O. (2011). Electromagnetically induced transparency and slow light with optomechanics, *Nature*, **472**(7341), 69–73.

66. Hryciw, A. C., and Barclay, P. E. (2013). Optical design of split-beam photonic crystal nanocavities, *Optics Letters*, **38**(10), 1612–1614.

67. Pernice, W. H. P. (2013). Photonic crystal dumbbell resonators in silicon and aluminum nitride integrated optical circuits, *Journal of Nanophotonics*, **7**(1), 073095.

68. Ollier, E. (2002). Optical MEMS devices based on moving waveguides, *IEEE Journal of Selected Topics in Quantum Electronics*, **8**(1), 155–162.

69. Pruessner, M. W., Siwak, N., Amarnath, K., Kanakaraju, S., Chuang, W.-H., and Ghodssi, R. (2006). End-coupled optical waveguide MEMS devices in the indium phosphide material system, *Journal of Micromechanics and Microengineering*, **16**(4), 832–842.

70. Zinoviev, K., Dominguez, C., Plaza, J. A., Busto, V. J. C., and Lechuga, L. M. (2006). A novel optical waveguide microcantilever sensor for the detection of nanomechanical forces, *Journal of Lightwave Technology*, **24**(5), 2132–2138.

71. Li, M., Pernice, W. H. P., and Tang, H. X. (2009). Broadband all-photonic transduction of nanocantilevers, *Nature Nanotechnology*, **4**(6), 377–382.

72. Roels, J., De Vlaminck, I., Lagae, L., Maes, B., Van Thourhout, D., and Baets, R. (2009). Tunable optical forces between nanophotonic waveguides, *Nature Nanotechnology*, **4**(8), 510–513.

73. Van Thourhout, D., and Roels, J. (2010). Optomechanical device actuation through the optical gradient force, *Nature Photonics*, **4**(4), 211–217.

74. Sauer, V. T. K., Diao, Z., Freeman, M. R., and Hiebert, W. K. (2014). Optical racetrack resonator transduction of nanomechanical cantilevers *Nanotechnology*, **25,** 055202. doi:10.88/0957-4484/25/5 /055202.

75. Basarir, Onur, Bramhavar, S., Basilio-Sanchez, G., Morse, T., and Ekinci, K. L. (2010). Sensitive micromechanical displacement detection by scattering evanescent optical waves, *Optics Letters*, **35**(11), 1792–1794.

76. Basarir, O, Bramhavar, S., and Ekinci, K. L. (2012). Motion transduction in nanoelectromechanical systems (NEMS) arrays using near-field optomechanical coupling, *Nano Letters*, **12**(2), 534–539.

77. Pernice, W. H. P., Li, M., and Tang, H. X. (2009). Theoretical investigation of the transverse optical force between a silicon nanowire waveguide and a substrate, *Optics Express*, **17**(3), 1806.

78. Basarir, O., Bramhavar, S., and Ekinci, K. L. (2012). Monolithic integration of a nanomechanical resonator to an optical microdisk cavity, *Optics Express*, **20**(4), 4272.

79. Eichenfield, M., Camacho, R., Chan, J., Vahala, K. J., and Painter, O. (2009). A picogram- and nanometre-scale photonic-crystal optomechanical cavity, *Nature*, **459**(7246), 550–555.

80. Pernice, W. H. P., Li, M., and Tang, H. X. (2009). Optomechanical coupling in photonic crystal supported nanomechanical waveguides, *Optics Express*, **17**(15), 12424.

81. Lee, K. K., Lim, D. R., Luan, H.-C., Agarwal, A., Foresi, J., and Kimerling, L. C. (2000). Effect of size and roughness on light transmission in a Si/SiO_2 waveguide: Experiments and model, *Applied Physics Letters*, **77**(11), 1617.

82. Dumon, P. (2007). *Ultra-Compact Integrated Optical Filters in Silicon-on-insulator by Means of Wafer-Scale Technology*, Ghent University.

83. Ekinci, K. L., Yang, Y. T., and Roukes, M. L. (2004). Ultimate limits to inertial mass sensing based upon nanoelectromechanical systems, *Journal of Applied Physics*, **95**(5), 2682.

84. Cleland, A. N., and Roukes, M. L. (1999). External control of dissipation in a nanometer-scale radiofrequency mechanical resonator, *Sensors and Actuators A: Physical*, **72**(3), 256–261.

85. Li, M., Tang, H. X., and Roukes, M. L. (2007). Ultra-sensitive NEMS-based cantilevers for sensing, scanned probe and very high-frequency applications, *Nature Nanotechnology*, **2**(2), 114–20.

86. Mile, E., Jourdan, G., Bargatin, I., Labarthe, S., Marcoux, C., Andreucci, P., Hentz, S., Kharrat, C., Colinet, E., and Duraffourg, L. (2010). In-plane nanoelectromechanical resonators based on silicon nanowire piezoresistive detection, *Nanotechnology*, **21**(16), 165504.

87. Ekinci, K L. (2005). Electromechanical transducers at the nanoscale: actuation and sensing of motion in nanoelectromechanical systems (NEMS), *Small*, **1**(8–9), 786–797.

88. Tobing, L. Y. M., Lim, D. C. S., Dumon, P., Baets, R., and Chin, M.-K. (2008). Finesse enhancement in silicon-on-insulator two-ring resonator system, *Applied Physics Letters*, **92**(10), 101122.

Chapter 11

Integrated Silicon Optomechanical Transducers and Their Application in Atomic Force Microscopy

Jie Zou,[a,b] Marcelo Davanco,[a,c] Yuxiang Liu,[a,b,d] Thomas Michels,[a,b] Kartik Srinivasan,[a] and Vladimir Aksyuk[a]

[a]*Center for Nanoscale Science and Technology, National Institute of Standards and Technology, Gaithersburg, MD 20899, USA*
[b]*Maryland Nanocenter, University of Maryland, College Park, MD 20742, USA*
[c]*Department of Applied Physics, California Institute of Technology, Pasadena, CA 91125, USA*
[d]*Department of Mechanical Engineering, Worcester Polytechnic Institute, Worcester, MA 01609, USA*

vladimir.aksyuk@nist.gov

11.1 Introduction

Physics experiments measuring unknown quantities by transducing them to mechanical motion have a long and distinguished history.[1] The advent of micro- and now nano-fabricated mechanical transducers has continued this trend over the last two decades, where miniaturization enabled better coupling and measurement of microscopic physical phenomena, from electron and nuclear spins[2,3] to individual vortices in superconductors,[4] from quantum vacuum

Nanocantilever Beams: Modeling, Fabrication, and Applications
Edited by Ioana Voiculescu and Mona Zaghloul

ISBN 978-981-4613-23-1 (Hardcover), 978-981-4613-24-8 (eBook)
www.panstanford.com

fluctuations in nanoscale optical cavities[5,6] to shapes and masses of individual molecules.[7] While the micromechanical rotation and acceleration sensors are now ubiquitous in both cars and cell phones, in the physics laboratory the nanoscale mechanical devices have now been cooled to their quantum mechanical ground state[8,9] and continue to enable measurements with unprecedented degree of precision and control. Far from being confined to a narrow set of unique applications, the micromechanical measurement is at the center of the atomic force microscope (AFM),[10,11] with its wide variety of operation modes and applications in physics, biology, and industrial inspection.[12]

Fabrication techniques for nanobeams and more complicated mechanical transducers with one or more critical dimensions below 100 nm are now well established. Due to their small size and low mass, such devices can achieve a unique combination of high speed, high sensitivity, and strong and local coupling to nanoscale systems and phenomena. Mechanical resonance frequencies above 100 MHz can be achieved without sacrificing mechanical compliance and force sensitivity.[2–7] The ultimate limit on the performance of such transducers is imposed by the fundamental thermodynamic mechanical force noise in accordance with the fluctuation dissipation theorem. It is therefore critical to minimize the mechanical dissipation in the transducer. With proper materials and fabrication techniques very high mechanical quality factor (Q_m) in the 10^4 to 10^6 range can be achieved in vacuum environment. In combination with the very low mass and stiffness, phenomenally low intrinsic noise can be achieved, even at room temperature. Although when operated in ambient, the air damping significantly reduces the Q_m, the absolute value of the damping coefficient and the corresponding force sensitivity can scale favorably compared to larger mechanical transducers.

However, one of the most significant roadblocks to realizing the full potential of nanomechanical sensing is the readout of the motion of the small transducer with high sensitivity, high bandwidth, and without excess power dissipation. Electrical means of motion readout,[13] such as capacitive, magnetomotive, piezoresistive, and piezoelectric have been successfully employed, but all of them suffer from various combinations of poor scaling with reduced size, power dissipation limitations, magnetic field and materials requirements, and all introduce thermal Johnson noise into the readout signal. On the other hand, optical techniques do not suffer from thermal noise, do not in principle need to dissipate any power at the transducer, and have an enormous measurement bandwidth. By introducing optical cavities, the light is trapped and made to interact longer with the mechanical transducer, allowing matching the bandwidth of optical readout to the mechanical transduction bandwidth.[14–16] The reduction in the readout bandwidth is traded for a drastic increase in the readout gain.

Effective coupling of a microscopic transducer to an external, bulk optical cavity is technically challenging due to alignment, drift and the need for high reflectivity coating on the transducer.[14] Because of the diffraction limit, in such a system the light cannot be focused to a spot much smaller than the optical wavelength. When the transducer (cantilever, nanobeam, etc.) size is reduced below the wavelength of light, the coupling and the sensitivity deteriorate dramatically. Here, these challenges are overcome by integrating the nanomechanical transducer in the near field of a nanophotonic cavity microfabricated together on the same chip.[17–19] The cavity can be self-aligned to the nanobeam with the accuracy of a few nanometers, and the mechanical beam can interact strongly with the optical field in the cavity without mechanical contact. The optical quality of the cavity can be maintained as well. This fully integrated, stable, and practical optomechanical device can be fiber connectorized and implement the readout of the mechanical motion of the nanoscale transducer with GHz bandwidth and the imprecision near the standard quantum limit with only microwatts of optical power.

The combined sensor can achieve the fundamental thermomechanical noise floor not only near the mechanical resonance frequency, but over a broad frequency range from near DC to several times the mechanical resonance.[19] Furthermore, the dynamics of the mechanical transducer can be modified and tailored to specific needs by either or both using optical forces or a suitable other actuation mechanism, such as an integrated electrostatic actuator. Both a classical feedback control scheme with virtually no excess noise injected into the transducer, and a quantum control scheme through optical forces, whereby the transducer acts directly on the cavity optical field, can be implemented. Such schemes can be used to widely tune the mechanical frequency, implement a regenerative oscillator or cold-dampen a high mechanical Q_m, low noise transducer and produce a flat transfer function for dynamic operation over a broad frequency range.

This chapter is arranged as follows: First, we discuss the device design, the transduction scheme, and the measurement result of a typical device. Then the numerical simulations are explained in details. In Section 11.4, we show the progress towards the application of our devices as AFM probes. Finally, we conclude with a summary and outlook.

11.2 Optomechanical Transduction and Device Fabrication

11.2.1 Design and Transduction Scheme

Typical optical readout approaches for measuring cantilever motion, such as beam deflection[20] and laser interferometry,[14] rely upon free-space optics and make measurements in the far field. In contrast, our approach (Fig. 11.1a) utilizes near-field interaction as a sensitive probe of nanocantilever motion. Our optical readout tool is a silicon microdisk resonator, which is a device that supports "whispering gallery" optical modes. These modes circulate around the microdisk edge and have evanescent tails that extend out into the surrounding air cladding. Introducing a cantilever into this evanescent region induces a shift in the optical mode frequencies ω_{opt}, with the amount of shift depending on the specific location of the cantilever with respect to the disk. Thus, as the cantilever vibrates, the resonant frequency of a given optical mode varies, and this in turn can be mapped to a varying optical intensity in a number of ways. One straightforward approach[15] is to use a laser that is tuned to the shoulder of the cavity mode optical resonance, as shown in Fig. 11.1b. Alternately, measurements of phase on resonance with the cavity mode can be used.

This approach is characterized by several parameters that determine the optical readout sensitivity. The first is the optomechanical coupling parameter $g_{OM} = d\omega_{opt}/dx$, which represents the change in the microdisk mode's optical frequency ω_{opt} per unit change in the disk-cantilever separation x. This parameter thus determines the amount of frequency shift induced in the optical mode when the cantilever vibrates. Next, the optical quality factor (Q_{opt}) of the mode is important. Qualitatively, one expects the minimum detectable frequency shift to be set by (a fraction of) the cavity mode optical linewidth. More precisely, the cavity mode lineshape determines the conversion between the frequency modulation created by the cantilever motion and the intensity modulation produced when the input laser is tuned to the shoulder of the microdisk optical mode (Fig. 11.1b). Other important parameters include the out-coupled optical power from the microdisk-cantilever sensor and the noise equivalent power of the photodetector used.

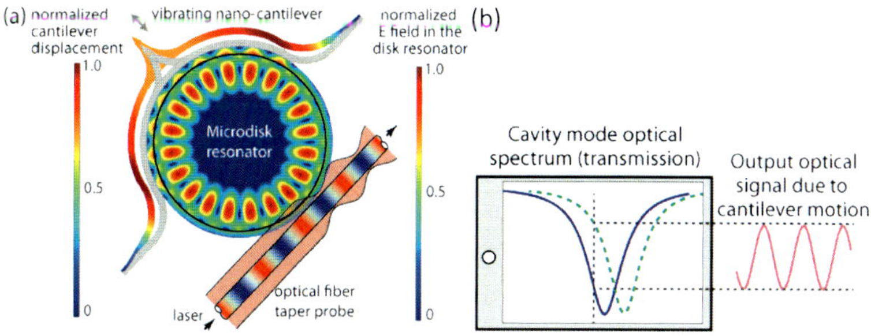

Figure 11.1 (a) Working principle of the disk-cantilever optomechanical sensor. The cantilever's equilibrium position is depicted in gray. The colored cantilever shows the FEM-calculated deformed shape (with an exaggerated amplitude) of the first order, in-plane, even-symmetry mechanical mode, for a system with disk diameter of 2.5 μm, cantilever width of 125 nm, and cantilever thickness of 260 nm. The color map in the microdisk resonator represents the absolute value of the FEM-calculated electric field amplitude of the relevant optical mode. The left scale bar is for the cantilever displacement, the right one for the electric field amplitude in the microdisk, and the electric field in the fiber probe is not in scale. (b) Using a tunable laser whose wavelength is aligned to the shoulder of an optical mode enables fluctuations in the cavity mode optical frequency due to cantilever motion to be mapped to an intensity-modulated optical signal. Adapted with permission from ref. 18.

Our sensor geometry has been chosen to optimize parameters such as g_{OM} and Q_{opt}. Using a semicircular cantilever shape increases the interaction length between the microdisk optical mode and the cantilever's mechanical mode with respect to what can be achieved with a straight cantilever. This is verified by finite element method simulations (see Section 11.3 for more details),[18] reproduced in Fig. 11.2a, where g_{OM} for the curved cantilever geometry is seen to exceed the value calculated for a straight cantilever by about one order of magnitude. Typical values for our system are $g_{OM}/2\pi \approx 1$ GHz/nm to 10 GHz/nm at gaps $G \approx 100$ nm. In addition, semicircular cantilever shape largely preserves the low optical loss possible in Si microdisks,[21] so that $Q_{opt} \geq 10^4$ can be readily achieved. We note that the requirement of high Q_{opt} sets a limit on the cantilever width we can use. As the width increases above ≈300 nm, the cantilever changes from having a perturbative effect on the microdisk mode to having a much stronger influence. In particular, once the cantilever width becomes wide enough to support optical modes, optical quality factors of the microdisk deteriorate dramatically.

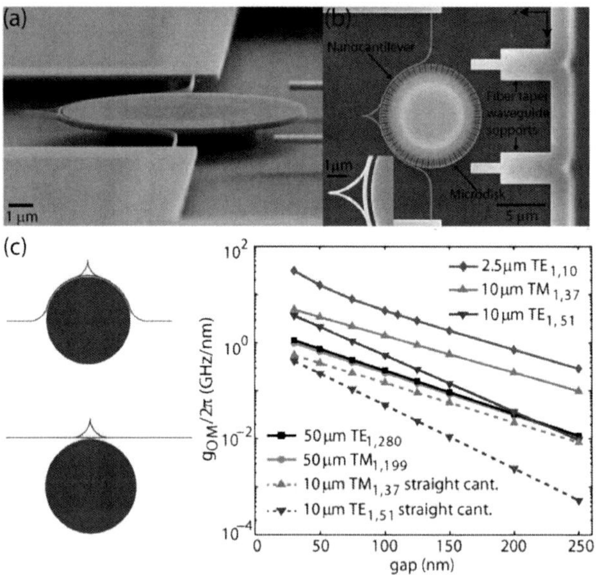

Figure 11.2 Scanning electron micrograph of fabricated device (D = 10 μm) from (a) a tilted angle and (b) the top view. The inset: a close-up of the gap and width of the cantilever near the tip. (c) Calculated optomechanical coupling parameter g_{OM} as a function of disk-cantilever gap, for different optical modes ($TM_{1,37}$, $TE_{1,51}$) and the two different cantilever shapes (curved and straight) shown in the diagram to the left. Panels (a, b) are adapted with permission from ref. 17. Copyright (2011) American Chemical Society. Panel (c) is adapted from ref. 18.

11.2.2 Fabrication

In experiments focused on establishing the basic optomechanical transduction mechanism, a simplified fabrication process was used, consisting of electron beam lithography (needed to establish ≈100 nm gaps and cantilever widths), inductively coupled plasma reactive ion etching of the silicon layer, hydrofluoric acid (HF) wet etching of the sacrificial oxide layer, and a liquid CO_2 critical point drying. These devices, shown in Fig. 11.2a, are suitable for understanding the thermal noise spectrum of different cantilever geometries, and the sensitivity with which the cantilever motion can be optically read out by the microdisk resonator. As they do not have access waveguides fabricated on-chip, optical fiber taper waveguides are used to couple light into and out of the devices. Experiments utilizing the device in a scanning probe configuration require a more extensive fabrication process, which will be elaborated in Section 11.4.

11.2.3 Detection Setup

A detection setup for characterizing these devices is shown in Fig. 11.3a. Light from a tunable laser is coupled into the device using, for example, an optical fiber taper waveguide. The fiber taper waveguide was fabricated by heating and stretching the optical fiber down to ≈1 μm in diameter. A local indentation ("dimple") with ≈10 μm radius of curvature is formed within the thinnest region of the fiber,[22] allowing for selective probing of devices within two-dimensional arrays. A swept-wavelength laser with a wavelength range of 1520 to 1630 nm was used as the light source, and was sent into a polarization controller before going into the fiber taper waveguide, allowing for polarization adjustment to maximize the coupling depth of the desired optical mode before recording data. Light circulates within the microdisk resonator hundreds or thousands of times (depending on the cavity's finesse) before exiting back through the same fiber taper waveguide. The output of the fiber is detected with a low bandwidth photodetector, and the transmission spectrum of the device is recorded, revealing the spectral location and spectral width of the cavity's optical modes. As described above, motion of the cantilever results in a frequency modulation of the optical cavity modes, which can be translated into an intensity modulation by probing these modes on the side of their resonance minima. The information obtained from the transmission spectra is thus used to determine the wavelength position to set the laser. The output signal exiting the device, which now carries the imprint of the mechanical motion as an intensity fluctuation, is detected on a high-bandwidth photodetector before being sent to an electronic spectrum analyzer to reveal the spectrum of mechanical modes.

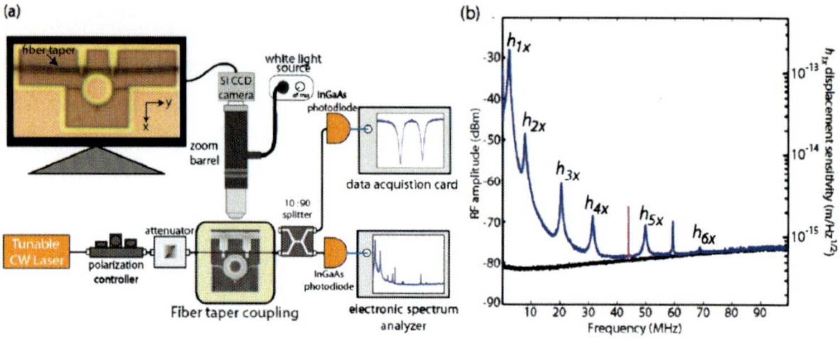

Figure 11.3 (a) Typical experimental setup. Light from a tunable laser is sent through a polarization controller and variable optical attenuator before being coupled into the device, in this case using an optical fiber taper waveguide. The light exiting the device is coupled back out into the same optical fiber taper waveguide, and then split into two channels with a fiber coupler. The first channel is used with a low bandwidth photodetector, and is used to measure the optical cavity's transmission spectrum in situations in which the laser wavelength is swept. The second channel is used with a high-bandwidth photodetector, and is used in situations in which the laser wavelength is fixed on the shoulder of the optical cavity mode. Here, the output of the photodetector is fed to an electronic spectrum analyzer to resolve the cantilever mechanical modes. (b) The thermomechanical noise spectrum is shown for a typical disk-cantilever device (D = 10 μm, w = 65 nm, G = 100 nm). Six in-plane mechanical modes are identified. The red spike corresponds to an electrical driving signal that is applied for a calibration purpose. Reprinted with permission from ref. 17. Copyright (2011) American Chemical Society.

Figure 11.3b displays the thermomechanical noise spectrum of one typical device with the microdisk diameter D = 10 μm. The thermal motion of the first six in-plane mechanical modes (see next section for the simulated mode shapes) is observed and the results are summarized in Table 11.1. The FEM simulation was used to calculate theoretical mechanical resonance frequency, f_m, and the effective mass, m_{eff}, based on the measured dimensions. Q_{opt} is obtained from the experimental optical spectra. The experimental f_m and the mechanical quality factor Q_m were obtained by fitting Lorentzian functions to the experimentally acquired mechanical spectra around the resonance peaks. The experimental and simulation results are in excellent agreement.

Table 11.1 Experimentally measured and calculated properties for in-plane mechanical modes of a typical disk-cantilever device (D = 10 μm, w = 65 nm, G = 100 nm)

Mode	k (N/m)	m_{eff} (pg)	Exp. f_m(MHz)	Cal. f_m(MHz)	Q_m
h_{1x}	0.14	0.73	2.23	2.35	4.9
h_{2x}	1.41	0.58	7.82	7.89	13.1
h_{3x}	5.72	0.35	20.37	20.51	38.5
h_{4x}	12.43	0.32	31.17	31.36	44.4
h_{5x}	41.95	0.44	49.36	49.86	61.2
h_{6x}	74.05	0.40	68.13	68.71	91.0

Source: Adapted with permission from ref. 17. Copyright (2011) American Chemical Society.

11.3 Numerical Simulation

Calculation of the important parameters, such as ω_{opt}, Q_{opt}, f_m, and g_{OM}, helps us optimize the microdisk-cantilever design. Optical and mechanical resonances of an optomechanical microcavity can be calculated by solving the appropriate eigenvalue equations, respectively, for the electromagnetic field and mechanical displacement. Then one can obtain the optomechanical coupling g_{OM} from the solutions of the electromagnetic field and mechanical displacement. In this section, we briefly describe the numerical methods and its application to our geometries.

11.3.1 Optical Resonances

For optical resonances, the eigenvalue equation is

$$\nabla \times \nabla \times \mathbf{E} = \varepsilon(\mathbf{r})\left(\frac{\omega}{c}\right)^2 \mathbf{E}, \tag{11.1}$$

assuming a time-harmonic electric field $\mathbf{E} = \mathbf{E}(\mathbf{r})\exp(i\omega t)$. In Eq. 11.1, $\varepsilon(\mathbf{r})$ is the permittivity distribution, c is the speed of light and ω is the resonance frequency. The permittivity distribution $\varepsilon(\mathbf{r})$ is known and essentially defines the optical resonator geometry. Solving Eq. 11.1 produces a discrete set of solutions with frequencies ω_i and corresponding field distributions (eigenvectors) $\mathbf{E}_i$, which are the optical cavity resonances. While in some simplified geometries Eq. 11.1 can be solved analytically, generally a numerical partial differential equation solving method, such as finite difference and finite elements,[23] is employed. Finite difference and finite elements are the most popular—but not the only—methods for PDE solving, and are widely available commercially or free.

An important aspect of optical cavities is that in general the cavities are "open," meaning that the optical field extends past the physical boundaries of the cavity. In fact the field extends infinitely into the region surrounding the geometry. Therefore, ideally Eq. 11.1 must be solved over the entire (infinite) space. A variety of methods have been devised to reduce the solution domain to practically realizable sizes, generally involving the enforcement of special field conditions at the boundaries of the (finite) computational window, that locally approximate the form of the real solution. Currently, however, the most popular method for limiting the size of the computational window is the perfectly matched layer (PML) method.[24] Essentially, it consists of surrounding the computational window with a thin layer of an artificial material that is both absorbing and optically matched to the medium external to the optical cavity. Because the PML is optically matched to the medium it surrounds, waves leaking from the cavity and impinging upon it are not reflected, and are absorbed once inside the PML.

Power leakage from an optical cavity is characterized by the cavity quality factor Q_{opt}, defined as $Q_{\mathrm{opt}} = \omega_{\mathrm{opt}} \cdot \tau_{\mathrm{p}}$, where τ_{p} is the cavity photon lifetime. This is essentially a measure of how many resonance cycles are necessary for a photon to leave the cavity,[25] or for the energy stored in the cavity to drop by $1/e$. The cavity Q_{opt} can be obtained from Eq. 11.1 by solving it with appropriate boundary conditions, as discussed above, which simulate open domains. In this case, complex eigenfrequencies $\tilde{\omega}$ are obtained, and (given the time-harmonic form of the electric field) the quality factor can be obtained as $Q_{\text{opt}} = \text{Re}(\tilde{\omega})/2\text{Im}(\tilde{\omega})$.

An alternative, very popular technique for calculating optical modes is the finite difference time domain method.[26] Here, one solves a discretized (finite difference) version of Maxwell's curl equations in time domain, in contrast to solving the frequency-domain eigenvalue problem of Eq. 11.1. The curl equations for the electric and magnetic fields are mutually coupled, and both **E** and **H** are calculated simultaneously on a grid at each time-step. This typically is done using the Yee algorithm (after K. Yee, who pioneered the technique[26,27]) in which a grid for **E** and **H** is defined such that the **E** components are surrounded by circulating **H** components and vice versa.

Resonant modes can be calculated by exciting the cavity with a spectrally broad source and monitoring the field decay in time. These modes are long lived and appear as peaks in the Fourier transform of the time evolution of the field. Typically, once a resonant mode is identified, subsequent simulations are performed in which the same mode is excited with a spectrally narrow source centered at its frequency, so that the field evolution is given only by the target resonance. Same as in the frequency-domain case, the computational domain is truncated, and PMLs are typically used to simulate open spaces. The radiation-limited quality factor of the cavity can be obtained either by monitoring the decay of the energy in the cavity and the power loss through the computational window boundaries, all of which can be calculated from the electromagnetic field, or also by fitting the time-evolution of the one or more field components at specific points within the cavity.

Simulation results presented in this chapter were obtained using the finite element method.

11.3.2 Mechanical Resonances

Mechanical resonances are solutions to the eigenvalue equation for the mechanical displacement $Q(\mathbf{r})$[28,29]

$$\nabla \cdot \left[\mathfrak{C} : \frac{(\nabla + \nabla^T)}{2} \mathbf{Q} \right] = \rho \omega^2 \mathbf{Q}, \tag{11.2}$$

where ρ is the mass density and $\mathfrak{C}$ is the elasticity tensor, which is of rank four. Both $\mathfrak{C}$ and $\mathbf{Q}$ are spatially variant and effectively define the geometry. For isotropic materials,

$$\frac{1}{\mathfrak{C}}=\frac{1}{E}\begin{bmatrix} 1 & -\nu & -\nu & 0 & 0 & 0 \\ -\nu & 1 & -\nu & 0 & 0 & 0 \\ -\nu & -\nu & 1 & 0 & 0 & 0 \\ 0 & 0 & 0 & 2\cdot(1+\nu) & 0 & 0 \\ 0 & 0 & 0 & 0 & 2\cdot(1+\nu) & 0 \\ 0 & 0 & 0 & 0 & 0 & 2\cdot(1+\nu) \end{bmatrix}, \quad (11.3)$$

where E is the Young's modulus, $-\nu$ is the Poison ratio, and the double-dot product corresponds to a fourth-rank to second-rank tensor product. The eigenvalue equation (Eq. 11.2) produces a discrete set of frequencies ω_{m} and corresponding displacement fields $\mathbf{Q}(\mathbf{r})$, which are the mechanical modes of the simulated structure. Equation 11.2 is typically solved with the finite element method. Using a commercial FEM software, we can calculate the mode shapes [relating to $\mathbf{Q}(\mathbf{r})$] and the mechanical frequencies f_{m}. Figure 11.4 displays the mode shapes of the first four in-plane modes with the displacement mainly along the x direction.

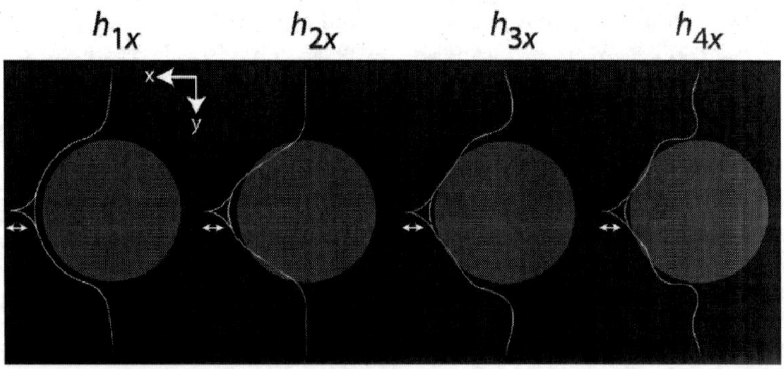

Figure 11.4 Numerically calculated mechanical mode shapes (amplitude exaggerated for clarity) with dominant displacement along the x axis for a device with D = 10 μm and w = 65 nm. Reprinted with permission from ref. 17. Copyright (2011) American Chemical Society.

11.3.3 Optomechanical Coupling Rate

The shift in the frequency ω_{opt} of a particular optical resonance due to the displacement of the nanostructure boundaries produced by a mechanical resonance at frequency f_{m} is quantified by the optomechanical coupling $g_{\text{OM}} = \partial\omega_{\text{opt}}/\partial x = \omega_{\text{opt}}/L_{\text{OM}}$; here, x is the cavity boundary displacement and L_{OM} is an effective optomechanical interaction length [1]. For the disk-cantilever work presented in this chapter, the following procedure was followed for calculating g_{OM}. First, the mechanical modes of the cantilever were simulated to determine the frequency and shape of the mode of interest (in-plane even-symmetry mode). The cantilever was then deformed with the mechanical mode shape, until the gap between the disk and the center point of the cantilever reached the specific gap value. This was to simulate the real motion of the cantilever that we measured in the experiment. The resonant optical modes in the disk with the deformed cantilever were simulated by solving the eigenvalue problem of the optical field. After obtaining the resonant frequencies of a specific optical mode at different gaps, we find g_{OM} as the slope of the fitted frequency-gap curve. We focus on the results obtained with 1st radial order optical modes, as these modes have the highest radiation-limited optical Q_{opt} and are predicted to couple well to the cantilever mechanical mode of interest.

Alternatively, we can employ $g_{\text{OM}} = \omega_{\text{opt}}/L_{\text{OM}}$ and estimate the effective length L_{OM} via the perturbative expression[28]

$$L_{\text{OM}} = \frac{2\int \varepsilon |\mathbf{E}|^2 \, dV}{\int (\mathbf{Q}\cdot\mathbf{n})\left[\Delta\varepsilon |\mathbf{E}_{\parallel}|^2 - \Delta\varepsilon^{-1} |\mathbf{D}_{\perp}|^2\right] dA}, \tag{11.4}$$

where $\mathbf{E}$ and $\mathbf{D}$ are the modal electric and electric displacement fields, respectively, $\Delta\varepsilon = \varepsilon_{\text{diel}} - \varepsilon_{\text{air}}$ and $\Delta\varepsilon^{-1} = (\varepsilon_{\text{diel}})^{-1} - (\varepsilon_{\text{air}})^{-1}$, where $\varepsilon_{\text{diel}}$ and ε_{air} are the permittivities of the (dielectric) microdisk material and air, respectively. The mass displacement due to the mechanical resonance is given by $\mathbf{Q}$, and the normal surface displacement at the structure boundaries is $\mathbf{Q}\cdot\mathbf{n}$, where $\mathbf{n}$ is the surface normal. In Eq. 11.4, $\mathbf{Q}$ is normalized to the maximum displacement: $\mathbf{Q} \rightarrow \mathbf{Q}/|\mathbf{Q}_{max}|$. The integral in the denominator is performed over the entire surface of the nanostructure.

11.4 Towards Optomechanical AFM Probes

11.4.1 Wide Range of Spring Constants

Commercial AFM cantilevers cover a wide range of spring constants k, from 0.01 N/m to over 1000 N/m. Cantilevers with soft spring constants usually find applications in imaging soft samples, such as biological materials, while the cantilevers with hard spring constants have proven useful in ultrahigh (atomic or even sub-atomic) resolution imaging. In Section 11.2.2, we have shown a cantilever in our integrated optomechanical system with a spring constant on the order of magnitude of 0.1 N/m. It is worthwhile investigating if the very soft and very hard regimes of typical spring constants are compatible with our disk-cantilever optomechanical configuration. By varying the microdisk diameter from 2.5 to 50 μm and the corresponding cantilever lengths accordingly (Fig. 11.5), we achieve spring constants ranging from 0.01 to 300 N/m (Table 11.2). Meanwhile, the resonant frequency of the fundamental mode of such a cantilever is much higher than the conventional cantilever with a comparable spring constant. In particular, the fundamental mechanical frequencies (0.265 to 111.4 MHz) are significantly higher than those of the conventional AFM probes (10 to 1.1 MHz).[30] This high frequency range, enabled by the small mass of the nanoscale cantilevers, may increase the imaging acquisition rate, decrease thermal drifts, reduce ambient vibration and acoustic noise, and enhance force sensitivity. More importantly, the displacement sensitivity is not greatly sacrificed for the cantilever with a very soft or very hard spring constant. The displacement sensitivity remains in the fm/Hz$^{1/2}$ range across the full range of cantilever stiffness. This is comparable to other state-of-the-art transduction schemes.

It is interesting to note that the device with D = 2.5 μm has a low optical Q_{opt} (10^3 to 10^4) due to the radiation loss; however, its optomechanical coupling g_{OM} is much higher than the one with D = 50 μm, because of a more confined mode volume (for D = 2.5 μm). These two effects balance each other and the displacement sensitivities for D = 2.5 μm and 50 μm are similar.

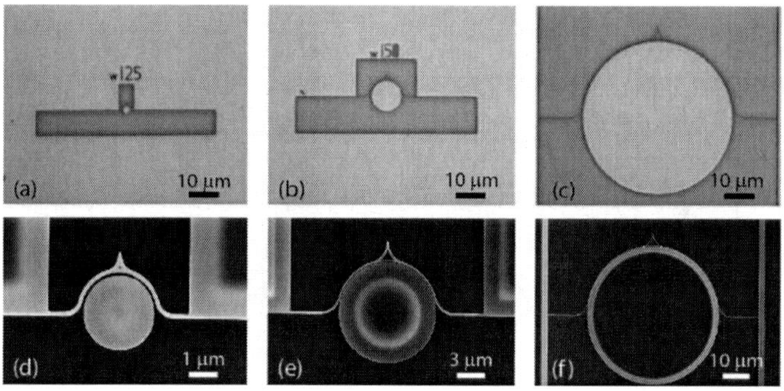

Figure 11.5 Optical microscope images (a–c) and SEM images (d–f) of fabricated cantilever-disk devices. The disk diameter, D, and cantilever width, w, in the devices are: (a), (d) D = 2.5 µm, w = 132 nm ± 6 nm; (b), (e) D = 10 µm, w = 172 nm ± 5 nm; and (c), (f) D = 50 µm, w = 155 nm ± 7 nm. Reprinted with permission from ref. 18.

Table 11.2 Experimentally measured and calculated properties of the disk-cantilever devices

D (µm)	ω(nm)	Q_{opt}	Exp. f_m (MHz)	Cal. f_m (MHz)	m_{eff} (pg)	k_{eff} (N/m)	Q_m	Typical disp sens. (fm/Hz$^{1/2}$) ± 15%
2.5	106 ± 8	1.6×10^4	57.65	61	0.28	36	28	2.0
	132 ± 6	7.2×10^3	68.78	75	0.34	64	66	
	158 ± 7	3.1×10^3	83.70	90	0.40	110	55	
	205 ± 7	5.2×10^3	100.0	120	0.51	200	69	
	238 ± 12	1.2×10^4	111.4	140	0.60	290	80	
10	124 ± 3	7.7×10^4	3.97	4.3	1.4	0.99	10	0.2
	149 ± 3	3.7×10^4	4.77	5.1	1.7	1.5	13	
	172 ± 5	9.9×10^4	5.62	5.9	1.9	2.4	22	
	224 ± 3	1.4×10^4	7.30	7.7	2.5	5.3	18	
	256 ± 4	8.7×10^3	8.17	8.8	2.9	7.6	21	
	271 ± 5	1.5×10^5	8.87	9.3	3.0	9.4	27	
50	107 ± 5	1.6×10^4	0.265	0.27	3.8	0.011	1.1	1.0
	128 ± 5	4.2×10^4	0.375	0.33	4.5	0.025	1.9	
	155 ± 7	3.4×10^4	0.433	0.40	5.5	0.041	1.7	
	210 ± 5	3.4×10^4	0.538	0.54	7.4	0.085	2.5	
	233 ± 10	6.2×10^4	0.615	0.60	8.2	0.12	4.3	

Source: Reprinted with permission from ref. 18.

Note: The typical photodetector-limited displacement sensitivity numbers are taken for representative devices within each disk diameter range (D = 2.5 µm, ω = 205 nm; D = 10 µm, ω = 172 nm; and D = 50 µm, ω = 155 nm).

11.4.2 Towards Optomechanical AFM

The integrated silicon transducer we have discussed so far requires an external fiber taper waveguide and positioners to align the fiber taper in order to coupled light into and out of the device. This makes it difficult to be integrated with a scan system for the AFM application. Furthermore, the tip of the cantilever is in-plane and located far away from the edges. Thus, it cannot be moved to the proximity of an off-chip surface under study. Moreover, the tip is not very sharp compared with the conventional cantilever. In this section, we describe our recent development towards optomechanical AFM probes and show preliminary results where the optomechanical AFM probe works in the contact mode.[31]

The optomechanical AFM probe comprises of a nanoscale cantilever, a microdisk, and on-chip waveguides (Fig. 11.6a). These elements are nanofabricated on an integrated chip. Batch fabrication at a wafer level is under development. Optical fibers are pigtailed, making external positioners for the fiber unnecessary. The tip of the cantilever is fabricated about 20 μm away from the edge and protected by oxide. Using focused ion beam, we exposed the tip to be overhanging on the edge and sharpened it before the final release by hydrofluoric acid. As displayed in Fig. 11.6b, the radius of the tip is found to be about 20 nm. The silicon microdisk acts as a whispering-gallery-mode optical cavity. The highest optical Q factor of the particular microdisk resonator (D = 10 μm) in this experiment is observed to be about 57,000, probably limited by imperfect plasma etching in this particular sample. Q factors up to 1 million have been demonstrated for microdisks with similar designs.[17,19] As we discussed in Sec. 1.2, the motion of the cantilever changes the effective optical length and thereby modulates the resonant wavelength of the microdisk cavity. Fixing the laser wavelength at the shoulder of the resonance, we can measure the displacement of the cantilever based on the change in the transmission intensity. The Brownian motion of the cantilever (ω = 100 nm and G = 100 nm) is detected and the resonant frequency of the fundamental in-plane mode is about 3.5 MHz with a typical quality factor about 20.

Next we demonstrate engagement of the probe to a sample surface under investigation. The sample (high-purity gold on mica) is mounted on a piezo scanner. The scanner sits on a stack of manual translational stages and slip-stick positioners. During the engagement, the laser wavelength is maintained on the shoulder of an optical resonance and the transmission intensity is monitored. Assuming we stay in the linear regime, the force between the tip and the surface is proportional to the change in the transmission intensity.

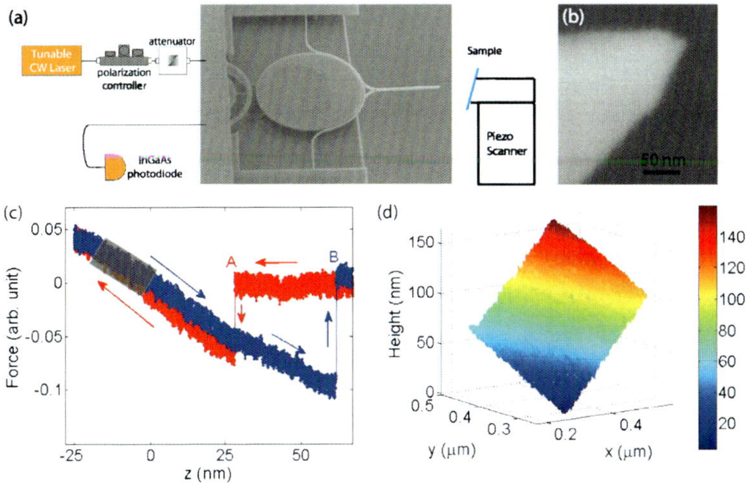

Figure 11.6 (a) Schematics of the experimental setup. (b) Scanning electron micrograph of a sharpened tip. The radius is estimated to be about 20 nm. (c) A typical force-distance curve. "Snap-in" (marked by A) and "pull-off" (marked by B) occur when the tip jumps onto and off from the sample surface. The gray shadow indicates the regime where the tip is placed when operating in the contact mode. (d) The scan image of a slightly tilted flat surface of the gold-on-mica sample. © 2013 NSTI http://nsti.org. Reprinted and revised, with permission, from ref. 31.

It is worth noting that our probe is more fragile than conventional cantilevers because of the nanoscale gap (≈100 nm) between the cantilever and the disk. If the approach procedure does not stop in time after the tip touches the surface, the cantilever will be translated too close to the microdisk and it may adhere to the sidewall of the microdisk even after the tip is separated from the surface. This phenomenon is known as stiction. After stiction happens, we need to utilize a micro manipulator to separate the cantilever from the disk. This procedure is time-consuming and not always successful. In the experiment, in order to avoid this issue at our best, we use an in-situ scanning electron microscope (SEM) to perform the initial course positioning and employ the standard "woodpecker" approach in a slow and careful manner.

After the surface is located, we record force-distance curves (Fig. 11.6c). "Snap-in" (marked by A) and "pull-off" (marked by B) correspond to the tip jumping onto and off from the sample. Small dissipation is observed, as is expected for the gold-on-mica sample. In order to image the topography of the sample, the position of the cantilever is then set to the repulsive-force regime (gray area in Fig. 11.6c). The optomechanical transducer works as an AFM probe in the contact mode and x/y piezo scanner is turned on to scan the surface. The measured displacement of the cantilever reflects nanoscale height changes on the surface. The topography image in Fig. 11.6d captures the slightly curved nature of the gold-on-mica sample. The fluctuation in the image is originated mostly from the noise in the AFM setup rather than the surface roughness. Further investigation indicates that the mechanical noise from the home-made scanner system dominates the noise performance. Implementation of the optomechanical probes in commercial AFM systems is under way to suppress the mechanical noise from the mechanical setup.

11.5 Summary and Perspective

This chapter describes the basic design, simulation, and fabrication for fully Si integrated, waveguide coupled, optomechanical force and displacement sensors. The approach of full Si integration of all stationary nanophotonic components with mechanically separated movable components creates the opportunity of independently engineering these two parts for a variety of MEMS and NEMS sensing applications, which requires high precision, high bandwidth, and small footprint. The further integration of actuators for static and dynamic actuation is also possible.

The next development steps will be a system modification for scanning probe microscopy (SPM), which is a major application for mechanical motion detection. As the preferred method in SPM is based on the measurement of the mechanical probe frequency change, it is necessary to drive the probe oscillation. In most commercially available scanning probe microscopes, this is realized by an external driving piezoelectric element mounted near the cantilever.[32] The excitation takes place through mechanical stimulation by shaking the complete cantilever holder. This method becomes inapplicable in liquid environments, because it generates standing waves through the acoustic stimulation of the medium. Furthermore, it is challenging to excite narrow high frequency resonances in the MHz regime using this method. These drawbacks can be overcome by directly driving the cantilever with an integrated actuator. The most promising candidates to achieve an active oscillation control and full compatibility with commercial available scanning probe microscopes are electrostatic or thermal actuation, due to their relatively simple design integration and compatibility with the existing fabrication process.

We have already demonstrated the integration of an electrostatic actuator to tune the optical cavity resonance by 5.54 nm, useful for adjusting the device to operate with a fixed optical laser source, such as a compact and relatively inexpensive stabilized laser diode.[19] The optomechanical coupling and the readout sensitivity are tunable by two orders of magnitude for optimizing the sensor gain and dynamic range.

Alternatively, a thermal actuator can be realized with a bimorph structure, which is exposed to a temperature difference. Thermal actuation (also called bimorph actuation) requires two combined materials with a mismatch of the coefficient of thermal expansion (CTE). If heat is applied locally, the bimetal beam experiences a deflection. This can be achieved through a focused laser beam[33,34] or Joule heating.[35] The advantage of bimorph actuation is the simple implementation in the fabrication process and independence of the substrate doping level.

Another interesting aspect of the device is the demonstrated cold damping of the mechanical degree of freedom by more than three orders of magnitude. Cold damping of the mechanical mode is realized by applying an electrical feedback signal, which is derived from the optical readout signal, to the integrated electrostatic actuator. While not reducing the input- referred force noise, i.e., the Langevin force acting on the probe, the near-critical damping stabilizes the probe position and also flattens the frequency dependent sensor gain, allowing us to use the sensor effectively over a very broad frequency range without severe dynamic range constrains. The ability to dampen the mechanical noise can strongly reduce the backaction of the sensor onto a system being measured. Moreover, we achieved a noise level ≈2.3 times the standard quantum limit (SQL)[16,36] for our mechanical system, approaching the fundamental readout limits. With future parameter improvement, it may be possible to cool the sensor to the quantum mechanical ground state while maintaining the high readout bandwidth[37] (i.e., in the "bad" cavity limit).

In summary, we have demonstrated a novel class of fully integrated cavity optomechanical transducers for mechanical position and motion sensing with high precision, high bandwidth, and small footprint.

Acknowledgment

JZ, YL, and TM acknowledge support under the Cooperative Research Agreement between the University of Maryland and the National Institute of Standards and Technology Center for Nanoscale Science and Technology, Award 70NANB10H193, through the University of Maryland.

References

1. Cavendish, H. (1798). Experiments to determine the density of the earth, *Philos. Trans. R. Soc. Lond.*, **88**, 469–526.
2. Rugar, D., Budakian, R., Mamin, H. J., and Chui, B. W. (2004). Single spin detection by magnetic resonance force microscopy, *Nature*, **430**, 329–332.
3. Poggio, M., and Degen, C. L. (2010). Force-detected nuclear magnetic resonance: Recent advances and future challenges, *Nanotechnology*, **21**, 342001.
4. Bolle, C. A., Aksyuk, V., Pardo, F., Gammel, P. L., Zeldov, E., Bucher, E., Boie, R., Bishop, D. J., and Nelson, D. R. (1999). Observation of mesoscopic vortex physics using micromechanical oscillators, *Nature*, **399**, 43–46.
5. Lamoreaux, S. K. (1997). Demonstration of the Casimir force in the 0.6 to 6 μm range, *Phys. Rev. Lett.*, **78**, 5–8.
6. Mohideen, U., and Roy, A. (1998). Precision measurement of the Casimir force from 0.1 to 0.9 μm, *Phys. Rev. Lett.*, **81**, 4549–4552.

7. Hanay, M. S., Kelber, S., Naik, A. K., Chi, D., Hentz, S., Bullard, E. C., Colinet, E., Duraourg, L., and Roukes, M. L. (2012). Single-protein nanomechanical mass spectrometry in real time, *Nat. Nanotechnol.,* **7**(9), 602–608.
8. Teufel, J. D., Donner, T., Li, D., Harlow, J. W., Allman, M. S., Cicak, K., Sirois, A. J., Whittaker, J. D., Lehnert, K. W., and Simmonds, R. W. (2011). Sideband cooling of micromechanical motion to the quantum ground state, *Nature,* **475**(7356), 359–363.
9. Chan, J., Mayer Alegre, T. P., Safavi-Naeini, A. H., Hill, J. T., Krause, A., Groblacher, S., Aspelmeyer, M., and Painter, O. (2011). Laser cooling of a nanomechanical oscillator into its quantum ground state, *Nature,* **478**, 89–92.
10. Binnig, G., Quate, C. F., and Gerber, C. (1986). Atomic force microscope, *Phys. Rev. Lett.,* **56**, 930–933.
11. Albrecht, T. R., Akamine, S., Carver, T. E., and Quate, C. F. (1990). Microfabrication of cantilever styli for the atomic force microscope, *J. Vac. Sci. Technol. A*, **8**, 3386–3396.
12. Giessibl, F. (2003). Advances in atomic force microscopy, *Rev. Mod. Phys.*, **75**, 949–983.
13. Ekinci, K. L. (2005). Electromechanical transducers at the nanoscale: Actuation and sensing of motion in nanoelectromechanical systems (NEMS), *Small*, **1**, 786.
14. Rugar, D., Maimin, H. J., and Guenther, P. (1989). Improved fiber-optic interferometer for atomic force microscopy, *Appl. Phys. Lett.,* **55**, 2588–2590.
15. Kippenberg, T. J., and Vahala, K. J. (2007). Cavity opto-mechanics, *Opt. Express*, **15**(25), 17172–17205.
16. Anetsberger, G., Arcizet, O., Unterreithmeier, Q. P., Scjliesser, A., Weig, E. M., Kotthaus, J. P., Kippenberg, K. J. (2009). Near-field cavity optomechanics with nanomechanical oscillators, *Nat. Phys.,* **5**, 909–914.
17. Srinivasan, K., Miao, H., Rakher, M. T., Davanço, M., and Aksyuk, V. (2011). Optomechanical transduction of an integrated silicon cantilever probe using a microdisk resonator, *Nano Lett.*, **11**(2), 791–797.
18. Liu, Y., Miao, H., Aksyuk, V., and Srinivasan, K. (2012). Wide cantilever stiffness range cavity optomechanical sensors for atomic force microscopy, *Opt. Express*, **20**(16), 18268–18280.
19. Miao, H., Srinivasan, K., and Aksyuk, V. (2012). A micromechanically controlled cavity optomechanical sensing system, *N. J. Phys.*, **14**, 075015.

20. Meyer, G., and Amer, N. M. (1988). Novel optical approach to atomic force microscopy, *Appl. Phys. Lett.*, **53**, 1045–1047.

21. Borselli, M., Johnson, T. J., and Painter, O. (2005). Beyond the Rayleigh scattering limit in high-Q silicon microdisks: Theory and experiment, *Opt. Express*, **13**(5), 1515–1530.

22. Michael, C. P., Borselli, M., Johnson, T. J., Chrystal, C., and Painter, O. (2007). An optical fiber-taper probe for wafer-scale microphotonic device characterization, *Opt. Express*, **15**(8), 4745–4752.

23. Jin, J. (2002). *The Finite Element Method in Electromagnetics*, John Wiley & Sons, New York.

24. Berenger, J. (1994). A perfectly matched layer for the absorption of electromagnetic waves, *J. Comput. Phys.*, **114**(2), 185–200.

25. Jackson, J. D. (1999). *Classical Electrodynamics, 3rd ed.*, Wiley, New York, p. 371.

26. Taflove, A., and Hagness, S. (2000). *Computational Electrodynamics: The Finite-Difference Time-Domain Method*, Artech House, Norwood, MA.

27. Yee, K. (1966). Numerical solution of initial boundary value problems involving Maxwell's equations in isotropic media, *IEEE Trans. Antennas Propagation*, **14**(3), 302–307.

28. Eichenfield, M., Chan, J., Camacho, R. M., Vahala, K. J., Painter, O. (2009). Optomechanical crystals, *Nature*, **462**, 78–82.

29. Auld, B. A. (1973). *Acoustic Fields and Waves in Solids, vol. I and II*, Wiley-Interscience, New York.

30. Bhushan, B., and Marti, O. (2010). Scanning probe microsopy—principle of operation, instrumentation, and probes, in *Springer Handbook of Nanotechnology, 3rd ed.* (Bhushan, B., ed.) Springer, Heidelberg, pp. 573–612.

31. Zou, J., Miao, H., Michels, T., Aksyuk, V., A Novel Integrated Optomechanical Transducer and Its Application in Atomic Force Microscopy, *Technical Proceedings of the 2013 NSTI Nanotechnology Conference & Expo—Nanotech 2013*, vol. 2, 173–176.

32. Tamayo, J., Humphris, A. D. L., Malloy, A. M., and Miles, M. J. (2001). Chemical sensors and biosensors in liquid environment based on microcantilevers with amplified quality factor, *Ultramicroscopy*, **86**, 167–173.

33. Ilic, B., Krylov, S., Craighead, H. G. (2010). Theoretical and experimental investigation of optically driven nanoelectromechanical oscillators, *J. Appl. Phys.*, **107**, 034311.

34. Ziegler, D., Bahri, D., Klaassen, A., Ashby, P. D. (2012). *Encased cantilevers for ultra-low-noise force spectroscopy*, paper presented at 56th Biophysical Society Annual Meeting, San Diego, CA.

35. Michels, T., Guliyev, E., Klukowski, M., Rangelow, I. W. (2012). Micromachined self-actuated piezoresistive cantilever for high speed SPM, *Microelectron. Eng.,* **97**, 265–268.

36. Teufel, J. D., Donner, T., Castellanos-Betran, M. A., Harlow, J. W. Lehnert, K. W. (2009). Nanomechanical motion measured with an imprecision below that at the standard quantum limit, *Nat. Nanotechnol.,* **4**, 820–823.

37. Genes, C., Vitali, D., Tombesi, P., Gigan, S., Aspelmeyer, M. (2008) Ground-state cooling of a micromechanical oscillator: Comparing cold damping and cavity-assisted cooling schemes, *Phys. Rev. A,* **77**, 033804.

Chapter 12

Nanocantilever Beam for Gas-Sensing Applications

Ritu Bajpai,[a,b] Mona Zaghloul,[a] Abhishek Motayed,[b,c] and Albert Davydov[b]

[a]*Department of Electrical and Computer Engineering, The George Washington University, Washington, DC 20052, USA*
[b]*Material Measurement Laboratory, National Institute of Standards and Technology (NIST), Gaithersburg, MD 20899, USA*
[c]*Institute for Research in Electronics and Applied Physics, University of Maryland, College Park, MD 20742, USA*

ritu.bajpai@gmail.com

Cantilever and beam structures have been widely used as sensing devices due to their distinct advantages which have been reviewed in literature.[1–3] Several transduction methods can be applied to cantilever and beam type devices making them usable for wide variety of applications. Some of the commonly used transduction techniques are (i) frequency change due to additional mass loading, (ii) bending of bimetallic cantilever for temperature sensing, (iii) stress sensing by measuring the cantilever bending due to stress applied on one side, and (iv) chemical sensing by change in resistivity of the cantilever. The evolution of cantilever and beam structures from macro- to micromachined structures and now nanowire (NW) and other nanostructure-based devices have

Nanocantilever Beams: Modeling, Fabrication, and Applications
Edited by Ioana Voiculescu and Mona Zaghloul

ISBN 978-981-4613-23-1 (Hardcover), 978-981-4613-24-8 (eBook)
www.panstanford.com

made them further attractive for sensing applications. Nanoscale cantilever and beam devices are widely used for sensing applications because of large surface area, miniature size, low power consumption, and nanomoles-per-mole (ppb) sensitivity.[4] This chapter focuses on chemiresistive-type NW and hybrid nanoparticle (NP)/NW devices that have been recently demonstrated.[5,6]

In a chemiresistor-type NW sensor, which is fabricated as a two-terminal electrical device, the sensitivity and selectivity depend on the interaction of different chemical analytes with the NW surface. Constrained by the surface properties of the NW material, most NW sensors can detect only specific type of analytes. In order to make a nanosensor array for a wide range of analytes, there is a need to tune the device sensitivity and selectivity toward different chemicals. In the hybrid NP/NW devices discussed in this chapter, GaN NWs with relatively inactive surface properties (i.e., with no chemiresistive sensitivity to different classes of organic vapors) have been used and have been functionalized with analyte-dependent active metal–oxides. Photoconductive metal–oxide-semiconductors are attractive as a functionalizing material due to their active surface properties and possible.

12.1 Metal Oxides for Gas Sensing

Metal oxides have long been used for gas sensing. An extensive review of various metal oxides used for sensing different gases is available in literature.[7–9] Semiconducting metal oxides thin films are used for chemical sensing[7,10] because their transport properties readily respond to the adsorbed analytes. SnO_2 is one of the most widely used semiconducting metal oxide for gas sensing followed by ZnO.[9] Gas-sensing properties of SnO_2 and ZnO have been widely studied.[11,12] Nowadays, nanowires and nanostructures of these semiconducting metal oxides are being heavily used for sensing applications. Recently, nanowires and other nanostructures of these materials have been shown to exhibit even better sensing characteristics due to increased surface area, reduction in size and power consumption, and nmol/mol ppb sensitivity.[13,14] For this reason, there has been vast research in the growth and properties of nanostructures of ZnO[15] and SnO_2 along with other metal oxides for the fabrication of nano-noses[16] and sensor arrays.

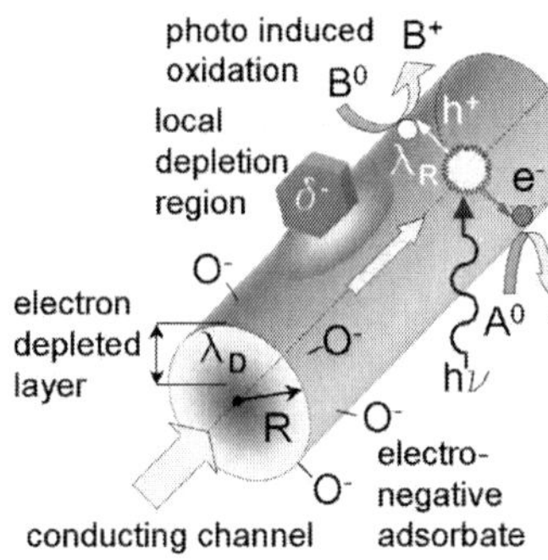

Figure 12.1 A summary of a few of the electronic, chemical, and optical processes occurring on metal oxides that can benefit from reduction in size to the nanometer range.[23]

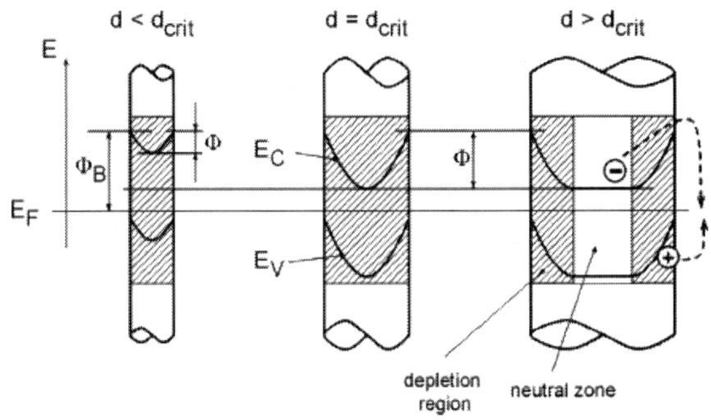

Figure 12.2 Dependence of depletion region (shaded), shape of conduction (E_c) and valence band edges (E_v), and recombination barrier Φ on the nanowire diameter d. The relative energetic locations of E_c, E_v, and E_f are not on scale. The detail on the right shows the surface recombination mechanism of the photoexcited carriers.[32]

What makes these oxides special for gas sensing? As discussed elsewhere,[17] various processes may take place on the surface of metal oxides as a result of the gas–semiconductor interaction, such as reduction/oxidation processes of semiconductor, adsorption of the chemical species directly on the semiconductor, and/or adsorption by reaction with surface states associated with pre-adsorbed ambient oxygen, catalytic effects and in general chemical reactions between the different adsorbed chemical species. Important properties of metal oxides for gas sensing are summarized as follows[17]:

- Density of surface states: For the surface of the metal oxide to be able to respond to the analytes by modulation in surface band bending, it is important that the concentration of native surface states be minimized.
- Electronic structure of gas-sensing metal oxides: Post-transition metal oxides have been found to be most suitable for chemiresistive sensing. Pre-transition metal oxides such as magnesium oxide magnesium oxide (MgO) and aluminum oxide (Al_2O_3) are expected to be inert because they have a large bandgap and can be neither reduced nor oxidized easily. Many transition metal oxides, on the other hand, have small bandgap and have metallic behavior.
- Adsorption/desorption parameters: It has been shown that for a bigger gas sensor effect, the activation energy of chemisorption on the oxide surface should be small and the activation energy for the desorption of the species should be large. Therefore, for the chemisorptions-type sensors, a material with optimal activation energy of desorption is needed.
- Catalytic activity: The high catalytic activity of the surface and, in particular, selectivity of this reaction to the analyte to be detected is an advantage for gas sensing. Many metal oxides that cannot be used as a chemiresistive sensor by themselves serve as a good catalyst for the chemiresistive sensing, such as cerium oxide, Fe_2O_3, and NiO.

For SnO_2 and ZnO, it has been established that the surface properties of these semiconducting metal oxides, due to adsorbed surface oxygen, make them sensitive to chemical species. At room temperature, the adsorbed oxygen on the surface of these oxides is most likely to exist as O_2^- species.[18,19] Surface-adsorbed oxygen captures a negative charge from the oxide [$O_2(g) + e^- \rightarrow O_2^-$ (ads.)] and creates the surface depletion layer and upward band bending at the surface. Depending on the oxide layer thickness, the depletion of the layer could partial or complete. Exposure to reducing gases decreases the width of depletion region by injection of electrons into the oxide, whereas exposure to oxidizing gases further increases the width of depletion region by trapping more number of electrons from the oxide.[20] Most semiconducting metal oxide sensors are operated at high temperature, usually between 200—and 400°C to assist in the process of sensing by oxygen desorption.

12.2 3D vs. 2D Sensors

There has been a trend of synthesizing nanostructures of metal oxides, which have been widely used for gas sensing, for example, tin oxide, indium oxide and zinc oxide.[21] Their gas-sensing behavior has widely been demonstrated in the recent years.[22] Properties of nanostructures that establish their advantages over 3D structures have been summarized as follows[23]:

(1) Large surface-to-volume ratio for the nanostructures provides larger response because sensing in chemiresistive sensors is a surface phenomenon.
(2) The width of the surface depletion region is comparable to the diameter of the nanostructures, which implies that the electronic properties can be strongly influenced by the processes at the surface.
(3) The average time for the photogenerated carriers to reach the surface ($\sim 10^{-12}$ to 10^{-10} s) is reduced compared to the electron-hole pair recombination time ($\sim 10^{-9}$–10^{-8} s). This implies that the surface photo-induced redox reactions with quantum yield close to 1 are possible on nanowires.
(4) Semiconducting oxide nanowires are usually stoichiometrically better defined and have a greater level of crystallinity than the multigranular oxides currently used in sensors, potentially reducing the instability associated with percolation or hopping conduction.
(5) Nanowires can easily be configured as field-effect transistors (FETs). It is possible to vary the position of the Fermi level within the bandgap using gate voltage and therefore alter the surface processes electronically.

Some of the electronic, chemical, and optical processes occurring on metal oxides that can benefit from reduction in size to the nanometer range have been schematically presented in Fig. 12.1. Additionally, nanostructures have a potential of massive multiplexing in small devices. Nanowire devices have been shown to exhibit faster recovery time than thin-film sensors.[23]

Chemical dopants and catalytic additives have been introduced to nanowires to enhance their selectivity and efficiency.[24,25] In other cases, surface nanoclusters and nanoparticles have been used as transducers for the chemically inert nanowires. For example, it has been shown that GaN NWs can be surface functionalized to induce sensitivity toward specific analytes.[26–28] These sensors provide the flexibility to start with an analyte-inert nanowire material and make them selectively sensitive based on the surface functionalization.

12.3 Properties of GaN, ZnO, and SnO_2 Nanostructures

GaN nanowires functionalized with ZnO and SnO_2 nanoparticles and ZnO nanowires have been used in this work to fabricate alcohol gas sensors. In this section, we will review the important properties of these materials, which made them a suitable choice for this work. Most semiconductor surfaces possess electronic surface states, and the impact of these surface states on electronic properties of the semiconductor has been studied.[29] Depending on the nature of the surface states, they may carry positive or negative charge that is screened by the opposite charge inside the semiconductor material known as the space charge. Distribution of the space charge is related to the curvature of the valence and conduction bands in the space charge region by Poisson's equation. This surface band bending causes photogenerated charge carrier separation when the light above the bandgap energy is incident on the device. For a semiconductor with positive space charge region, holes move toward the surface, whereas electrons have a tendency to remain in the bulk. This phenomenon is also observed in semiconducting nanowires, where the formation of the depletion region occurs on the entire surface of the nanowire forming a central conducting region surrounded by a depletion region.

12.3.1 Properties of ZnO and SnO_2 Nanostructures

When the metal oxide surface is illuminated with UV light, energetic enough to create photogenerated electron–hole pair, an increase in the conductivity is observed. This increase is not only because of the photogenerated charge carriers but also because photogenerated holes can remove the adsorbed oxygen from the oxide surface [$h^+ + O_2^-(ads.) \rightarrow O_2\ (g)$] and release the captured electron back into the oxide as explained for the metal oxide semiconductors and also demonstrated for the metal oxide NWs.[30,31] Therefore, the adsorbed oxygen is released in the presence of UV light and promotes photogenerated charge carrier separation in the ZnO and SnO_2. The holes migrate toward the surface and the unpaired electrons can be collected by the applied electric field to result in the increased conductivity of the oxide. An analyte that can undergo oxidation reaction in presence of adsorbed oxygen and UV light will also assist in removal of adsorbed oxygen and increase in the oxide conductivity. This property can be used for room-temperature gas sensing.

12.3.2 Properties of GaN Nanowires

Photoconductive behavior of GaN nanowires has been discussed.[32,33] GaN nanowires, like most other semiconducting nanowires, exhibit a depletion space charge layer on the surface with an extension of the order of the nanowire diameter as shown in Fig. 12.2. Depending on the diameter and doping of the nanowires, fully or partially depleted nanowires exist. In the presence of photoexcitation with optimal light intensity, the photogenerated charge carrier separation is governed by the depletion width. It has also been shown in that for an *n*-type background doping in the 10^{17} cm^{-3} range, the depletion space charge layers should have extensions of 50–100 nm into the bulk.[32] This can be calculated using the following equation:[34]

$$W_d = \sqrt{\frac{2\varepsilon_0 \varepsilon_r \psi_s}{qN_a}},$$

where W_d is the depletion width, ε_0 is the permittivity in vacuum, ε_r is the unspecific relative dielectric constant, Ψ_s is the surface potential with respect to bulk, q is the electronic charge, and N_a is the doping density of the material.

Therefore, nanowires with diameters less than 100 nm could be fully depleted in dark. As the diameter of the NWs used in this work is greater than 100 nm, we have a neutral conducting core in the center of the NWs and a depletion region at the surface. When a voltage is applied across the ends of the NW, we observe the dark current of the device. In the presence of light, an enhanced current is observed due to photogenerated charge carriers and their separation assisted by surface band bending. In our case, as the GaN nanowire surface is inert to chemical species, the photocurrent remains unaltered when the chemical vapors are passed over the bare GaN nanowire devices.

12.4 Device Fabrication

We discuss three types of sensor devices along with their fabrication and characterization details:

(1) ZnO nanoparticle (NP)/GaN NW alcohol sensors: Two-terminal suspended GaN nanowire devices were fabricated. The devices were functionalized with ZnO nanoparticles for alcohol detection.
(2) SnO_2 NP/GaN NW alcohol sensors: Two-terminal suspended GaN nanowire devices were fabricated. The devices were functionalized with SnO_2 nanoparticles for alcohol detection.
(3) ZnO NW alcohol sensors: These sensors were fabricated using ZnO nanowires. Unlike the other two sensor devices listed above, surface functionalization with nanoparticles was not performed in this case.

All the devices shared the same fabrication steps, except for the nanowire and nanoparticle material selection.

Vertically grown, *c*-axis oriented ZnO and GaN nanowires have been used in this work. The length of the ZnO nanowires varied between 5 and 15 um, whereas the diameter was between 80 and 150 nm. For the GaN nanowires, the length varied from 5 to 20 um, whereas the diameter was between 200 and 350 nm. Details about the growth and properties of GaN nanowires and ZnO nanowires used in this work have been given elsewhere.[35–37] The NWs were detached from the substrate by sonication in isopropanol and dielectrophoretically aligned across the prepatterned electrodes. The electrodes were fabricated using photolithography followed by deposition of a metal stack of Ti (40 nm)/Al (420 nm)/Ti (40 nm). Thick bottom electrodes ensure the free suspension of the NWs. For the formation of ohmic contacts to the NW ends, the top metal contacts were fabricated using a metal stack of Ti (70 nm)/Al (70 nm)/Ti (40 nm)/Au (40 nm).[38] Figure 12.3 shows SEM image of a suspended GaN nanowire bridge structure after deposition of top electrodes.

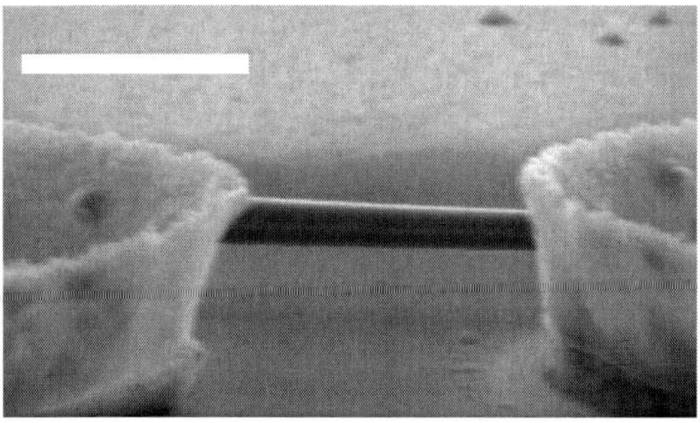

Figure 12.3 SEM image of a suspended GaN nanowire bridge structure after deposition of top electrodes.

For the first two types of sensor devices listed in the beginning of this chapter, ZnO NP/GaN NW and SnO_2 NP/GaN NW, GaN nanowires with relatively inactive surface properties (i.e., with no chemiresistive sensitivity to different classes of organic vapors) have been used. The selectivity and sensitivity of these devices for alcohol vapors has been induced by performing surface functionalization with ZnO or SnO_2 nanoparticles. A commonly used technique for the deposition of nanoparticles is sputtering.[26,39,40] Therefore, for the hybrid metal oxide nanoparticle/GaN nanowire sensors fabricated in this work, the nanoparticles were deposited using the RF sputtering tool (Denton Vacuum Discovery 550). The deposition time was optimized to ensure that the sputter deposited metal oxide formed isolated islands and that the nanoparticles did not coalesce to form a conducting film on the surface. Figures 12.4 and 12.5 show the SEM images of the sputter deposited ZnO and SnO_2 nanoparticles on the surface of GaN nanowire, respectively.

Rapid thermal anneal (RTA) was performed for 30 s at 700°C for the GaN nanowire.[26] The anneal temperature of 300°C for 30 s was found to be optimal for ZnO nanowire devices. In both cases, the temperature ramp rate was 100°C per minute and the process was performed under a 6000 sccm flow of ultra-high purity Ar. RTA assists the device in three ways: (i) It facilitates ohmic contact formation to NWs, (ii) it induces recrystallization of the metal oxide nanoclusters, and (iii) it reduces mechanical stresses in the metal stacks forming the alignment and top electrodes, which is an important consideration for wire bonding the devices.

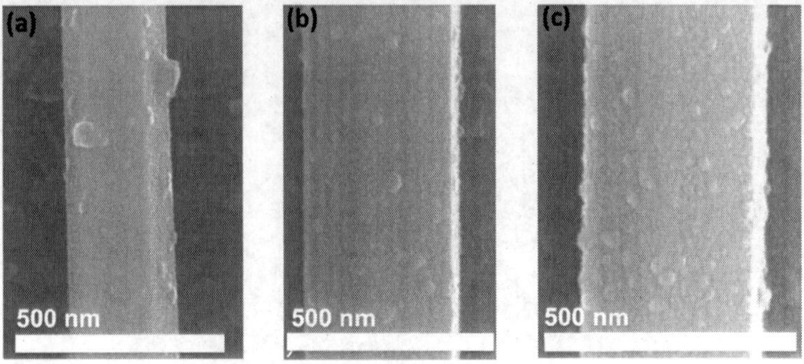

Figure 12.4 SEM image of sputter deposited ZnO nanoparticles on GaN nanowire surface with deposition times (a) 160 s, (b) 200 s, and (c) 250 s.

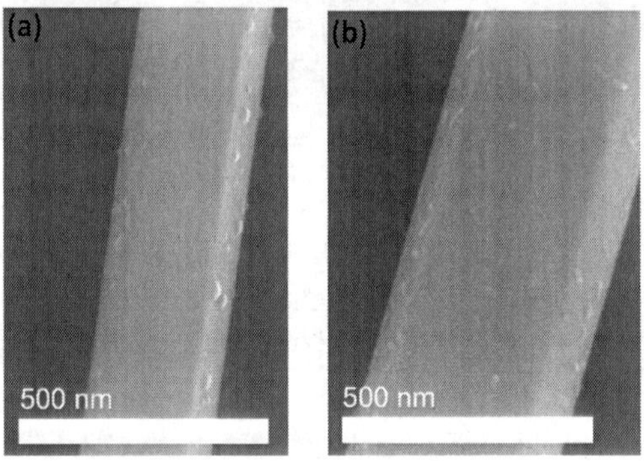

Figure 12.5 SEM image of sputter deposited SnO_2nanoparticles on GaN nanowire surface with deposition times (a) 7 min (b) 15 min.

12.5 Device Characterization

Electrical characterization of the devices was performed by measuring the current voltage (*I*–*V*) characteristics. All the electrical measurements were performed using Agilent B1500A semiconductor parameter analyzer. *I*–*V* characterization was repeatedly performed after (i) fabrication of the top electrodes, (ii) rapid thermal anneal of the devices, and (iii) device packaging.

Figures 12.6 and 12.7 show the *I–V* characteristic and X-ray spectrum for hybrid ZnO NP/GaN NW and SnO_2 NP/GaN NW devices, respectively. A DC probe station was employed for performing all the measurements before device packaging. These measurements assisted in studying the electrical properties of the nanowires and the contacts.

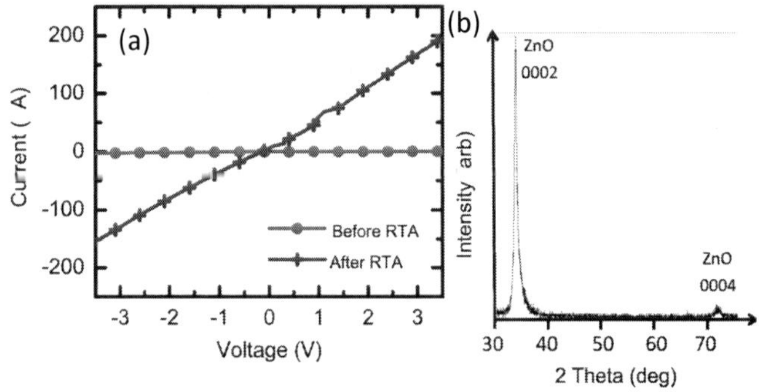

Figure 12.6 Electrical and morphological characteristics of ZnO NP/GaN NW device. (a) *I–V* characteristic of the device before and after RTA anneal (b) XRD Ω-2Θ scan of a 300 nm-thick ZnO film.

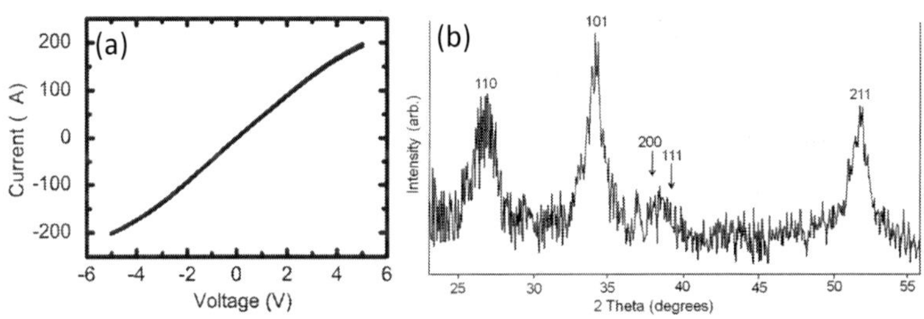

Figure 12.7 Electrical and morphological characteristics of SnO_2 NP/GaN NW device: (a) Dark *I–V* characteristic of the device, (b) XRD Ω-2Θ scan of a 40 nm-thick SnO_2 film after 700°C anneal in argon for 30s. Indices correspond to tetragonal SnO_2 phase (P42/mnm) with a = 4.758(3) Å and c = 3.175(2) Å lattice parameters.

Nanowire sensors were also characterized by SEM, HR-ATEM/STEM, and SAED. The specimens were analyzed in a FEI Titan 80–300 TEM/STEM microscope operating at 300 kV accelerating voltage. The instrument was also equipped with an EDAX Si/Li energy-dispersive X-ray spectrometer and high-angle annular dark-field (HAADF), bright-field (BF), and annular dark-field (ADF) STEM detectors to perform spot, line profile and areal compositional analyses. The identification of crystalline phases was also performed by XRD and EBSD. The XRD spectra were collected using a Bruker-AXS D8 scanning X-ray microdiffractometer equipped with a general area detector diffraction system (GADDS) using Cu-K radiation. The EBSD patterns were recorded using an HKL Nordlys II EBSD detector attached to the Hitachi S-4700 SEM

12.6 Results

Sensing test results were collected from several sensor devices. The naming convention used for the devices in this work is as follows: R (number) S (number)-E(number), where R stands for the recipe and the following number tells the recipe used for the fabrication of alignment electrodes (generation 1, generation 2, or generation 3), the number following S denotes the serial number of the device in the sequential order of fabrication, and the number following E denotes the particular electrode on the chip where the device is located. For example, R2S4-E6-4 used generation 2 recipe for the fabrication of alignment electrodes, it was the fourth sensor device to be fabricated (generation 1 devices were not counted because there was no yield) and the particular device is located on electrode E6-4 on the chip R2S4. The response of the sensor devices was tested toward breathing air, nitrogen, alcohols (methanol, ethanol, n-propanol, isopropanol, n-butanol, and isobutanol), water vapor, acetone, and hexane. The results for each category of devices have been presented in different sections. Section 12.6.1 describes the sensing results obtained from the ZnO NP/GaN NW devices. Sensing results of the SnO_2 NP/GaN NW devices have been described in Section 12.6.2. Finally, the sensing results obtained from the ZnO NW devices have been shown in Section 12.6.3. Table 12.1 gives the list of sensor devices whose results have been presented in this chapter.

12.6.1 Response of the ZnO NP/GaN NW Devices

This section gives the sensing test results of ZnO nanoparticle/GaN nanowire devices. Response (in presence of UV illumination) to breathing air, nitrogen, alcohols, and possible interfering gases with alcohol vapors (acetone, benzene, alkanes) has been presented.

12.6.1.1 Response to the breathing air

Breathing air was used as the carrier gas for all the sensing experiments performed in this work. Device conductivity at 40 sccm flow of breathing air was regarded as the sensor baseline. Figure 12.8 shows the response of ZnO NP/GaN NW device (R2S5-E8-2) to the different flow rates of breathing air. At time t = 0, UV illumination was turned off and the device photocurrent decayed until the illumination was turned on again. The device was allowed to attain a stable photocurrent and then the airflow was turned on. As seen from the figure, device conductance decreased upon exposure to the breathing air. Also, as the flow rate of the air increased, the conductivity decreased further. When the illumination was turned off, the current once again decayed toward the dark current value.

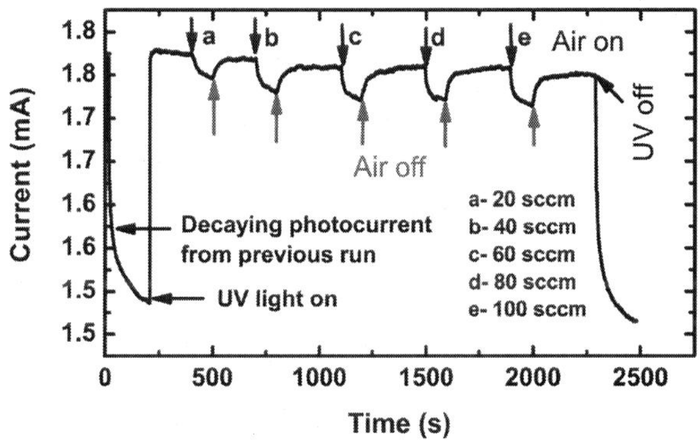

Figure 12.8 Device (R2S5-E8-2) response to different flow rates of breathing air.

12.6.1.2 Response to nitrogen

All the devices were also tested for a response toward nitrogen gas in presence of UV illumination. An opposite response to that toward the breathing air was observed, i.e., an increase in conductivity was observed when the devices were exposed to nitrogen. Additionally, a greater increase was observed with the increasing flow of nitrogen. Figure 12.9 shows the response of the device R2S5-E8-2 toward different flow rates of nitrogen.

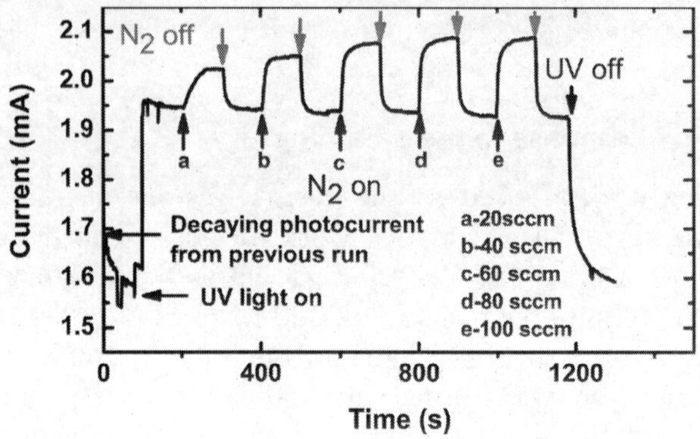

Figure 12.9 Device (R2S5-E8-2) response to different flow rates of nitrogen gas.

However, if the device conductivity in nitrogen was treated as the sensor baseline and gases were passed through the chamber using N_2 as the carrier gas in the presence of light, negligible device response was observed.

12.6.1.3 Response to alcohols

When exposed to alcohol vapors, the devices showed an increase in conductivity, with maximum sensitivity toward methanol. The response of ZnO NP/GaN NW device (R2S4-E6-4) toward 500 mol/mol (ppm) of the four alcohols—methanol, ethanol, n-propanol, and n-butanol—is shown in Fig. 12.10. As seen from the figure (marked by the dotted circle), when the flow of methanol and the air mixture was turned on, there was a residual effect of the air that was already present in the chamber, i.e., initially there was a decrease in conductivity (due to breathing air), but once a constant flow of methanol vapors had been established, an increase in conductivity (due to methanol + air mixture) was observed. Similarly, when the flow of methanol was turned off and the flow of breathing air was turned on, there was an initial increase in conductivity, which was followed by a decrease once all the methanol vapors had been flushed out of the chamber. For this reason, two consecutive measurements were made to record the sensor response to every concentration of the analyte. The first measurement ensured that a steady state had been established, whereas the second measurement was recorded as the actual data.

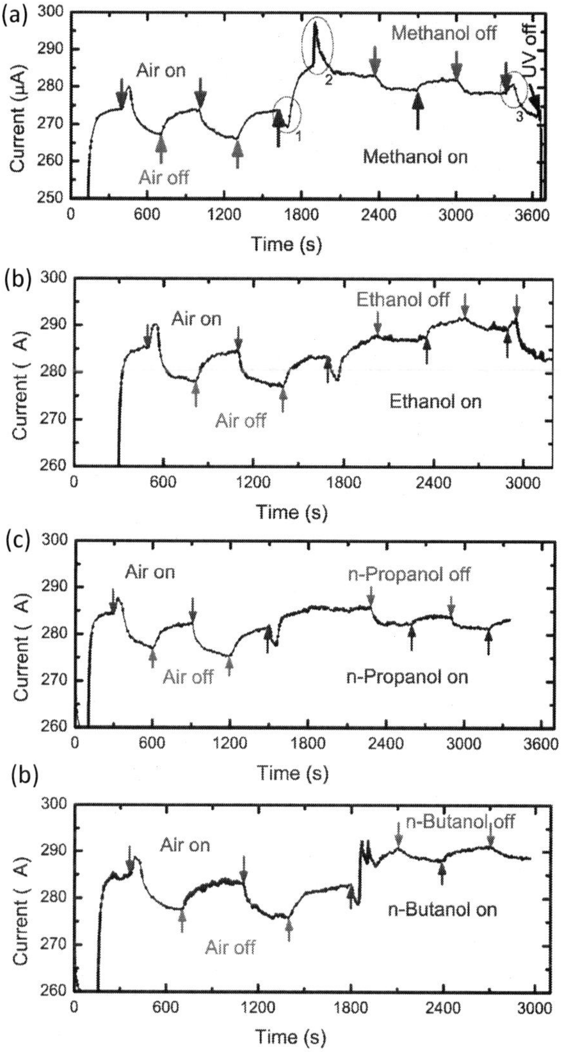

Figure 12.10 Response of the ZnO NP/GaN NW device (R2S4-E6-4) to 500 ppm of (a) methanol, (b) ethanol, (c) n-propanol, and (d) n-butanol.

Figure 12.11 shows the relative response of devices R2S4-E6-4 and R2S4-E8-2 to 500 mol/mol (ppm) of methanol and butanol vapors. In both the cases, a larger response is observed for methanol than butanol. Also, the response for n-butanol is found to be greater than isobutanol in each case.

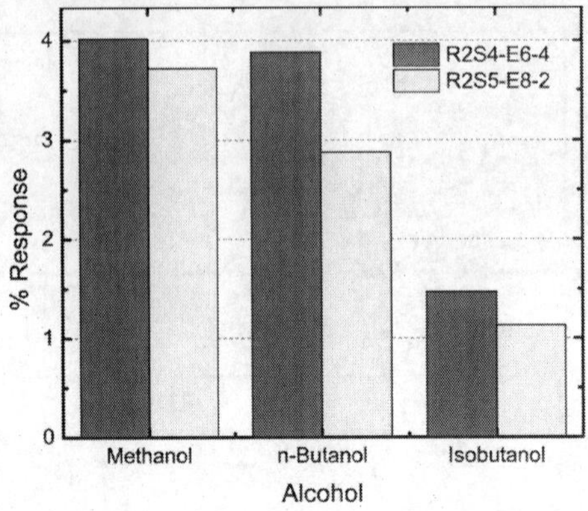

Figure 12.11 Percentage response of ZnO NP/GaN NW devices to 500 mol/mol (ppm) of methanol and butanol vapors.

12.6.1.4 Response to acetone, benzene and hexane

Figure 12.12 shows the relative response of device R2S4-E6-4 toward ethanol, acetone, hexane, and benzene for different concentrations. Acetone, benzene, and alkanes could be the common interfering chemicals with alcohol vapors. The devices show a negligible response to benzene and hexane, whereas the sensitivity toward 100 mol/mol (ppm) of ethanol is similar to the sensitivity toward 1000 mol/mol (ppm) of acetone. Ethanol vapor concentration down to 100 nmol/mol (ppb) has been successfully detected and the detection of even lower concentrations is only limited by our measurement setup.

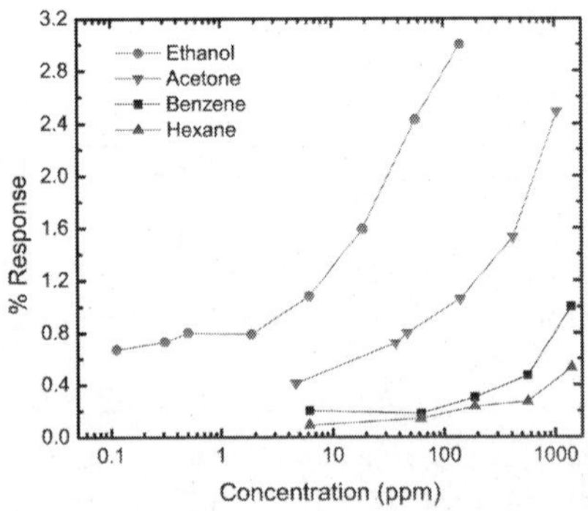

Figure 12.12 Response of the ZnO NP/GaN NW device (R2S4-E6-4) to different concentrations of ethanol, acetone, benzene, and hexane.

12.6.2 Response of the SnO_2 NP/GaN NW Devices

Similar set of tests were performed on SnO_2 NP/GaN NW devices. Once again, as observed from ZnO NP/GaN NW devices, these devices show an increased conductivity when exposed to nitrogen gas flow. However, no change in conductivity was observed when analyte vapors were passed through the chamber using nitrogen as the carrier gas. Figure 12.13a shows the response of device R2S23-E3-3 to 500 ppm of different alcohols. Figure 12.13b shows the response of the same device to different concentrations of methanol vapors. As expected, the response of the device increases with an increasing concentration of methanol. Additionally, in this case a clear trend of a decrease in the response with an increase in the length of the carbon chain in the alcohol molecule was observed.

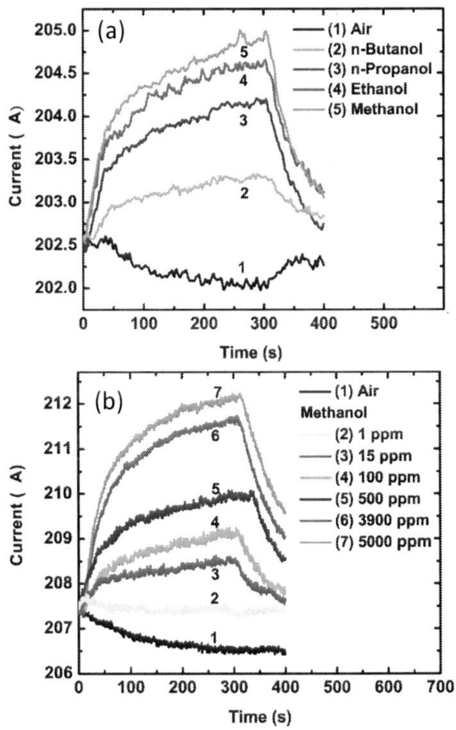

Figure 12.13 Response of the SnO_2 device (R2S23-E3-3) to (a) 500 mol/mol (ppm) of different alcohols (b) different concentrations of methanol.

Figure 12.14a shows the response of SnO_2 device (R2S23-E3-3) to 500 mol/mol (ppm) of acetone, and hexane vapors. As seen from the figure, the response toward acetone and hexane was comparable to the response toward isobutanol, the alcohol producing the least response from the device. Negligible response was observed toward benzene vapors. Response of the devices toward water was similar to the response toward alcohols, i.e., an increase in photoconductivity was observed with increasing concentration of water vapors as shown in Fig. 12.14b.

Response of ZnO NP/GaN NW and SnO_2 NP/GaN NW devices toward various gases has been listed in the Table 12.1.

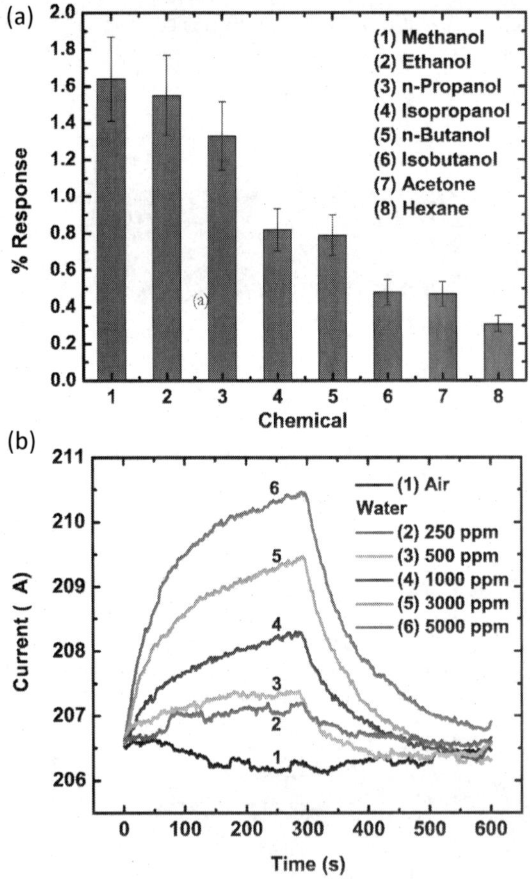

Figure 12.14 Response of the SnO_2 device (R2S23-E3-3) to (a) various chemicals and (b) water.

Table 12.1 Summary of the response of ZnO NP/GaN NW and SnO_2 NP/GaN NW devices

Gas	UV-assisted response
Breathing air	Decreased conductivity
Nitrogen	Increased conductivity
Alcohols (in air)	Increased conductivity (maximum response to methanol)
Water vapor (in air)	Increased conductivity
Acetone, benzene, and hexane (in air)	Increased conductivity (small compared to alcohols)

12.6.3 Response of ZnO NW Devices

ZnO nanowire devices were also tested for a response toward air, nitrogen, and alcohols. In addition to these tests, these devices were tested for variation in sensitivity with variation in light intensity.

12.6.3.1 Response to air

Response of the ZnO NW device (R3S30-E11) to 40 sccm flow of breathing air has been shown in Fig. 12.15. Similar to the response of ZnO nanoparticle and SnO_2 nanoparticle-coated GaN nanowires, these devices also exhibited a decrease in conductivity on exposure to airflow. However, the percentage response of these devices was much larger than that of nanoparticle-coated GaN nanowire devices.

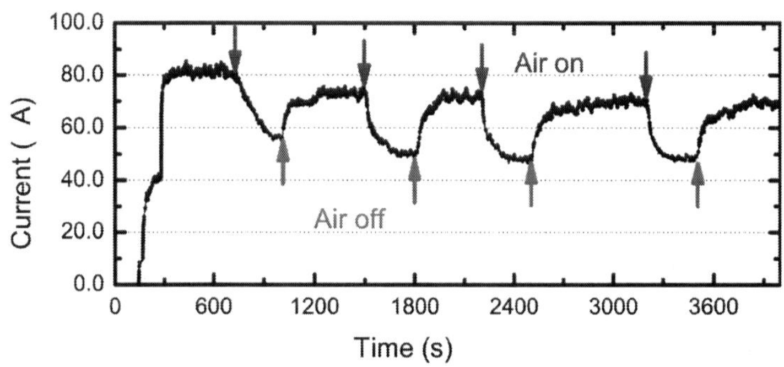

Figure 12.15 Response of the ZnO NW device (R3S30-E11) to 40 sccm flow of breathing.

12.6.3.2 Response to nitrogen

The response of ZnO NW device (R3S30-E11) toward nitrogen has been shown in Fig. 12.16. Once again, the response of these devices followed the trend of ZnO nanoparticle and SnO_2 nanoparticle-coated GaN nanowire devices, i.e., there was an increase in conductivity on exposure to the flow of nitrogen gas. As in the case of air, the response was greater than that seen with the nanoparticle-coated GaN devices.

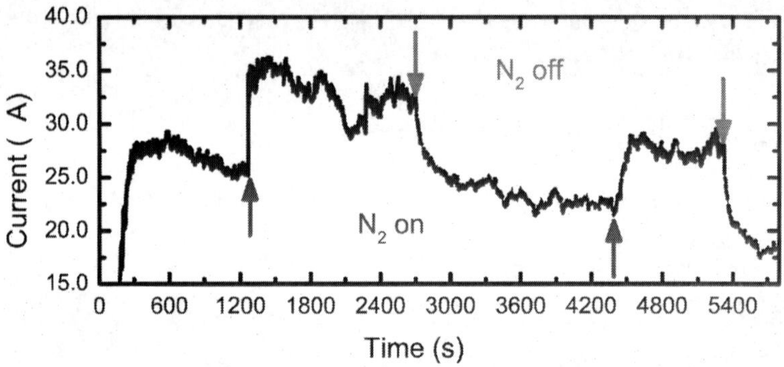

Figure 12.16 Response of the ZnO NW device (R3S30-E11) to 40 sccm flow of breathing.

12.6.3.3 Response to alcohols

The response of one of the ZnO NW devices (R3S25-E5-1) toward the series of alcohols has been shown in Fig. 12.17. From Fig. 12.17, it is observed that the tested alcohols induced an increase in conductivity of ZnO NW devices just as in the hybrid metal oxide (MO) NP/GaN NW devices. Additionally, it is observed that propanol shows a slow recovery response, which is not understood. It appears as if there is a fast recovery behavior followed by the slow recovery behavior of the device. Figure 12.18 shows the percentage response of the ZnO NW device R3S25-E5-1, toward different alcohols.

The response of the ZnO nanowire devices toward alcohols was measured at different light intensities using the neutral density filters. Figure 12.19 shows the response of ZnO NW device toward methanol, ethanol, n-propanol, and n-butanol at three different light intensities. It was observed that the response toward the alcohols decreased with the decreasing intensity of light.

Therefore, light intensity could be used to control the response magnitude of the sensor devices.

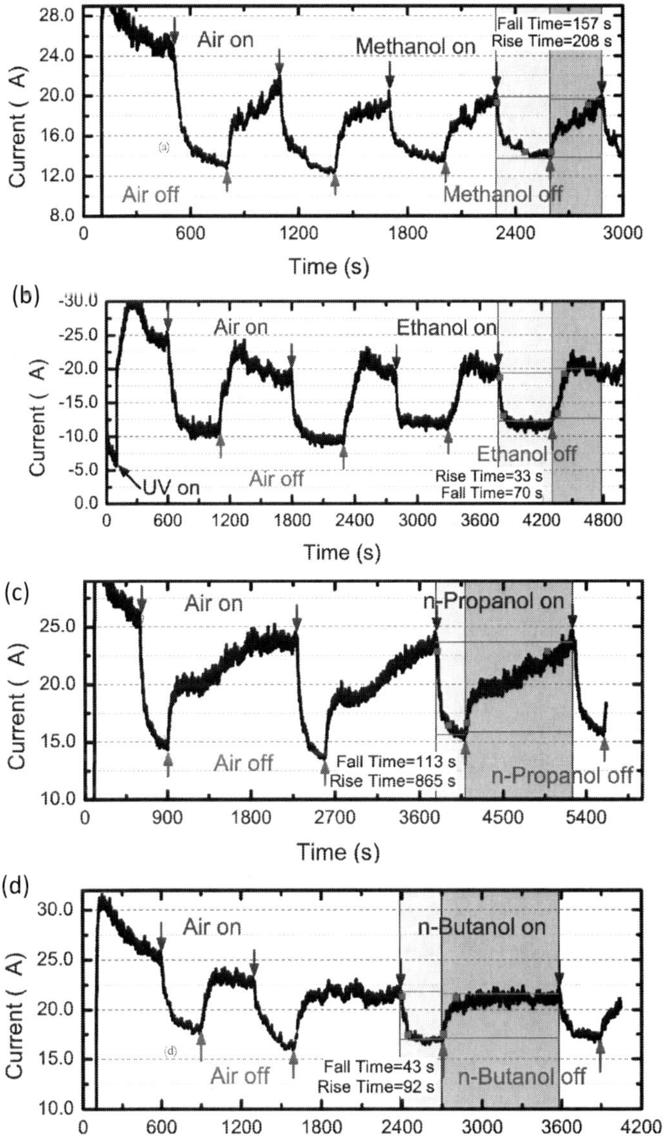

Figure 12.17 Response of the ZnO NW device (R3S25-E5-1) to 500 ppm of (a) methanol, (b) ethanol, (c) n-propanol, and (d) n-butanol.

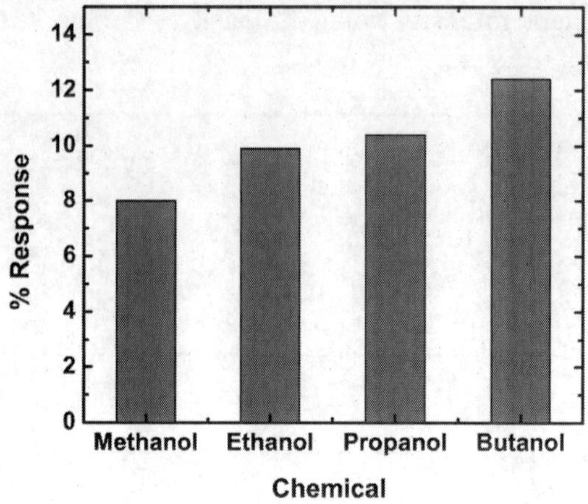

Figure 12.18 Relative response to different alcohols.

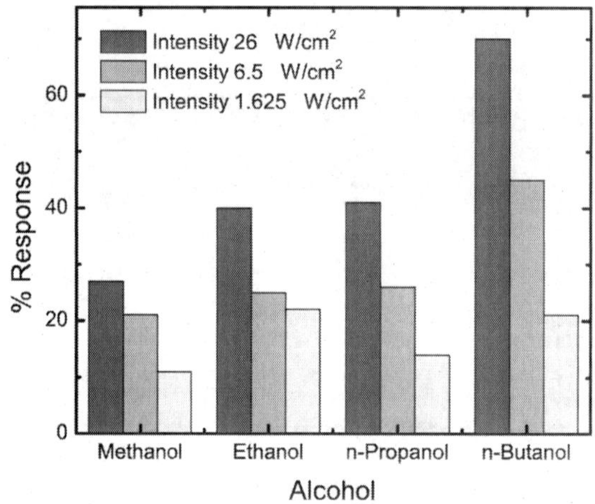

Figure 12.19 Response of the ZnO NW device to different alcohols with varying intensity.

12.6.3.4 Response to acetone, benzene, and hexane

Response of the ZnO nanowire device (R3S24-E11) to 500 ppm of acetone, benzene, and hexane vapors has been shown in Fig. 12.20. As seen from the figure, the response toward acetone and benzene is less than ethanol and the response toward hexane in negligible. Therefore, relative response of the interfering chemicals with respect to most responsive alcohol, n-butanol in this case is going to be even less than that compared to ethanol.

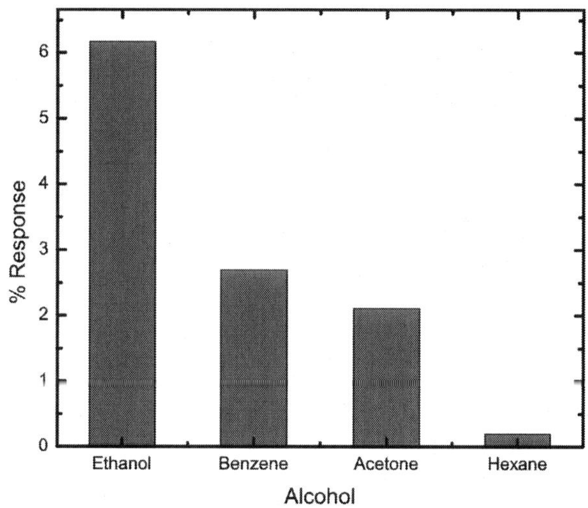

Figure 12.20 Relative response of ZnO NW devices to 500 ppm of acetone, benzene, and hexane.

Response of ZnO nanowire sensor devices to various gases has been listed in Table 12.2 and the important observations have been summarized below:

Table 12.2 Summary of the response of ZnO NW devices

Gas	UV-assisted response	Dark response
Breathing air	Decreased conductivity	Negligible response
Nitrogen	Increased conductivity	Negligible response
Alcohols (in air)	Increased conductivity (maximum response to methanol)	Negligible response
Acetone, benzene, and hexane (in air)	Increased conductivity (smaller response than alcohols)	Negligible response

12.6.4 Summary of Sensing Results

Similar to ZnO NP/GaN NW and SnO_2 NP/GaN devices, for ZnO NW devices

(1) A decrease in conductivity was observed when the devices were exposed to air.

(2) Exposure to nitrogen resulted in an increased conductivity.
(3) Passing alcohols through the chamber with air as the carrier gas also led to an increase in the conductivity of the devices.
(4) A weak response toward acetone, benzene, and hexane in form of increased conductivity was also observed.
(5) Once again, no response was observed toward the analytes when nitrogen was used as the carrier gas.

Additionally,

(1) Percentage response of these devices was larger than ZnO and SnO_2 nanoparticle-coated GaN nanowire devices. However, it should be noted that the measurements for these devices were performed at 365 nm wavelength with intensity 26 W/cm^2 whereas ZnO and SnO_2 nanoparticle-coated GaN nanowire devices were illuminated with a broadband source with intensity 375 nW/cm^2 at 365 nm wavelength.
(2) A relationship opposite to that recorded with the NP/NW devices was observed between the response toward an alcohol and the length of the carbon chain in the alcohol molecule. A bigger carbon chain gave a greater response.

12.7 Sensing Mechanism

Sensing behavior of ZnO nanowire devices has been examined first because of simplicity and a large amount of existing information in literature about the same. Next, a possible sensing mechanism has been proposed for the hybrid metal oxide nanoparticle-coated GaN nanowire sensors.

12.7.1 Sensing Properties of ZnO Nanowire Devices

As explained in Section 12.1, the sensing properties of ZnO nanowires are caused by the surface-adsorbed oxygen species. Removal of the adsorbed oxygen can be assisted by high temperature or UV illumination and results in an increase in the conductivity of ZnO. Most of the conventional metal oxide gas sensors are assisted by elevated temperature to facilitate oxygen de-sorption. UV light-assisted sensing operation at room temperature as used in this work, has been demonstrated as an alternate to high-temperature sensing in existing literature and an equivalence between temperature-assisted and light-assisted sensing of the metal oxide gas sensors has been established.[41–43] In this section, we discuss the possible mechanism for the response of ZnO NW sensor devices described in this work to different analytes.

12.7.1.1 Response to air

Presence of adsorbed oxygen species in the form of molecular O_2, and ionized O_2^- at room temperature has been shown for ZnO. This adsorbed oxygen captures the conducting electrons from ZnO and result in the formation of a depletion layer near the surface. When the oxide is exposed to light with photon energy greater than or equal to the bandgap of ZnO, photogenerated electron–hole pairs are produced in the material. The electrons move in the direction of the applied electric field, whereas the holes follow the potential slope created by the depletion region and move toward the surface. At the surface, these holes displace the adsorbed oxygen and help release the electrons that were trapped by the oxygen species back into the oxide. Oxygen can re-adsorb on the surface and can be again driven out by holes. This process continues until equilibrium is established between the adsorbed oxygen and the number of majority charge carriers, i.e., electrons in the metal oxide hence leading to a defined conductivity. Response of ZnO nanowires in presence of UV light has been discussed in earlier literature.[44] A schematic representation of photoinduced increase in the conductivity of ZnO nanowire devices has been shown in Fig. 12.21.

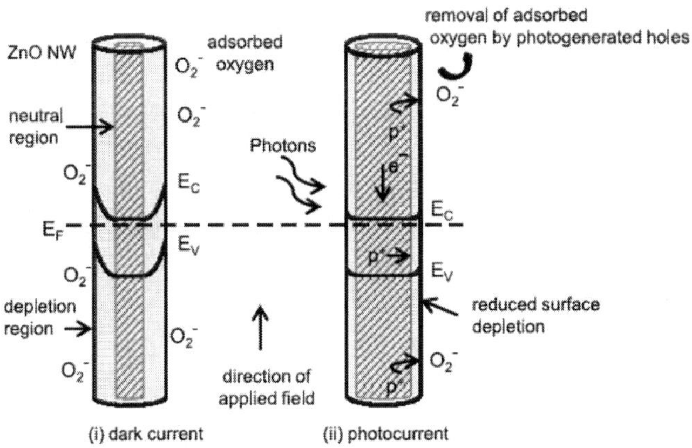

Figure 12.21 Schematic representation of increase in the conductivity of ZnO NWs on exposure to UV light.

For the sensor devices described here, the initial state of the sensor and hence the absolute conductivity with respect to constant flow of air is unknown because when the sensor is sitting under the ambient conditions the percentage coverage of the device surface with oxygen is unpredictable as it would depend on the room temperature, humidity and surface properties for the particular device. However, when the device is exposed to a known airflow, the relative surface coverage with respect to smaller or greater flow of air through the chamber is known.

Device conductivity reduces with increasing flow of air through the chamber because the coverage of the device surface with oxygen increases, which captures more conducting electrons, hence decreasing the conductivity. No response to air is observed in dark because the device has an established surface coverage by oxygen, which is in equilibrium and cannot be increased further. Additionally, this oxygen can be removed from the surface only when assisted by elevated temperature or UV light. When the light is turned on, depletion width is reduced to minimum possible equilibrium value at that intensity. Now on further exposure to oxygen, the depletion width can be increased leading to a decrease in the device conductivity.

12.7.1.2 Response to nitrogen

To the best of our understanding, nitrogen molecules displace adsorbed oxygen when supported by the photogenerated holes. Displacement of adsorbed oxygen from the oxide surface results in increased conductivity due to increased photogenerated charge carrier separation. However, no effect of nitrogen was observed in dark. This is consistent with the results achieved with ZnO nanowire UV photodetectors, where no change in dark current was observed in the vacuum before illumination, indicating that oxygen desorption was indeed supported by photogenerated holes.[31]

12.7.1.3 Response to alcohols

Effect of alcohols on the conductivity of ZnO NWs and thin films has been explained in existing literature. It has been observed that exposure to alcohols results in an increased conductivity of ZnO. Similar to the photogenerated holes, when chemical species interact with the adsorbed oxygen species, they may remove the O_2^-

from the surface causing the captured electron to be released back into the oxide nanoparticle

Alcohols can oxidize on the surface of ZnO forming aldehyde (most probable at room temperature in absence of catalysts) and water using the adsorbed oxygen and releasing back the conducting electrons in the oxide, which explains the increase in oxide conductivity. The above described mechanism of operation of ZnO alcohol sensors is widely accepted in literature.[45–48] A schematic representation of alcohol sensing action of ZnO is shown in Fig. 12.22.

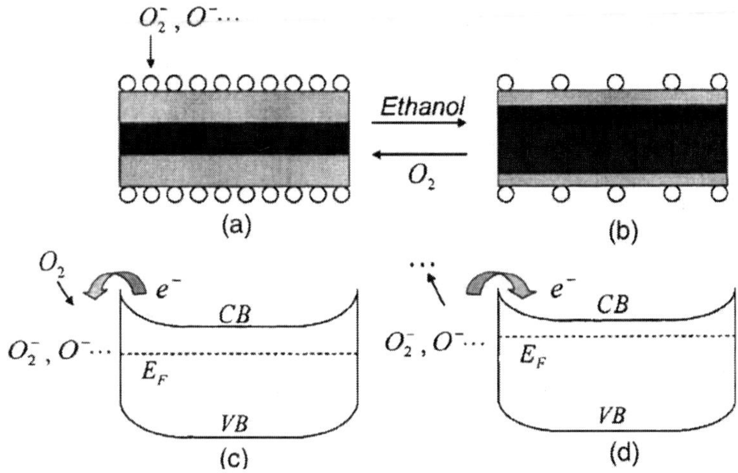

Figure 12.22 (a) The adsorption of O_2 traps some free electrons and widens the surface depletion width. (b) The reactions release some trapped electrons back to the conduction band and narrow the electron depletion width. (c) and (d) are the energy band diagrams of (a) and (b), respectively.[49]

However, it was observed that if nitrogen was used as a carrier gas, alcohols did not produce a change in device conductivity. This behavior can be understood from the initial response of the device toward nitrogen gas. As nitrogen removes the adsorbed oxygen from the device surface, when alcohols are introduced into the sensing chamber, adsorbed oxygen is not readily available for reaction with alcohols. Therefore, the alcohols do not exhibit a response when nitrogen is used as a carrier gas.

It is interesting to note the response trend of the device toward different alcohols. Maximum response is obtained from exposure to n-butanol molecules and the response decreases with decrease in the length of the carbon chain, i.e., minimum response is observed toward methanol vapors. This trend seems to correspond to the acidity of alcohols in gas phase, which increases with the increasing length of the carbon chain in the molecule.[50] The following order of relative acidities has been found: neopentyl alcohol > t-butyl > isopropyl > ethyl > methyl > water; and t-butyl ≈ n-pentyl ≈ n-butyl > n-propyl > ethyl. It should be noted that in contrast to the solution behavior, t-butoxide in gas phase is less basic than methoxide.

12.7.1.4 Response to other chemicals

The devices were also tested for a response to possible interfering chemicals with alcohol vapors like acetone, benzene, and hexane. Although the response toward these gases was small compared to alcohols, an increase in conductivity was observed when their vapors were introduced into the chamber. At this moment, we assume that the response due to these gases should also be related to physical or chemical de-sorption of oxygen; however, their details are not understood.

12.7.2 Sensing Properties of Hybrid NP/NW Sensors

The GaN nanowire surface in our sensor devices is coated with SnO_2 or ZnO nanoparticles. As described in Section 12.3.2, GaN nanowires have a surface depletion region, which affects the conductivity of these nanowires. In this work, the nanowire and the nanoparticle material are chosen such that they have closely matching bandgap energy. Hence, photogenerated electron-hole pairs are produced in both the oxide nanoparticles and the GaN nanowires. We speculate that in presence of UV light, an analyte that injects electrons in the oxide nanoparticles on the GaN nanowire surface promotes photogenerated carrier separation in the GaN nanowire by trapping/recombining with the photogenerated holes in the NW at the surface. This leads to a subsequent increase in the nanowire photocurrent.

The proposed sensing mechanism explains the response of the hybrid sensor to the various gases tested in this work. Schematic representation of the proposed sensing mechanism for the hybrid metal-oxide nanoparticle/GaN nanowire sensors has been shown in Fig. 12.23.

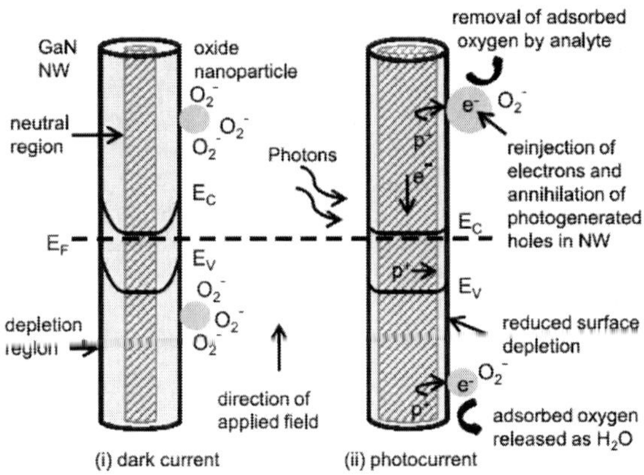

Figure 12.23 Schematic representation of the sensing mechanism of hybrid nanoparticle nanowire sensors.

12.7.2.1 Response to alcohols

The interaction of alcohols with ZnO surface has been discussed in Section 12.7.1.3. In this section, we will review the interaction of alcohol molecules with SnO_2 surface and then attempt to explain the response of nanoparticle/nanowire sensors toward alcohols.

The decomposition of alcohols on metal oxide surfaces in general, and SnO_2 in particular, has been widely studied. The adsorption and decomposition of alcohols on metal oxide surfaces could primarily proceed by (i) selective oxidation to aldehyde,[11,24] (ii) dehydration to form alkene and water,[11] and/or (iii) complete oxidation to carbon dioxide and water.[51,52] Selective oxidation to aldehyde seems to be the favored reaction on SnO_2 surface in the absence of catalytic additives. Thermal desorption spectroscopy of methanol on stoichiometric SnO_2 (110) surface detected formaldehyde and water as the only desorption products besides methanol.[53] A similar mechanism for dissociation of ethanol on the surface of single crystal and polycrystalline SnO_2 (110) to form acetaldehyde and water has been described elsewhere.[11] It has been suggested that gaseous ethanol can undergo a reversible dissociation to an ethoxy group and adsorbed hydrogen as

$$CH_3CH_2OH(gas) \rightarrow CH_3CH_2OH(ads.) \quad (12.1)$$

$$CH_3CH_2OH(ads.) \rightarrow CH_3CH_2O(ads.) + H(ads.) \tag{12.2}$$

$$CH_3CH_2O(ads.) \rightarrow CH_3CHO(ads.) + H(ads.) \tag{12.3}$$

$$CH_3CH_2O(ads.) \rightarrow CH_3CHO(gas) + H(ads.) \tag{12.4}$$

$$4H(ads.) + O^-_2(ads.) \rightarrow 2H_2O(gas) + e^- \tag{12.5}$$

Although the above reactions have been shown to proceed only at temperatures above 300 K, the possibility is that UV excitation in our case promotes these reactions on the device surface even at room temperature. Alternatively, alcohols could be dissociatively adsorbed as alkoxy group and hydrogen (step 2 above) without further dissociation to aldehyde at room temperature.[54] Through either route, hydrogen released in steps 2, 3, or 4 can react with the adsorbed oxygen to form water and desorb from the oxide surface as shown in step 5. This alcohol-assisted removal of adsorbed oxygen to release free electrons in the ZnO and SnO_2 nanoparticles manifests itself as the increased conductivity in the nanowire in the presence of alcohol vapors. When tested with various alcohol vapors, our devices show a sensitivity trend in the decreasing order of methanol > ethanol > n-propanol > isopropanol > n-butanol > isobutanol. This is opposite to the behavior observed with the SnO_2 thin films,[55] other metal oxide nanostructures such as tungsten oxide (WO_3) nanoplates[56] and ZnO nanowires as seen in this work, i.e., the sensitivity is the highest for n-butanol and the lowest for methanol.

At present, the reason for the reversal in trend from oxide thin film/nanowire to nanocrystalline coating is not understood. While the ZnO nanowire devices seem to follow the expected trend of decreasing sensitivity from butanol to methanol, we observe a reverse trend in the GaN NW devices functionalized with ZnO or SnO_2 nanoparticles.

12.7.2.2 Response to nitrogen

Response of the nanoparticle/nanowire devices toward nitrogen can be understood with the help of the response of the ZnO nanowire devices toward nitrogen. As observed in the case of ZnO nanowire devices, when ZnO and SnO_2 nanoparticles are exposed to nitrogen in presence of UV illumination, desorption of surface oxygen causes increased electron concentration in the nanoparticles hence increasing the photogenerated charge carrier separation in GaN nanowires. This is observed as an increase in the conductivity of the hybrid nanoparticle/nanowire devices. As the nanoparticles do not respond to nitrogen in dark, there is no change in the conductivity of the hybrid devices when exposed to nitrogen in absence of UV illumination.

12.7.2.3 Response to water

Various possible mechanisms for the interaction of water molecules with SnO_2 surface have been discussed in literature.[57] An increase in surface conductivity in SnO_2 thin films in presence of water has been experimentally proven.[58] The response of SnO_2 nanowires to H_2O at room temperature also shows an increase in conductivity when exposed to water vapors.[59] In our case, as with the SnO_2 nanowire, water should remove adsorbed oxygen on the surface of the metal oxide nanoparticles leading to an increased concentration of electrons in the nanoparticles and subsequent increase in the photocurrent of GaN NWs.

12.8 Modeling of Nanowire Devices

The bridge and the cantilever structures are the two of the most widely used structures; hence, modeling of nanowire cantilever and suspended bridge structures has been presented in this chapter. Modal and harmonic analysis has been performed for both the structures. Simulation and modeling has been used to study the possible application of these structures as frequency-based mass sensors.

12.8.1 Simulation of Resonant Frequency

Figures 12.24a,b show the model of ZnO nanowire cantilever and the bridge structure, respectively. The cantilever and the suspended bridge structure have been modeled as a nanowire of length 700 nm and the width and thickness of 100 nm each. To simplify the model, the ZnO nanowire is assumed to have a square instead of a hexagonal cross section. Manhattan mesh type as shown in Fig. 12.24 is used for the FEM analysis. The device is made on a silicon substrate using aluminum metal for the electrode structure.

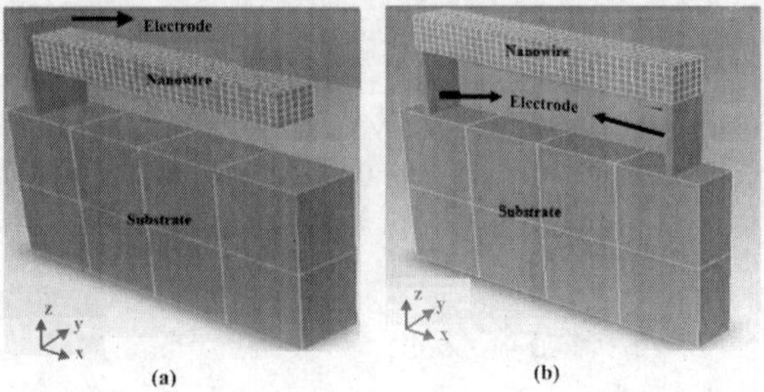

Figure 12.24 (a) Model of a nanowire cantilever structure with the FEM mesh. (b) Model of a nanowire bridge with the FEM mesh.

12.8.1.1 Modal analysis

Resonant frequencies for several modes of vibration for both cantilever and the bridge structure were evaluated using modal analysis. At these resonant frequencies, an undamped (lossless) mechanical structure responds to a bounded excitation with an unbounded response. Both the structures were actuated by using the piezoelectric property of ZnO. For the cantilever structure a 3 V voltage was applied between the top and the bottom surface of the nanowire, while for the bridge structure the voltage was applied at the two ends attached to the electrodes. The possible vibration modes for these structures could be longitudinal, torsional, and flexural. Some of the simulated modes of vibration for the bridge structure are shown in Fig. 12.25.

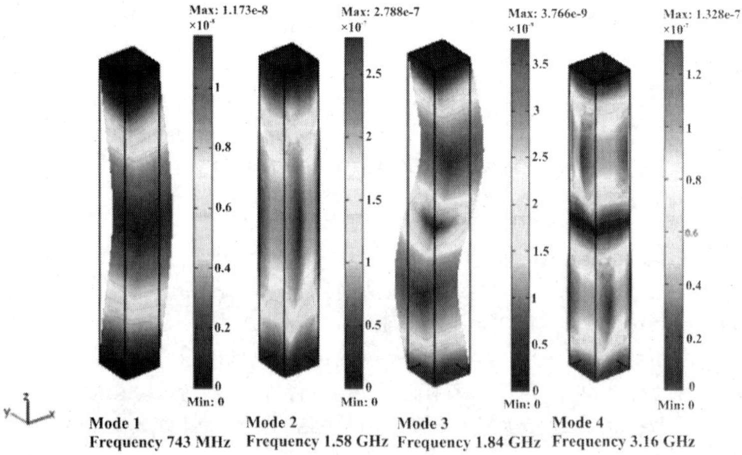

Figure 12.25 Modal shapes for the bridge structure.

12.8.1.2 Harmonic analysis

Harmonic analysis gives the dynamic response of the structure when loads with the sinusoidal time dependence are applied. In this case, harmonic voltage with amplitude 3 V, has been applied between the top and the bottom surface of the cantilever structure and at the two ends of the bridge structure. A plot for the displacement magnitude as a function of excitation frequency has been shown in Fig. 12.26 for the cantilever structure and Fig. 12.27 for the bridge structure. From the plot, we see that the resonant frequency is 1.7 and 1.74 GHz for the cantilever and the bridge structures, respectively.

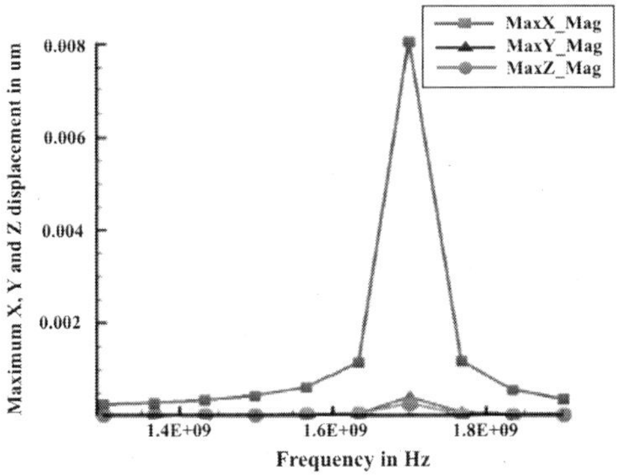

Figure 12.26 Harmonic response of the cantilever structure, length 700 nm.

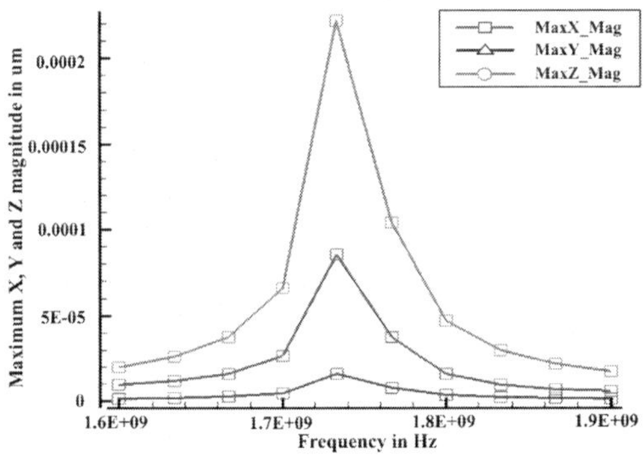

Figure 12.27 Harmonic response of the beam structure, length 800 nm.

From the frequency response, we also see that the dominant mode of vibration is longitudinal mode for the cantilever structure and transverse mode for the bridge structure.

12.8.2 Simulation of a Mass Sensor

The idea behind using such devices as a mass sensor is that as we add an external mass to the resonating structure, there should be a change in the fundamental resonant frequency of the device and this change should be dependent on the amount of deposited external mass. To simulate this effect, the cantilever and bridge structure model is modified by placing an external mass on the tip of the cantilever and in the middle of the bridge structure. Figures 12.28 and 12.29 show the frequency response of the device for the various amounts of mass loading for the cantilever and the bridge structure. As the quantity of externally deposited mass increases, the resonant frequency of vibration decreases and the amplitude of vibration increases.

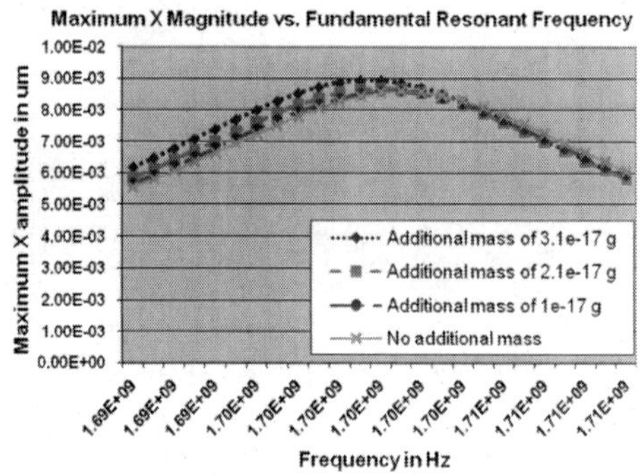

Figure 12.28 Frequency response of the cantilever structure with external mass loading.

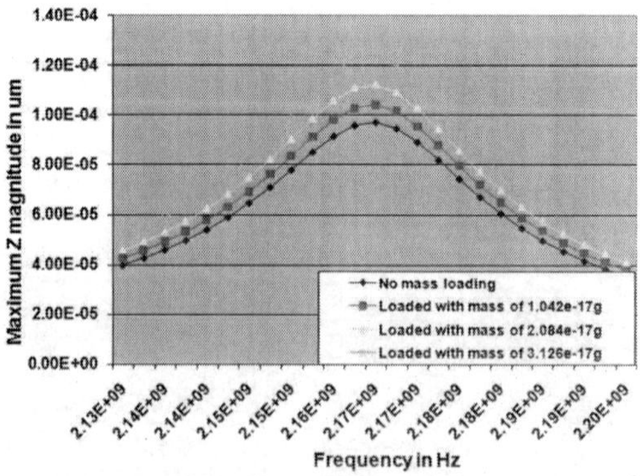

Figure 12.29 Frequency response of the suspended beam structure with external mass loading.

12.8.3 Theoretical Analysis

It has been shown that the nanoscale structures having dimensions in the order of hundreds of nanometers can be simulated using FEM modeling techniques as the principles of continuum mechanics still hold valid. Therefore, it should be possible to theoretically verify our results using the formulas from classical mechanics. In this section, we will theoretically verify the validity of the results obtained from simulation studies.

12.8.3.1 Fundamental resonant frequency calculation

Classical theory of mechanics is used to compute the fundamental resonant frequency of the nanowire cantilever and bridge structure. Longitudinal mode of vibration is the dominant mode of vibration for the cantilever structure as shown in the previous section. The resonant frequency of vibration for a cantilever in longitudinal mode is given by the following formula[60]:

$$f = \frac{2n-1}{4L}\sqrt{\frac{E_{zz}}{\rho}},$$

where n is a natural number indicating the mode number (i.e., $n = 1$ for the first mode, etc.), ρ (= 5343 kg/m^3) is the density and E_{zz} is Young's modulus (= 119.7 GPa). Substituting the given values and length, L of the nanowire as 700 nm, the value of the longitudinal resonance frequency is calculated to be 1.69 GHz, which is 0.6% off from the simulated frequency of 1.7 GHz as obtained from Fig. 12.26.

Fundamental resonant frequency for a beam with length L, thickness t, and material constants Young's modulus E and density ρ is given by[61]

$$f_0 = \frac{\lambda_0^2}{2\pi L^2}\sqrt{\frac{E}{12\rho_b}},$$

The value of coefficient λ_0 for bridge (both-end-supported beams) = 4.73. Density ρ_b for ZnO is 5680 kg/m^3[61] and Young's Modulus is 140 GPa.[62]

12.8.4 Mass Frequency Relationship

The fundamental resonance frequency can be related to the mass of the cantilever as[61]

$$f = \frac{1}{2\pi}\sqrt{\frac{k}{m^*}},$$

where m^* is the effective mass of the cantilever and k is the force constant for the cantilever. Theoretically, the relationship between change in frequency with addition of external mass can be expressed as[1]

$$\frac{\Delta f}{f} = -\frac{1}{2}\frac{\Delta m}{m}.$$

12.8.5 Comparison of Simulation and Theoretical Results

Comparing the shift in the fundamental resonant frequency, Δf with addition of external mass, $\Delta m = 1 \times 10^{-17}$ g, we see that for the cantilever structure $\Delta f = 0.22$ MHz from the calculations and it is 0.5 MHz from simulations. For the bridge structure, from the theoretical analysis we obtain $\Delta f = 0.279$ MHz, whereas from simulation $\Delta f = 0.125$ MHz. For both cases, the frequency shift Δf is found to be linear with respect to mass loading in simulation; however, a smaller sensitivity factor is observed. This sensitivity offset could be because the formula is for uniformly distributed mass on the structure, whereas we are placing the mass at the tip of the cantilever and on the center of the bridge.

12.8.5.1 Comparison of the cantilever and the bridge structure

Bridge structure has two advantages over the cantilever structure. The cantilever structure will need optical method for the detection of the output signal, whereas for the bridge structure the output signal can be measured directly in terms of an electrical signal using the network analyzer. Second, bridges show higher resonance frequency than cantilever structures for the given geometrical and material conditions.[61] Therefore the mass sensitivity of the bridge structures should be better than the cantilever structures. This can also be verified by simulation of the two structures in CoventorWare as shown in Fig. 12.30.

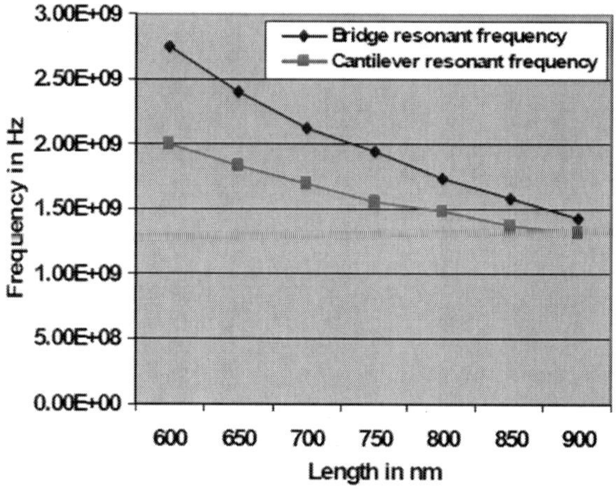

Figure 12.30 Frequency vs. length for the cantilever and the bridge structure.

Since the cantilever has only one end fixed while the other end is free, voltage is applied to the top and the bottom surface of the cantilever as compared to the bridge structure where we apply the voltage between the two ends of the nanowire, which makes it easier to measure the output signal electrically when the device is fabricated. Due to piezoelectric property of the ZnO when the voltage is applied to the top and bottom surface of the cantilever the fundamental frequency f_c is the longitudinal vibration frequency given by[63]

$$f_c = \frac{1}{4L}\sqrt{\frac{E}{\rho_b}}. \tag{12.6}$$

When the voltage is applied to the top and the bottom surface of the bridge structure, the fundamental resonant frequency is the flexural vibration frequency given by Eq. 12.6. As seen in Fig. 12.29 the difference between the fundamental frequency of the bridge and the cantilever structure will be more for the smaller dimensions, as the bridge frequency depends inversely on length square, whereas the cantilever frequency is inversely proportional to length. Since the smaller length will give us higher frequency and hence higher sensitivity, bridge structure is a better choice for this kind of proposed design of a mass detector.

12.9 Conclusion

This chapter presented the current role of semiconductor sensors in field of chemical sensing along with an insight into how the current research could produce some of the desired improvements in the gas-sensing technology. Finally, we summarize the important highlights and look into the possible applications of alcohol sensors presented in this work.

(1) Two-terminal NW devices can be used for conductivity-based gas sensing. They can also be used for making nanoresonators and frequency-based sensors.

(2) Alcohol sensors have been fabricated by functionalizing GaN nanowires with ZnO and SnO_2 metal oxide nanoparticles. This fabrication technique separates the sensing and transduction process in the sensor. The sensing task is performed by the nanoparticles, whereas the electrical transduction is achieved by the GaN nanowires. The device selectivity of the hybrid NP/NW gas sensor devices can be tuned by changing the functionalizing oxide without changing any of the fabrication steps. This makes these hybrid NP/NW sensors highly suitable candidates for making nanosensor arrays for a wide variety of chemical analytes.

(3) Room-temperature operation of semiconductor sensor devices provides greater flexibility in choosing the process steps during device fabrication compared to the sensors that require integrated on-chip heater for temperature-assisted operation. UV-assisted devices discussed in this work are low-power and highly integrable. They can be incorporated into portable electronic devices by on-chip integration of a UV led.

(4) Variation in the response of ZnO nanowire devices with variation in light intensity UV has been observed. It has been shown that the light intensity can be used to tune to sensitivity of UV-assisted sensor devices.

12.10 Applications

Alcohol sensors find major application in traffic safety and environment industry. For traffic safety, alcohol sensors are used in breathalyzers and in breath alcohol ignition interlock devices (BAIIDs). The ongoing research on traffic safety supported by Automotive Coalition for Traffic Safety and National Highway Traffic Safety Administration (NHTSA), is geared toward developing and demonstrating non-invasive in-vehicle alcohol detection technologies that can quickly and accurately measure a driver's blood alcohol content (BAC).[64] In environment industry, alcohol sensors are important for establishing the indoor and outdoor air quality.

References

1. Ziegler, C. Cantilever-based biosensors. *Anal. Bioanal. Chem.*, **379,** 946–959 (2004).
2. Lavrik, N. V., Sepaniak, M. J., and Datskos, P. G. Cantilever transducers as a platform for chemical and biological sensors. *Rev. Sci. Instrum.*, **75,** 2229–2253 (2004).
3. Raiteria, R., Grattarolaa, M., Buttb, H.-J., and Skládalc, P. Micromechanical cantilever-based biosensors. *Sens. Actuators B Chem.*, **79**, 115–126 (2001). doi:10.1016/S0925-4005(01)00856-5.
4. Huang, X.-J., and Choi, Y.-K. Chemical sensors based on nanostructured materials. *Sens. Actuators B Chem.*, **122,** 659–671 (2007).
5. Bajpai, R., et al. UV-assisted alcohol sensing using SnO_2 functionalized GaN nanowire devices. *Sens. Actuators B Chem.*, **171–172,** 499–507 (2012).
6. Bajpai, R., Motayed, A., Davydov, A. V., Bertness, K. A., and Zaghloul, M. E. UV-assisted alcohol sensing with zinc oxide-functionalized gallium nitride nanowires. *IEEE Electron Device Lett.*, **33,** 1075–1077 (2012).
7. Kanan, S. M., El-Kadri, O. M., Abu-Yousef, I. A., and Kanan, M. C. Semiconducting metal oxide based sensors for selective gas pollutant detection. *Sensors,* **9,** 8158–8196 (2009).
8. Meixner, H., and Lampe, U. Metal oxide sensors. *Sens. Actuators B Chem.*, **33,** 198–202 (1996).

9. Eranna, G., Joshi, B. C., Runthala, D. P., and Gupta, R. P. Oxide materials for development of integrated gas sensors—A comprehensive review. *Crit. Rev. Solid State Mater. Sci.*, **29,** 111–188 (2004).
10. Kohl, D. Function and applications of gas sensors. *J. Phys. Appl. Phys.*, **34,** R125 (2001).
11. Kohl, D. Surface processes in the detection of reducing gases with SnO_2-based devices. *Sens. Actuators,* **18,** 71–113 (1989).
12. Heiland, G. Homogeneous semiconducting gas sensors. *Sens. Actuators,* **2,** 343–361 (1981).
13. Tomchenko, A. A., Harmer, G. P., and Marquis, B. T. Detection of chemical warfare agents using nanostructured metal oxide sensors. *Sens. Actuators B Chem.*, **108**, 41–55 (2005).
14. Sheila, M. P., and Holland, C. *Inclusive Housing in An Ageing Society: Innovative Approaches* (2007). doi:10.1111/j.1741-6612.2007.00228.x.
15. Wang, Z. L. Zinc oxide nanostructures: growth, properties and applications. *J. Phys. Condens. Matter*, **16**, R829 (2004).
16. Chen, P.-C., Shen, G., and Zhou, C. Chemical sensors and electronic noses based on 1-D metal oxide nanostructures. *IEEE Trans. Nanotechnol.,* **7,** 668–682 (2008).
17. Korotcenkov, G. Metal oxides for solid-state gas sensors: What determines our choice? *Mater. Sci. Eng. B,* **139**, 1–23 (2007).
18. Chang, S.-C. Oxygen chemisorption on tin oxide: Correlation between electrical conductivity and EPR measurements. *J. Vac. Sci. Technol.*, **17**, 366–369 (1980).
19. Yoon, D. H., and Choi, G. M. Microstructure and CO gas sensing properties of porous ZnO produced by starch addition. *Sens. Actuators B Chem.*, **45**, 251–257 (1997).
20. Barsan, N., and Weimar, U. Conduction model of metal oxide gas sensors. *J. Electroceram.*, **7**, 143–167 (2001).
21. Sberveglieri, G., et al. Synthesis and characterization of semiconducting nanowires for gas sensing. *Sensors Actuators B Chem.*, **121**, 208–213 (2007).
22. Huang, J., and Wan, Q. Gas sensors based on semiconducting metal oxide one-dimensional nanostructures. *Sensors,* **9**, 9903–9924 (2009).
23. Kolmakov, A., and Moskovits, M. Chemical sensing and catalysis by one-dimensional metal-oxide nanostructures. *Ann. Rev. Mater. Res.*, **34**, 151–180 (2004).

24. Batzill, M., and Diebold, U. The surface and materials science of tin oxide. *Prog. Surf. Sci.,* **79,** 47–154 (2005).

25. Cabot, A., Diéguez, A., Romano-Rodríguez, A., Morante, J., and Bârsan, N. Influence of the catalytic introduction procedure on the nano-SnO_2 gas sensor performances: Where and how stay the catalytic atoms? *Sens. Actuators B Chem.,* **79,** 98–106 (2001).

26. Aluri, G. S., et al. Highly selective GaN-nanowire/TiO_2-nanocluster hybrid sensors for detection of benzene and related environment pollutants. *Nanotechnology,* **22,** 295503 (2011).

27. Dobrokhotov, V., et al. Principles and mechanisms of gas sensing by GaN nanowires functionalized with gold nanoparticles. *J. Appl. Phys.,* **99,** 104302–104302-7 (2006).

28. Lim, W., et al. Room temperature hydrogen detection using Pd-coated GaN nanowires. *Appl. Phys. Lett.,* **93,** 072109–072109-3 (2008).

29. Lüth, H. *Surfaces and Interfaces of Solid Materials.* Springer (1995).

30. Kind, H., Yan, H., Messer, B., Law, M., and Yang, P. Nanowire ultraviolet photodetectors and optical switches. *Adv. Mater.,* **14**, 158–160 (2002).

31. Soci, C., et al. ZnO Nanowire UV photodetectors with high internal gain. *Nano Lett.,* **7,** 1003–1009 (2007).

32. Calarco, R., et al. Size-dependent photoconductivity in MBE-grown GaN–nanowires. *Nano Lett.,* **5,** 981–984 (2005).

33. Sanford, N. A., et al. Steady-state and transient photoconductivity in c-axis GaN nanowires grown by nitrogen-plasma-assisted molecular beam epitaxy. *J. Appl. Phys.,* **107,** 034318 (2010).

34. Keller, D. Nano Field Effect Transistors as basic building blocks for sensing. PhD Thesis, University of Basel (2008).

35. Bertness, K. A., Roshko, A., Mansfield, L. M., Harvey, T. E., and Sanford, N. A. Mechanism for spontaneous growth of GaN nanowires with molecular beam epitaxy. *J. Cryst. Growth,* **310,** 3154–3158 (2008).

36. Bertness, K. A., Roshko, A., Mansfield, L. M., Harvey, T. E., and Sanford, N. A. Nucleation conditions for catalyst-free GaN nanowires. *J. Cryst. Growth,* **300,** 94–99 (2007).

37. Huang, C.-C., Pelatt, B. D., and Conley Jr, J. F. Directed integration of ZnO nanobridge sensors using photolithographically patterned carbonized photoresist. *Nanotechnology,* **21,** 195307 (2010).

38. Motayed, A., et al. Electrical, thermal, and microstructural characteristics of Ti/Al/Ti/Au multilayer Ohmic contacts to n-type GaN. *J. Appl. Phys.,* **93,** 1087 (2003).

39. Xu, M., Ng, V. M., Huang, S. Y., Long, J. D., and Xu, S. Growth of SiC nanoparticle films by means of RF magnetron sputtering. *IEEE Trans. Plasma Sci.*, **33**, 242–243 (2005).
40. Okumu, J., et al. Photochromic silver nanoparticles fabricated by sputter deposition. *J. Appl. Phys.*, **97**, 094305–094305-6 (2005).
41. De Lacy Costello, B. P. J., Ewen, R. J., Ratcliffe, N. M., and Richards, M. Highly sensitive room temperature sensors based on the UV-LED activation of zinc oxide nanoparticles. *Sens. Actuators B Chem.*, **134**, 945–952 (2008).
42. Peng, L., et al. Light induced enhancing gas sensitivity of copper-doped zinc oxide at room temperature. *Sens. Actuators B Chem.*, **131**, 660–664 (2008).
43. Prades, J. D., et al. Equivalence between thermal and room temperature UV light-modulated responses of gas sensors based on individual SnO_2 nanowires. *Sens. Actuators B Chem.*, **140**, 337–341 (2009).
44. Li, Y., Della Valle, F., Simonnet, M., Yamada, I., and Delaunay, J.-J. Competitive surface effects of oxygen and water on UV photoresponse of ZnO nanowires. *Appl. Phys. Lett.*, **94**, 023110 (2009).
45. Wan, Q., et al. Fabrication and ethanol sensing characteristics of ZnO nanowire gas sensors. *Appl. Phys. Lett.*, **84**, 3654–3656 (2004).
46. Bie, L.-J., Yan, X.-N., Yin, J., Duan, Y.-Q., and Yuan, Z.-H. Nanopillar ZnO gas sensor for hydrogen and ethanol. *Sens. Actuators B Chem.*, **126**, 604–608 (2007).
47. Sun, Z.-P., Liu, L., Zhang, L., and Jia, D.-Z. Rapid synthesis of ZnO nano-rods by one-step, room-temperature, solid-state reaction and their gas-sensing properties. *Nanotechnology*, **17**, 2266 (2006).
48. Hsueh, T.-J., Hsu, C.-L., Chang, S.-J., and Chen, I.-C. Laterally grown ZnO nanowire ethanol gas sensors. *Sens. Actuators B Chem.*, **126**, 473–477 (2007).
49. Feng, P., Wan, Q., and Wang, T. H. Contact-controlled sensing properties of flowerlike ZnO nanostructures. *Appl. Phys. Lett.*, **87**, 213111–213111-3 (2005).
50. Brauman, J. I., and Blair, L. K. Gas-phase acidities of alcohols. Effects of alkyl groups. *J. Am. Chem. Soc.*, **90**, 6561–6562 (1968).
51. Jinkawa, T., Sakai, G., Tamaki, J., Miura, N., and Yamazoe, N. Relationship between ethanol gas sensitivity and surface catalytic property of tin oxide sensors modified with acidic or basic oxides. *J. Mol. Catal. Chem.*, **155**, 193–200 (2000).

52. Park, S.-S., Zheng, H., and Mackenzie, J. D. Ethanol gas sensing properties of SnO_2-based thin-film sensors prepared by the sol-gel process. *Mater. Lett.,* **17**, 346–352 (1993).

53. Victoria, A., and Gercher, D. F. C. Oxygen-vacancy-controlled chemistry on a metal oxide surface: Methanol dissociation and oxidation on SnO_2(110). *Surf. Sci.,* 279–293 (1994). doi:10.1016/0039-6028(94)90072-8.

54. Bowker, M., and Madix, R. J. XPS, UPS and thermal desorption studies of alcohol adsorption on Cu(110): II. Higher alcohols. *Surf. Sci.,* **116**, 549–572 (1982).

55. Gong, H., Wang, Y. J., Teo, S. C., and Huang, L. Interaction between thin-film tin oxide gas sensor and five organic vapors. *Sens. Actuators B Chem.,* **54**, 232–235 (1999).

56. Chen, D., et al. The enhanced alcohol-sensing response of ultrathin WO_3 nanoplates. *Nanotechnology,* **21**, 035501 (2010).

57. Barsan, N., Schweizer-Berberich, M., and Göpel, W. Fundamental and practical aspects in the design of nanoscaled SnO_2 gas sensors: a status report. *Fresenius J. Anal. Chem.,* **365**, 287–304 (1999).

58. Henrich, V. E., and Cox, P. A. *The Surface Science of Metal Oxides.* Cambridge University Press, UK, 480 pp. (1996).

59. Kuang, Q., Lao, C., Wang, Z. L., Xie, Z., and Zheng, L. High-sensitivity humidity sensor based on a single SnO_2 nanowire. *J. Am. Chem. Soc.,* **129,** 6070–6071 (2007).

60. Lee, W., Jeong, M.-C., and Myoung, J.-M. Fabrication and application potential of ZnO nanowires grown on GaAs(002) substrates by metal–organic chemical vapour deposition. *Nanotechnology,* **15**, 254 (2004).

61. Kim, B. H., Kern, D. P., Raible, S., and Weimar, U. Fabrication of micromechanical mass-sensitive resonators with increased mass resolution using SOI substrate. *Microelectron. Eng.,* **61–62**, 947–953 (2002).

62. Sun, C. T., and Zhang, H. Size-dependent elastic moduli of platelike nanomaterials. *J. Appl. Phys.,* **93**, 1212–1218 (2003).

63. Reichel, E. K., Weiß, B., Riesch, C., Jachimowicz, A., Jakoby, B. A novel micromachined liquid property sensor utilizing a doubly clamped vibrating beam, *Proc. XX Eurosensors*, 17–20, (2006).

64. Driver Alcohol Detection System For Safety (DADSS). *Century Counc.* Available at http://www.centurycouncil.org/judicial-guide/driver-alcohol-detection-system-safety-dadss.

Chapter 13

Bimaterial Nanocantilever Beam Calorimeter for Atmospheric Pressure and Liquid Applications

Ioana Voiculescu,[a] Masaya Toda,[b] Takahito Ono,[b] and Fei Liu[a]

[a]*Mechanical Engineering Department, City College of New York, Convent Ave. at 138th street, New York, NY 10031, USA*
[b]*Department of Mechanical Systems and Design, Tohoku University, 6-6-01 Aramaki-Aza-Aoba, Aoba-ku, Sendai 980-8579, Japan*

voicules@ccny.cuny.edu

13.1 Introduction

A simple cantilever beam fixed at one side and free at the other side could be used as a calorimeter when it is fabricated from two different materials with different coefficients of thermal expansion. Materials used for cantilever beam calorimeter's fabrication are: silicon or silicon nitride and thin metallic films such as gold or silver. When the temperature changes the different values of the coefficient of thermal expansion of metal and silicon cause the sensor to deflect. The heat could be applied at the end of the calorimetric sensor or uniformly along the entire length of the sensor. The deflection of the bimaterial beam is a function of temperature. In order to enhance the cantilever beam sensitivity

Nanocantilever Beams: Modeling, Fabrication, and Applications
Edited by Ioana Voiculescu and Mona Zaghloul

ISBN 978-981-4613-23-1 (Hardcover), 978-981-4613-24-8 (eBook)
www.panstanford.com

to very small temperature variations the beam has to be very thin with the thickness at nanoscale dimensions. The dimensions of bimaterial cantilever beams are in the range: length 200 to 750 μm, width 20 to 40 μm and thickness 400 to 600 nm.

The bimaterial cantilever beam was first introduced as a calorimeter to measure the heat generated by chemical reactions with power and energy resolutions in the range of 1 nW and 1 pJ, respectively.[1,2] Another interesting application of bimetalic cantilever beams is as IR detector.[3–7] In this case, the bimaterial cantilever beam was demonstrated to measure power of 100 pW or energy of 150 fJ. Other applications include spatially resolved calorimetry,[8–12] ferromagneticresonance,[13] scanning thermal imaging probes,[14] thermomechanical data storage,[15–16] nanolithography,[17–20] and thermal actuators.[21–23] The cantilever beam calorimeter could operate at atmospheric pressure, in liquid, or in vacuum. In this chapter, the cantilever beam calorimeter behavior at atmospheric pressure and in liquid will be discussed. Formulas for the calculation of the nanocantilever beam deflection as a function of incident power applied at the free end or on the entire length of the nanocantilever beam operated in air and in the liquid will be presented.

13.2 Fabrication of Cantilever Beam Calorimeter

Possible fabrication process of composite cantilever beam calorimeter could start with a Si wafer with thin layers of SiO_2 deposited by wet thermal oxidation and SiN_x grown by low-pressure chemical vapor deposition (LPCVD).[24] Very thin layers of Cr and Au were deposited on the SiN_x layer by thermal evaporation. The thickness of the Au layer was 90 nm. The Cr layer with the thickness of 5 nm was used to enhance the adhesion between the Au layer and the SiN_x surface. To define the bimaterial cantilever beam structure, the Cr–Au layers were subsequently patterned by wet-etching method. To release the nanocantilever beam, the backside of the Si wafer was deep reactive ion etched (RIE) using CF_4 and O_2 gases. The last fabrication step was to detach the cantilever beam from the SiO_2 support using wet etching with buffered hydrofluoric acid (BHF) solution. A scanning electron microscopy (SEM) image of the bimaterial cantilever beam fabricated using this procedure is shown in Fig. 13.1. This long cantilever beam is curved upward, because of the imbalanced residual stresses (strains) in the different layers that form the cantilever beam.

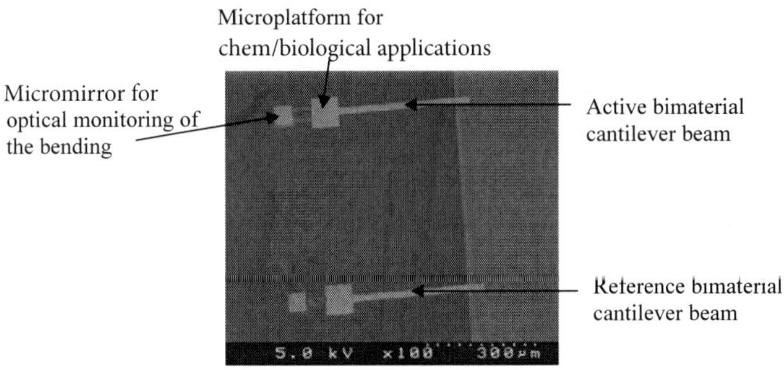

Figure 13.1 SEM image of the bimaterial cantilever beam calorimeter, fabricated from SiN_x and Cr–Au. Each cantilever beam was conceived with a microplatform for active measurements and a secondary microplatform for optical measurements of the deflection.[24]

Commercial atomic force microscope (AFM) triangular bimaterial cantilever beam (from Nano And More GmbH, European distributor for AFM probes)[25] was also successfully used as a calorimeter.[1] This AFM cantilever beam is presented in Fig. 13.2 and was fabricated from layers of Si_3N_4 and aluminum (Al). The dimensions are as follows: length 200 μm, width 20 μm each leg, and thickness 600 nm.

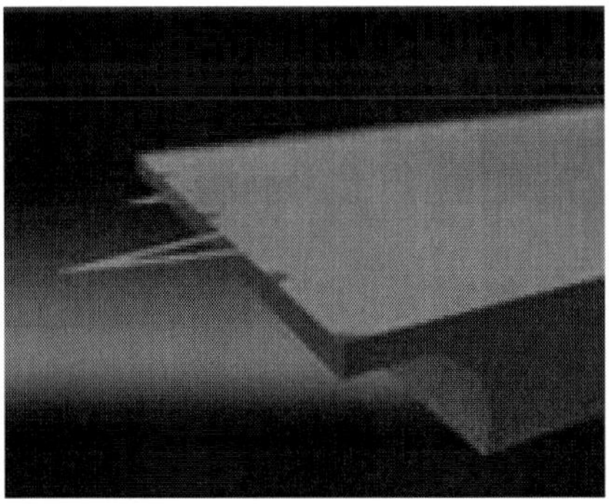

Figure 13.2 SEM image of a triangular commercial bimaterial cantilever beam fabricated from SiNx and Al and used for AFM applications.[25]

13.3 Thermal Model at Atmospheric Pressure with Heat Locally Applied at the Free End

The heat could be applied locally at the bimetalic cantilever beam tip or uniformly over the entire area of the cantilever beam. A laser diode could be used to apply a heat spot near the free end of the calorimeter. The cantilever beam deflection due to the heat applied only at the cantilever beam tip could be observed with the microscope.

The heat transfer analysis of the nanocantilever beam could be approximate as a fin conduction model. Equation 13.1 represents the governing equation. The temperature, T_∞ is the ambient temperature and x is the distance along the cantilever beam length.

$$\frac{d^2T}{dx^2} - m^2(T - T_\infty) = 0 \tag{13.1}$$

$$\text{where } m^2 = \frac{2\bar{h}}{\lambda_1 t_1 + \lambda_2 t_2} \tag{13.2}$$

In Eq. 13.2, $\bar{h}$ is the natural convective heat transfer coefficient, λ_1 and λ_2 are the thermal conductivity of Au and SiN_x, respectively, and t_1 and t_2 are the thicknesses of the Au layer and SiN_x layer, respectively. When Biot number, Bi < 0.1, the temperature distribution along the cross sections of a cantilever beam could be neglected and only the temperature distribution along the beam axis (1D) is considered. Due to the very small dimensions of the cantilever beam, the Rayleigh number is equal to 10^{-5}.[26,27] Thus most of the energy lost from the nanocantilever beam is through conduction into the air and not through convection. Therefore, the energy loss from the cantilever beam in this study was modeled as conduction into the surrounding air.

The boundary conditions for Eq. 13.1 are given by Eqs. 13.3 and 13.5:

$$-\bar{\lambda} w(t_1 + t_2)\frac{dT(0)}{dx} = P \tag{13.3}$$

$$\text{where } \bar{\lambda} = \frac{\lambda_1 t_1 + \lambda_2 t_2}{t_1 + t_2} \tag{13.4}$$

$$T(L) = T_0 \tag{13.5}$$

The solution to Eq. 13.1 can be expressed as follows:

$$T(x) - T_\infty = C_1 e^{-mx} + C_2 e^{mx}, \tag{13.6}$$

where C_1 and C_2 are constants and m is calculated using Eq. 13.2 and P is the power absorbed by the free end of the cantilever beam.

The constants C_1 and C_2 could be calculated by applying Eqs. 13.3 and 13.5 into Eq. 13.6, as:

$$C_1 = \frac{T_0 - T_\infty}{e^{-mL} + e^{mL}} + \frac{Pe^{2mL}}{\bar{\lambda} w(t_1 + t_2) m(1 + e^{2mL})} \tag{13.7}$$

$$C_2 = \frac{T_0 - T_\infty}{e^{-mL} + e^{mL}} - \frac{P}{\bar{\lambda} w(t_1 + t_2) m(1 + e^{2mL})}, \tag{13.8}$$

where, T_0 is the temperature at the clamped side of the cantilever beam, w and L are the width and the length of the cantilever beam, respectively.

Assuming $T_0 = T_\infty$, the temperature of the cantilever beam at the clamped side is identical to the ambient temperature. Then Eqs. 13.7 and 13.8 can be simplified as follows:

$$C_1 = \frac{P}{2m(\lambda_1 t_1 + \lambda_2 t_2)w} \frac{e^{mL}}{\cosh(mL)} \tag{13.9}$$

$$C_2 = \frac{-P}{2m(\lambda_1 t_1 + \lambda_2 t_2)w} \frac{e^{-mL}}{\cosh(mL)} \tag{13.10}$$

The temperature distribution along the bimaterial cantilever beam situated in a uniform ambient temperature bath and heated at the free end is expressed by Eq. 13.11.

$$T(x) - T_\infty = \frac{P}{m(\lambda_1 t_1 + \lambda_2 t_2)w} \frac{\sinh[m(L - x)]}{\cosh(mL)} \tag{13.11}$$

The ambient temperature, T_∞, was chosen equal to 293 K. The power applied at the free end was P = 4 mW. The natural convective heat transfer coefficient was $\bar{h}$ = 3900 W/m^{2}K. The temperature obtained at the free end using this model was 143°C as illustrated in Fig. 13.3.

Using the beam theory,[28,29] the deflection of the bimaterial strip with different expansion coefficients could be calculated with Eq. 13.12, which relates the beam deflection with the temperature variations.

$$\frac{d^2z}{dx^2} = 6(\gamma_2 - \gamma_1)\frac{t_1 + t_2}{t_2^2 K}[T(x) - T_0], \tag{13.12}$$

$$\text{where } K = 4 + 6\left(\frac{t_1}{t_2}\right) + 4\left(\frac{t_1}{t_2}\right)^2 + \frac{E_1}{E_2}\left(\frac{t_1}{t_2}\right)^3 + \frac{E_2}{E_1}\left(\frac{t_2}{t_1}\right) \tag{13.13}$$

The parameters γ_1 and γ_2 are the thermal expansion coefficients corresponding to the Au layer and SiN_x layer, respectively, and E_1 and E_2 are Young's modulus of the Au layer and SiN_x layer, respectively. The boundary conditions necessary to solve Eq. 13.13 are given by Eqs. 13.14 and 13.15.

$$z(L) = 0 \tag{13.14}$$

$$\frac{dz(L)}{dx} = 0 \tag{13.15}$$

Using Eqs. 13.4, 13.13, 13.14 and 13.15, the deflection of the cantilever beam free end could be calculated with Eq. 13.16.

$$z(0) = \frac{RP}{m^2(\lambda_1 t_1 + \lambda_2 t_2) w \cosh(mL)}\left[\frac{\sinh(mL)}{m} - L\right], \tag{13.16}$$

$$\text{where } R = 6(\gamma_2 - \gamma_1)\frac{t_1 + t_2}{t_2^2 K} \tag{13.17}$$

Using Eq. 13.16 the deflection z corresponding to different values of the power applied at the free end when the cantilever beam is operated in the air could be calculated. The power of the laser beam was calibrated. In this way, the operated power applied at the free end of the cantilever beam was known. Young's modulus decreases with the temperature, and for this reason, the variation $\Delta E/\Delta T$ of Young's modulus was considered in this calculation.

The value of the natural convective heat transfer coefficient $\bar{h}$ was approximated from literature.[27,28] In the thermal modeling, the value of the natural convective heat transfer coefficient $\bar{h}$ was chosen 1670 W/m²K at applied temperature close to room temperature and 3900 W/m²K at higher temperature. The values were chosen because they provided a good fit for the experimental measurements.

For the experimental measurements, the deflection z was calculated with Eq. 13.18:

$$z \cong \frac{L^2}{2R}, \tag{13.18}$$

where R is the radius of curvature of the cantilever beam. The bimaterial cantilever beam radius of curvature R was measured by fitting the cantilever beam curvature using the graphic program Digital Curve Tracer and Matlab.

In the thermal modeling of the temperature profile along the bimaterial cantilever beam, only the convection was considered. The cantilever beam length was 400 μm. In this study, the maximum power 4 mW applied at the free end of the bimaterial cantilever beam raised the device temperature to 143°C at the free end (Fig. 13.3). For this reason, the radiation was not considered in the governing Eq. 13.1 used for the temperature model.

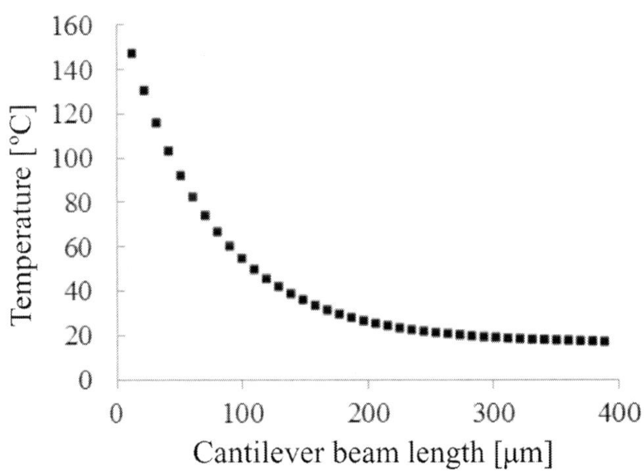

Figure 13.3 The temperature profile along the bimaterial cantilever beam. The ambient temperature, T_∞ was 20°C. The power applied at the free end was P = 4 mW. The natural convective heat transfer coefficient $\bar{h}$ = 3900 W/m²K. The temperature obtained at the free end using this model was 143°C.

13.4 Thermal Model at Atmospheric Pressure with Heat Uniformly Distributed on the Cantilever Beam Length

There are applications of cantilever beam calorimeter where the cantilever beam is uniformly heated on the entire length. One example is a commercial AFM cantilever beam heated evenly over the entire upper surface.[1] For experimental measurements the light was applied on the AFM cantilever beam with a 1.6 m WHeNe laser at 633 nm. The optical fiber used to guide the light onto the sensor had a diameter of 200 μm and the cantilever beam length was 200 μm. In this way, the entire sensor was uniformly illuminated and absorption occurred over its full length. When the power was uniformly distributed over the entire sensor, the temperature profile could be expressed with Eq. 13.19[1]:

$$(T-T_0)(x)=(l^2-x^2)\frac{P}{2lw(\lambda_1 t_1+\lambda_2 t_2)} \tag{13.19}$$

The beam deflection is formulated by Eq. 13.20:

$$\frac{d^2z}{dx^2}=\frac{5}{4}(\gamma_2-\gamma_2)\frac{t_1+t_2}{t_2^2K}\frac{l^3}{(\lambda_1 t_1+\lambda_2 t_2)w}P \tag{13.20}$$

To simplify the mathematical analysis, this AFM cantilever beam was modeled as a rectangular composite structure, although the real geometry was triangular with two narrow legs (illustrated in Fig. 13.2). The sensor was fabricated from S_3N_4 and Al. The parameters γ_1 and γ_2 are the thermal expansion coefficients, corresponding to Al and S_3N_4 layer, respectively. P is the total power absorbed by the sensor and in this case is distributed evenly over the entire sensor. The parameter K is a device parameter and λ is the thermal conductivity of the layers 1 and 2, and l and w are the length and width of the cantilever beam, respectively.

Experimentally, the power was uniformly applied on the cantilever beam sensor using the 1.6 mW HeNe laser at 633 nm and an optical fiber with the diameter of 200 μm. The absorbed power was 1.9 μW and the sensor deflection was experimentally measured as 115 Å. The experimental value of the sensitivity was 0.061 Å/nW. When a material absorbs a photon, a fraction of the energy may be transformed into heat. A measurement of the photothermal heating as a function of a wavelength can provide an absorption spectrum of the material. A very thin bimetalic microcantilever beam could be used as a sensor capable of detecting heat changes of picojoules. The small size of the sensor allows picograms quantities of materials to be studied. The spectroscopic method is highly sensitive and could use commercial cantilever beam fabricated for the AFM. A very small sample was deposited on the aluminum surface and produced heat during optical absorbtion. The experiments reported in the literature used latex spheres covered with fluorescent dye molecules and deposited on an AFM cantilever beam.[1] The mass of this sample was very small in the order of tens of picograms. This sample was illuminated by light and the heat was generated when the sample was illuminated with light at the absorption wavelength. The photothermal spectrum of the sample situated on the AFM cantilever beam displays a sharp pick at 460 nm.[1] This is in agreement with the absorption energy observed from the bulk sample that was studied with a spectrophotometer.

13.5 Thermal Model with the Cantilever Beam Immersed in the Liquid with Heat Locally Applied at the Free End

The bimorph cantilever beam calorimeter has been demonstrated as highly sensitive and sophisticated sensor for chemical and biological detection in liquid. To study the temperature distribution on the cantilever beam when it is operated in liquid, the bimorph cantilever beam was totally immersed in the liquid and heated with a microwire heater. The experimental values of the deflection were observed. In analogy with mammalian cell temperature measurement, the heater was situated at a distance of 15 μm from the free end of the cantilever beam. This experimental setup could mimic a mammalian cells floating in the cell media near the cantilever beam calorimeter. The thermomechanical response of the cantilever beam was used to determine the heat transfer coefficient from the cantilever beam to the liquid. The heat transfer coefficient has different values when the heater temperature varies from 28 to 72°C. At low temperatures, the value was 54 W/m^{2}K and at higher temperatures, 780 W/m^{2}K.

The experimental setup used for the study of the heat transfer along the bimaterial cantilever beam immersed in liquid is illustrated in Fig. 13.4. In order to clearly observe the cantilever beam's curvature produced by the temperature variations, the bimaterial cantilever beam was mounted on one side, inside a small experimental chamber containing water. The microheater (Veeco model 1610-00) fabricated from a platinum-rhodium micro wire with the diameter 5 μm, was positioned at 15 μm distance from the cantilever beam free end (Fig. 13.4). A commercial thermometer was also immersed in the experimental chamber in order to monitor the temperature of the liquid during the experiment. For this experimental setup, the chamber containing the cantilever beam was situated under an optical microscope with a CCD camera. The experimental setup and the microscope were placed in a room with controlled temperature, to avoid environmental temperature variations during the experiment. The microheater was mounted on a precision holder with the capability of micro positioning the heater in *x*, *y*, and *z* directions. The cantilever beam curvature was recorded with the video camera of the system. A side view micrograph of the cantilever beam used in this study and heated at the free tip with the microheater is shown in Fig. 13.4.

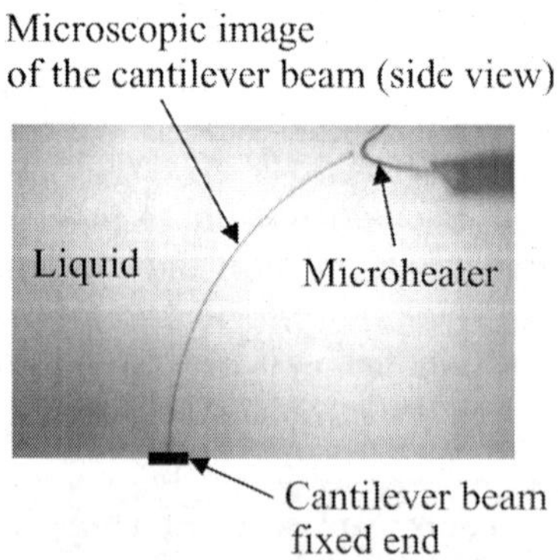

Figure 13.4 Microscopic imaging of the cantilever beam calorimeter heated with the microheater. The cantilever beam was immersed in water at room temperature and was positioned on one side. The heater was mounted on a micropositioner, for precise manipulation in *x*, *y*, *z* directions and positioned at 15 μm near the cantilever beam free end.

A schematic of the cantilever beam is illustrated in Fig. 13.5. The x-axis is set along the cantilever beam. The y-axis and z-axis are oriented in the directions of the width and thickness of the cantilever beam, respectively. To simulate the temperature profile along the cantilever beam (x-axis), we considered the following assumptions: The temperature profile of in the cantilever beam varies only along the length ($T = T(x)$), the fluid is at constant temperature (room temperature T_0) and the temperature of the cantilever beam fixed side is considered to be the same as the fluid temperature, the heat source is situated at $x = X$, the heat transfer away from the cantilever beam is characterized by the heat transfer coefficient, $\bar{h}$, which is independent of the position along the cantilever beam, and the heat transfer in the z direction can be ignored. With these assumptions, the cantilever beam can be modeled as a thermal fin; a similar concept was used for the model in the air.

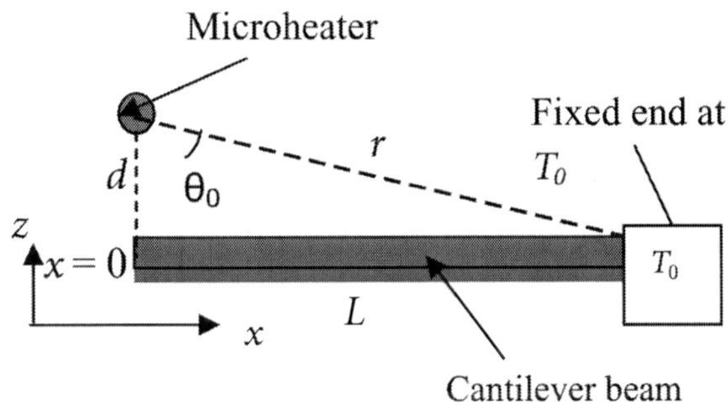

Figure 13.5 Illustration used for the theoretical calculation of the cantilever beam immersed in the liquid and the heater. The microheater is positioned closed to the free end of the cantilever beam. The distance between the microheater and the cantilever beam is d. The cantilever beam was immersed in the water and its deflection was observed on z direction.

The cantilever length was marked L. The distance d between the heat source and the cantilever beam is constant and equal to 15 μm.

As illustrated in Fig. 13.5, the cantilever beam was heated on the upper side by heat coming from the surrounding water that was heated by the heater. On the bottom side, the cantilever beam exchanges heat by convection.

Assuming that the heater is a point source, and the power P generated by the heater is constant, the heat conduction was expressed in spherical coordinates. The temperature distribution in the water, $T_{wo}(r)$ is expressed by Eq. 13.21.

$$T_{wo}(r) = \frac{P}{4\pi k_1 r} \tag{13.21}$$

$$\text{where } r = \sqrt{d^2 + x^2} \tag{13.22}$$

Here k_1 is the heat conductance of water. T_0 is the water temperature, when the water was not heated. In this study, T_0 was 23°C. Substituting Eq. 13.22 in Eq. 13.21, we get Eq. 13.23, which is the heat profile in the water. The temperature applied on the cantilever beam by the microwire heater located at a certain distance d from the cantilever beam tip, $T_{wo}(x)$, decreased along the cantilever length as expressed by Eq. 13.23. As we note, the temperature profile $T_{wo}(x)$ is expressed without the cantilever beam existence. The value X represents the distance between the heater and the tip of the cantilever beam.

$$T_{wo}(x) = \frac{P}{4\pi k_1 \sqrt{d^2 + (x - X)^2}} + T_0 \tag{13.23}$$

First, we will consider the situation that the heat source is situated at the apex of the cantilever beam ($X = 0$). Considering a small volume x of a length dx along the cantilever ($x = 0$ means the apex position on the cantilever), the conserved heat transfers inside, $\dot{E}_{in}$, and outside, $\dot{E}_{out}$, is expressed by Eqs. 13.24 and 13.25. Here, q_x and dq_c are the heat conduction in the cantilever beam and heat loss in water by convection at x position on the cantilever. On the other hand, dq_w is the heat transfer from water at x on the cantilever.

$$\dot{E}_{in} = q_x + dq_w \tag{13.24}$$

$$\dot{E}_{out} = q_x + \frac{\partial q_x}{\partial x} dx + dq_c \tag{13.25}$$

According to Fourier's law, the heat conduction could be expressed with Eq. 13.26:

$$q_x = -\bar{\lambda} A_c \frac{dT(x)}{dx} \tag{13.26}$$

where $\bar{\lambda}$ and A_c are the gravity center of thermal conductivity and the cross sectional area of the cantilever beam, respectively. When the temperature profile in the water near the cantilever is assumed as equivalent to the temperature profile in water, dq_w can be expressed by Fourier's law, Eq. 13.27.

$$dq_w = -k_1 \frac{dT_w(r)}{dr} wdx = \frac{P}{4\pi r^2} wdx = \frac{P}{4\pi(d^2 + x^2)} wdx \tag{13.27}$$

According to Newton's law, the temperature profile in water $T_{wo}(x)$ is introduced in the following equation of the temperature profile along the cantilever $T(x)$ using the heat transfer coefficient h.

$$dq_c = h(T(x) - T_{wo}(x))wdx \tag{13.28}$$

Using Eqs. (13.24–13.28), the governing equation is obtained as follows (Eq. 13.29):

$$\bar{\lambda} A_c \frac{d^2T(x)}{dx^2} + \frac{dq_w}{dx} - h(T(x) - T_{wo}(x))w = 0 \tag{13.29}$$

Equation 13.27 is introduced in Eq. 13.29, and the temperature profile on the cantilever is expressed in Eq. 13.30. Equation 13.30 is solved using a numerical method.

$$\frac{d^2T(X)}{dx^2} + \frac{P}{4\pi\bar{\lambda}t(d^2 + x^2)} - \frac{h}{\bar{\lambda}t}(T(x) - T_{wo}(x)) = 0 \tag{13.30}$$

The boundary conditions are as follows:

$$T(0) = T_w(d) = \frac{P}{4\pi k_1 d} + T_0 \tag{13.31}$$

$$T(\mathrm{L}) = T_0 \tag{13.32}$$

The beam deflection produced by temperature variations could be calculated with Eq.13.33, already demonstrated in Chapter 13.3.

$$\frac{d^2z(x)}{dx^2} = 6(\gamma_2 - \gamma_1)\frac{t_1 + t_2}{t_2^2 K}[T(x) - T_0] \tag{13.33}$$

$$\text{where } K = 4 + 6\left(\frac{t_1}{t_2}\right) + 4\left(\frac{t_1}{t_2}\right)^2 + \frac{E_1}{E_2}\left(\frac{t_1}{t_2}\right)^3 + \frac{E_2}{E_1}\frac{t_2}{t_1} \tag{13.34}$$

The cantilever beam deflection and the temperature distribution along the cantilever beam immersed in the are calculated with Eqs. 3.20 and 3.30, respectively. The values of the thermal conductivity in this research were as follows: for gold 320 $Wm^{-1}K^{1}$, for SiN_x 32 $Wm^{-1}K^{-1}$, and for water 0.6 $Wm^{-1}K^{-1}$. When the heater temperature was 63.4°C, according to Eq. 13.21, the temperature at 15 μm distance from the heater was equal to 37.5°C. This temperature was applied at the tip of the cantilever beam when Eq. 13.30 was used for the calculation of the temperature profile on the cantilever beam. Fig. 13.6a illustrates the temperature profile on the cantilever beam, $T(x)$, when 37.5°C was applied at the free end of the cantilever beam. The cantilever beam deflection corresponding to 37.5°C applied on the free end was 14.13 μm, $z(x)$ is shown in Fig. 13.6b. Here, the deflection of 15 μm is calculated at the apex of the cantilever, $z(0)$. The measured deflection of the apex was experimentally determined to be 15 μm, measured by the optical microscope.

In order to precisely calibrate the cantilever beam calorimeter, it is important to estimate the heat transfer coefficient $\bar{h}$, which characterizes the cantilever beam calorimeter operated in the liquid. To find the values for the heat transfer coefficient, $\bar{h}$, during the experiment, six different temperatures were applied on the microcantilever beam with the microheater situated at the distance of 15 μm from the cantilever beam tip. The temperatures produced by the microheater were 88, 74.5, 63.4, 53.5, 45, and 38.1°C (see Table 13.1). The experimental deflections corresponding to these temperatures were measured using the microscopic image of the cantilever beam side view and the graphic program Digital Curve Tracer. The experimental deflection values are given in Table 13.1. The values of the heat transfer coefficient $\bar{h}$ were tuned in order to approximate the experimental values and are presented in Table 13.1. The differences between the calculated deflection and the experimental deflection were insignificant.

At the temperature of 28°C applied on the free end of the cantilever beam, the $\bar{h}$ value was 178 W/m^{2}K. The temperature of 45°C was the highest temperature applied on the cantilever beam free end, and the corresponding $\bar{h}$ value was 1350 W/m^{2}K. Since this cantilever beam calorimeter was designed for biological experiments, temperatures higher than 45°C were not used in the experiment.

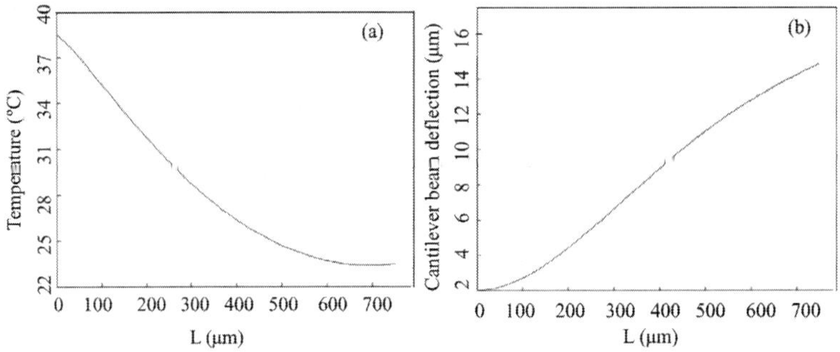

Figure 13.6 Thermal model of the heat temperature profile on the cantilever beam and the cantilever beam deflection when the heater temperature was 63.4°C, and the heater was positioned at 15 μm from the cantilever beam. (a) The temperature profile on the cantilever beam, when the temperature at the free end was 37.5°C. (b) Cantilever beam deflection generated by the temperature of 37.5°C applied at the free end was 14.13 μm.

Table 13.1 Estimated values of the heat transfer coefficient $\bar{h}$

T applied by the heater (°C)	*T* at free end of the beam (°C)	Experimental deflection (μm)	*h* (W/m^{2}K)
88.0	45	20	1350
74.5	41	15.40	1040
63.4	37.5	14.10	750
53.5	33.5	12.70	473
45.0	31	11.10	330
38.1	28	9	178

Note: The heater temperature was chosen between 38.1 and 88°C. The heater was situated 15 μm from the cantilever beam free end. The temperatures applied on the free end of the beam varied from 28 to 45°C. For the temperature range of this experiment, the heat transfer coefficient $\bar{h}$ values were between 178 and 1350 W/m^{2}K.

13.6 Conclusions

Bimetalic cantilever beams could be used as calorimeters. This type of cantilever beam calorimeter has the length at the microscale dimensions and thickness at the nanoscale dimensions in order to generate large deflection and be sensitive to very small temperature variations. They are considered nanocantilever beam. Because of its nanoscale dimensions, the cantilever beam calorimeter is considered in the category of nanocantilever beam and included in this book. This calorimeter could sense heat applied at the cantilever beam tip or on the entire area of the cantilever beam. In this chapter, the cantilever beam calorimeters operated at the atmospheric pressure and in the liquid were discussed. This technique has many applications such as temperature measurements, study of bacteria or individual cells reaction to a biochemical, and gas sensing when the heat is produced by specific gases to be detected, which react with the particular coating on the calorimeter. This type of calorimeter could be also used as a diagnostic sensor, in which the heat is generated by the cellular metabolic processes associated with the cell-antigen reactions.

When the cantilever beam was evaluated in the liquid, the cantilever beam calorimeter was immersed into water and the heater was situated at 15 μm from the free end of the cantilever beam. The heater temperature varied from 38.1 to 88°C and the cantilever beam deflection was measured using the microscopic picture and a graphic program. In biological applications, the temperature could not reach high values. That is why in the thermal modeling of the temperature profile along bimaterial cantilever beam, only the convection was considered. For the experimental testing, the applied temperature values were low and the heat radiation was not considered in the governing equation used for the temperature model. For the temperature range of this experiment, the values of the heat transfer coefficient $\overline{h}$ in the water were estimated to be between 178 and 1350 W/m^{2}K. Since the bimaterial cantilever is sensitive to small temperature variation applied by a heater positioned at a distance from the cantilever beam, we consider that this instrument could be used in the future to measure the temperature variations produced by a single cell when the cell is floating near the cantilever beam or is situated on the cantilever beam. In the air, the values of the heat transfer coefficient $\bar{h}$ were much larger and equal to 1670 W/m^{2}K at temperature close to room temperature and 3900 W/m^{2}K at higher temperature with values up to 200°C.

References

1. Barnes, J. R., Stephenson, R. J., Woodburn, C. N., et al. (1994). A femtojoule calorimeter using micromechanical sensors, *Rev. Sci. Instrum.*, **65** (12), 3793–3798.
2. Barnes, J. R., Stephenson, R. J., Welland, M. E., Gerber, C., and Gimzewski, J. K. (1994). Photothermal spectroscopy with femtojoule sensitivity using a micromechanical device, *Nature,* **372**, 79–81.
3. Varesi, J., Lai, J., Perazzo, T., Shi, Z., and Majumda, A. (1997). Photothermal measurements at picowatt resolution using uncooled micro-optomechanical sensors, *Appl. Phys. Lett.*, **71**, 306–308.
4. Datskos, P. G., Lavrik, N. V., and Rajic, S. (2004). Performance of uncooled microcantilever thermal detectors, *Rev. Sci. Instrum.* **75**(13.4), 1134–1148.
5. Datskos, P. G., Oden, P. I., Thundat, T., et al. (1996). Remote infrared radiation detection using piezoresistive microcantilever, *Appl. Phys. Lett.*, **69**, 2986–2988.
6. Van Neste, C. W., Senesac, L. R., Yi, D., and Thundat, T. (2008). Standoff detection of explosive residues using photothermal microcantilevers, *Appl. Phys. Lett.*, **92**, 134102.
7. Wig, A., Arakawa, E. T., Passian, A., Ferrel, T. L., and Thundat, T. (2006). Photothermal spectroscopy of Bacillus anthracis and Bacillus cereus with microcantilevers, *Sens. Actuators B Chem.*, **114**, 206–211.
8. Nonnenmacher, M., and Wickramasinghe, H. K. (1992). Scanning probe microscopy of thermal conductivity and subsurface properties, *Appl. Phys. Lett.*, **61**, 168.
9. Abedinov, N., Grabiec, P., Gotszalk, T., et al. (2001). Micromachined piezoresistive cantilever array with integrated resistive microheater for calorimetry and mass detection, *J. Vacuum Sci. Technol. A,* **19**, 2884.
10. Gotsmann, B., and Durig, U. (2004). Thermally activated nanowear modes of a polymer surface induced by a heated tip, *Langmuir*, **20**, 1495.
11. Berger, R., Lang, H., Gerber, C., et al. (1998). Micromechanical thermogravimetry, *Chem. Phys. Lett.,* **294**, 363.

12. Pinnaduwage, L. A., Wig, A., Hedden, D. L., et al. (2004). Detection of trinitrotoluene via deflagration on a microcantilever, *J. Appl. Phys.*, **95**, 5871.

13. Wallis, T. M., Moreland, J., Riddle, B., and Kabos, P. (2005). Microwave power imaging with ferromagnetic calorimeter probes on bimaterial cantilevers, *J. Magn. Magn. Mater.*, **286**, 320.

14. Nakabeppu, O., Chandrachood, M., Wu, Y., Lai, J., and Majumda, A. (1995). Scanning thermal imaging microscopy using composite cantilever probes, *Appl. Phys. Lett.*, **66**, 694–696.

15. King, W. P., Kenny, T. W. K. Goodson, K. E., et al. (2001). Atomic force microscope cantilevers for combined thermomechanical data writing and reading, *Appl. Phys. Lett.*, **78**, 1300.

16. Mamin, H. J., Fan, L. S., Rugar, D., and Hoen, S. (1995). Tip-based data storage using micromechanical cantilevers, *Sens. Actuators A: Phys.*, **48**, 215–219.

17. Wang, D. A., Bullen, J., Zou, C., Liu, J., and Mirkin, C. (2004). Thermally actuated probe array for parallel dip-pen nanolithography, *J. Vacuum Sci. Technol. B*, **22**, 2563.

18. Gotsmann, B., Duerig, U., Frommer, J., and Hawker, C. J. (2006). Exploiting chemical switching in a Diels-Alder polymer for nanoscale probe lithography and data storage, *Adv. Funct. Mater.*, **16**, 1499.

19. Sheehan, P. E., Whitman, L. J., King, W. P., and Nelson, B. A. (2004). Nanoscale deposition of solid inks via thermal dip-pen nanolithography, *Appl. Phys. Lett.*, **85**, 1589.

20. Nishimura, S., Takemura, Y., and Shirakashi, J. (2007). SPM local oxidation nanolithography with active control of cantilever dynamics, *J. Phys. Conf. Ser.*, **61**, 1066.

21. Lalinsky, T., Burian, E., Drzik, M., et al. (2000). Thermal actuation of a GaAs cantilever beam, *J. Micromechan. Microeng.*, **10**, 293.

22. Grigorov, A. V., Davis, Z. J., Rasmussen, P., and Boisen, A. (2004) A longitudinal thermal actuation principle for mass detection using a resonant micro-cantilever in a fluid medium, *Microelectron. Eng.*, **73**, 881.

23. Sulchek, T., Minne, S. C., Adams, J. D., et al. (1999). Dual integrated actuators for extended range high speed atomic force microscopy, *Appl. Phys. Lett.*, **75**, 1637.

24. Toda, M., Ono, T., Liu, F., and Voiculescu, I. (2010). Evaluation of bimaterial cantilever beam for heat sensing at atmospheric pressure, review of scientific instruments, *Rev. Sci. Instrum.*, **81**, 055104.

25. Commercial cantilever beam from Nanoworld: http://www.nanoworld.com/pyrex-nitride-triangular-silicon-nitride-tipless-cantilever-afm-tip-pnp-tr-tl.

26. Incropera, F. P., DeWitt, D. P., Bergman, T. L., and Lavine, A. S. (2006). *Fundamentals of Heat and Mass Transfer*, 6th ed., Wiley & Sons.

27. Serrano, J. R., Phinney L. M., and Rogers, J. W. (2009). Temperature amplification during laser heating of polycrystalline silicon microcantilevers due to temperature-dependent optical properties, *Int. J. Heat Mass Transfer*, **52**, 2255–2264.

28. Shen, S., Narayanaswamy, A., Goh, S., and Chen, G. (2008). Thermal conductance of bimaterial microcantilevers, *Appl. Phys. Lett.*, **92**, 063509.

29. Roark, R. J., and Young, W. C. (1975). *Formulas for Stress and Strain*, 5th ed., McGraw-Hill, New York.

Chapter 14

Advances and Challenges to Bring Nanomechanical Biosensors to Biochemistry Labs and Clinical Use

Priscila M. Kosaka, Javier Tamayo, and Monsterrat Calleja

Bionanomechanics Lab, Institute of Microelectronics of Madrid, CSIC, Isaac Newton 8 (PTM), Tres Cantos, 28760 Madrid, Spain

priscila@imm.cnm.csic.es

14.1 Introduction

The ability to measure the abundance of many proteins sensitively and simultaneously in real samples is vital to disease detection and monitoring. Immunoassays based on antibodies have been used over 30 years and are still among the most important diagnostic tools, which is widely used in medical diagnostics.

The need of label-free techniques with higher levels of sensitivity has boosted the development of an increasing number of label-free techniques based on nanotechnology such as optical whispering gallery microcavity sensors, plasmon resonance sensors based on nanostructures, electrical nanowire sensors, suspended microchannel resonators, and micro- and nanocantilever sensors, to name a few.[1–6] These developments were generally driven by

Nanocantilever Beams: Modeling, Fabrication, and Applications
Edited by Ioana Voiculescu and Mona Zaghloul

ISBN 978-981-4613-23-1 (Hardcover), 978-981-4613-24-8 (eBook)
www.panstanford.com

the necessity of low sample consumption, robustness, cost and portability. Moreover, the necessity of high sensitive techniques is needed for the early detection of cancer or other pathologies, because the early stages of disease are typically treated with the greatest probability of success. High-sensitivity will allow less invasive disease diagnosis and will offer less discomfort to the patients as well.

After the first work of Fritz and co-workers[7] showing the potential of microcantilevers as a relevant label-free clinical diagnostic technique in the year 2000, many papers were published showing, as a proof of concept, the label-free detection of proteins and DNA and the ability to sequence DNA by hybridization.[7–11]

Similarly, the cantilever array sensor can play a relevant role in the study of the tremendous complexity of the glycome. The identification and quantification of carbohydrate interactions with other biomolecules is important to understand the biological function of carbohydrates. To this end, Gruber and co-workers[12] described the development of cantilever array biosensors with a self-assembling carbohydrate-based sensing layer that selectively and sensitively detects carbohydrate–protein binding interactions.

Recently, arrays of microcantilevers coated with a specific oligonucleotide detected malignant melanoma. Huber and colleagues have focused on the detection of BRAFV600E mutations in melanoma. However, it is clear that this cantilever-based approach is applicable to mutations in other types of cancer such as gastrointestinal stromal tumors.[13] Another recent example of an interesting application of microcantilevers as biosensors was shown by Longo and co-workers[14]; they showed that the response of bacteria to antibiotics can be quickly assessed by monitoring the fluctuations of cantilevers coated with the.

The works mentioned were very useful and essential to understand and establish this new technology, but nanomechanical sensing has not yet been established as a practical alternative to well-established bioanalytical techniques such as ELISA, microarrays or electrophoresis methods. Finally, after almost 15 years, it is time to approach the microcantilever biosensors to the clinical use. In this chapter, we will discuss the progress made by the scientific community and the challenges that must be overcome to can find the nanomechanical sensors in every biochemistry laboratory and as a point-of-care technique in the future.

14.2 Biosensors Based on Nanomechanical Systems

Advances in micro- and nanofabrication technologies are enabling to achieve incredibly smaller mechanical transducers with micro- and nanosized moving parts whose deformation and vibration are sensitively modified upon molecular adsorption. This subcategory of mechanical biosensors is referred to as nanomechanical biosensors.[15] These devices can be mass produced by adopting well-established microfabrication techniques used in the semiconductor industry which results in a large number of identical devices in a relatively low cost and the small sensor area guarantees low reagent and sample volumes.

14.2.1 Modes of Operation

The nanomechanical sensors can be split into surface stress or static mode and dynamic mode (Figs. 14.1a,b, respectively). The resulting mechanical response due to the interaction with the targeted molecule is either a deformation or a resonance frequency shift, respectively.

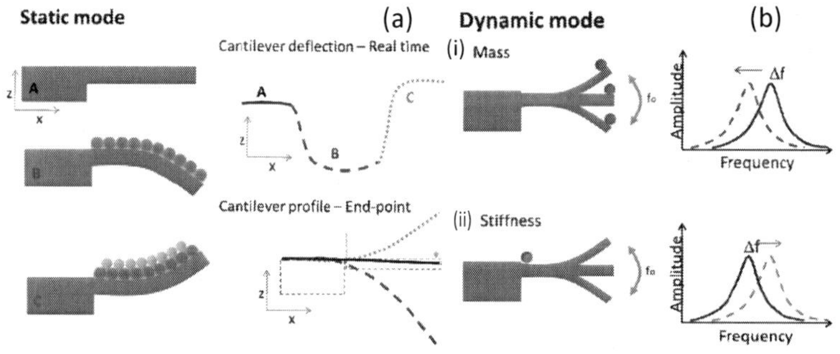

Figure 14.1 Main operation modes of nanomechanical sensors: (a) static mode and (b) dynamic mode. Both modes can operate in real time or as end-point detection assays.

Experiments with nanomechanical sensors can be designed to follow the mechanical response in real time (Fig. 14.1a) or, alternatively, the mechanical signals can be measured before and after incubation of the chips in the test sample to perform end-point detection of an analyte (Fig. 14.1b). Detection in real time provides information about the kinetics of the biomolecular interactions, while end-point detection is usually the preferred choice for clinical applications because of its simplicity, possibility of high-throughput and suitability for their use at the point-of-care.[16]

Surface stress biosensors are based on the deformation of the cantilever surface that can are usually related to changes in the surface stress arising from intermolecular forces and charge reorganization.[17] In surface stress sensing applications the functionalization of the cantilever surface with a sensitive layer for molecular recognition must be performed on just one side of the device. Thus, the asymmetrically structure will bend when the target molecules interact with the sensitized surface of the cantilever, resulting in a measurable mechanical deflection of the cantilever. Biomolecular recognition generates small surface stress (0.001–0.01 N m^{-1}), which is equivalent to cantilever deflection of few nanometers that can be easily measured by optical and electrical methods. The expansion of a surface is defined as compressive surface stress and contraction as tensile surface stress.

In the dynamic mode, cantilevers as excited close to their resonance frequency. The resonance frequency change shifting occurs due to the added mass of the analyte on any of the sensor surfaces, usually a flexural-mode beam resonator, and in many cases it can also be influenced by changes in the mechanical stiffness of the adsorbate-beam system.[18]

14.2.2 Immobilization of Receptors

To turn a nanomechanical system into a biosensor, it has to be coated with a biorecognition element. These are usually biomolecules such as nucleic acids, proteins, and carbohydrates. The functionalization of the cantilever surface comprises (i) the chemically modification of the surface if it does not already contain appropriate reactive functional groups to bind the biorecognition element, (ii) the biorecognition element that must be immobilized onto the surface by reaction with surface chemical groups and should ensure the natural activity of the receptors, its bioavailability to targets and the formation of stable, robust, homogeneous and reproducible sensing layers and, finally, (iii) if bare areas remain on the cantilever surface, they must be blocked to avoid non-specific adsorption of undesired molecules (Review Tamayo).

When working with nanomechanical biosensors there are some other requirements necessary to ensure the success of the recognition of the targeted molecule. The biorecognition layer on the cantilever surfaces should be (i) thin, to avoid changes in the mechanical properties of the cantilever, (ii) uniform, to generate a uniform stress, (iii) compact, to avoid interactions with the underneath solid substrate and (iv) stable and physically strong against changes in buffer and temperature.[7]

Many approaches can be used to immobilize the biorecognition agents to the microcantilever surface. The most employed methods are self-assembled monolayers (SAM) on gold surfaces and organosilanization of silicon surfaces.

Intensive literature deals with the formation, properties, and applications of SAM of alkylthiolates on metals. The high affinity of thiols (R-SH) for metals (gold, silver, copper, palladium, platinum, and mercury) drive the formation of well-organized surfaces with useful chemical functionalities displayed at the exposed interface for biosensor applications. Gold is usually the surface material of choice due to its chemical inertness and the well-defined structure of the obtained film densely packed and ordered array of long chain molecules. Gold does not oxidize at room temperature, does not react with atmospheric oxygen, and is biocompatible. Normally, the top cantilever surface is coated with a thin layer of chromium or titanium (2–3 nm) and 20–30 nm of gold by either sputtering or thermal evaporation. Chromium and titanium act as adhesion layers for the gold. The gold surface on the top side of the cantilevers is then easily modified by thio-labeled nucleic acids or proteins exposing cysteins at their surfaces. Proteins that do not bear cysteine residues can be genetically modified to engineered site-specific thiol groups distal from the active sites of the molecule.[19]

Silicon oxide is present on the surface of the most common materials used for fabrication of nanomechanical systems such as silicon and silicon nitride. Silicon oxide surface avidly adsorbs water molecules producing silanol groups (Si-OH) with high density.[20,21] These groups are used in silanization methods as connection sites that react with organosilicon molecules forming a silicon-oxygen-silicon carbon linkage. This bonding is thermodynamically and hydrolytically stable. The organosilicon molecules are composed of a small organic group such as methyl, a reactive group such as chloride, methoxy, and ethoxy, and a terminal group that can be replaced with different functional groups like amino, carboxyl, or thiol group that provides the desired properties for the modified surface.[22] The simplicity of the reaction and the high stability of the product make this method a customary surface modification route. The silanization reaction is usually performed dipping the surfaces into a solution containing the organosilicon molecule and both sides of the cantilever are functionalized. Therefore, this method is more suitable for biosensors based on nanomechanical resonators that do not require the functionalization asymmetry needed in nanomechanical biosensors based on the surface stress. As the asymmetry of the functionalized top surface and the bottom surface is especially important for the static mode, then one must think in a way to confine the biorecognition molecules on only one side of the structure.

Since cantilevers are sensitive to the slightest side effects such as temperature variations, ionic strength of a bath, and mechanical bending by hydrodynamic forces, it is difficult to clearly correlate the deflection signal of a single cantilever with specific binding events.[23] The best strategy to circumvent the competitive effects is to use an array of cantilevers where one or more cantilever surfaces can be coated with different molecules and at least one of them will be used as reference cantilever. One should take into account that the choice of coating for the reference cantilever should not only be governed by its inertness with the reaction of interest, but also by the similarity of its elastic properties to the elastic properties of the sensing layer.[24] To analyze the data, the nonspecific signal from the reference cantilever is subtracted from the specific signal from the sensing cantilever; this is called differential sensor response. By determining the differential deflections, effects such as nonspecific interactions are significantly minimized and the sensitivity can be optimized.

14.3 High Throughput Measurements

An important bottleneck in the nanomechanical sensing technology lays in the improvement of its throughput. The lack of large datasets acquired with nanomechanical sensors is present in most of the already demonstrated applications. Large numbers of experiments are necessary to push the technology beyond the mere proof of concept.[16] An important step in this direction was when Fritz and co-workers in the year 2000 used an array of eight cantilevers to detect different DNA sequences.[7] Unfortunately, after almost 15 years, the throughput of nanomechanical biosensing technologies has hardly gone much further. Arrays of eight cantilevers or less are still the first option used by scientists worldwide to follow applications expanding from gas detection to polymer studies and microbiology.

The difficulty to perform high-throughput experiments with cantilevers arises not only in the sensor fabrication, but also in sensor incubation in diverse test samples and their massive parallel readout. Achieve high throughput for the measurements using nanomechanical biosensor can enable the evaluation of large numbers of samples in a short time and different targeted analytes can be assessed in a single assay. In this section, we will discuss the advances made by scientific community to improve throughput and multiplexing capability for the nanomechanical sensing technology.

14.3.1 Functionalization of Microcantilever Arrays

The method used for cantilever coating should be fast, reproducible, and reliable and allow differential coating of the cantilevers in an array. Commercial microcantilever chips can be easily handled in serial experiments and the most straightforward and used approach has been the incubation of each single chip containing an array of 2–8 cantilevers in a microcapillary. A few other solutions have been proposed for the functionalization of cantilever sensors with a biolayer and they will also be mentioned here.

Using the microcapillaries, each microcantilever of the array is inserted into disposable glass microcapillaries filled with the liquid containing the probe molecules (Fig. 14.2a) and the entire array is functionalized in parallel. This method has been largely applied for the deposition of thiol-functionalized single-strand DNA[7,10,25,26] and in a few cases for proteins.[27] However, it is challenging to avoid evaporation and retraction of the liquid from the cantilevers for typical immersion duration of 20 min.[23] Another disadvantage of this technique is the manual alignment of the microcantilever array and the microcapillary functionalization tool, so the technique is not viable for coating a large number of cantilever arrays. Additionally, the reaction chemistry of the fluid in the microcapillaries is not designed to react only with one region of the cantilever. The two sides of the cantilever are wetted using this approach, making this technique not suitable when the cantilever surface is modified using the organosilanization. For surface stress mode sensors, this is a limitation, since the asymmetry between the top and bottom surfaces is required.

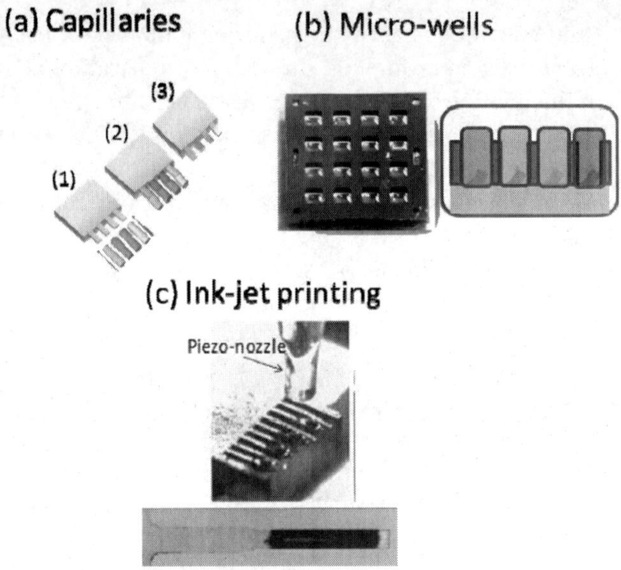

Figure 14.2 Sample dispensing and surface functionalization strategies for nanomechanical sensors. (a) Capillaries used to individually, (b) microwells matching the design of an array of 128 microcantilevers in 16 groups of 8 cantilevers each, and (c) ink-jet printing for cantilever functionalization. Different parts of the figure are adapted with permission © *Nature Nanotechnology* 2009, 4, 179, copyright 2009.

More recently, a chip consisting of 16 separate wells (Fig. 14.2b) was designed with pitch dimensions matching the SBS (Society for Biomolecular Sciences) standards for compatibility with existing automated dispensing devices. The wells here were also large enough for alternative hand micropipetting of solutions. Each well can be filled with approximately 2 μL of solution. The design was matching a two-dimensional cantilever array comprising 128 cantilevers distributed in the 16 reaction wells.[28] Using this chip, the user can study 16 different samples, make trials with duplicates, and triplicate or perform statistical studies in a single experiment.

A technique that coats many cantilever sensor arrays in a rapid and reliable way is the inkjet spotting[23,29] (Fig. 14.2c). In this technique, an *x-y-z* positioning system allows the positioning of a fine nozzle with an accuracy of 10 μm over a cantilever. Individual drops can be dispensed at the desired region of the microcantilever, or if the droplets are spotted with a pitch smaller than 0.1 mm, they merge and form a continuous film. By adjusting the number of droplets spotted on the cantilever, the film thickness can be controlled precisely, and as a result, a very homogeneous and reproducible film can be obtained within seconds. Inkjet spotting affords efficient and controlled functionalization of only one side of the cantilever, which is very important when cantilevers need to be coated with proteins or other layers that adsorb non-specifically on both sides when immersed. Prado and co-workers[30] showed that inkjet print with small modifications is suitable for the deposition of proteins. If necessary, the inkjet system can be operated in a controlled humidity environment, decreasing the evaporation rate of the deposited drops and allowing longer adsorption times. Also, a controlled humidity environment is desired for maintaining the hydration of the adsorbed layer and opens the possibility of generating inexpensive and dense bioarrays such as phage libraries, bacterial artificial chromosome libraries, or cell-based sensors.[30] The success of the inkjet technique for the coating of self-assembled single-stranded DNA[23,29] and for a more accurate way to position cells on microcantilevers for the microbial growth detection for clinical application[31] has been demonstrated. In conclusion, inkjet spotting has proved to be a very efficient, accurate, and versatile method for functionalization that can be used to coat any flat surface and can be applied to all shape sensors.

14.3.2 Read-Out of Microcantilever Arrays

Regarding the readout of arrays of sensors, nanomechanical sensor can be measured by optical and electrical techniques.

Among electrical techniques, piezoresistive detection is the most widespread. This detection requires the integration of a piezoresistive element during the fabrication of the nanomechanical system, which is usually located near the clamping regions. The electrical detection techniques impose high complexity in the fabrication and major restrictions in the size, geometry, and materials of the sensors. However, electrical detection provides an impressive degree of miniaturization, which is crucial for point-of-care applications. Unfortunately, electrical readout has not been extensively used in liquid, due to the difficulty of encapsulation of the sensor circuitry.

The optical lever method is the most widely used in biosensing applications of nanomechanical sensors.

Both electrical and optical detection techniques at the state of the art can achieve comparable displacement sensitivities in the range of 1–1000 fm $Hz^{-1/2}$ and bandwidths of hundreds of MHz. The electrical detection techniques impose higher complexity in the fabrication and major restrictions in the size, geometry, and materials of the sensors. However, electrical detection provides an impressive degree of miniaturization, which is crucial for point-of-care applications. The method consists of measuring the deflection of a laser beam reflected off the nanomechanical system surface by means of a position-sensitive photodetector. In 2005, an alternative laser-leveraging readout was proposed.[32] It is based on mounting either the laser or the sample in voice-coil actuators, so the displacement of the beam on the surface in two dimensions allows to attain a three dimensional picture of the surface and therefore, the deformation of the cantilevers with respect to the clamping surface in a variety of environmental conditions such as measurements in liquids or in mixtures of gases.

In 2010, an instrument able to address cantilever arrays of any pitch and cantilever number in a fast and reliable manner was developed.[28] Here the alignment of the laser beam is made automatically and can detect the position of the chip and the correction for any misalignment. Besides, the instrument provides a readout rate of tens of cantilevers per second, and it can be applied to any pitch and size of the cantilevers. The absolute displacement of any point on the cantilever surface can be recorded with respect to the fixed clamping region, so comparative successive measurements can be performed after removal of cantilever chips from the measurement platform. It is therefore possible to record the real profile with respect to the clamping region and repeat the measurements after further sensitization or molecular recognition experiments are performed outside the measurement set-up. This allows the ex-situ analysis, which is the preferred choice for point-of-care applications. Recently, the relative displacement of the laser beam and the surface has also served to read cantilever arrays with a DVD head.[33]

For further information, we refer the reader to several excellent reviews.[17,35]

14.4 Specificity and Sensitivity of Nanomechanical Biosensors

During the past decades, researchers sought the miniaturization of nanomechanical devices and the reduction of their noise levels to the thermomechanical noise floor. The thermomechanical noise determines the ultimate detection limits of nanomechanical biosensors in their different operation modes: static[36] and dynamic.[37] Although thoroughly understanding the physics behind the nanomechanical detection and improvement of the sensor performance are both key tasks, looking for the perfect nanomechanical biosensor may be sterile if the inherent variability of biological processes is disregarded.[16] One should be aware that when nanomechanical biosensors functionalized with bioreceptors are dipped into complex solutions such as urine or serum to detect the presence of biomarkers the ultimate detection limit can be far from the real-detection limits.

In this section, we will discuss the results obtained in a comprehensive analysis of the surface stress generated on the surface of microcantilevers as a consequence of the analyte recognition. We will explore the key points that should be taken into account for the development of nanomechanical biosensors with high level of confidence.

14.4.1 A Statistical Approach to Achieve Highly Sensitive and Specific End-Point Detection of Immunoreactions by Nanomechanical Biosensors

Recently we have performed for the first time a statistical study of the potential of nanomechanical biosensors based on surface stress for the label-free and end-point immunoassays.[38] The vision was to join the features that these devices demonstrated in liquid with the simplicity and high-throughput of end-point bioassays, which is the preferred choice in the biomedical field. Horseradish peroxidase (HRP) was chosen as the model antigen because of its small size (about 4 nm) to assess the capability of the technique for specific detection of small molecules. The study comprised the analysis of 1012 microcantilevers and more than 3000 values of surface stress were generated during the functionalization, detection, and control experiments. The measurement of a significant number of points was performed to define the noise level of the system, which will depend on the receptor layer of choice, the background medium, the device intrinsic mechanical properties, and surface characteristics. It is important to highlight that no microcantilever used in this extensive study was discarded.

The focus of this study is the variability of the sensor response rather than on the determination of impressive detection limits. In fact, until now any work in the literature has not addressed the rate of false positives and false negatives of nanomechanical biosensors.

14.4.1.1 Choosing the best blocking strategy

Ideally, the blocking molecule should: (i) show negligible interactions with other non-sought molecules present in a biological fluid, (ii) avoid the non-specific adsorption of the antigen, (iii) not hinder the capture of the analyte by the antibody, and (iv) strongly bind to the substrate to prevent exchange reactions. Bovine serum albumin (BSA) and poly(ethylene glycol) (PEG) are one of the most used blocking molecules and were used as models to find the best blocking strategy to avoid non-specific interaction on nanomechanical biosensor. BSA is a blocking molecule traditionally used in label-based immunoassays with a reported blocking efficiency of about 90%. PEG is a synthetic polymeric molecule that recently has been used for preventing nonspecific adsorption and it is considered among blocking agents as one of the most resistant to protein adsorption.

Before analyzing the effect of the molecular recognition between the surface-anchored antibodies and the antigen, let us first analyze the statistics of surface stress values obtained in the control experiments and can find the best blocking strategy to avoid nonspecific adsorption. This is a crucial step for optimizing the sensitivity and specificity of the assay, and it is particularly vital in label-free biosensors, in which the transduction physical signal may interpret non-specific adsorption as molecular recognition events (false positive).

Two control assays were tested. In the first control, an antibody nonspecific to HRP was immobilized on the surface. In the second control, the anti-HRP was immobilized on the surface and the sample was exposed an enzyme smaller that the HRP.

The cantilever response to non-specific antibody-antigen interactions (data not shown) is of -2.8 ± 6.4 mN/m for the blocking based on BSA, whereas the cantilever response is significantly smaller, -0.86 ± 2.05 mN/m, for the cantilevers blocked with PEG. The results indicate a three times better blocking efficiency of the PEG with respect to the BSA. This feature is a result of two effects: (i) BSA is a larger molecule (prolate ellipsoid with dimensions of 14 nm × 4 nm) than the HRP and it is expected the generation of gaps of few nanometers during the random chemisorption of BSA on the cantilever surface where small molecules can non-specifically bind and (ii) PEG achieves a grafting density of the order of 10^{13} cm^{-2} and adopts a brush conformation that efficiently repels non-specific adsorption.[39]

14.4.1.2 The influence of antibody surface density on the specificity of nanomechanical biosensors

The density of biomolecular receptors plays a crucial role in the sensitivity and specificity of the cantilever bending response to the subsequent molecular recognition reactions.[7,28,38,40] The cantilever response for three antibody concentrations (5, 50 and 250 μg/mL) was characterized and the surface density for each concentration was measured by radiolabelling. For the lowest concentration (5 μg/mL) the antibody surface density found was of about 2200 molecules per μm^2, whereas for 50 μg/mL the surface density increased to 6400 molecules per μm^2. For the highest antibody concentration (250 μg/mL), the surface density slightly increased indicating that the saturation of the surface was achieved.

Figure 14.3 shows the mean and standard deviation values of the surface stress induced due to the recognition of 1 μg/mL of HRP and the control experiments for the BSA and PEG blocking strategies. The results offer information that will be discussed in detail below:

- The molecular recognition of HRP is evidenced as a surface stress distribution wider than that of the control experiments.

However, the surface stress distribution for the HRP and control experiments partially overlaps. This overlapping is much higher when BSA-based blocking strategy is used, as a consequence of the significant nonspecific adsorption discussed in the previous section.

- The second piece of information that gets our attention is the fact that the surface stress induced by molecular recognition exhibits a high variability and can be either positive or negative for similar assays. A significant source of noise, so-called adsorption surface stress noise, may play a dominant role in some experimental situations. This noise arises from the randomness of the adsorption processes and the steep distance-dependence of the intermolecular interactions when the molecules are near contact. This source of noise, so far ignored, must be accounted for in order to develop reliable label-free surface stress biosensors. Moreover, since the surface density undergoes fluctuations between experiments, the steep variation of the mean surface stress with the surface density at high-packing (that is the situation that one finds to attain a highly responsive surface stress biosensor) amplifies the noise of the surface stress induced by physisorption and molecular recognition processes.[38]
- For the assays in which the lowest antibody concentration was used (5 μg/mL), the cantilever response was not significantly different from that of the control assays especially when BSA blocking is used. This can be related to the low surface density of antibodies that causes negligible changes in the intermolecular interactions on the cantilever surface when the antigen is captured. When the antibody concentration is increased to 50 μg/mL, the highest antigen response was found for both blocking strategies. The results indicate that a dense packing of the antibodies near, but below saturation, is an indispensable condition to obtain a good cantilever response to the antigen binding. As expected, saturation of the surface with antibodies enormously degrades the performance of the biosensor due to the crowding or overlapping of antibody fragments, hence reducing the antigen binding due to the steric hindrance but a dense packing of the antibodies is still necessary to attain measurable surface stress upon antigen binding.

- The specificity of the end-point nanomechanical immunodetection critically depends on the surface density of receptors and blocking strategy. The highest specificity to the HRP was found in cantilevers functionalized with an antibody concentration of 50 μg/mL and subsequently blocked with PEG. This can be due to the ability of PEG to fill the voids between the antibodies generating a high osmotic pressure that assists in the required "elastic" connectivity between the molecular receptors on the cantilever. This percolation mechanism enables a significant cantilever bending upon molecular recognition as a consequence of the collective and cooperative deformation of the biological layer formed on the cantilever surface.

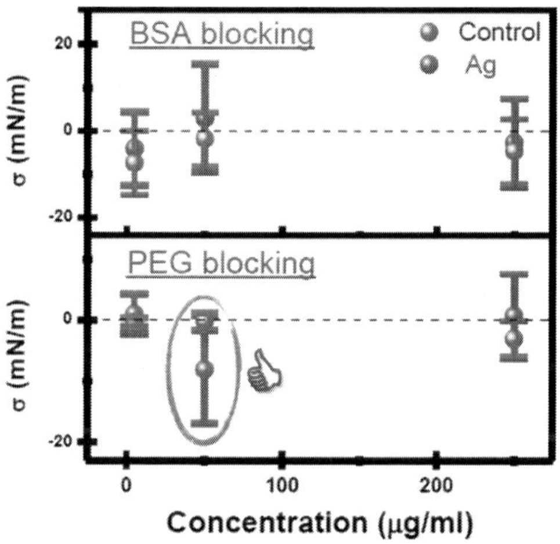

Figure 14.3 Mean and standard deviation of the surface stress induced by the specific HRP binding (red) and control experiments (grey) for cantilevers blocked with BSA and PEG, respectively. Reproduced with minor modifications from *Analyst*, 2013, **138**, 863–872, DOI: 10.1039/C2AN36192B (Ref. 38).

The false positive rate found when BSA was used as blocking molecule was so high that makes its use unreliable for the development of an end-point label-free immunoassay based on cantilevers. When the PEG blocking was used the false negatives are much lower, but it is still high for a reliable immunoassay. The results found until now can be a promising baseline for further improvement by developing immobilization methods that enable a major control on the nanometer-scale configuration of the antibodies on the cantilever surface.[41]

14.4.1.3 Reliability of the nanomechanical biosensors and development of a quality control test

Despite the important challenges of attaining reproducible and reliable assays with nanosensors, little attention has been paid to the variability of nanosensor responses in the last decade. Receiving Operating Curves (ROC curves) can be used to evaluate the accuracy of diagnostic tests and classify individuals into health and disease groups. Here we will propose the use of ROC curves as a tool to quantify the reliability of nanomechanical biosensors. The ROC curve will be plotted for the microcantilevers immobilized using 50 μg/mL of antibody and blocked with PEG, the optimized functionalization conditions settled before (see Sections 14.4.1.1 and 14.4.1.2). In Section 14.4.2 you will find a brief introduction to ROC curves.

The sensor response to the antigen binding was defined as[38]

$$R = \frac{\sigma - \langle \sigma_c \rangle}{\Delta \sigma_c}, \tag{14.1}$$

where σ is the obtained surface stress, $\langle \sigma_c \rangle$ is the mean surface stress in the control experiments and $\Delta\sigma_c$ is the surface stress standard deviation in the control experiments. The ROC curve in Fig. 14.4a was plotted several cut-off values of R and with no data rejection. The value found for the area under the curve is 0.842, which is a good value considering that the area under the curve can vary from 0.5 (insensitive biosensor) to 1 (perfect biosensor).

The analysis of the pattern of the surface stress generated by the antibody immobilization, blocking step, and antigen binding for each cantilever can provide additional insights for the improvement of immobilization strategies to enhance the sensitivity and selectivity of immunoassays based on micro and nano-cantilevers.[38] Moreover, a quality test that enables the exclusion of cantilevers in which the immobilization characteristics do not lead to sensitive and specific antigen response can be delineated. The quality test can conduct to the development of label-free immunoassays based on cantilevers with a high level of confidence in addition to high sensitivity and specificity.

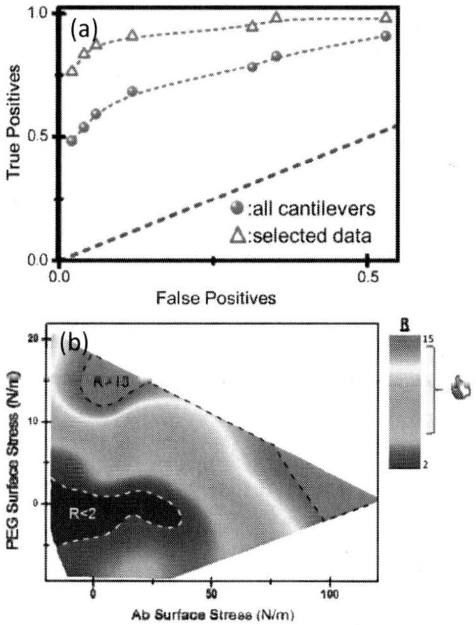

Figure 14.4 (a) ROC curve for all the data from cantilevers functionalized with 50 mg mL^{-1} of anti-HRP and subsequently blocked with PEG (circles), and for selected data based on the statistical criteria described in the text (triangles) and (b) Antigen response (R) map as a function of the surface stress induced by the functionalization steps. Reproduced with minor modifications from *Analyst*, 2013, **138**, 863–872, DOI: 10.1039/C2AN36192B (Ref. 38).

Figure 14.4b shows a color-intensity two-dimensional map of the antigen response as a function of the surface stress induced by the two previous steps in the assay, the antibody immobilization and the surface blocking. The antigen response map allows delineating the conditions in which the highest antigen response is obtained. The conditions that guarantee the optimal detection of the antigen with a high degree of confidence are: (i) tensile surface stress for the antibody immobilization higher than 40 mN/m and (ii) blocking-induced surface stress higher than 3 mN/m. The first found condition indicates that the optimum packing densities are those for which the antibodies are separated by nanometer-scale gaps and the intermolecular interactions are attractive, and rapidly vary with the gap distance. The second condition indicates that a high packing of blocking molecules in the voids between antibodies contributes to an effective "mechanical" percolation[28,42] of the antibody film that results into a high nanomechanical change of the biological layer upon antigen binding.

Now, let us check the validity of the quality control designed above by plotting the ROC curve for the cantilevers that a sensor response (R) specific to the antigen binding is expected (Fig. 14.4a). The value of the area under the curve found for the cantilevers that approved the quality test increased from 0.842 to 0.936. This new value is much closer to the value of a perfect biosensor.

The study presented in this chapter was much more focused on the variability of the nanomechanical biosensors rather than on the determination of detection limits. However, microcantilever biosensors have shown enormous potential for highly sensitive label-free detection of proteins and DNA[7,10,12,18,33,43,44] and the limit of detection (LOD) of any biosensor is one of the most important characteristic of any biosensor.

So, what is the limit of detection (LOD)? The definition of LOD given by the American Chemical Society (ACS) and by the International Union of Pure and Applied Chemistry (IUPAC), respectively, are:

(i) The limit of detection is the lowest concentration of an analyte that the analytical process can reliably detect.[45]

(ii) The limit of detection, expressed as a concentration of quantity, is derived from the smallest measure that can be detected with reasonable certain for a given analytical procedure.[46,47]

Adopting the quality control described above, we measured the LOD of the current nanomechanical immunoassay and the LOD achieved was 10 ng/mL. This value is at least two orders of magnitude smaller than that obtained with well-established label-free biosensors, QCM and SPR, with similar samples in direct assays.[48,49]

14.4.2 Receiving Operating Characteristics Analysis

A receiver operating characteristics (ROC) graph is a technique for visualizing, organizing and selecting classifiers based on their performance.[50] The performance of a diagnostic test or the reliability of a test to discriminate disease cases from normal cases is evaluated using the ROC curve analysis.[51] The ROC curve can also be used to compare the performance of two or more diagnostic test. Although ROC graphs are apparently simple, there are some common misconceptions and pitfalls when using them in practice. Here we will give just a brief introduction to ROC curves and you can find excellent detailed literature on the use of ROC curves for diagnostic testing.[50,52,53]

Analyzing the results of a diagnostic test in two populations, one population with a disease and another one without the disease, it will be rarely observed a perfect separation between these two groups. Indeed, the distribution of the test results will overlap, as shown in the Fig. 14.5a.

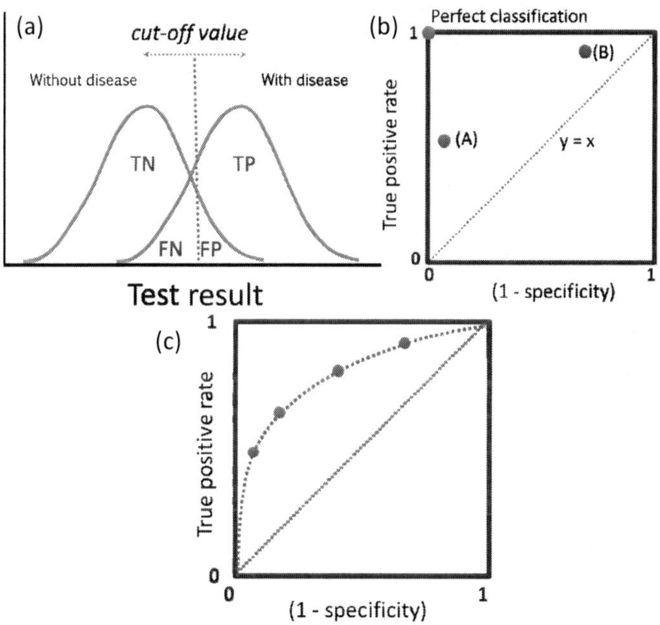

Figure 14.5 (a) Two hypothetical distributions of a quantity on which decisions are based, showing one possible cut-off value, (b) a basic ROC curve with different classifiers, and (c) typical ROC curve.

For any possible cut-off point selected to discriminate between the two populations, there will be some cases that the disease will be correctly classified as positive (TP = true positive), but there will be some positive cases that will be wrongly classified as negative (FN = false negative). Moreover, some cases without the disease will be correctly classified as negative (TN = true negative), but some cases without the disease will be classified as positive (FP = false positive).

The ROC curve is defined as a plot of a test rue positive rate (TPR) or sensitivity as the *y* coordinate versus its false positive rate (FPR) or 1-specificity as the *x* coordinate (Figs. 14.5b and 14.5c). The true positive rate (TPR) can be defined as:

$$\mathrm{TPR} = \frac{\mathrm{TP}}{\mathrm{All\ positives}}. \tag{14.2}$$

The false positive rate (FPR) is

$$\mathrm{FPR} = \frac{\mathrm{FN}}{\mathrm{Total\ negatives}}. \tag{14.3}$$

The specificity is defined as

$$\mathrm{Specificity} = \frac{\mathrm{TN}}{(\mathrm{FP} + \mathrm{TN})}, \tag{14.4}$$

$$\text{or Specificity} = 1 - \mathrm{FPR} \tag{14.5}$$

Some points in a ROC graph are important to note (see Fig. 14.5b). The lower left point (0,0) and the upper right point (1,1) represent insensitive biosensors. The point (0,1) represents perfect classification, 100% of true positives and no false negatives. The point A, which is closer to the *x* axis than the point B, makes positive classification only with strong evidence. On the other hand, the point B classifies nearly all positives correctly, but it has higher false positive rate than the point A. Good tests approach the top left corner of the ROC curve.

The points on the diagonal line $y = x$ may be said to have no information about the class (random guessing) and the points in the lower right triangle performs worse than random guessing. This region of the ROC graph is usually empty.

When the result of a test falls into one or two obviously defined categories, such as the presence or absence of a disease, the test has only one pair of sensitivity and specificity values. A single pair of sensitivity and specificity values is not sufficient to describe the full range of a diagnostic performance. What is done to solve this problem is to choose different cut-off levels to define a positive test result and as a result, we will have a ROC graph with different classifiers as shown in Fig. 14.5c. If the multiple pairs of sensitivity and specificity defined for different cut-off values are connected, a ROC curve can be estimated from these points. One of the most popular measures is the area under the ROC curve (AUC). The AUC is a measurement of the overall performance of a diagnostic test and can be understood as the average value of sensitivity for all possible values of specificity. The closer the value of the AUC is to 1, the better is the overall diagnostic performance of the test. The perfect test will have an AUC equal to 1.

14.5 Conclusions

In the past decade, microcantilever biosensors have shown their capability for highly sensitive label-free detection of nucleic acid and proteins. Despite the enormous advances, the promise of applications of this technology in the biomedical field has been delayed because of its low reproducibility. In this chapter, we discussed the advances that can achieve high-throughput analysis and the guidelines to minimize the deviations in the nanomechanical biosensor response the reproducibility issue in microcantilever biosensors and provide the guidelines to minimize the deviations in the biosensor response between different assays. We foresee further innovative ways to take advantage of nanomechanics for biosensing in the coming years.

References

1. Ferrari, M. (2005). Cancer nanotechnology: Opportunities and challenges, *Nat. Rev. Cancer,* **5**(3), 161–171.

2. Cheng, M. M. C., Cuda, G., Bunimovich, Y. L., Gaspari, M., Heath, J. R., Hill, H. D., Mirkin, C. A., Nijdam, A. J., Terracciano, R., Thundat, T., and Ferrari M. (2006). Nanotechnologies for biomolecular detecion and medical diagnostics, *Curr. Opin. Chem. Biol.,* **10**(1), 11–19.

3. Arlett, J. L., Myers, E. B., and Roukes M. L. (2011). Comparative advantages of mechanical biosensors, *Nat. Nanotechnol.,* **6**(4), 203–215.

4. Burg, T. P., Godin, M., Knudsen, S. M., Shen, W., Carlson, G., Foster, J. S., Babcock, K., and Manalis S. R. (2007). Weighinig of biomolecules, single cells and single nanoparticles in fluid, *Nature,* **446**(7139), 1066–1069.

5. Melli, M., Scoles, G., and Lazzarino, M. (2011). Fast detection of biomolecules in diffusion-limites regime using micromechanilcal pillars, *ACS Nano*, **5**(10), 7928–7935.

6. Joo, J., Kwon, D., Yim, C., and Jeon, S. (2012). Highly sensitive giagnostic assay for the detection of protein biomarkers using microresonators and multifuncional nanoparticles, *ACS Nano,* **6**(5), 4375–4381.

7. Fritz, J., Baller,M. K., Lang, H. P., Rothuizen, H., Vettiger, P., Meyer, E., Güntherodt, H., Gerber, C., and Gimzewski, J. K. (2000). Translating biomolecular recognition into nanomaterial, *Science,* **288**(5464), 316–318.

8. Wu, G., Datar, R. H., Hansen, K. M., Thundat, T., Cote, R. J., and Majumdar, A. (2001). Bioassay of prostate-specific antigen (PSA) using microcantilevers, *Nat. Biotechnol.,* **19**(9), 856–860.

9. Arnts, Y., Seeling, J. D., Lang, H. P., Zhang, J., Hunziker, P., Ramseyer, J. P., Meyer, E., Hegner, M., and Gerber, C. (2003). Label-free protein assay based on a nanomechanical cantilever array, *Nanotechnology,* **14**(1), 86–90.

10. McKendry, R., Zhang, J., Arntz, Y., Strunz, T., Hegner, M., Lang, H. P., Baller, M. K., Certa, U., Meyer, E., Guntherodt, H. J., and Gerber, C. (2002). Multiple label-free biodetection and quantitative DNA-binding assay on a nanomechanical cantilever array, *Proc. Natl. Acad. Sci. U. S. A.,* **99**(15), 9783–9788.

11. Wu, G. Hansen, K., Thundat, T., Datar, R., Cote, R., Hagan, M. F., Chakraborty, A. K., and Majumdar A. (2001). Origin of nanomechanical cantilever motion generated from biomolecular interactions, *Proc. Natl. Acad. Sci. U. S. A.,* **98**(4), 1560–1564.

12. Gruber, K., Horlacher, T., Castelli, R., Mader, A., Seeberger, P. H., and Hermann, B. A. (2011). Cantilever array sensors detect specific carbohydrate-protein interactions with picomolar sensitivity, *ACS Nano*, **5**(5), 3670–3678.

13. Huber, F., Lang, H. P., Backmann, N., Rimoldi, D., and Gerber, Ch. (2013). Direct detection of a *BRAF* mutation in total RNA from melanoma cells using cantilever arrays, *Nat. Nanotechnol.,* **8**(2), 125–129.

14. Longo, G., Alonso-Sarduy, L., Marques Rio, L., Bizzini, A., Trampuz, A., Notz, J., Dietler, G., and Kasas, S. (2013). Rapid detection of bacterial

resistance to antibiotics using AFM cantilevers as nanomechanical sensors, *Nat. Nanotechnol.,* **8**(7), 522–526.

15. Tamayo, J., Kosaka, P. M., Ruz, J. J., San Paulo, A., and Calleja, M. (2013). Biosensors based on nanomechanical systems, *Chem. Soc. Rev.,* **42**(3), 1287–1311.
16. Calleja, M., Kosaka, P. M., San Paulo, A., and Tamayo, J. (2012). Challenges for nanomechanical sensors in biological detection, *Nanoscale,* **4**(16), 4925–4938.
17. Godin, M., Tabard-Cossa, V., Miyahara, Y., Monga, T., Williams, P. J., Beaulieu, L. Y., Bruce Lennox, R., Grutter, P. (2010). Cantilever-based sensing: The origin of surface stress and optimization strategies, *Nanotechnology,* **21**(7), 075501.
18. Waggoner, P. S., and Craighead, H. G. (2007). Micro- and nanomechanical sensors for environmental, chemical, and biological detection, *Lab Chip,* **7**(10), 1238–1255.
19. Firestone, M. A., Shank, M. L., Sligar, S. G., and Bohn, P. W. (1996). Film architecture in biomolecular assemblies. Effect on linker on the orientation of genetically engineered surface-bound proteins, *J. Am. Chem. Soc.,* **118**(38), 9033–9041.
20. Rye, R. R., Nelson, G. C., Dugger, M. T. (1997). Mechanistic aspects of alkylchlorosilane coupling reactions, *Langmuir,* **13**(11), 2965–2972.
21. Haensch, C., Hoeppener, S., and Schubert, U. S. (2010). Chemical modification of self-assembled silane based monolayers by surface reactions, *Chem. Soc. Rev.,* **39**(6), 2323–2334.
22. Pesek, J. J., and Matyska, M. T. (1997). Methods for the modification and characterization of oxide surfaces, *Interface Sci.,* **5**(2–3), 103–117.
23. Bietsch, A., Zhang, J., Hegner, M., Lang, H. P., and Gerber, Ch. (2004). Rapid functionalization of cantilever array sensors by inkjet printing, *Nanotechnology,* **15**(8), 873–880.
24. Sushko, M. L. (2009). Nanomechanics of organic/inorganic interfaces : A theoretical insight, *Faraday Discuss,* **143**, 63–80.
25. Shu, W., Liu, D., Watari, M., Riener, C. K., Strunz, T., Welland, M. E., Balasubramanian, S., McKendri, R. A. (2005). DNA molecular motor driven micromechanical cantilever arrays, *J. Am. Chem. Soc.,* **127**(48), 17054–17060.
26. Zhang, J., Lang, H. P., Huber, F., Bietsch, A., Grange, W., Certa, U., McKendry, R.,Güntherodt, H. J., Hegner, M., and Gerber, Ch. (2006). Rapid and label-free nanomechanical detection of biomarker transcripts in human RNA, *Nat. Nanotechnol.,* **1**(3), 214–220.

27. Backmann, N., Zahnd, C., Huber, F., Bietsch, A., Plückthun, A., Lang, H. P., Güntherodt, H. J., Hegner, M., and Gerber, C. (2005). A label-free immunosensor array using single-chain antibody fragments, *Proc. Natl. Acad. Sci. U. S. A.,* **102**(41), 14587–14592.

28. Martinez, N. F., Kosaka, P. M., Tamayo, J., Ramirez, J., Ahumada, O., Mertens, J., Hien, T. D., Rijn, C. V., and Calleja, M. (2010). High throughput optical readout of dense arrays of nanomechanical systems for sensing, *Rev. Sci. Instrum.*, **81**(12), 125109.

29. Bietsch, A., Hegner, M., Lang, H. P., Gerber, C. (2004). Inkjet deposition of alkanethiolate monolayers and DNA oligonucleotides on gold: Evaluation of spot uniformity by wet etching, *Langmuir,* **20**(12), 5119–5122.

30. Pardo, L., Cris Wilson Jr., W., and Boland, T. (2003). Characterization of patterned self-assembled monolayers and protein arrays generated by the ink-jet method, *Langmuir,* **19**(5), 1462–1466.

31. Lukacs, G., Maloney, N., and Hegner, M. (2012). Ink-jet printing: Perfect tool for cantilever array sensor preparation for microbial growth detection, *J. Sens.,* **2012**, 561259.

32. Alvarez, M., and Tamayo, J. (2005). Optical sequential readout of microcantilever arrays for biological detection, *Sens. Actuators B Chem.,* **106**(2), 687–690.

33. Bosco, F. G., Hwu, E. T., Chen, C. H., Keller, S., Bache, M., Jakobsen, M. H., Hwang, I. S., and Boisen, A. (2011). High throughput label-free platform for statistical biomolecular sensing, *Lab Chip,* **11**(14), 2411–2416.

34. Boisen, A., Dohn, S., Keller, S. S., Schmid, S., and Tenje, M. (2011). Cantilever-like micromechanical sensors, *Rep. Prog. Phys.,* **74**, 036101.

35. Goeders, K. M., Colton, J. S., and Bottomley, L. A. (2008). Microcantilevers: Sensing chemical interactions via mechanical motion, *Chem. Rev.,* **108**(2), 522–542.

36. Alvarez, M., Tamayo, J., Plaza, J. A., Zinoviev, K., Dominguez, C., and Lechuga, L. M. (2006). Dimension dependence of the thermo-mechanical noise of microcantilevers, *J. Appl. Phys.,* **99**, 024910.

37. Cleland, A., and Roukes, M. (2002). Noise processes in nanomechanical resonators, *J. Appl. Phys.,* **92**(5), 2758–2769.

38. Kosaka, P. M., Tamayo, J., Ruz, J. J., Puertas, S., Polo, E., Grazu, V., de la Fuente, J. M., and Calleja, M. (2013). Tackling reproducibility in microcantilever biosensors: A statistical approach for sensitive and specific end-point detection of immunoreactions, *Analyst,* **138**(3), 863–872.

39. Sharma, S., Johnson, R. W., Desai, T. A. (2004). Evaluation of the stability of nonfouling ultrathin poly(ethylene glycol) films for silicon-based microdevices, *Langmuir,* **20**(2), 348–356.

40. Kelling, S., Paoloni, F., Huang, J., Ostanin, V. P., Elliott, S. R. (2009). Simultaneous readout of multiple microcantilever arrays with phase-shifiting interferometric microscopy, *Rev. Sci. Instrum.,* **80**(9), 093101.

41. Jin, G., Meng, Y., Liu, L., Niu, Y., Chen, S., Cai, Q., and Jiang, T. (2011). Development of biosensor based on imaging ellipsometry and biomedical applications, *Thin Solid Films,* **519**(9), 2750–2757.

42. Tamayo, J., Pini, V., Kosaka, P., Martinez, N. F., Ahumada, O., and Calleja, M. (2012). Imaging the surface stress and vibration modes of a microcantilever by laser beam deflection microscopy, *Nanotechnology,* **23**(31), 315501.

43. Noy, P., Steiner, R., Voelkle, J., Hegner, M., and Fattinger, C. (2012). Instrument for label-free detection of noncoding RNAs, *J. Sens.*, **2012**, 208079.

44. Zhang, J., Lang, H. P., Yoshikawa, G., and Gerber, C. (2012). Optimization of DNA hybridization efficiency by pH-driven nanomechanical bending, *Langmuir*, **28**(15), 6494–6501.

45. ACS Committee on Environmental Improvement (1980). Guidelines for data acquisition and data quality evaluation in environmental chemistry, *Anal. Chem.*, **52**, 2242–2249.

46. IUPAC, Analytical Chemistry Division, Nomenclature, Symbols, Units and their usage in spectrochemical analysis—II (1978) *Spectrochim Acta Part B*, **33**, 242.

47. IUPAC, Analytical Chemistry Division (1976). Nomenclature, Symbols, Units and their usage in spectrochemical analysis—II. Data Interpretation, *Pure Appl. Chem.*, **45**(2), 99–103.

48. Arntz, Y., Seelig, J., Lang, H., Zhang, J., Hunziker, P., Ramseyer, J., Meyer, E., Hegner, M., and Gerber, C. (2003). Label-free protein assay based on a nanomechanical cantilever array, *Nanotechnology*, **14**(1), 86–90.

49. Qiu, D., Cosgrove, T., and Howe, A. M. (2007). Steric interactions between physically adsorbed polymer-coated colloidal particles: Soft or hard?, *Langmuir*, **23**(2), 475–481.

50. Fawcet, T. (2006). An introduction to ROC analysis, *Pattern Recognit. Lett.,* **27**(8), 861–874.

51. Taffé, P., Halfon, P., Ghali, W. A., and Burnand, B. (2012). Test result-based sampling: An efficient design for estimating the accuracy of patient safety indicators, *Med. Decis. Making*, **32**(1), E1–E12.

52. Swets, J. A., Dawes, R. M., Monahan, J., (2000). Better decisions through science, *Sci. Am.*, **283**(4), 82–87.

53. Park, S. H., Goo, J. M., Jo, C. H. (2004). Receiver operating characteristics (ROC) curve: Practical review for radiologists, *Korean J. Radiol.*, **5**(1), 11–18.

Chapter 15

Nanocantilever Beams as Biological Sensors

Ankit Jain and Muhammad Ashraful Alam

School of Electrical and Computer Engineering, Purdue University, West Lafayette, IN 47907, USA

jankiit@gmail.com, alam@purdue.edu

15.1 Introduction

Detection of biological molecules such as viruses, proteins, DNA, etc., are essential for food safety, early warning of biological attack, early stage diagnosis of cancer, and genome sequencing. Nanoscale devices are widely regarded as a potential candidate for ultra-sensitive, low-cost, label-free detection of bio-molecules and are considered as a technology alternative to the existing chemical or optical detection schemes. Label-free schemes detect bio-molecules using their intrinsic properties, namely, size, mass, or charge of the molecule, instead of using extrinsic optical or magnetic labels attached to the target molecule. Among the various label-free technologies, significant research has focused on developing ultra-sensitive biological sensors based on nanocantilevers.[1,2]

The use of a cantilever as a sensor dates back to 1943 when Norton proposed a hydrogen gas sensor based on such a system.[3]

Nanocantilever Beams: Modeling, Fabrication, and Applications
Edited by Ioana Voiculescu and Mona Zaghloul

ISBN 978-981-4613-23-1 (Hardcover), 978-981-4613-24-8 (eBook)
www.panstanford.com

The opportunity to develop the cantilever as a highly sensitive biosensor, however, had to wait the invention and widespread adoption of the atomic force microscope (AFM).[4] An AFM measures the forces between the tip of a cantilever and the sample surface by recording the tip deflection of a static cantilever (contact mode AFM) or changes in the resonance frequency of a vibrating cantilever (dynamic mode AFM). As we will see in Section 15.2, nanocantilever based biosensors operate on a closely related principle, where interaction with biological molecules changes the bending (static mode) or resonance frequency (dynamic mode) of the cantilever.[5] Note that these mechanical sensors offer an advantage of detecting both charged and neutral bio-molecules. In contrast, their electronic counterpart, e.g., ion-sensitive field-effect transistors (ISFETs)[6] can only detect charged molecules.

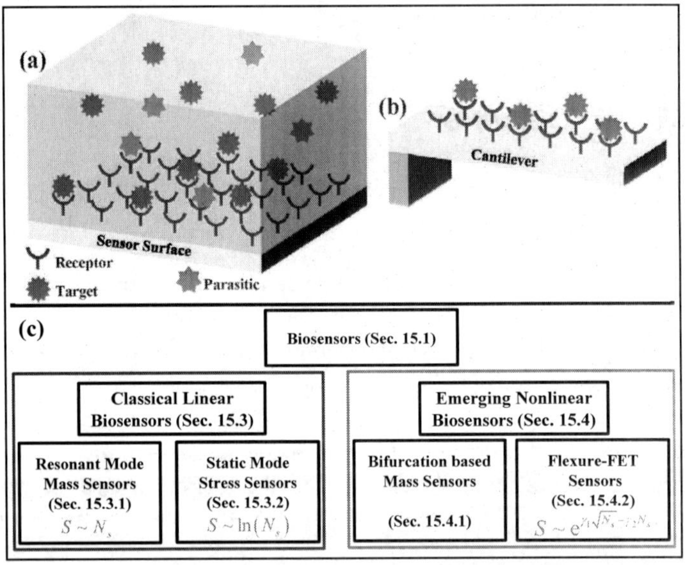

Figure 15.1 (a) Schematic of a generic biological sensor operating in a fluidic environment. The receptor, target, and parasitic molecules are also shown. (b) Drawing of a nanocantilever in which the capture of target molecules changes the cantilever's static or dynamic response through change in the mass, stiffness and/or surface stress. (c) A chart showing various linear and nonlinear cantilever biosensors to be discussed in this chapter. S is the sensitivity of respective biosensor and N_s is the areal density of captured bio-molecules on the cantilever surface.

Today, cantilever-based devices find applications in a broad range of fields such as communication, computation, optics, scanning probe microscopy, and sensing. Specific examples include RF-MEMS capacitive/ohmic switches,[7] varactors, tunable oscillators,[8] NEMS relays,[9] NEMFET,[10] deformable mirrors,[11] displays,[12] accelerometers,[13] and chemical/gas sensors. In this chapter, however, we focus on the use of cantilever as a biological sensor and discuss both classical and emerging modes of biosensor operation.

Figures 15.1a,b show the schematic of a sensor surface operating in a fluidic environment. The sensor surface is functionalized with receptor molecules that will subsequently conjugate to the target molecules (without any optical or magnetic labels) contained in the fluid. For example, if the target is an antibody, then the receptor is the corresponding antigen, or if the target is a DNA base (e.g., Adenine), then receptor is its conjugate base (e.g., Cytosine). The specific binding of the target and receptor molecules changes the sensor characteristics and the change is measured as a signature of detection. For example, the binding may induce an additional surface stress or may change the mass or stiffness of the cantilever, resulting in bending or change in the resonance frequency.[5]

There are three key metrics of any sensing scheme (electronic or mechanical), namely, response time,[14] selectivity,[15] and sensitivity.[16] *Response time* is the time required to capture a certain number of target molecules to get a detectable output signal. Interestingly, response time depends on the geometry of the sensor surface and cylindrical geometry allows the smallest response time.[14] On the other hand, *selectivity* is associated with the problem of non-specific binding, i.e., binding of parasitic molecules with receptor molecules, producing a "false-positive" signal. In a highly selective sensor, receptor molecules should only bind to the target molecules and not to any other molecules in the solution. Finally, *sensitivity* can be defined in number of ways; in general, it measures the change in the sensor characteristics (e.g., resonance frequency of a cantilever[5] or drain current of ISFET[17]) in response to the capture of a given number of target molecules. Note that, response time and selectivity of a sensor do not depend on the sensing scheme, whereas sensitivity does. Therefore, in this chapter, we only discuss sensitivity-related issues of nanocantilever-based biological sensors. Note that, sensitivity of the sensor depends whether it is operated in linear or nonlinear regime. Figure 15.1c summarizes different types of linear and nonlinear cantilever biosensors to be discussed in the following sections.

The rest of the chapter is organized as follows. In Section 15.2, we discuss the spring-mass model of cantilever and illustrate classical mass- and stress-based sensors in Section 15.3. We then present the emerging nonlinear biosensors, such as, bifurcation-based mass sensors and Flexure-FET in Section 15.4. We finally conclude in Section 15.5.

15.2 Cantilever-Based Sensors as Spring Mass System

The static as well as dynamic response of cantilever-based sensors is best described by Euler–Bernoulli beam equation.[18] In this chapter, however, we use a lumped parameter, spring-mass system (Fig. 15.2b) of a cantilever to illustrate its key features. The equation of motion of the lumped parameter spring-mass system is given by

$$m\frac{d^2y}{dt^2}+\frac{m\omega_0}{Q}\frac{dy}{dt}-k(y_0-y)-k'(y_0-y)^3=F_{\text{ext}}, \tag{15.1}$$

where m is the effective mass of the cantilever, y is the position of vibrating beam, t is time, ω_0 is the natural frequency of cantilever, Q is the quality factor, k is the effective spring constant of the cantilever such that $\omega_0 = \sqrt{k/m}$, y_0 is the position of the cantilever in its rest position, and k' is the constant associated with cubic nonlinearity of the spring. Note that $k = \frac{\alpha EWH^3}{(1-\nu)L^3}$ is the spring constant of the cantilever where α is a geometrical factor, E is Young's modulus, ν is Poisson's ratio, W is the width, H is the thickness, and L is the length. F_{ext} is the external force acting on the cantilever, e.g., surface forces, electrostatic forces, etc. Historically, the cantilever-based sensors have been operated without applying any external force (i.e., $F_{\text{ext}} = 0$) and in the linear response regime i.e., $k' \approx 0$ (Section 15.3). We will explore the nonlinear $F_{\text{ext}} \neq 0$ and $k' \neq 0$ operation in Section 15.4.

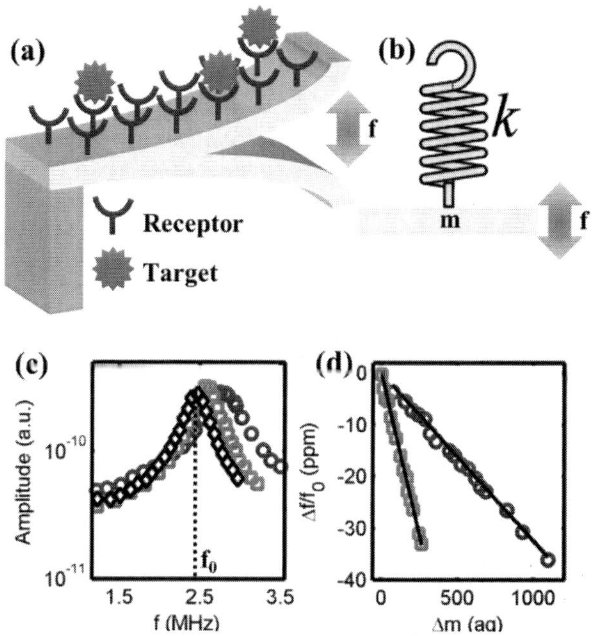

Figure 15.2 Dynamic resonant mode sensing using nanocantilever. (a) Schematic of a vibrating cantilever. (b) Equivalent spring-mass model of the vibrating cantilever. (c) Amplitude *vs.* frequency spectrum of a freestanding cantilever (○), after receptor molecules attachment (□), and after capture of target molecules (◇).[19] (d) Change in resonance frequency as a function of added mass of the bio-molecules for two different cantilevers.[20]

15.3 Classical Linear Biosensors

15.3.1 Resonant Mode Mass Sensors

In resonant mode sensing, a vibrating nanocantilever can be used as a microbalance and bio-molecules can be detected by observing the change in dynamic response of the cantilever.[1,5] Fundamentally, dynamic response of a cantilever is governed by its resonance frequency f_0 that is given by (using Eq. 15.1 with $k' = 0$ and $F_{ext} = 0$)

$$f_0 = \frac{\omega_0}{2\pi} = \frac{1}{2\pi}\sqrt{\frac{k}{m}}. \tag{15.2}$$

Once the target molecules are captured, the change in m (and/or k) shifts the resonance frequency (Δf) to indicate the capture of bio-molecules. Experimentally, the resonance frequency of the cantilever can be obtained by measuring the amplitude-frequency spectrum of vibrating cantilever using optical techniques. Figure 15.2c shows amplitude-frequency spectrum of a vibrating cantilever for three different conditions: before functionalizing with receptor molecules (blue circles), after functionalizing (red squares), and after capture of target molecules (black diamond).[19] By definition, the peak in the amplitude-frequency spectrum corresponds to the resonance frequency f_0. As expected, the resonance frequency decreases following the attachment of receptor molecules due to the added mass on the cantilever. It decreases further after capture of target molecules by receptor molecules due to further increase in the mass. The change in the resonance frequency due to the adsorption of the molecules can be obtained using (Eq. 15.2), and is given by

$$\frac{\Delta f}{f_0} \approx -\frac{\Delta m}{2m} + \frac{\Delta k}{2k}, \tag{15.3}$$

where Δm is the mass of added molecules and Δk is the change in stiffness. Figure 15.2d shows Δf as a function of Δm for two different cantilevers[20] and linear dependence of Δf on Δm confirms Eq. 15.3 (assuming $\Delta k = 0$). Equation 15.3 suggests that the sensitivity $S \equiv \Delta f / f_0$ of nanocantilever biosensors can only vary linearly with Δm (assuming $\Delta k = 0$) and therefore $S \propto N_s$ with N_s being the areal density of captured bio-molecules on cantilever surface. We emphasize that these biosensors—with careful design and appropriate instrumentation—can be extraordinarily sensitive; indeed, zeptogram mass detection has been reported.[21] It is also important to realize that the linear sensitivity with Δm is achieved only if the change in stiffness (Δk) due to capture of bio-molecules is negligible (Eq. 15.3). In general, the capture of target molecules increases k.[19] If Δk compensates Δm, (Eq. 15.3) suggests that the resonance frequency may not shift at all (i.e., $\Delta f \sim 0$) and the sensitivity could be vanishingly small. Therefore, one must independently measure the change in k to decouple the "mass effect" from "stiffness effect," so that the mass of the adsorbed molecule can be correctly estimated.[22,23]

The sharpness of the peak (or the width of the amplitude-frequency spectrum) vibrating cantilever is characterized by its quality factor (Q) (Eq. 15.1) and depends on the damping due to the surrounding medium. As Q increases, resonance peak becomes sharper and the width of the spectrum is reduced. Unfortunately, the value of minimum detectable Δf increases as Q is reduced.[24] Therefore, measurements in vacuum or air can be more sensitive (capable of resolving small Δf and therefore smaller Δm) as compared to detection in a fluidic environment.[24] This dependence of Δf on Q has inspired design of suspended microchannel resonators[25–27] that do not suffer from Q degradation due to the surrounding fluid. Such resonators are operated either in vacuum or in air and the fluid containing the target bio-molecules flows through the microchannel, embedded within the cantilever itself.

To summarize, the response of resonant mode nanocantilever-based biosensors is linear with respect to the added mass of bio-molecules. Minimum detectable mass depends on the quality factor of the vibrating cantilever. And suspended microchannel resonators can detect lower masses due to their high quality factors.

15.3.2 Stress-Based Static Mode Sensors

Another class of nanocantilever sensor involves operation in the static mode, in which capture of target molecules introduces a surface stress.[5] Change in the surface stress can be the result of an adsorption process or electrostatic interaction between charged molecules on the surface or conformational changes of the adsorbed molecules. This change in the surface stress bends the cantilever as shown in Fig. 15.3a. The deflection of the tip of the cantilever Δy is then measured as a signature of bio-molecules capture. Stoney's equation[28] relates the deflection Δy with the change in the surface stress $\Delta\sigma$ as follows:

$$\Delta y = \frac{3L^2(1-\nu)}{EH^2}\Delta\sigma, \tag{15.4}$$

where L is the length, ν is Poisson's ratio, E is Young's modulus, and H is the thickness of the cantilever. Note that, Eq. 15.4 can be obtained from Eq. 15.1 (with time derivatives and $k' = 0$) with appropriately chosen $F_{\text{ext}} = -\frac{3\alpha WH}{L}\Delta\sigma$.

Deflection Δy of the cantilever tip can be measured optically (e.g., using a laser and photodiode) or electrically (e.g. using an integrated piezoresistor). Figure 15.3b shows deflection Δy as a function of the target bio-molecules concentration in the solution for two different cantilevers having different geometrical dimensions.[29] The response Δy is *sub-linear* with respect to the concentration and it depends on the geometrical dimensions of the cantilever. Using Eq. 15.4 and data shown in Fig. 15.3b, $\Delta\sigma$ can be calculated and is shown in Fig. 15.3c. Interestingly, $\Delta\sigma$ for two different cantilevers follow a single curve (Fig. 15.3c), suggesting that $\Delta\sigma$ only depends on the concentration of the bio-molecules and not on the cantilever properties.

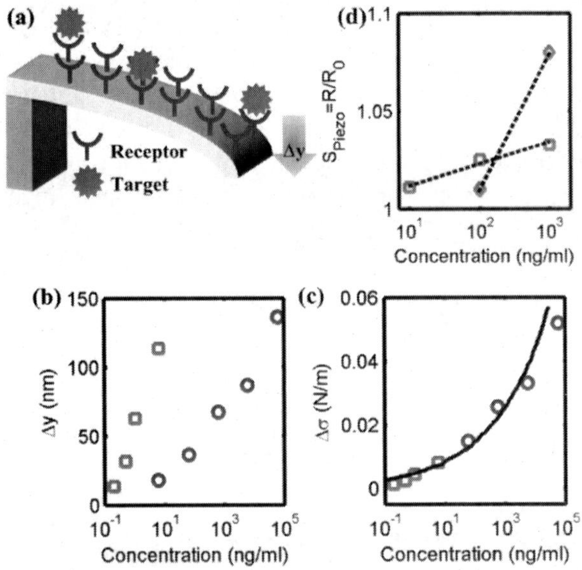

Figure 15.3 Stress-based sensing using the static response of a nanocantilever. (a) Schematic of a bent cantilever due to capture of bio-molecules on its surface. (b) Displacement of the tip of the cantilever as a function of bio-molecule concentration for two different cantilevers.[29] (c) Corresponding change in the surface stress of the cantilever. Symbols denote experimental data and solid line is a guide to the eye. (d) Ratio of the resistance of piezoresistive material attached to the cantilever after (R) and before (R_0) capture of bio-molecules.[30] Symbols denote experimental data and dotted line is a guide to the eye.

Instead of optical measurement of Δy or $\Delta\sigma$ through laser-photodiode system, one can measure $\Delta\sigma$ by measuring the change in the resistance of a piezoresistor attached to the cantilever.[30] For these piezoresistive-based cantilever biosensors, the sensitivity is defined as the ratio of resistance after (R) and before (R_0) the capture of bio-molecules, i.e., $S_{\text{Piezo}} \equiv R/R_0$. Figure 15.4d shows S_{Piezo} for two different cantilevers, suggesting that response S_{Piezo} varies logarithmically with bio-molecules concentration. We, therefore, conclude that these static mode nanocantilever sensors respond only sub-linearly with the analyte concentration i.e., $S \sim \ln(N_0)$.

15.4 Emerging Nonlinear Biosensors

In the previous section, we discussed classical linear biosensors that can either be operated in static or dynamic mode. Now, we discuss a new class of emerging nonlinear biosensors that utilize inherent instability of nanocantilever's static/dynamic response to achieve better sensitivity towards bio-molecules capture.

15.4.1 Bifurcation-Based Mass Sensors

As discussed above, classical resonant mode biosensors rely on the change in resonance frequency due to the capture of bio-molecules. Note that when operated in the linear regime (under small amplitude limit), the amplitude-frequency spectrum is symmetric and bell-shaped, as shown in Fig. 15.4a. In this case, detection of bio-molecules is achieved by observing the shift in the peak frequency (i.e., Δf, see Fig. 15.4a), as discussed in Section 15.3.1.

In the large amplitude nonlinear response regime, however, higher order spring nonlinearities (i.e., $k' \neq 0$ in Eq. 15.1) distort the response, and amplitude-frequency spectrum is no longer symmetric.[31] Figure 15.4b shows one such amplitude-frequency spectrum with softening nonlinearity ($k' < 0$ and $F_{\text{ext}} = F_0 \sin(2\pi f_{\text{ext}})$ in Eq. 15.1) with F_0 being the excitation amplitude and f_{ext} is the excitation frequency) for a Duffing-like resonator. Interestingly, spectrum exhibits sudden jumps at points P and S representing saddle-node bifurcations. The hysteretic behavior shown in Fig. 15.4b is achieved, when $F_0 > F_c$ with F_c being a critical threshold.

Kumar et al., has proposed a bifurcation-based mass sensor that utilizes these sudden jumps and rely on the shift in the amplitude (but not on the shift in the frequency) to signal bio-molecules capture.[32] In bifurcation-based sensing, the resonator is operated near one of the critical points (say P). Capture of the bio-molecules reduces the fundamental frequency f_0 and increases f_{ext}/f_0 resulting in the sudden change in the amplitude of oscillation ΔA, as shown in Figs. 15.4b–c. Measurement of ΔA (using laser Doppler vibrometer (LDV)) is then used as the signature of capture of bio-molecules. It should be appreciated that this sensing scheme is very sensitive to small quantities of added molecules due to the amplification offered by inherent instability of mechanical system.

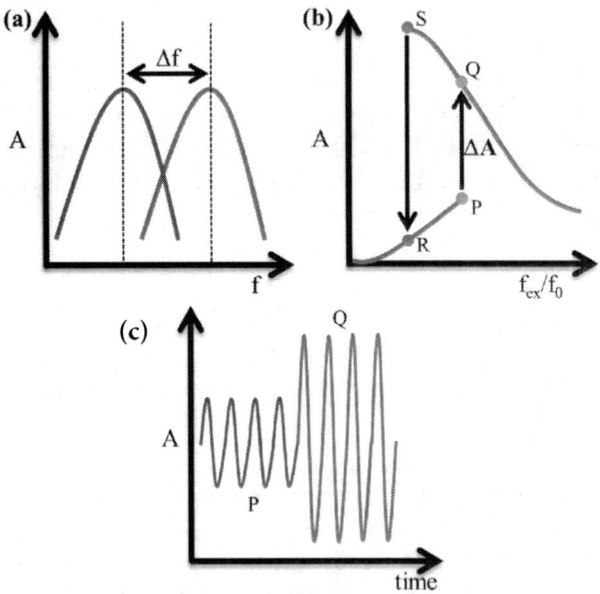

Figure 15.4 A comparison of classical linear mass sensors with nonlinear bifurcation-based mass sensors. Amplitude-frequency spectrum of (a) classical linear and (b) bifurcation-based mass sensors proposed in ref. [32]. (c) Amplitude as a function of time in bifurcation-based mass sensors. Classical sensors rely on the change in frequency Δf, whereas bifurcation-based sensors rely on change in the amplitude ΔA due to capture of bio-molecules.

15.4.2 Electromechanical Coupling-Based Flexure-FET Biosensors

We have discussed both linear and nonlinear cantilever-based biosensors that rely on optical readout of y or f_0. Now we discuss a new class of nonlinear biosensors called Flexure-FET[16] that utilize the electromechanical coupling between a suspended beam and a field-effect transistor to achieve much higher sensitivity compared to traditional biosensors. A Flexure-FET consists of a channel biased through a thin-film suspended gate (Fig. 15.5a). While the structure is similar to that of a suspended-gate FET,[33] NEMFET[10] or resonant gate transistor,[34] we call the sensor Flexure-FET to emphasize its distinctive nonlinear operation specifically optimized for ultrasensitive detection of bio-molecules. In a Flexure-FET, any change in the mechanical property of the suspended gate is directly reflected in the change of the drain current of the integrated field-effect transistor and thereby enables electrical readout. As shown in Fig. 15.5b, ultra high sensitivity arises from the coupling of two nonlinear responses, namely (i) spring-softening[35] in which stiffness decreases nonlinearly with the applied gate bias V_G and vanishes at the pull-in point (for detailed discussions on pull-in instability, see Refs. [36,37]), and (ii) sub-threshold electrical conduction[38] in which current depends exponentially on the surface potential. Such nonlinear electro-mechanical coupling enables exponentially high sensitivity for Flexure-FET sensors, which is fundamentally unachievable by exclusive use of existing nanoscale electronic or mechanical biosensors.

It should be noted that from a mechanical perspective, Flexure-FET operates close to pull-in instability, a critical point. Similar critical point sensing has also been reported for vapor sensors that operates close to bucking-instability[39,40] and for mass sensor that operates close to saddle-node bifurcation[32] (discussed in Section 15.4.1) and their higher sensitivity have been confirmed experimentally. However, beyond the critical point sensing, the integrated transistor-action in the sub-threshold regime provides the Flexure-FET an *additional* exponential sensitivity (and simpler DC read-out) that could not be achieved by other nonlinear sensor schemes.

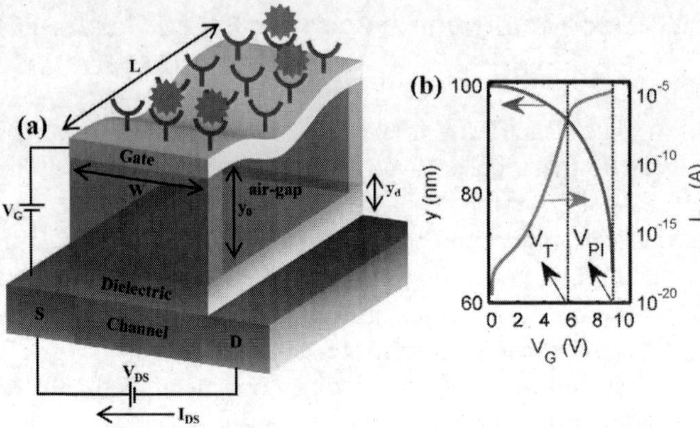

Figure 15.5 Static stiffness-based nonlinear electromechanical sensing with electrical readout. (a) Schematic of Flexure-FET biosensor and (b) y and I_{DS} vs. gate voltage characteristic of Flexure-FET. V_T is the threshold voltage and V_{PI} is the pull-in voltage of Flexure-FET.

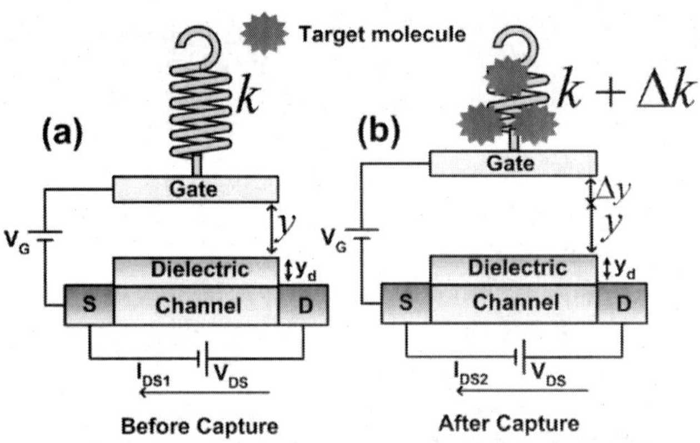

Figure 15.6 Equivalent spring-mass model of Flexure-FET. Stiffness changes from k to $k + \Delta k$ after the capture of bio-molecules, and therefore, position of gate changes from y to $y + \Delta y$, which results in the modulation of drain current from I_{DS1} to I_{DS2}.

The operating principle of Flexure-FET can be understood based on a spring-mass system coupled to electrostatic actuation (see Fig. 15.6).[10,34] With the application of gate bias V_G, the gate moves downward toward the dielectric (y vs. V_G curve in Fig. 15.5b) and the corresponding increase in gate capacitance is reflected in the increased drain current I_{DS}, as shown in Fig. 15.5b. The static behavior of the device is dictated by the balance of spring and electrostatic forces (Eq. 15.1) with time derivatives zero and $k' = 0$), i.e.,

$$k(y_0 - y) = F_e \equiv \frac{1}{2}\varepsilon_0 E_{air}^2 A, \tag{15.5}$$

where E_{air} is the electric field in the air, and $A = WL$ is the area of the gate electrode. The electric field below the membrane E_{air} is equal to $\varepsilon_s E_s(\psi_s)$, where ε_s is the dielectric constant of the substrate, and

$$E_s(\psi_s) = \sqrt{\frac{2qN_A}{\varepsilon_0\varepsilon_s}}\left[\psi_s + \left(e^{-\frac{q\psi_s}{k_BT}} - 1\right)\frac{k_BT}{q} - \left(\frac{n_i}{N_A}\right)^2\left(\psi_s - \left(e^{-\frac{q\psi_s}{k_BT}} - 1\right)\frac{k_BT}{q}\right)\right]^{\frac{1}{2}}, \tag{15.6a}$$

where $E_s(\psi_s)$ is the electric field at the substrate-dielectric interface (see Ref. [38], page 64, for a detailed derivation of Eq. 15.6a), (ψ_s) is the surface potential, q is the charge of an electron, N_A is the substrate doping, k_B is the Boltzmann constant, T is the absolute temperature, and n_i is the intrinsic carrier concentration in the substrate. The voltage drop in air $(y\varepsilon_s E_s(\psi_s))$, dielectric $\left(\frac{y_d}{\varepsilon_d}e_s E_s(\psi_s)\right)$, and substrate (ψ_s) can be related to the applied gate bias V_G as follows:

$$V_G = \left(y + \frac{y_d}{\varepsilon_d}\right)\varepsilon_s E_s(\psi_s) + \psi_s, \tag{15.6b}$$

where y_d is the dielectric thickness. Equations 15.5–15.6 are solved self-consistently for y and ψ_s at each V_G. The corresponding inversion charge density (Q_i) in the channel and drain current (I_{DS}) are given by

$$Q_i = \frac{qn_i^2}{N_A}\int_0^{\psi_s}\frac{e^{\frac{q\psi}{k_BT}} - 1}{E_s(\psi)}d\psi, \tag{15.7}$$

$$I_{DS} = \mu_n L Q_i \frac{V_{DS}}{W}, \tag{15.8}$$

where μ_n is the channel mobility for electrons, V_{DS} is the applied drain to source voltage. Figure 15.5b shows the steady-state response of Flexure-FET as a function of biasing voltage V_G, obtained from the numerical simulations of Eqs. 15.5–15.8.

For transduction, Flexure-FET biosensor utilizes the change in suspended-gate stiffness from k to $k + \Delta k$, due to the capture of bio-molecules. The change in stiffness due to the capture of bio-molecules has been demonstrated by several recent experiments of mass sensing using nanocantilever-based resonators.[19,41–43] This well-known observation of stiffness change has been attributed to the change in the membrane thickness, Young's modulus, and/or surface stress of the beam.[19,41–43] Indeed, Craighead[44] suggests its use as a basis of a new class of mechanical biosensor.

In the following analysis, we model change in k by the change in the effective thickness H of the gate (ΔH), *although it should be stressed that the conclusions do not depend on the particular hypothesis regarding* Δk. For now, we ignore the details of the spatial distribution of molecules associated with random sequential adsorption,[15] and assume a uniform distribution of adsorbed molecules on the sensor surface. Therefore, the conservation of volume suggests $\Delta H = N_s A_t H_t$, where N_s is the areal density, A_t is the effective cross-sectional area, and H_t is the effective thickness of the target molecule. Using the fact that $k = \frac{\alpha EWH^3}{(1-\nu)L^3}$, change in stiffness Δk due to ΔH ($<<H$) can be related to adsorbed molecule density N_s as follows:

$$\frac{\Delta k}{k} \approx \frac{3N_s A_t H_t}{\mathrm{H}}. \tag{15.9}$$

It can be shown that if Flexure-FET is operated close to pull-in and in sub-threshold regime, sensitivity S (using Eqs. 15.5–15.9, see Ref. [16] for details of derivation) is given by

$$S_{\text{Flexure}} \equiv \frac{I_{\text{DS1}}}{I_{\text{DS2}}} \approx \exp(\gamma_1 \sqrt{N_s} - \gamma_2 N_s), \tag{15.10}$$

where γ_1/γ_2 are two-sensor geometry-dependent constants. Equation 15.10 confirms the exponential sensitivity of Flexure-FET toward bio-molecules capture.

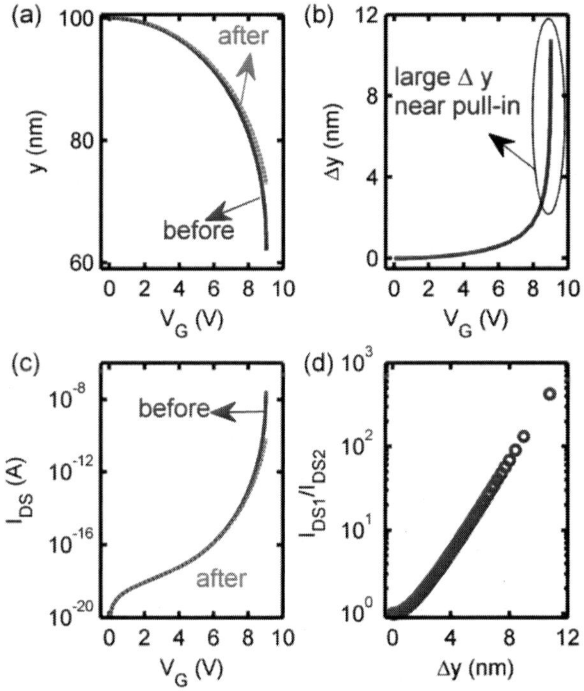

Figure 15.7 Change in the sensor characteristics due to capture of target molecules on the surface of the gate, (a) y vs V_G before and after capture, and (b) corresponding change in the position of gate electrode Δy vs. V_G. Δy increases rapidly near pull-in due to spring-softening effect. The capture of target molecules is directly mirrored in the change in I_{DS}. (c) I_{DS} vs. V_G for before and after capture, and (d) corresponding ratio of the two currents I_{DS1} (before capture) and I_{DS2} (after capture) as a function of Δy. Orders of magnitude change in I_{DS} can be easily achieved for typical surface density of $N_s = 5 \times 10^{12}$ cm^{-2}, projected area of the bio-molecule, $A_t = \pi R_t^2$ with R_t = 1 nm, and H_t = 5.1 nm. These parameters translate to just an equivalent $\Delta k \sim 6\%$. The device considered has the following typical parameters: L = 4 μm, W = 1 μm, H = 40 nm, E = 200 GPa, y_0 = 100 nm, y_d = 5 nm, ε_s = 11.7, ε_d = 3.9, $N_A = 6 \times 10^{16}$ cm^{-3}.

The results for the change in sensor characteristics due to the capture of bio-molecules are summarized in Fig. 15.7. For example, Fig. 15.7a shows y vs. V_G before and after capture of target molecules. After the capture, gate moves up (for a fixed V_G) due to increased restoring spring force (because of increase in k (see

Fig. 15.7a). Interestingly, the change in gate position Δy is maximum close to pull-in due to spring-softening effect, as shown in Fig. 15.7b. The change in gate position Δy is directly reflected in change in I_{DS}. Figure 15.7c shows I_{DS} vs. V_G before and after capture of bio-molecules. Interestingly, I_{DS} decreases after capture due to increased separation between the gate and the dielectric (hence decreased capacitance). The corresponding ratio of the currents I_{DS1} (before capture) and I_{DS2} (after capture) increases exponentially with Δy (Fig. 15.7d), and becomes maximum near pull-in.

15.5 Conclusions

In this chapter, we have discussed various ways of detecting bio-molecules using nanocantilevers. Classical resonant mode biosensors rely on the change in the resonance frequency of vibrating cantilever and require complex optical instrumentation for detection, especially when very high sensitivity is desired. Stress-based static mode sensors detect the deflection of the tip of the cantilever and responds linearly to change in the stress. To achieve better sensitivity than achieved by classical linear biosensors, critical-point nonlinear bio-sensors have started to appear in the literature. For example, we discussed bifurcation-based mass sensors that operate close to a saddle-node bifurcation. Finally, we have discussed Flexure-FET biosensor that integrates a transistor for direct electrical readout and utilizes nonlinear electromechanical coupling for its exponential sensitivity. We believe that these critical point nonlinear biosensors with electrical readout will offer opportunity to integrate highly sensitive sensors in low-cost point-of-care applications.

References

1. J. L. Arlett, E. B. Myers, and M. L. Roukes, Comparative advantages of mechanical biosensors, *Nat. Nanotechnol.*, **6**, 203–215, 2011.
2. J. Tamayo, P. M. Kosaka, J. J. Ruz, A. San Paulo, and M. Calleja, Biosensors based on nanomechanical systems, *Chem. Soc. Rev.*, **42**(3), 1287–1311, 2013.
3. F. J. Norton, Gas analyzer, U.S. Patent 2,307,8001943.

4. G. Binnig and C. F. Quate, Atomic force microscope, *Phys. Rev. Lett.*, **56**(9), 930–933, Mar. 1986.

5. N. V. Lavrik, M. J. Sepaniak, and P. G. Datskos, Cantilever transducers as a platform for chemical and biological sensors, *Rev. Sci. Instrum.*, **75**, 2229–2253, 2004.

6. P. Bergveld, Thirty years of ISFETOLOGY—What happened in the past 30 years and what may happen in the next 30 years, *Sens. Actuators B-Chem.*, **88**, 1–20, 2003.

7. G. M. Rebeiz and J. B. Muldavin, RF MEMS switches and switch circuits, *IEEE Microwave Mag.*, **2**, 59–71, 2001.

8. J. J. Yao, RF MEMS from a device perspective, *J. Micromechan. Microeng.*, **10**, R9–R38, 2000.

9. V. Pott, H. Kam, R. Nathanael, J. Jeon, E. Alon, and T.-J. King Liu, Mechanical computing redux: Relays for integrated circuit applications, *Proc. IEEE*, **98**(12), 2076–2094, Dec. 2010.

10. H. Kam, D. T. Lee, R. T. Howe, and T.-J. King, A new nano-electro-mechanical field effect transistor (NEMFET) design for low-power electronics, presented at *IEEE International Electron Device Meeting*, 2005. http://ieeexplore.ieee.org/xpl/login.jsp?tp=&arnumber=1609380&url=http%3A%2F%2Fieeexplore.ieee.org%2Fiel5%2F10701%2F33791%2F01609380.pdf%3Farnumber%3D1609380

11. T. Bifano, Adaptive imaging: MEMS deformable mirrors, *Nat. Photon.*, **5**(1), 21–23, Jan. 2011.

12. Q. M. Technologies, Value Proposition of mirasol Displays, 2011. https://www.qualcomm.com/media/documents/files/mirasol-value-proposition.pdf.

13. L. M. Roylance and J. B. Angell, A batch-fabricated silicon accelerometer, *IEEE Trans. Electron Devices*, **26**, 1911–1917, 1979.

14. P. R. Nair and M. A. Alam, Performance limits of nanobiosensors, *Appl. Phys. Lett.*, **88**(23), 233120, Jun. 2006.

15. P. R. Nair and M. A. Alam, Theory of 'Selectivity' of label-free nanobiosensors: A geometro-physical perspective, *J. Appl. Phys.*, **107**(6), 64701, Mar. 2010.

16. A. Jain, P. R. Nair, and M. A. Alam, Flexure-FET biosensor to break the fundamental sensitivity limits of nanobiosensors using nonlinear electromechanical coupling, *Proc. Natl. Acad. Sci. U. S. Am.*, **109**(24), 9304–9308, Jun. 2012.

17. P. R. Nair and M. A. Alam, Screening-limited response of nanobiosensors, *Nano Lett.*, **8**, 1281–1285, 2008.

18. C. Ziegler, Cantilever-based biosensors, *Anal. Bioanal. Chem.*, **379**(7–8), 946–959, 2004.

19. A. K. Gupta et al., Anomalous resonance in a nanomechanical biosensor, *Proc. Natl. Acad. Sci. U. S. A.*, **103**, 13362–13367, 2006.

20. K. L. Ekinci and M. L. Roukes, Nanoelectromechanical systems, *Rev. Sci. Instrum.*, **76**(6), 2005. http://scitation.aip.org/content/aip/journal/rsi/76/6/10.1063/1.1927327.

21. Y. T. Yang, C. Callegari, X. L. Feng, K. L. Ekinci, and M. L. Roukes, Zeptogram-scale nanomechanical mass sensing, *Nano Lett.*, **6**, 583–586, 2006.

22. F. Sadeghian, H. Goosen, A. Bossche, and F. van Keulen, Application of electrostatic pull-in instability on sensing adsorbate stiffness in nanomechanical resonators, *Thin Solid Films*, **518**, 5018–5021, 2010.

23. E. Gil-Santos et al., Nanomechanical mass sensing and stiffness spectrometry based on two-dimensional vibrations of resonant nanowires, *Nat. Nanotechnol.*, **5**, 641–645, 2010.

24. K. L. Ekinci, Ultimate limits to inertial mass sensing based upon nanoelectromechanical systems, *J. Appl. Phys.*, **95**(5), 2682, Mar. 2004.

25. T. P. Burg et al., Weighing of biomolecules, single cells and single nanoparticles in fluid, *Nature*, **446**(7139), 1066–1069, May 2007.

26. J. Lee, W. Shen, K. Payer, T. P. Burg, and S. R. Manalis, Toward attogram mass measurements in solution with suspended nanochannel resonators, *Nano Lett.*, **10**(7), 2537–2542, Jul. 2010.

27. J. L. Arlett and M. L. Roukes, Ultimate and practical limits of fluid-based mass detection with suspended microchannel resonators, *J. Appl. Phys.*, **108**(8), 084701, Oct. 2010.

28. G. G. Stoney, The tension of metallic films deposited by electrolysis, *Proc. R. Soc. Lond. A*, **82**, 172, 1909.

29. G. H. Wu, R. H. Datar, K. M. Hansen, T. Thundat, R. J. Cote, and A. Majumdar, Bioassay of prostate-specific antigen (PSA) using microcantilevers, *Nat. Biotechnol.*, **19**, 856–860, 2001.

30. K. W. Wee et al., Novel electrical detection of label-free disease marker proteins using piezoresistive self-sensing micro-cantilevers, *Biosens. Bioelectron.*, **20**, 1932–1938, 2005.

31. H. A. C. Tilmans, M. Elwenspoek, and J. H. J. Fluitman, Micro resonant force gauges, *Sens. Actuators A: Phys.*, **30**(1–2), 35–53, Jan. 1992.

32. V. Kumar et al., Bifurcation-based mass sensing using piezoelectrically-actuated microcantilevers, *Appl. Phys. Lett.*, **98**, 2011.

33. N. Abele, R. Fritschi, K. Boucart, F. Casset, P. Ancey, and A. M. Ionescu, Suspended-Gate MOSFET: Bringing new MEMS functionality into solid-state MOS transistor, presented at *IEEE International Electron Device Meeting,* 2005. http://ieeexplore.ieee.org/xpl/login.jsp?tp=&arnumber=1609384&url=http%3A%2F%2Fieeexplore.ieee.org%2Fiel5%2F10701%2F33791%2F01609384
34. H. C. Nathanson, W. E. Newell, R. A. Wickstrom, and J. R. Davis, The resonant gate transistor, *Trans. Electron Devices*, **14**(3), 117–133, 1967.
35. H. Torun, K. K. Sarangapani, and F. L. Degertekin, Spring constant tuning of active atomic force microscope probes using electrostatic spring softening effect, *Appl. Phys. Lett.*, **91**, 2007.
36. S. Krylov and R. Maimon, Pull-in dynamics of an elastic beam actuated by continuously distributed electrostatic force, *J. Vibration Acoustics-Trans. Asme*, **126**, 332–342, 2004.
37. S. Krylov, Lyapunov exponents as a criterion for the dynamic pull-in instability of electrostatically actuated microstructures, *Int. J. Non-Linear Mechan.*, **42**, 626–642, 2007.
38. T. Yuan and N. T. H., *Fundamentals of Modern VLSI Devices.* Cambridge University Press.
39. D. R. Southworth, L. M. Bellan, Y. Linzon, H. G. Craighead, and J. M. C.-A. 163503 Parpia, Stress-based vapor sensing using resonant microbridges, *Appl. Phys. Lett.*, **96**, 2010.
40. D. J. Joe, et al., Stress-based resonant volatile gas microsensor operated near the critically buckled state, *J. Appl. Phys.*, **111**(10), 104517, May 2012.
41. J. Tamayo, D. Ramos, J. Mertens, and M. Calleja, Effect of the adsorbate stiffness on the resonance response of microcantilever sensors, *Appl. Phys. Lett.*, **89**, 2006.
42. P. S. Waggoner and H. G. Craighead, The relationship between material properties, device design, and the sensitivity of resonant mechanical sensors, *J. Appl. Phys.*, **105**, 2009.
43. D. Ramos, J. Tamayo, J. Mertens, M. Calleja, and A. Zaballos, Origin of the response of nanomechanical resonators to bacteria adsorption, *J. Appl. Phys.*, **100**, 2006.
44. H. Craighead, Nanomechanical systems—Measuring more than mass, *Nat. Nanotechnol.*, **2**(18–19), 2007.

Chapter 16

Micro/Nano-Mechanical Cantilevers for Cancer Diagnosis

Kilho Eom[a] **and Taeyun Kwon**[b]

[a]*Biomechanics Laboratory, College of Sport Science, Sungkyunkwan University, Suwon 440-746, Republic of Korea*
[b]*SKKU Advanced Institute of Nano Technology (SAINT), Sungkyunkwan University (SKKU), Suwon 440-746, Republic of Korea*

kilhoeom@skku.edu., taeyunkwon@gmail.com

16.1 Introduction

Cantilever has served as a theoretical model for more than a century for understanding the fundamental principles of mechanics,[1] which can be applicable for designing the engineering structures such as building, automobile, aircraft, and so forth. In particular, a cantilever system has been employed to quantitatively characterize the mechanical deformation (e.g., bending, torsional, and axial deformation) of the system due to a mechanical load (force). The fundamental principles of such deformations using a cantilever system have dated back to the 1920s, in which Timoshenko[2] had largely contributed to the development of elasticity theory using a simple elastic model such as the beam

Nanocantilever Beams: Modeling, Fabrication, and Applications
Edited by Ioana Voiculescu and Mona Zaghloul

ISBN 978-981-4613-23-1 (Hardcover), 978-981-4613-24-8 (eBook)
www.panstanford.com

model (e.g., cantilever). Currently, the computational techniques such as finite element analysis, which has emerged in the 1970s, have been widely utilized along with classical elasticity theory (sometimes based on a simple model such as a cantilever system) to design the functional structures (e.g., building, automobile, etc.) that can sustain the load. Now, the cantilever system has appeared in the textbook for engineering undergraduate students,[1] since the cantilever is a simple but rigorous model that helps engineering students build up the fundamental concepts of mechanics.

The cantilever, which opened an avenue in classical mechanics (particularly, elasticity) about a century ago as stated above, has recently paved a new way for characterizing the mechanical deformation behavior of nanoscale structures and materials. In particular, an atomic force microscopy (AFM) that allowed Binnig[3] to become a Nobel laureate, was developed as an imaging toolkit in 1980. The basic principle of AFM imaging is to measure the interaction between AFM force probe and a sample surface that is desired to be imaged, where a cantilever is used as an AFM force probe to estimate the interaction force (between the sample and AFM tip).[4] The AFM imaging technique along with theories developed based on the cantilever system has enabled the researchers to acquire the high-resolution images of atomic/molecular surfaces with nanometer precisions.[5]

In a past decade, the cantilever that was previously employed as a force probe to image the surface at atomic resolution, has enabled the label-free detection of biological or chemical molecules, which are the signature of specific diseases like cancers, based on the chemomechanics transduction mechanism.[6–8] Specifically, the adsorption of biological/chemical molecules onto the surface of a cantilever leads to a change in the surface energy of a cantilever and, consequently, generation of a surface stress[9] acting on the surface of the cantilever. This surface stress change due to molecular adsorption, which occurs on the cantilever's surface, results in the bending deflection of a cantilever. This detection scheme based on measuring the deflection change of a cantilever due to molecular adsorption was first employed by Gerber and coworkers,[10] who measured the surface stress driven by adsorption of alkanethiol onto a gold surface. This detection scheme has been widely employed for highly sensitive, label-free detection of biological molecules such as DNA, RNA, and proteins with their low concentrations (e.g., ~1 pM). The sensitive detection of specific biological molecules is of high significance for early diagnosis of diseases such as cancers, because at the early stage of cancer development the amount of marker proteins (or enzymes) expressed on a cancer cell is difficult to be sensed by a conventional bioanalytical toolkit such as enzyme-linked immunosorbent assay (ELISA). Recent studies have shown the potential of cantilever sensor in the sensitive identification of early stage of cancer by measuring the minute amount of marker proteins (or enzymes) expressed on cancerous cells (for example, see Section 16.3.4).

More recently, instead of measuring the deflection change of a cantilever due to molecular adsorption, the label-free detection has been implemented by estimating the resonant frequency behavior of a cantilever in response to molecular adsorption onto the cantilever.[11–13] The detection scheme based on the resonance of a cantilever has been receiving a significant attention due to the fact that such a detection scheme may allow the miniaturization of a cantilever. Specifically, since the resonant frequency (ω) of a cantilever is inversely proportional to the square of the length of a cantilever,[14] i.e., $\omega \propto L^{-2}$, the miniaturization of a cantilever increases its resonant frequency and, consequently, the frequency shift due to the molecular adsorption (i.e., the detection sensitivity of a cantilever). In recent years, a nanocantilever has been utilized to sense and detect the specific chemical/biological molecules even at single-molecule resolution[15]; this implies that the nanocantilever can be useful in detecting a single or few biological/chemical molecules, which may open a new window for the development of nanosensors whose detection limit approaches the single-molecule resolution.

This chapter is aimed toward presenting the current state of the art of cantilever-based label-free detection for future applications in early diagnosis of cancer, and the underlying principles for a cantilever-based detection. Section 16.2 describes physical principles, which are useful in delineating the mechanical behavior (e.g., bending and/or resonance behaviors) of a cantilever as well as understanding the underpinning mechanisms of a cantilever-based label-free detection. Section 16.3 is dedicated to recent efforts that have been made to develop the label-free biological detection based on measurement of resonance behavior of the cantilever in response to biomolecular adsorption. Moreover, in Section 16.3, we also discuss fundamental mechanisms in cantilever-based biological detection by using physical models described in Section 16.2. Finally, in Section 16.4, we present the future perspectives in cantilever-based biological detection for future cancer diagnosis.

16.2 Physical Principles

This section reviews the fundamental principles of cantilever-based diagnosis. The early diagnosis of cancer via cantilever assay is attributed to the ability of a cantilever sensor to detect extremely minute amount of disease-specific protein molecules, DNA molecules, or RNA molecules. The detection principle of a cantilever sensor is the transduction of chemical interactions on a cantilever's surface into the change of mechanical behavior for a cantilever. In particular, when target molecules are captured on a cantilever's surface, where probe molecules having binding affinity with target molecules are functionalized, the molecular binding between probe molecules and target molecules lead to the change of physical properties for a cantilever, such as surface stress or the effective mass of a cantilever. The change of these physical properties can be estimated by measuring the cantilever's bending deflection or resonant frequency. Moreover, in this section, we particularly pay attention to physical models that enable the characterization of the vibrational properties of a cantilever, since the resonating cantilever is more useful for highly sensitive detection. Here, it should be noted that the physical models described in this section are restricted to continuum mechanics model that may fail to capture the nanoscale effect, which affects the vibrational behavior of nanoscale resonator (or cantilever), since the details of molecular interactions or atomistic effects (e.g., finite size effect) are lacking in the continuum mechanics model. If the reader is more interested in multiscale models that couple atomistic model to continuum model for gaining insight into the vibrational characteristics of nanoscale resonator, they may read the review article[13] that properly describes the issue of multiscale modeling as well as future perspectives in modeling that can be coupled with theory and experiments.

16.2.1 Continuum Mechanics: Bending Deflection

For the past decade, a cantilever has been utilized for studying the surface chemistry, particularly chemical interaction between molecule and surface. For example, Gerber and coworkers[10] quantitatively studied the kinetics of alkanethiol adsorption onto a gold surface by measuring the cantilever's bending deflection. The fundamental principle of cantilever-based characterization of molecular adsorption onto a surface is that when molecules are adsorbed onto the surface of a cantilever, the surface stress is generated due to the change of surface energy driven by molecular adsorption. This surface stress results in the bending deflection of a cantilever (see Fig. 16.1).

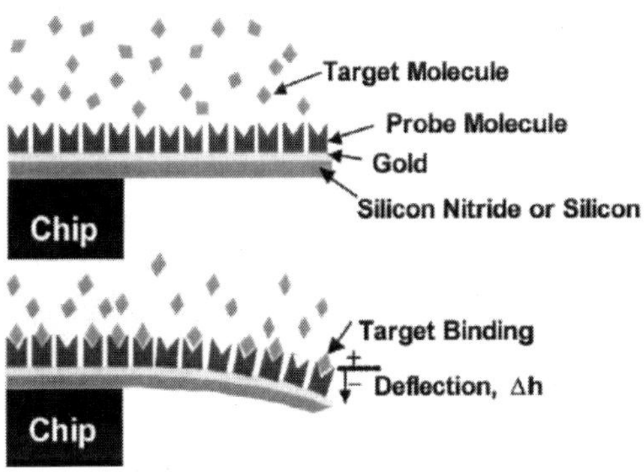

Figure 16.1 Detection principle of cantilever-based sensing based on measuring the bending deflection of a cantilever due to biomolecular adsorption. When target molecules are specifically bound to receptor molecules that were functionalized on the surface of a cantilever, the biomolecular binding results in the surface stress change that leads to the bending deflection motion of the cantilever. Reprinted with permission from ref. 18.

This principle was mathematically expressed via Stoney's formula,[16] which provides a relationship between the surface stress and the deflection of a cantilever such as

$$\tau = \frac{Et^2}{6R(1-\nu)} = \frac{E}{3(1-\nu)}\left(\frac{t}{L}\right)^2 \Delta w, \tag{16.1}$$

where τ is the surface stress, E, ν, t, and L represent the elastic modulus, Poisson's ratio, thickness, and length of a cantilever, respectively, and Δw is the bending deflection change of a cantilever due to the surface stress. Here, it is assumed that a cantilever undergoes the pure bending.

In last decades, the detection principle based on Stoney's formula given in Eq. 16.1 has been employed for developing the cantilever-based molecular detection. For the label-free detection of biological molecules, the surface of a cantilever has to be chemically modified in order for a cantilever to specifically capture the biological molecules such as proteins, DNA, or RNA. Specifically, the chemical modification of the surface of a cantilever is implemented by immobilizing receptor molecules (that can specifically capture target molecules) on the surface of the cantilever. This detection principle was used for studying the DNA hybridization[17,18]—the effect of ionic strength or hydration force on DNA hybridization was quantitatively characterized by measuring the deflection change of a cantilever driven by DNA hybridization.[19] Recent studies[20,21] report the possibility of not only detecting DNA molecules with their short nucleotide lengths but also sensing the single-nucleotide polymorphism of DNA molecules by measuring the deflection change of a cantilever due to DNA hybridization. Besides DNA detection, the cantilever bioassay was also reported for sensing disease-specific proteins (e.g., see Section 16.3.2).

Even though this detection principle was widely employed for detecting biological molecules as described above, this principle is subjected to following restrictions. First, the length of a cantilever should be at least 100 μm for reliable measurement of a cantilever's bending deflection change driven by molecular interaction on a cantilever's surface. In particular, as described in Eq. 16.1, the bending deflection change due to molecular interaction-driven surface stress is proportional to the square of a cantilever's length, i.e., $\Delta w \propto L^2\tau$. This indicates that this detection principle does not allow a cantilever to be scaled down to nanoscale. Second, it is difficult to quantify the amount of adsorbed molecules onto a cantilever's surface. These two restrictions have researchers to take into account detection principle based on measuring the resonant frequency behaviors of a cantilever in response to molecular interaction. The resonance-based detection principle is delineated in Section 16.2.3.

16.2.2 Continuum Mechanics Theory: Resonant Frequency

In this section, we describe physical theories (or models) that explain the resonance behavior of various resonant systems such as piezoelectric cantilever and doubly clamped nano/micro-scale structures. Here, it is essential to characterize the resonance behavior of a cantilever for understanding the frequency shift of a cantilever due to biomolecular adsorption, since the frequency shift is closely related to the resonance behavior of the cantilever. Moreover, we have also delineated the role of medium surrounding a cantilever or mechanical tension applied to the cantilever in its resonance behavior, which may be correlated with the sensing performance of the cantilever, by using continuum mechanics models.

16.2.2.1 Theoretical models

16.2.2.1.1 Theory: Euler–Bernoulli beam model

To characterize the vibrational properties such as resonant frequency of a cantilever, we consider the continuum mechanics theory such as Euler–Bernoulli beam theory, which is useful in depicting the mechanical deformation of a one-dimensional micro/nanoscale structure. In particular, when the length of the structure is much larger than other transverse dimensions (i.e., thickness and width), the Euler–Bernoulli beam theory is acceptable for modeling such one-dimensional micro/nanoscale structures. The vibrational behavior of micro-electromechanical system (MEMS) or nano-electromechanical system (NEMS) devices can be dictated by following governing equation for Euler–Bernoulli beam model.[22]

$$EI\partial_x^4 w(x,t) + c\partial_t w(x,t) + \rho A\partial_t^2 w(x,t) = 0, \qquad (16.2)$$

where $w(x, t)$ is the transverse deflection of a cantilever, E, I, ρ, and A indicate the elastic modulus, cross-sectional moment of inertia, mass density, and cross-sectional area of a MEMS/NEMS cantilever, respectively, and c is a damping coefficient that arises from the friction between a cantilever and the molecules of medium surrounding the cantilever. Here, it should be noted that the governing equation given by Eq. 16.2 is only applicable to the harmonic oscillation of a device. In other words, the nonlinear bending deflection of a device cannot be characterized by Eq. 16.2. For a theoretical convenience, we consider a case in which a device operates in a vacuum, that is, the damping effect can be ignored. For a vibrating device, the transverse deflection can be represented in the form of $w(x, t) = q(x) \times \exp[j\omega t]$, where ω and $q(x)$ are the resonant frequency of a device and its corresponding deflection eigenmode, respectively, and j is the unit of a complex number. As a consequence, the governing equation given by Eq. 16.2 becomes the eigenvalue problem as follows:

$$L[q] = \lambda q, \tag{16.3a}$$

where $L[\bullet]$ and λ are the differential operator and eigenvalue such as

$$L[\bullet] = EI\partial_x^4(\bullet) \text{ and } \lambda = \omega^2 \rho A \tag{16.3b}$$

When a specific boundary condition is prescribed, it is straight forward to characterize the vibrational behavior of a NEMS/MEMS device by solving the eigenvalue problem given in Eq. 16.3.

16.2.2.1.2 *Piezoelectric cantilever*

In a last decade, the piezoelectric cantilevers have been developed in such a way that such cantilevers consist of two layers such as piezoelectric layer and substrate layer. Here, piezoelectric layer is utilized to actuate a cantilever, while the substrate layer is used for patterning. In recent years, Kwon and colleagues[23,24] developed a novel piezoelectric film, i.e., $0.1Pb(Zn_{0.5}W_{0.5})$-$0.9Pb(Zn_{0.5}Ti_{0.5})O_3$ (PZW-PZT) film, which is useful to actuate MEMS/NEMS devices. The resonant frequency of piezoelectric cantilever made of two layers such as piezoelectric and substrate layers can be predicted based on the linear elastic Euler–Bernoulli beam model. In particular, the bending rigidity of the piezoelectric cantilever is given by[25,26]

$$\kappa = \frac{1}{12} E_p h_p^3 f(\alpha, \beta), \tag{16.4a}$$

where κ is the bending rigidity of a piezoelectric cantilever, E_p and h_p are the elastic modulus and thickness of a piezoelectric film, respectively, and $f(\alpha, \beta)$ is a dimensionless function defined as

$$f(\alpha,\beta)=\frac{1+\alpha^2\beta^4+\alpha\beta(2+2\beta^2+3\beta)}{1+\alpha\beta}. \tag{16.4b}$$

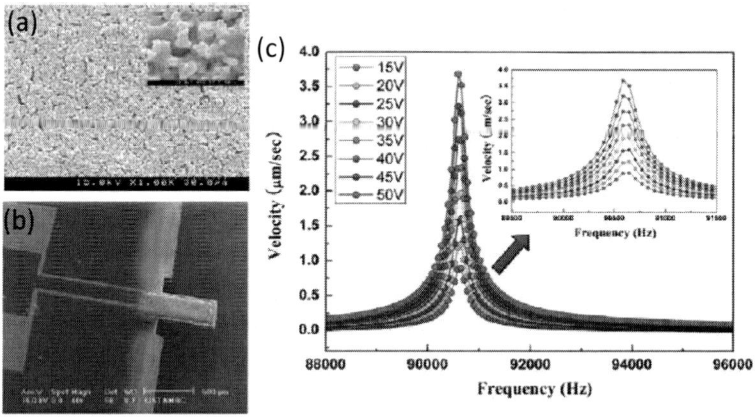

Figure 16.2 (a) Microscope image of piezoelectric film. (b) Scanning electron microscope (SEM) image of a fabricated piezoelectric microcantilever. (c) Resonance curves of piezoelectric microcantilever as a function of actuation voltage. Reprinted with permission from ref. 13.

Here, α and β are dimensionless parameters given by $\alpha = E_s/E_p$ and $\beta = h_s/h_p$, where E_s and h_s are the elastic modulus and thickness of a substrate, respectively. The effective mass per unit length of a piezoelectric cantilever is given as $m^* = \rho_p h_p + \rho_s h_s$, where ρ_p and ρ_s are the mass densities of piezoelectric film and substrate, respectively. Consequently, the resonant frequency of a piezoelectric cantilever operating in vacuum or air can be represented in the form of

$$\omega_n = \frac{\lambda_n^2}{2\pi L^2}\sqrt{\frac{\kappa}{m^*}}, \tag{16.5}$$

where ω_n is the resonant frequency at n-th flexural mode, λ_n is the n-th eigenvalue for cantilevered boundary condition. Table 16.1 shows that the experimentally measured resonant frequencies of piezoelectric cantilevers can be fitted with theoretical model described in Eq. 16.5. Figure 16.2 shows the frequency-amplitude behaviors of piezoelectric cantilevers.

Table 16.1 Experimentally measured and theoretically calculated resonant frequencies of piezoelectric microcantilevers with their various dimensions

Cantilever dimension (width × length) (μm × μm)	Experimentally measured resonant frequency (Hz)	Theoretically predicted resonant frequency (Hz)
400 × 380	154,857	154,770–155,130
400 × 480	99,250	99,052–99,281
400 × 580	68,875	68,786–68,945

16.2.2.1.3 *Doubly clamped micro/nanostructures*

Recently, doubly clamped nanoscale structures such as nanowires[27] and/or nanotubes[28] fabricated from bottom-up or top-down approaches have been employed as a resonant device that exhibits the high-frequency dynamic ranges even up to giga (10^9) hertz. The vibrational behavior of a doubly clamped beam is depicted by a governing equation as follows:

$$EI\partial_x^4 w(x,t) - [N + (EA/2L)\int_0^L \{\partial_x w(x,t)\}^2 dx]\partial_x^2 w(x,t) + \rho A \partial_t^2 w(x,t) = 0, \tag{16.6}$$

where N is an axial force applied to the clamped ends of a beam. Here, a second term in the parenthesis indicates the geometric nonlinear effect due to clamped boundary condition. For small deformation, the resonant frequency of a doubly clamped beam without any application of axial force N is given as

$$\omega_n = \frac{\xi_n^2}{2\pi L^2}\sqrt{\frac{EI}{\rho A}}, \tag{16.7}$$

where ξ_n is the n-th eigenvalue for doubly clamped boundary condition, i.e., $\xi_1 = 4.73$ (for the first harmonic flexural mode), $\xi_2 = 7.85$ (for the second harmonic mode), etc. A recent study by Roukes and coworkers[29] reports that the resonance behavior of a doubly clamped piezoelectric GaAs beam is well described by a Euler–Bernoulli beam model delineated in Eq. 16.7. The resonant frequency of piezoelectric GaAs beam can be tuned by using DC bias voltage V_{dc}, which results in the production of axial strain to GaAs beam due to piezoelectric effect. In particular, the axial force due to piezoelectric effect is given by $N = \sigma_r A = EAd_{3j}V_{dc}$, where σ_r is a residue stress driven by piezoelectric effect, and d_{3j} is an anisotropic piezoelectric constant. Accordingly, the resonant frequency of a piezoelectric GaAs beam is written as

$$\omega_n^* = \frac{\mu_n^2}{2\pi L^2}\sqrt{\frac{EI}{\rho A}\left(1 + \frac{d_{3j}V_{dc}L^2}{n^2\pi^2 I^2}\right)}, \tag{16.8}$$

When the frequency shift due to DC bias voltage is small such that $d_{3j}V_{DC}L^2/n^2\pi^2 I << 1$, the fundamental resonant frequency shift is measured as

$$\Delta\omega_1 = \omega_1^* - \omega_1 \approx -\frac{d_{3j}V}{2\pi h^2}\sqrt{\frac{3E}{\rho}}. \tag{16.9}$$

Here, h is the thickness of a piezoelectric beam. As shown in Fig. 16.3, the frequency shift due to DC bias voltage for a piezoelectric beam is well fitted to continuum mechanics model given by Eq. 16.9.

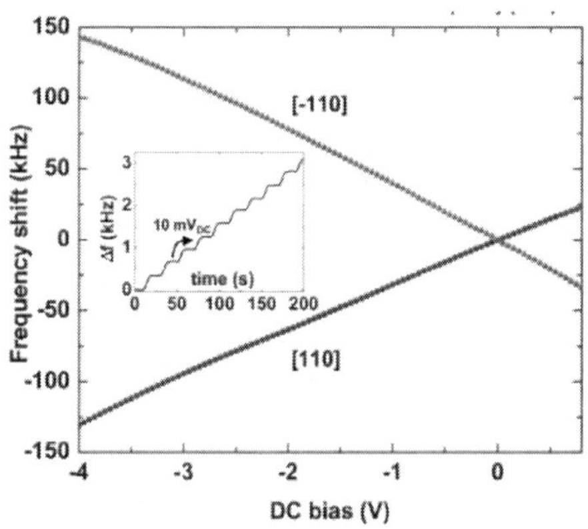

Figure 16.3 Frequency shift of a piezoelectric cantilever due to applied DC bias voltage. Here, the bias voltage induces the residual stress that serves as an axial load that can affect the resonant frequency of the cantilever. Reprinted with permission from ref. 29.

16.2.2.2 Effect of mechanical tension

In recent years, efforts have been made to improve the dynamic behavior of resonant devices such as doubly clamped nanobeam. In particular, Craighead and coworkers[30] have reported that a mechanical tension is a control parameter that is useful in tuning the resonance behavior of a doubly clamped nanostructure. It is shown that a mechanical tension applied to a doubly clamped beam allows for increasing not only the resonant frequency of a doubly clamped beam but also its quality factor that represents the rate at which the energy of a resonant device is lost per vibrational period due to interaction between a device and environment or due to intrinsic flaw or defect of a device. Here, we only consider the resonant frequency of a device, since the label-free detection of atoms or molecules has been implemented by measuring the frequency behavior of a device in response to atomic or molecular adsorption onto a device. The resonance characteristic of a doubly clamped nanobeam operating in vacuum or air under applied mechanical stress σ_0 can be dictated by a governing equation given as

$$EI\partial_x^4 w(x,t) - \sigma_0 A\partial_x^2 w(x,t) + \rho A\partial_t^2 w(x,t) = 0. \tag{16.10}$$

Here, E, I, ρ, and A represent the elastic modulus, cross-sectional moment of inertia, mass density, and cross-sectional area of a nanobeam, respectively. By solving Eq. 16.10, the resonant frequency of a nanobeam under applied mechanical stress is obtained as

$$\omega_n = \omega_n^0 \sqrt{1 + \Gamma_n}, \tag{16.11a}$$

where ω_n^0 is the resonant frequency of a nanobeam without any application of mechanical stress, and Γ_n is the dimensionless mechanical tension defined as

$$\Gamma_n = \frac{1}{\alpha_n}\left(\frac{\sigma_0}{E}\right)\left(\frac{L}{h}\right)^2, \tag{16.11b}$$

where L and h are the length and thickness of a nanobeam, respectively, and α_n is a constant depending on the mode index n. Here, it should be noted that the dimensionless tension Γ_n is a ratio of mechanical tension to a critical load that is the buckling load, i.e., $\Gamma_n = \sigma_0 A/P_{cr}$, where $P_{cr} = \pi^2 EI/L^2$. This indicates that when the amount of mechanical tension is comparable to a critical buckling load, the resonant frequency will be significantly enhanced (see Fig. 16.4). On the other hand, the compressive mechanical force will critically reduce the resonant frequency of a nanobeam.

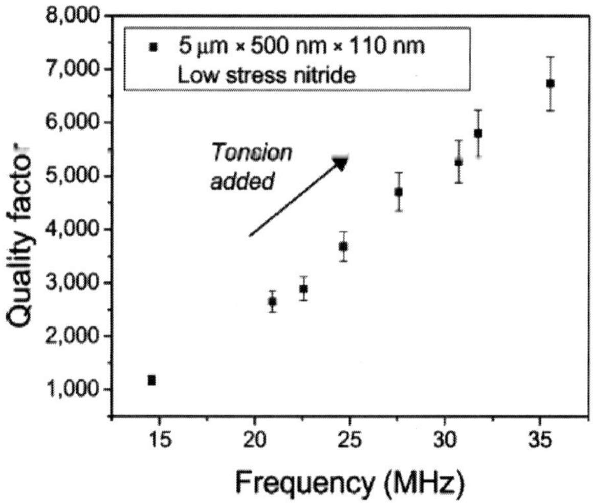

Figure 16.4 Effect of mechanical tension on the resonant frequency and quality factor of nanomechanical cantilever. Reprinted with permission from ref. 30.

16.2.2.3 Effect of hydrodynamic force

In order for a cantilever sensor to serve as a biosensor, it has to operate in an aqueous environment rather than vacuum or air, since the biological function of target biomolecules will be lost when they are exposed to vacuum or air. When a cantilever vibrates in aqueous environment, the resonance behavior of the cantilever will be affected by interactions between the cantilever and water molecules surrounding the cantilever. This effect of water molecules surrounding a cantilever on its resonance behavior is referred to as *hydrodynamic effect.* The hydrodynamic effect on the vibrational behavior of a cantilever was theoretically analyzed in Sader's model,[31] which was verified by numerical simulations by Paul and Cross,[32] finite element simulation by Raman and colleagues,[33] and experiments by Chon et al.[34]

The equation of motion for a cantilever operating in aqueous environment is described by a governing equation as follows:

$$EI\partial_x^4 w(x,t) + \rho A \partial_t^2 w(x,t) = f_H(x,t), \tag{16.12}$$

where $f_H(x, t)$ is a hydrodynamic force acting on a cantilever due to water molecules surrounding the cantilever. By using a Fourier transform, the equation of motion given by Eq. 16.12 becomes

$$\frac{EI}{L^4}\frac{d^4W(\xi,\omega)}{d\xi^4} - \omega^2 \rho A W(\xi,\omega) = F_H(\xi,\omega), \tag{16.13a}$$

where ξ is a rescaled coordinate defined as $\xi = x/L$, where L is the length of a cantilever, and Fourier transform is defined as

$$G(x,\omega) = \int_{-\infty}^{\infty} g(x,t) e^{-j\omega t} dt \tag{16.13b}$$

for any function $g(x, t)$ with j being the unit of a complex number. The hydrodynamic force can be computed from Navier–Stokes equation, and represented in the form[31]

$$F_H(x,\omega) = \frac{\pi}{4}\rho_f \omega^2 b^2 \Gamma(\omega) W(x,\omega). \tag{16.14}$$

Here, ρ_f is the mass density of a fluid surrounding a cantilever, b is the width of a cantilever, and $\Gamma(\omega)$ is the dimensionless hydrodynamic function. Accordingly, from Eqs. 16.13 and 16.14, the equation of motion represented in Fourier space is given by

$$\frac{d^4W(\xi,\omega)}{d\xi^4} - [\Lambda(\omega)]^4 W(\xi,\omega) = 0, \tag{16.15a}$$

where

$$\Lambda(\omega) = \lambda_1 \sqrt{\frac{\omega}{\omega_0}\left[1 + \frac{\pi \rho_f b^2}{4\rho A}\Gamma(\omega)\right]}. \tag{16.15b}$$

Here, λ_1 is the eigenvalue for a fundamental mode ($\lambda_1 = 1.87$ for cantilevered boundary condition), and ω_0 is a resonant frequency measured in a vacuum, i.e., $\omega_0 = (\lambda_1/L)^2(EI/\rho A)^{1/2}$. When the cantilever vibrates due to an excitation driven by Brownian motion of fluid, the hydrodynamic function can be regarded as a constant, and consequently, the function $\Lambda(\omega)$ can be written as[31]

$$\Lambda(\omega) \approx \Lambda_n(\omega) = \lambda_n \sqrt{\frac{\omega}{\omega_n^0}} \left[1 + \frac{\pi \rho_f b^2}{4\rho A} \left\{\Gamma_R(\omega_n^R) + j\Gamma_I(\omega_n^R)\right\}\right], \quad (16.16)$$

where ω_n^0 is the resonant frequency at the n-th flexural mode measured in a vacuum, and ω_n^R is the resonant frequency for flexural mode n in the absence of dissipative effect, i.e., $\Gamma_R >> \Gamma_I$. Consequently, the resonant frequency of a cantilever immersed in a fluid without any dissipative effect is represented in the form

$$\frac{\omega_n^R}{\omega_n^0} = \left[1 + \frac{\pi \rho_f b^2}{4\rho A} \Gamma_R(\omega_n^R)\right]^{-1/2}. \quad (16.17)$$

It should be noted that when the Reynolds number of a fluid becomes infinite (i.e., $Re = \rho\omega_1^0 b^2/\eta \to \infty$), the real component of hydrodynamic function approaches a unit, and thus the resonant frequency is obtained as

$$\frac{\omega_n^R}{\omega_n^0} = \left[1 + \frac{\pi}{4}\left(\frac{\rho_f}{\rho}\right)\left(\frac{b}{h}\right)\right]^{-1/2}, \quad (16.18)$$

where h is the thickness of a cantilever. From Eq. 16.18, it is straightforward to note that the hydrodynamic loading dictated by the density of a fluid plays a critical role on the resonance behavior of a cantilever immersed in the fluid (see Fig. 16.5).

16.2.3 Detection Principles

In the previous section, we present a theoretical background that enables the understanding of the vibrational behavior of cantilevers operating in various environments such as vacuum (or air) or fluid. Based on this theoretical background, we need to establish the framework that reveals detection principle for cantilever-based sensing. Here, we consider only the principles of cantilever-based detection based on measurement of frequency change for a cantilever due to atomic (molecular) adsorption on the cantilever's surface. Here, we take into account three major effects—(i) mass of adsorbed molecules, (ii) surface stress driven by molecular adsorption, and (iii) the elastic property of adsorbed molecules—playing a role in developing the detection principles.

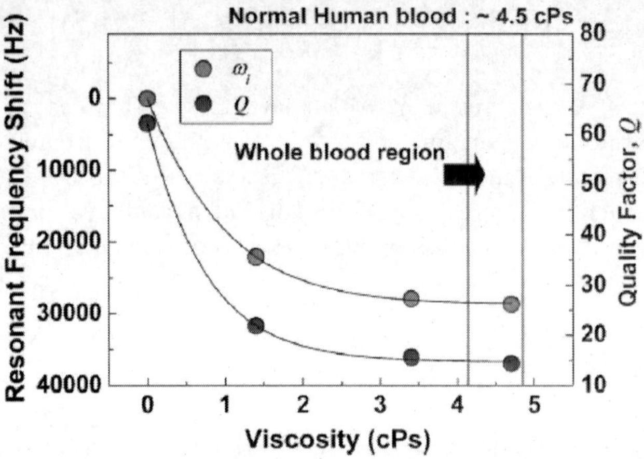

Figure 16.5 The change of resonant frequency for a microcantilever when it was immersed in a viscous fluid is shown. The viscous fluid affects the resonant frequency and quality factor of piezoelectric microcantilever due to hydrodynamic effect. Reprinted with permission from ref. 36.

16.2.3.1 Mass effect

The detection sensitivity of a resonating cantilever is related to its dimension, particularly its length, since the length of a cantilever determines its resonant frequency. In case of a cantilever vibrating in vacuum or air, the frequency change of a cantilever due to molecular adsorption onto the cantilever's surface can be represented in the form[35]

$$\frac{\Delta\omega_0}{\omega_0} \approx -\frac{1}{2}\frac{\Delta m}{m}. \tag{16.19}$$

Here, $\Delta\omega_0$ is the resonant frequency shift of a cantilever operating in vacuum (or air) due to adsorbed molecules with their total mass of Δm, while ω_0 and m are the resonant frequency and mass of a bare cantilever, respectively. It should be noted that a negative sign in Eq. 16.19 indicates that molecular adsorption decreases the resonant frequency of a cantilever, and that Eq. 16.19 holds only for $\Delta m << m$. Moreover, we note that a detection principle depicted in Eq. 16.19 is suitable for a case, where the elastic modulus of adsorbed molecules is much smaller than that of a cantilever. This assumption is appropriate for biomolecular detection using a resonant cantilever operating in air, since the elastic modulus of proteins or cells is measured in the order of 10 MPa to 1 GPa, which is much smaller than the elastic modulus of metal cantilever.

For application of a cantilever in the label-free detection of biological molecules, the detection principle given in Eq. 16.19 is insufficient, since the principle holds for the case in which a cantilever vibrates in vacuum or air rather than aqueous environment such as water. The inapplicability of Eq. 16.19 for understanding the frequency shift of a cantilever immersed in a fluid due to biomolecular adsorption is attributed to the hydrodynamic effect, which was described in Section 16.2.2.3. In particular, when proteins are adsorbed onto the surface of a cantilever immersed in water, the water molecules bound to proteins may affect the resonance behavior of the cantilever. The resonant frequency shift of a cantilever operating in a fluid owing to biomolecular adsorption is given by[36]

$$\frac{\Delta\omega}{\omega} = \frac{1}{2}\frac{\Delta m}{m}\theta + \frac{1}{2}\frac{\Delta m_{\mathrm{H}}}{m_{\mathrm{H}}}(1-\theta) \tag{16.20}$$

where $\Delta\omega$ and ω are the frequency shift and reference frequency measured in a fluid, m_{H} is the hydrodynamic loading acting on a cantilever due to water molecules, and Δm_{H} is the change of hydrodynamic loading due to biomolecular interaction on the surface of a cantilever vibrating in a fluid. Here, dimensionless parameter θ is defined as $\theta = m/(m + m_{\mathrm{H}})$ with a relation of

$$\frac{m_{\mathrm{H}}}{m} = \left(\frac{b}{h}\right)\left(1 + \frac{4}{(\lambda_{\mathrm{i}}b/L)\sqrt{b^2\omega_{\mathrm{i}}^0/v}}\right)\left(\frac{\rho_{\mathrm{F}}}{\rho}\right), \tag{16.21}$$

where b, h, and L are the width, thickness, and length of a cantilever, respectively, ω_{i}^0 is the resonant frequency of a cantilever operating in vacuum, v is a kinetic viscosity, ρ_{F} and ρ are the mass densities of a fluid and a cantilever, respectively.

16.2.3.2 Surface stress effect

As described in Section 16.2.1, when target molecules are bound to the functionalized surface of a cantilever, this molecular adsorption changes the surface state of a cantilever, specifically the surface energy of a cantilever, which results in generation of a surface stress that may affect the resonance behavior of a cantilever. Moreover, a scaling down of a cantilever leads to emergence of the surface stress, which is attributed to eminent imbalance between surface atoms and bulk atoms for a cantilever. Here, we review the current physical models that can explain the role of surface stress on the resonance behavior of a cantilever.

16.2.3.2.1 *One-dimensional model: Beam model*

In the earlier work of Thundat and coworkers,[37] it was found that the total mass of adsorbed biomolecules is insufficient to depict the resonant frequency shift of a cantilever due to such biomolecular adsorption. They conjectured that the frequency shift due to molecular adsorption may be ascribed to not only the total mass of adsorbed molecules but also the surface stress driven by molecular adsorption. The equation of motion for a cantilever with molecular adsorption was described as[38–40]

$$EI\partial_x^4 w(x,t) - N\partial_x^2 w(x,t) + \rho A\partial_t^2 w(x,t) = 0. \tag{16.22}$$

Here, N is an axial force induced by a surface stress such as $N = \tau_0 A$, where τ_0 is a constant surface stress driven by molecular adsorption. This theoretical model depicted in Eq. 16.22 was utilized in recent works by McFarland et al.,[39] Hwang et al.,[38] and Dorignac et al.[40] for studying the frequency response of a cantilever to (bio)molecular adsorption. The frequency shift due to a constant surface stress driven by molecular adsorption is represented in the form

$$\frac{\Delta\omega}{\omega_0} \approx \frac{\tau_0 L^3}{\pi EI}. \tag{16.23}$$

However, it should be noted that a theoretical model given by Eq. 16.22 violates Newton's law of motion, particularly Newton's third law—the law of action and reaction. Specifically, when a surface stress was produced due to molecular adsorption, there should be a residual stress that is a reaction to the surface stress. As described in the original work of Gurtin[41] and other recent studies,[42,43] it has been shown that a constant surface stress cannot affect the resonant frequency of a cantilever. This suggests that a phenomenological model dictated by Eq. 16.22 has to be modified in order to satisfy the physical law—Newton's laws of motion.

16.2.3.2.2 *Three-dimensional model: Plate model*

Sader and Lachut[44] revisited the issue of a role that a constant surface stress plays in the resonant frequency behavior of a cantilever by considering a three-dimensional elastic model such as a plate model. For gaining insight into a relationship between a constant surface stress and frequency shift due to such a surface stress, they used a scaling law rather than solving the analytically intractable equation of motion for the plate model.

First, let us consider a plate bearing the surface stress without any boundary conditions (i.e., essential boundary conditions). The force equilibrium with assuming no traction boundary conditions at the edges of the plate provides an in-plane stress state (σ_x, σ_y) given as $\sigma_x = \sigma_y = \tau_0/h$, where τ_0 is the total surface stress, and h is the thickness of a cantilever. Here, the coordinate x is defined along the longitudinal direction of the plate, while the coordinate y is defined along the width of the plate. Accordingly, the displacement field (u_x, u_y) of a plate is obtained as $u_x = -(1 - \nu) \times \tau_0 x/Eh$ and $u_y = -(1 - \nu)\tau_0 y/Eh$, where ν is the Poisson's ratio of the plate.

For a cantilevered plate with surface stress, the in-plane stress of a plate is no longer uniform as in the case of an unconstrained plate due to essential boundary condition. For $x < O(b)$ where b is the width of a cantilever, the in-plane stress exists, otherwise the in-plane stress becomes zero. Sader and Lachut assumed that the in-plane stress is uniform for $x < O(b/L)$. Using the equation of motion for a plate with in-plane stress, the effective bending rigidity D_{eff} of the plate exerting the surface stress is given by a relationship of

$$\frac{D_{\text{eff}} - D_0}{D_0} \approx \frac{(1-\nu)\tau_0}{Eh}\left(\frac{b}{L}\right)\left(\frac{b}{h}\right)^2, \tag{16.24}$$

where D_0 is the bending rigidity of a bare plate (i.e., without any surface stress) given as $D_0 = Eh^3/12(1 - \nu^2)$. As a consequence, the frequency shift of a cantilevered plate due to a constant surface stress is represented in the form of[44]

$$\frac{\Delta\omega}{\omega_0} = f(\nu)\frac{(1-\nu)\tau_0}{Eh}\left(\frac{b}{L}\right)\left(\frac{b}{h}\right)^2. \tag{16.25}$$

This does not contradict Gurtin's argument[41] that the resonant frequency of a cantilever is not affected by a constant surface stress. In particular, Eq. 16.25 demonstrates that a constant surface stress will not affect the resonant frequency of a cantilever, when b approaches zero (as assumed in the model of elastic beam).

16.2.3.3 Stiffness effect

In the theoretical models described in Sections 16.2.3.1 and 16.2.3.2, it is assumed that the elastic stiffness of adsorbed biomolecules (or cells or particles) is negligible when frequency shift due to biomolecular adsorption is taken into account. However, if the elastic modulus of a cantilever is comparable to that of adsorbed molecules (or cells or particles), then the frequency shift driven by adsorption is ascribed to not only the mass of adsorbed molecules but also their elastic modulus.

Tamayo and coworkers[45] recently developed a theoretical model that enables the quantitative descriptions on the frequency shift due to molecular adsorption as a function of the mass of adsorbed molecules as well as their elastic modulus. In their model, a single particle (or a single cell or protein monolayer) was adsorbed onto the location $l_1 < x < l_2$. The equation of motion for a cantilever with adsorption is given by

$$\mu(x)\partial_t^2 w(x,t) + \partial_x^2\left[D(x)\partial_x^2 w(x,t)\right] = 0. \tag{16.26}$$

Here, the parameters $\mu(x)$ and $D(x)$ are given as $\mu(x) = \rho_c b h_c + \rho_a b h_a[H(x - l_1) - H(x - l_2)]$ and $D(x) = E_c I_c^0[\{1 - H(x - l_1)\} + H(x - l_2)] + [E_c I_c(x) + E_a I_a(x)]\ [H(x - l_1) - H(x - l_2)]$, where ρ_c, b, h_c, E_c, and I_c^0 represents the mass density, width, thickness, elastic modulus, and cross-sectional moment of inertia for an unloaded cantilever, respectively, ρ_a, h_a, E_a, and $I_a(x)$ indicate the mass density, thickness, elastic modulus, and cross-sectional moment of inertia for adsorbed molecules, respectively, $I_c(x)$ is the cross-sectional moment of inertia for a cantilever onto which molecules are adsorbed, and $H(x)$ is the Heaviside unit step function. For a cantilever that vibrates, the transverse deflection is represented in the form of $w(x, t) = \phi(x)\cdot\exp[j\omega t]$, where ω is the resonant frequency of a cantilever with molecular adsorption, and $\phi(x)$ is the deflection eigenmode of unloaded cantilever, i.e., $\phi(x) = \sin(\lambda x/L) - \sinh(\lambda x/L) + [(\sin\lambda + \sinh\lambda)/(\cos\lambda + \cosh\lambda)] \times [\cosh(\lambda x/L) - \cos(\lambda x/L)]$ with λ being an eigenvalue such as $\lambda = 1.86$. By using Rayleigh–Ritz method,[14,22] the resonant frequency of molecule-loaded cantilever can be obtained from a relation of

$$\omega^2 = \frac{\int_0^L D(x)[\phi''(x)]^2 dx}{\int_0^L \mu(x)[(x)]^2 dx}. \tag{16.27}$$

When the molecules are uniformly adsorbed over the entire surface of a cantilever, the frequency shift of a cantilever due to molecular adsorption is represented in the form of[45]

$$\frac{\Delta\omega}{\omega_0} = \alpha\left(\frac{h_a}{h_c}\right) + \beta\left(\frac{h_a}{h_c}\right)^2, \tag{16.28a}$$

where

$$\alpha = \frac{1}{2}\left(3\frac{E_a}{E_c} - \frac{\rho_a}{\rho_c}\right) \tag{16.28b}$$

$$\beta = \frac{3}{8}\left[\left(\frac{\rho_a}{\rho_c}\right)^2 + 2\frac{E_a}{E_c}\left(1 - \frac{\rho_a}{\rho_c}\right) - 7\left(\frac{E_a}{E_c}\right)^2\right] \tag{16.28c}$$

Equation 16.28 demonstrates that the frequency shift due to molecular adsorption is determined from both the mass of adsorbed molecules and their elastic modulus.

16.2.4 Langmuir Kinetic Model

Cantilever sensor enables the real-time monitoring of biomolecular interactions by measuring the in situ frequency shift of a cantilever due to biomolecular interactions that occur on the surface of the cantilever. This implies that the cantilever sensor allows for quantitative characterization of the kinetics of biomolecular interaction. In this section, we describe a theoretical model that provides the relationship between the kinetics of biomolecular interaction and the in situ frequency shift due to biomolecular interaction.

First, as described in our previous work,[46] we consider a case in which target molecules are bound to the receptor molecules that are immobilized on the surface of a cantilever. We take into account the Langmuir kinetic model for binding between receptor (R) molecules and target (T) molecules such as

$$R + T \underset{k_u}{\overset{k_b}{\rightleftharpoons}} RT, \tag{16.29}$$

where k_b is a rate constant for molecular binding between receptor and target molecules, k_u is the rate constant of molecular unbinding (dissociation), and RT represents a complex via binding between receptor and target molecules. The Langmuir kinetic model provides a rate equation for molecular binding such as

$$\frac{dN_b}{dt} = k_b \left(N_{tot} - N_b\right)\xi - k_u N_b, \tag{16.30a}$$

where N_{tot} is the total number of receptor molecules, N_b is the number of receptor molecules that capture the target molecules, and ξ is the concentration of target molecules. With introduction of dimensionless parameter θ such as $\theta = N_b/N_{tot}$, the rate equation becomes

$$\frac{d\theta}{dt} = k_b(1-\theta)\xi - k_u\theta. \tag{16.30b}$$

Here, $d\theta/dt$ indicates the rate of an increase in the number of receptor molecules bound to target molecules, the first term in the right-hand side (r.h.s.) of Eq. 16.30b represents the number of receptor molecules incorporated to binding with target molecules, and the second term in r.h.s. shows the number of receptor molecules involved in molecular dissociation. By solving the rate equation given by Eq. 16.30, the dimensionless parameter is

$$\theta(t)=\theta_{\mathrm{eq}}\left[1-\exp(-t/\tau)\right], \tag{16.31}$$

where θ_{eq} is the equilibrium value of θ given by $\theta_{\mathrm{eq}} = k_{\mathrm{b}}\xi/(k_{\mathrm{b}}\xi + k_{\mathrm{u}})$, and τ is a time constant given as $\tau = k_{\mathrm{b}}\xi - k_{\mathrm{u}}$. Since the total mass of target molecules adsorbed onto the functionalized surface of a cantilever is linearly proportional to the dimensionless parameter θ, the resonant frequency shift of a cantilever due to molecular adsorption as a function of time can be represented in the form of

$$\Delta\omega(t)=-\Delta\omega_{\mathrm{eq}}\left[1-\exp(-t/\tau)\right], \tag{16.32}$$

where $\Delta\omega_{\mathrm{eq}}$ is the resonant frequency shift at equilibrium state. The negative sign in Eq. 16.32 indicates that molecular binding decreases the resonant frequency of a cantilever. This equation elucidates that the in situ frequency shift due to molecular adsorption with respect to time can be fitted to a exponential decaying function. In other words, the molecular adsorption leads to the exponential decay in the resonant frequency of a cantilever.

Now, let us consider the molecular dissociation such as enzymatic cleavage (or termed as proteolysis) that occurs on the surface of a cantilever. The rate equation for enzymatic cleavage of peptide chains is written as[47]

$$\frac{dN(t)}{dt}=-k_{\mathrm{p}}N(t), \tag{16.33}$$

where $N(t)$ is the number of intact peptides (i.e., peptide chains that are not cleft by enzymes) at time t, and k_{p} is a rate constant for enzymatic cleavage. Here, the rate equation is written for fixed concentration of enzymes. The number of enzymatically cleft peptide chains is thus obtained as $N_{\mathrm{p}}(t) = N_0[1 - \exp(-k_{\mathrm{p}}t)]$, where N_0 is the total number of peptide chains functionalized on the surface of a cantilever. The enzymatic cleavage of peptide chains leads to a decrease in the effective mass of a peptide-functionalized cantilever, resulting in an increase in the resonant frequency of the cantilever. Since the frequency shift of a cantilever owing to enzymatic cleavage is linearly proportional to the number of enzymatically cleft peptide chains, the resonant frequency shift can be expressed as

$$\Delta\omega(t) = \Delta\omega_{eq}[1 - \exp(-k_p t)]. \quad (16.34)$$

Here, $\Delta\omega_{eq}$ is the equilibrium value of resonant frequency shift. As described in Eq. 16.34, the enzymatic cleavage results in the exponential increase of the resonant frequency of a cantilever.

16.3 Cantilever-Based Label-Free Detection

16.3.1 DNA/RNA Detection

Gerber and coworkers[17] first reported the ability of a cantilever sensor to detect short DNA chains with their length of 12 nucleotide sequence without any labeling by measuring the deflection change of a cantilever due to DNA hybridization. In particular, when the target single-stranded DNA (ssDNA) chains are hybridized to the probe ssDNA chains that are functionalized on the surface of a cantilever, this hybridization changes the surface energy of a cantilever, and consequently, produces the surface stress leading to the bending deflection motion of a cantilever. More importantly, they showed that cantilever-based detection of DNA hybridization enables the quantitative descriptions on the Gibbs free energy related to DNA hybridization using Langmuir isotherm model.[48] Recently, they have also provided that a cantilever sensor allows for label-free detection of short RNA chains that can be used as a signature for a specific disease such as cancer.[49] Moreover, they have validated the ability of a cantilever to serve as a cancer diagnostic toolkit by showing that cantilever sensor is able to sensitively detect the short RNA chains that were expressed on a cancerous cell.[50] In particular, when the lysate of cancerous cells as a clinical sample was used for cantilever bioassay, the cantilever is able to sensitively capture the RNA chains that were contained in the lysate of cancer cells.

Although the detection scheme of a cantilever using its bending deflection change allows for highly sensitive detection of disease-specific DNA/RNA molecules, this detection scheme is inappropriate for quantifying the number of DNA molecules that were captured by a cantilever. However, the label-free detection based on estimation of frequency behavior of a cantilever in response to biomolecular interactions as described in Section 16.2.3 enables one to quantify the DNA molecules specifically adsorbed onto a cantilever. For instance, Graighead and coworkers[51] first reported the ability of nanoscale cantilever to detect a single DNA chain or to enumerate the number of few DNA chains by measuring the frequency change of a cantilever owing to DNA hybridization on the surface of a cantilever. Here, they used a cantilever that exhibits a high frequency in the order of 100 MHz for sensing a single DNA chain.

In the earlier work of Dravid and colleagues,[52] it is shown that the detection sensitivity of a resonant cantilever can be improved by using a nanoparticle as a mass amplifier, which can be attached to hybridized DNA molecules. The nanoparticle-based improvement of detection sensitivity for a resonant cantilever allows a cantilever sensor to have detection limit of 0.05 nM. However, this detection scheme using a nanoparticle as an amplifier is inappropriate for quantifying the number of DNA molecules specifically bound to the cantilever. This suggests that it is more appropriate to improve the detection sensitivity of cantilever by miniaturization rather than using mass amplifier (i.e., nanoparticle).

In recent years, Eom and coworkers[53] theoretically reported the resonance behavior of a cantilevered carbon nanotube in response to adsorption of DNA chain. Here, it should be noted that a carbon nanotube is able to capture a DNA molecule due to π–π interactions between the surface of a carbon nanotube and DNA molecule, and that a carbon nanotube is able to exhibit the high-frequency dynamic range up to 100 MHz and even up to 1 GHz, which implies that a resonant carbon nanotube exhibits the detection limit of single-molecule resolution. In this theoretical work,[53] it is found that a carbon nanotube is able to not only capture a single DNA chain but also identify the sequence of a DNA chain.

We note that a small-scale cantilever (e.g., nanoscale cantilever or cantilevered carbon nanotube), which can bear a high-frequency dynamic range in vacuum or air, is not favorable when it operates in aqueous environment.[13] This is attributed to the critical energy dissipation of a small-scale resonating cantilever due to interaction between water surrounding the cantilever and it, which leads to significant degradation of the quality factor of a cantilever, and consequently, results in the difficulty of measuring the frequency shift due to biomolecular interaction (e.g., DNA hybridization) in a fluid. We have previously shown that the piezoelectric microscale cantilever is able to exhibit a relatively high quality factor, which is appropriate in measuring the frequency shift due to biomolecular interaction in aqueous environment.[36] In particular, the piezoelectric cantilever allows for in situ observation of DNA hybridization by measuring the frequency shift in a fluid. More remarkably, it is shown that the in situ frequency shift due to DNA hybridization is well fitted to Langmuir kinetic model (as described in Section 16.2.4),[46] which suggests that the piezoelectric cantilever is able to quantitatively characterize the kinetics of DNA hybridization.

16.3.2 Protein Detection

In 2001, Majumdar and coworkers[54] first reported the ability of a cantilever to serve as a biosensor, which can perform the label-free detection of protein that is a signature of a specific disease such as cancer. In particular, they employed a cantilever for sensing prostate-specific antigen (PSA) protein that is a marker of prostate cancer by measuring the bending deflection change of the cantilever due to PSA antigen–antibody binding on the cantilever's surface. In their work,[54] the detection sensitivity of a cantilever was reported to be ~1 ng/ml.

In 2005, Lee et al.[55] considered a resonant cantilever for detecting marker proteins with high sensitivity. In particular, they have measured the frequency shift of a cantilever in air due to adsorption of proteins onto the cantilever. They reported that the resonant cantilever achieves the detection limit of 10 pg/ml, which is much lower than that of a static cantilever (i.e., detection based on measuring the deflection change of a cantilever due to biomolecular adsorption). Moreover, they found that the frequency shift due to protein adsorption onto a cantilever is attributed to not only the mass of adsorbed proteins but also the change of surface stress due to adsorbed proteins.[56] Hwang et al.[38] employed the theoretical model described in Section 16.2.3.2.1 for quantitative descriptions on the role that surface stress due to protein adsorption plays in the resonant frequency shift of a microcantilever owing to such adsorption. However, as Sader and Lachut argued,[44,57] the theoretical model employed by Hwang et al.[38] is inappropriate in that the model violates Newton's third law. Moreover, as Sader et al.[44] reported, the surface stress is still insufficient to describe the frequency shift of a cantilever owing to biomolecular adsorption. This suggests the necessity of developing a theoretical model that can explain the physical origin of the frequency shift of a cantilever due to biomolecular adsorption.

Bashir and colleagues[58] showed that biomolecular adsorption can increase the resonant frequency of a thin cantilever, which is attributed to the elastic property of adsorbed biomolecular monolayer (see Fig. 16.6). In particular, when the thickness of

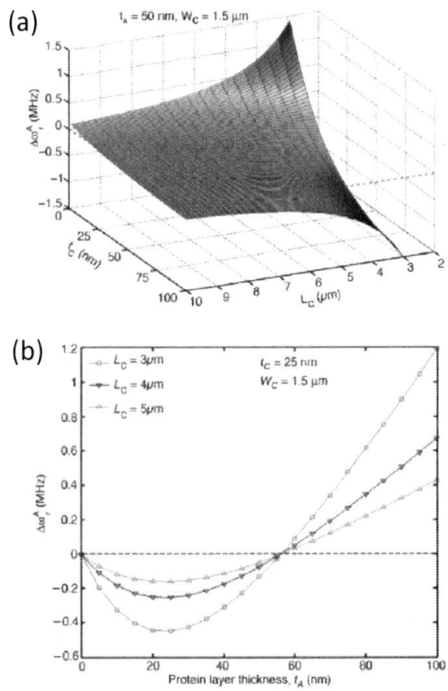

Figure 16.6 Frequency shift of nanoscale cantilever due to biomolecular adsorption. (a) Frequency shift due to biomolecular adsorption (with thickness of protein monolayer being 50 nm) was calculated as a function of the thickness and length of the cantilever. (b) Frequency shift driven by biomolecular adsorption was computed as a function of thickness of adsorbed protein monolayer as well as the length of a cantilever. Reprinted with permission from ref. 58.

a cantilever is comparable to the size of biomolecular layer adsorbed onto the cantilever, the frequency shift due to such adsorption is critically determined from the elastic modulus of adsorbed biomolecular layer. In a similar manner, Tamayo and coworkers[45] have employed a thin polymer cantilever for label-free detection of proteins. As described in Section 16.2.3.3, the frequency shift due to adsorption of protein monolayer onto a thin polymer cantilever is well described by a theoretical model depicted in Eq. 16.28. This indicates that the frequency shift due to biomolecular adsorption is critically affected by the elastic modulus of adsorbed biomolecular monolayer, if the thickness or elastic modulus of a cantilever is comparable to those of biomolecular monolayer.

In our previous report,[36] we provided that piezoelectric microcantilever, which exhibits a relatively high quality factor in a viscous fluid, allows for in situ monitoring of protein–protein interactions (e.g., protein antigen–antibody binding) in real time. It was shown that the frequency shift of piezoelectric microcantilever driven by protein–protein interactions is ascribed to the mass of adsorbed proteins as well as the change of hydrodynamic loading due to protein–protein interactions as has been elucidated by a theoretical model described in Eq. 16.20. Moreover, the in situ frequency shift due to protein–protein interactions enables the quantitative descriptions on the kinetics of such protein–protein interactions based on Langmuir kinetic model depicted in Section 16.2.4 (see Fig. 16.7).

In recent years, in order to improve the detection sensitivity of a cantilever even up to single-molecule resolution, researchers have made an effort to scaling down the cantilever even to nanoscale. Specifically, Roukes and colleagues[15] first reported the possibility of sensing a single protein molecule by using a resonant nanoscale cantilever. However, despite the ability of nanoscale cantilever to perform single-molecule detection, it is still unfavorable to sense and detect single protein molecules in aqueous environment, since the performance of nanoscale cantilever is severely depreciated in a liquid environment. It is still challenging to develop a nanoscale cantilever that can bear excellent vibrational characteristics (e.g., high quality factor) in a viscous fluid for future applications to in situ biosensing.

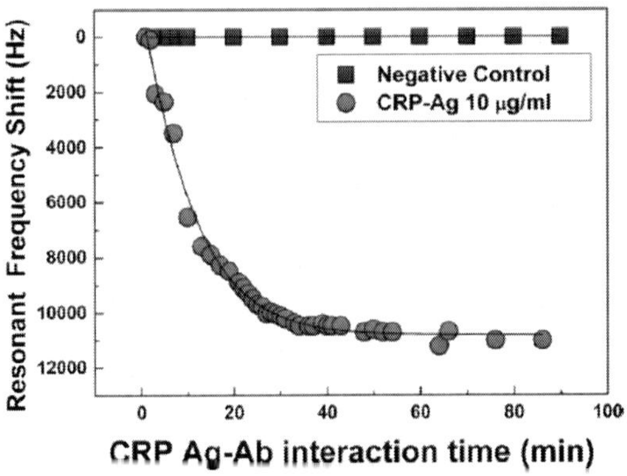

Figure 16.7 In situ frequency shift (indicated by dots), measured in a real-time, due to protein antigen–antibody binding on the surface of a microcantilever is well fitted to Langmuir kinetic model (represented by a solid line). Reprinted with permission from ref. 36.

16.3.3 Detection of Enzymatic Activity

The interactions between enzyme and protein, that is, enzyme specifically cleaves a protein (or a peptide chain), have played a vital role in cellular signaling transduction, which results in the malfunction of cell, leading to expression of diseases such as cancer.[59,60] This suggests that it is of high importance to quantitatively understand the interactions between enzyme and protein. Moreover, in recent decades, an inhibitor molecule as a drug has been developed in such a way that the inhibitor molecule is able to prevent the interaction between enzyme and protein such as enzymatic cleavage. This implies that it is essential to gain a quantitative insight into the interactions between enzyme, protein, and inhibitor for effective drug screening.

In the past decade, the interactions between enzyme and protein have been studied by measuring the bending deflection motion of a microcantilever due to enzymatic cleavage of protein chains. Specifically, the bending deflection of a cantilever, whose surface is chemically functionalized with enzymes or peptide chains (or protein chains), was monitored when such a cantilever is exposed to enzymes that can cleave peptide chains (or protein chains). The cantilever-based characterization of enzymatic cleavage has been well described in the literature.[7,61,62] However, it should be noted that the quantitative understanding of enzymatic cleavage by measuring the deflection change of a cantilever is restrictive, since the label-free detection based on measurement of the deflection change does not allow for quantifying the amount of peptide chains (or protein chains) that were reacted with enzymes.

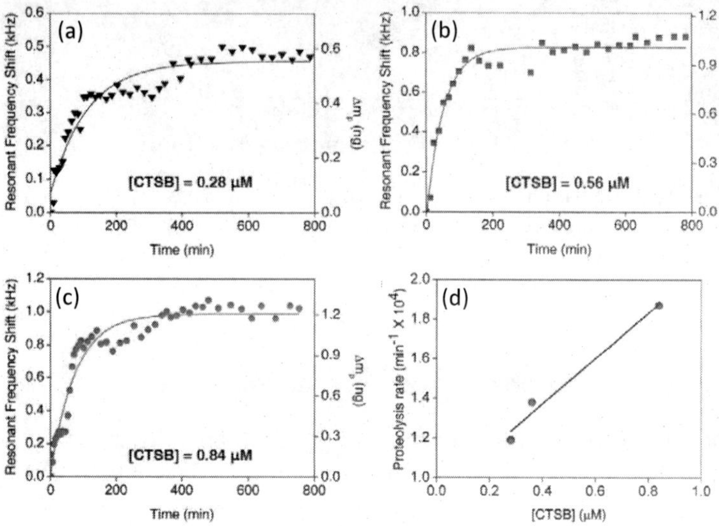

Figure 16.8 In situ frequency shift of a microcantilever due to proteolysis (i.e., enzymatic cleavage) of peptide chains functionalized on the surface of the cantilever is measured as a function of the concentrations of enzymes (here, cathepsin B) such as (a) 0.28 μM, (b) 0.56 μM, and (c) 0.84 μM. (d) Rate constant for proteolysis as a function of enzyme concentration. Reprinted from ref. 47 (under Creative Common License Agreement).

Recently, we have experimentally studied the interactions between enzyme and peptide chains by using a resonant microcantilever immersed in a liquid (see ref. 47). In particular, we measure the frequency shift of the cantilever, which was functionalized with peptide chains that can specifically interact with cathepsin B (CTSB), due to the CTSB-driven cleavage of peptide chains on the surface of the cantilever. Based on the measured frequency shift, we estimate the amount of peptide chains that were cleft by CTSB by using a theoretical model described by

Eq. 16.20. Moreover, in order to gain insight into the kinetics of enzymatic cleavage, we observe the real-time frequency shift of a cantilever, which was immersed in a buffer solution, due to enzymatic cleavage. Based on the Langmuir kinetic model delineated in Section 16.2.4 (see Eq. 16.32), we can estimate a rate constant for CTSB-induced enzymatic cleavage. As shown in Fig. 16.8, the in situ, real-time frequency shift of the cantilever is well fitted to Langmuir kinetic model. We found that the rate constant of enzymatic cleavage increases as a function of the concentrations of enzymes, and that the increase rate of rate constant for enzymatic cleavage, $R \equiv dk_p/d$ [CTSB], is estimated as $R = 1.15\ \mu M^{-1} \cdot min^{-1}$. This suggests that the efficacy of enzymatic cleavage is controlled by the concentrations of enzymes.

16.3.4 Detection of Matrix Metalloproteinase Expressed on Cancer Cells

Membrane type-1 matrix metalloproteinase (MT1-MMP) expressed on cancerous cells has been found to play a critical role in the proliferation and metastasis of cancers.[63] Specifically, the breakdown mechanism of extracellular matrix due to MT1-MMP expressed on a cancer cell is different from that due to MT1-MMP manifested on a normal cell; this difference is responsible for metastasis[64] or abnormal cell division[65] in cancer cells. This difference is attributed to not only the expression level of MT1-MMP but also the activity of MT1-MMP, i.e., the kinetics of proteolysis (enzymatic cleavage) due to MT1-MMP. In general, cancerous cells exhibit the highly expressed and highly activated MT1-MMPs on the surface of cancerous cells. This indicates that for cancer diagnosis, it is a priori requisite to characterize the amount of expressed MT1-MMPs and their activity.

The detection principle of cantilever-based sensing of MT1-MMP and its activity is a direct transduction of MT1-MMP-driven proteolysis of peptide chains, which were functionalized on the surface of a cantilever, into its frequency shift owing to the mass of cleft peptide chains (for details, see ref. 66). Figure 16.9a shows the frequency shift of a cantilever due to proteolysis induced by MT1-MMPs as a function of their concentrations. It is shown that the in situ frequency shift is well fitted to Langmuir kinetic model, which provides a rate constant for MT1-MMP-induced proteolysis as shown in Fig. 16.9b. The increase rate of a kinetic rate for proteolysis is given as $R = 0.16 \times 10^{-9}\ \text{min}^{-1} \cdot \text{M}^{-1}$.

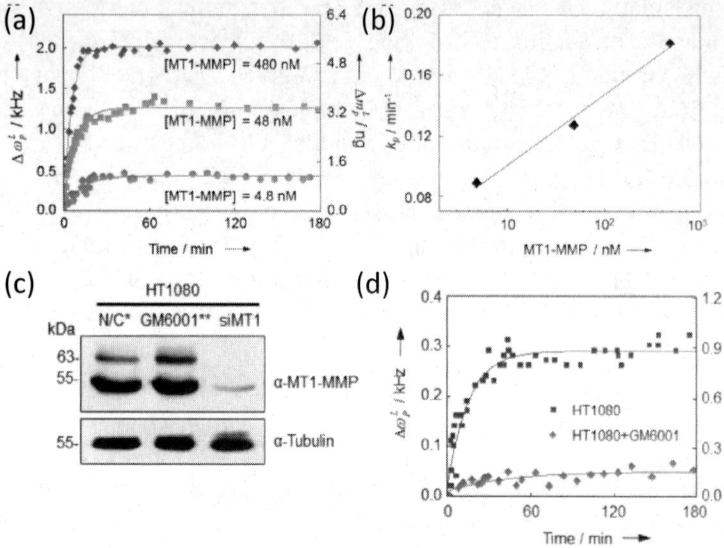

Figure 16.9 (a) In situ frequency shift due to MT1-MMP-induced proteolysis as a function of the concentrations of MT1-MMP. The dots indicate the experimental data, while the solid line represents the theoretical fitting to Langmuir kinetic model. (b) Rate constant for MT1-MMP-driven proteolysis as a function of MT1-MMP concentration. (c) Western blot analysis of HT1080 cell (N/C*), HT1080 cells treated with GM6001 (GM6001**), and HT1080 cells regulated by siRNA (siMT1). Western blot analysis cannot provide the GM6001-driven regulation of active MT1-MMPs, but enables an insight into the siRNA-directed regulation of MT1-MMP. (d) Frequency shift due to proteolysis driven by MT1-MMPs extracted from HT1080 cells (N/C*) and HT1080 cells treated with GM6001. It is shown that GM6001 critically decreases the resonant frequency shift due to MT1-MMP-driven proteolysis, indicating that GM6001 is able to effectively regulate the active MT1-MMPs expressed on cancerous cells. Reprinted with permission from ref. 66.

To validate the robustness of a cantilever sensor to serve as a biosensor for cancer diagnosis, we consider the cantilever-based detection of active MT1-MMPs that were extracted from the membranes of cancerous cells. In particular, we take into account a buffer solution that contains various proteins obtained from the whole cell lysate of fibrosarcoma HT1080 cells for cantilever bioassay; such buffer solution acquired from cell lysate was injected into a liquid cell, in which a vibrant cantilever is mounted. The expression of MT1-MMPs on the surface of HT1080 cells (or HT1080 cells treated with inhibitor GM6001) was confirmed by using western blot analysis. Western blot analysis shows that MT1-MMPs are highly expressed on tumor cells (i.e., HT1080 cells), but it is unable to provide the quantitative insight into the activity of MT1-MMP expressed on HT1080 cells. Moreover, as shown in Fig. 16.9c, western blot analysis does not provide insight into how GM6001 (inhibitor molecule) regulates the activity of MT1-MMP, while the western blot analysis is able to recognize the siRNA-directed regulation of MT1-MMP. On the other hand, as shown in Fig. 16.9d, the cantilever bioassay allows for quantitative characterization of the activity of MT1-MMPs expressed on HT1080 cells as well as the activity of MT1-MMPs that were regulated by inhibitor. Specifically, we can observe the frequency shift of a cantilever due to MT1-MMP-induced proteolysis of peptide chains, providing that the mass of cleft peptide chains is estimated as Δm_P = 0.9 ng, and that the rate constant of such proteolysis is measured as k_p = 0.066 min^{-1}. However, when we consider a buffer solution obtained from the lysate of HT1080 cells treated with GM6001, the frequency shift due to MT1-MMP-driven proteolysis is below 0.1 kHz, indicating that GM6001 is able to effectively regulate the activity of MT1-MMPs expressed on HT1080 cells. In particular, the mass of cleft peptides for GM6001-regulated HT1080 cells is measured as Δm_P = ~0.1 ng, and GM6001 decrease the rate constant of MT1-MMP-induced proteolysis to k_p = 0.023 min^{-1}. This suggests that a cantilever bioassay is able to quantitatively characterize the efficacy of drug molecules that regulates the activity of MT1-MMPs expressed on a cancerous cell, while such characterization has been made impossible using western blot analysis.

16.4 Future Perspectives

In this chapter, we describe the fundamental principles of cantilever-based label-free detection of biological molecules for future cancer diagnosis, and recent applications of cantilever sensor in biological/biochemical label-free detections. As elucidated in this chapter, the cantilever bioassay is very useful in a quantitative characterization of biomolecular interactions; such characterization cannot have been achieved in a classical bioassay toolkit such as ELISA. More remarkably, the cantilever bioassay allows for highly sensitive label-free detection of biomolecules with detection limit now being much lower than that of other classical assay toolkits such as ELISA and fluorescence-based detection.

Despite the ability of a resonant cantilever sensor to detect biomolecules with high detection sensitivity, the cantilever bioassay has to be extended to a clinical level by considering more realistic clinical samples. Until now, most of research works on cantilever bioassay has considered a buffer solution that contains biological molecules rather than clinically obtained samples such as blood (or tumors). In general, for effective cancer diagnosis, it is recommendable to use a clinical sample such as a whole blood that typically contains a variety of proteins, among which specific proteins have to be detected. As described in Section 16.3.4, we showed that cantilever bioassay can be utilized at a clinical level by using the whole cell lysate of cancer cells, indicating that cantilever bioassay can be employed for effective cancer diagnosis. We anticipate that cantilever bioassay may play a vital role in early diagnosis of cancer as long as we make an effort to advance the cantilever bioassay that can operate using a clinical sample.

References

1. J. M. Gere, *Mechanics of Materials*, Thomson Learning, 2003.
2. S. P. Timoshenko and J. N. Goodier, *Theory of Elasticity*, McGraw-Hill, 1970.
3. G. Binnig, C. F. Quate, and C. Gerber, *Phys. Rev. Lett.*, **56** (1986) 930.
4. D. J. Muller and Y. F. Dufrene, *Nat. Nanotechnol.*, **3** (2008) 261.
5. Y. Sugimoto, P. Pou, M. Abe, P. Jelinek, R. Perez, S. Morita, and O. Custance, *Nature*, **446** (2007) 64.
6. A. Boisen, S. Dohn, S. S. Keller, S. Schmid, and M. Tenje, *Rep. Prog. Phys.*, **74** (2011) 036101.
7. K. M. Goeders, J. S. Colton, and L. A. Bottomley, *Chem. Rev.*, **108** (2008) 522.

8. A. Boisen and T. Thundat, *Mater. Today*, **12** (2009) 32.
9. W. Haiss, *Rep. Prog. Phys.*, **64** (2001) 591.
10. R. Berger, E. Delamarche, H. P. Lang, C. Gerber, J. K. Gimzewski, E. Meyer, and H. J. Guntherodt, *Science*, **276** (1997) 2021.
11. P. S. Waggoner and H. G. Craighead, *Lab Chip*, **7** (2007) 1238.
12. M. Calleja, P. Monteiro Kosaka, A. San Paulo, and J. Tamayo, *Nanoscale*, **4** (2012) 4925.
13. K. Eom, H. S. Park, D. S. Yoon, and T. Kwon, *Phys. Rep.*, **503** (2011) 115.
14. L. Meirovitch, *Analytical Methods in Vibrations*, Macmillan, New York, 1967.
15. M. S. Hanay, S. Kelber, A. K. Naik, D., Chi, S., Hentz, E. C. Bullard, E., Colinet, L., Duraffourg, and M. L. Roukes, *Nat. Nanotechnol.*, **7** (2012) 602.
16. G. G. Stoney, *Proc. R. Soc. Lond. A*, **82** (1909) 172.
17. J. Fritz, M. K. Baller, H. P. Lang, H. Rothuizen, P. Vettiger, E. Meyer, H. J. Guntherodt, C. Gerber, and J. K. Gimzewski, *Science*, **288** (2000) 316.
18. G. H. Wu, H. F. Ji, K. Hansen, T. Thundat, R. Datar, R. Cote, M. F. Hagan, A. K. Chakraborty, and A. Majumdar, *Proc. Natl. Acad. Sci. U. S. A.*, **98** (2001) 1560.
19. J. Mertens, C. Rogero, M. Calleja, D. Ramos, J. A. Martin-Gago, C. Briones, and J. Tamayo, *Nat. Nanotechnol.*, **3** (2008) 301.
20. K. M. Hansen, H. F. Ji, G. H. Wu, R. Datar, R. Cote, A. Majumdar, and T. Thundat, *Anal. Chem.*, **73** (2001) 1567.
21. S. L. Biswal, D. Raorane, A. Chaiken, H. Birecki, and A. Majumdar, *Anal. Chem.*, **78** (2006) 7104.
22. W. Weaver, S. P. Timoshenko, and D. H. Young, *Vibration Problems in Engineering*, John Wiley & Sons, New York, 1990.
23. T. Y. Kwon, J. H. Park, Y. B. Kim, D. S. Yoon, C. I. Cheon, H. L. Lee, and T. S. Kim, *J. Cryst. Growth*, **295** (2006) 172.
24. T. Y. Kwon, Y. B. Kim, K. Eom, D. S. Yoon, H. L. Lee, and T. S. Kim, *Appl. Phys. A*, **88** (2007) 627.
25. J. W. Yi, W. Y. Shih, and W.-H. Shih, *J. Appl. Phys.*, **91** (2002) 1680.
26. J. H. Park, T. Y. Kwon, D. S. Yoon, H. Kim, and T. S. Kim, *Adv. Funct. Mater.*, **15** (2005) 2021.
27. A. Husain, J. Hone, H. W. C. Postma, X. M. H. Huang, T. Drake, M. Barbic, A. Scherer, and M. L. Roukes, *Appl. Phys. Lett.*, **83** (2003) 1240.

28. V. Sazonova, Y. Yaish, H. Ustunel, D. Roundy, T. A. Arias, and P. L. McEuen, *Nature*, **431** (2004) 284.
29. S. C. Masmanidis, R. B. Karabalin, I. De Vlaminck, G. Borghs, M. R. Freeman, and M. L. Roukes, *Science*, **317** (2007) 780.
30. S. S. Verbridge, D. F. Shapiro, H. G. Craighead, and J. M. Parpia, *Nano Lett.*, **7** (2007) 1728.
31. J. E. Sader, *J. Appl. Phys.*, **84** (1998) 64.
32. M. R. Paul and M. C. Cross, *Phys. Rev. Lett.*, **92** (2004) 235501.
33. S. Basak, A. Raman, and S. V. Garimella, *J. Appl. Phys.*, **99** (2006) 114906.
34. J. W. M. Chon, P. Mulvaney, and J. E. Sader, *J. Appl. Phys.*, **87** (2000) 3978.
35. T. Braun, V. Barwich, M. K. Ghatkesar, A. H. Bredekamp, C. Gerber, M. Hegner, and H. P. Lang, *Phys. Rev. E*, **72** (2005) 031907.
36. T. Y. Kwon, K. Eom, J. H. Park, D. S. Yoon, T. S. Kim, and H. L. Lee, *Appl. Phys. Lett.*, **90** (2007) 223903.
37. S. Cherian and T. Thundat, *Appl. Phys. Lett.*, **80** (2002) 2219.
38. K. S. Hwang, K. Eom, J. H. Lee, D. W. Chun, B. H. Cha, D. S. Yoon, T. S. Kim, and J. H. Park, *Appl. Phys. Lett.*, **89** (2006) 173905.
39. A. W. McFarland, M. A. Poggi, M. J. Doyle, L. A. Bottomley, and J. S. Colton, *Appl. Phys. Lett.*, **87** (2005) 053505.
40. J. Dorignac, A. Kalinowski, S. Erramilli, and P. Mohanty, *Phys. Rev. Lett.*, **96** (2006) 186105.
41. M. E. Gurtin, X. Markenscoff, and R. N. Thurston, *Appl. Phys. Lett.*, **29** (1976) 529.
42. P. Lu, H. P. Lee, C. Lu, and S. J. O'Shea, *Phys. Rev. B*, **72** (2005) 085405.
43. G.-F. Wang and X.-Q. Feng, *Appl. Phys. Lett.*, **90** (2007) 231904.
44. M. J. Lachut and J. E. Sader, *Phys. Rev. Lett.*, **99** (2007) 206102.
45. J. Tamayo, D. Ramos, J. Mertens, and M. Calleja, *Appl. Phys. Lett.*, **89** (2006) 224104.
46. T. Kwon, K. Eom, J. Park, D. S. Yoon, H. L. Lee, and T. S. Kim, *Appl. Phys. Lett.*, **93** (2008) 173901.
47. T. Kwon, J. Park, J. Yang, D. S. Yoon, S. Na, C.-W. Kim, J. S. Suh, Y. M. Huh, S. Haam, and K. Eom, *PLOS ONE*, **4** (2009) e6248.
48. R. McKendry, J. Y. Zhang, Y. Arntz, T. Strunz, M. Hegner, H. P. Lang, M. K. Baller, U. Certa, E. Meyer, H. J. Guntherodt, and C. Gerber, *Proc. Natl. Acad. Sci. U. S. A.*, **99** (2002) 9783.

49. J. Zhang, H. P. Lang, F. Huber, A. Bietsch, W. Grange, U. Certa, R. McKendry, H. J. Guntherodt, M. Hegner, and C. Gerber, *Nat. Nanotechnol.*, 1 (2006) 214.
50. F. Huber, H. P. Lang, N. Backmann, D. Rimoldi, and C. Gerber, *Nat. Nanotechnol.*, **8** (2013) 125.
51. B. Ilic, Y. Yang, K. Aubin, R. Reichenbach, S. Krylov, and H. G. Craighead, *Nano Lett.*, **5** (2005) 925.
52. M. Su, S. Li, and V. P. Dravid, *Appl. Phys. Lett.*, **82** (2003) 3562.
53. M. Zheng, K. Eom, and C. Ke, *J. Phys. D: Appl. Phys.*, **42** (2009) 145408.
54. G. H. Wu, R. H. Datar, K. M. Hansen, T. Thundat, R. J. Cote, and A. Majumdar, *Nat. Biotechnol.*, **19** (2001) 856.
55. J. H. Lee, K. S. Hwang, J. Park, K. H. Yoon, D. S. Yoon, and T. S. Kim, *Biosens. Bioelectron.*, **20** (2005) 2157.
56. J. H. Lee, T. S. Kim, and K. H. Yoon, *Appl. Phys. Lett.*, **84** (2004) 3187.
57. M. J. Lachut and J. E. Sader, *Phys. Rev. B*, **85** (2012) 085440.
58. A. K. Gupta, P. R. Nair, D. Akin, M. R. Ladisch, S. Broyles, M. A. Alam, and R. Bashir, *Proc. Natl. Acad. Sci. U. S. A.*, **103** (2006) 13362.
59. T. Kodadek, *Nature*, **453** (2008) 861.
60. J. S. Sebolt-Leopold and J. M. English, *Nature*, **441** (2006) 457.
61. Y. Weizmann, R. Elnathan, O. Lioubashevski, and I. Willner, *J. Am. Chem. Soc.*, **127** (2005) 12666.
62. Y. Weizmann, R. Elnathan, O. Lioubashevski, and I. Willner, *Nano Lett.*, **5** (2005) 741.
63. M. Seiki, *Cancer Lett.*, **194** (2003) 1.
64. N. Sugiyama, M. Varjosalo, P. Meller, J. Lohi, K. M. Chan, Z. Zhou, K. Alitalo, J. Taipale, J. Keski-Oja, and K. Lehti, *Proc. Natl. Acad. Sci. U. S. A.*, **107** (2010) 15786.
65. D. Ponti, A. Costa, N. Zaffaroni, G. Pratesi, G. Petrangolini, D. Coradini, S. Pilotti, M. A. Pierotti, and M. G. Daidone, *Cancer Res.*, **65** (2005) 5506.
66. G. Lee, K. Eom, J. Park, J. Yang, S. Haam, Y.-M. Huh, J. K. Ryu, N. H. Kim, J. I. Yook, S. W. Lee, D. S. Yoon, and T. Kwon, *Angew. Chem. Int. Ed.*, **51** (2012) 5837.

Index